Communications in Computer and Information Science 1391

More information about this series at http://www.springer.com/series/7899

Alexander Dudin · Anatoly Nazarov ·
Alexander Moiseev (Eds.)

Information Technologies and Mathematical Modelling

Queueing Theory and Applications

19th International Conference, ITMM 2020
Named after A.F. Terpugov, Tomsk, Russia, December 2–5, 2020
Revised Selected Papers

Springer

Editors
Alexander Dudin
Belarusian State University
Minsk, Belarus

Anatoly Nazarov
Tomsk State University
Tomsk, Russia

Alexander Moiseev
Tomsk State University
Tomsk, Russia

ISSN 1865-0929 ISSN 1865-0937 (electronic)
Communications in Computer and Information Science
ISBN 978-3-030-72246-3 ISBN 978-3-030-72247-0 (eBook)
https://doi.org/10.1007/978-3-030-72247-0

This Springer imprint is published by the registered company Springer Nature Switzerland AG
The registered company address is: Gewerbestrasse 11, 6330 Cham, Switzerland

Preface

The series of scientific conferences Information Technologies and Mathematical Modelling (ITMM) was started in 2002. In 2012, the series acquired international status, and selected revised papers have been published in *Communications in Computer and Information Science* since 2014. The conference series was named after Alexander Terpugov, one of the first organizers of the conference, an outstanding scientist of the Tomsk State University and a leader of the famous Siberian school on applied probability, queueing theory, and applications.

Traditionally, the conference has about ten sections in various fields of mathematical modelling and information technologies. Throughout the years, the sections on probabilistic methods and models, queueing theory, and communication networks have been the most popular ones at the conference. These sections gather many scientists from different countries. Many foreign participants come to this Siberian conference every year because of our warm welcome and serious scientific discussions. In 2020, the ITMM conference was held using technologies for online conferences due to the pandemic of COVID-19.

This volume presents selected papers from the 19th ITMM conference. The papers are devoted to new results in queueing theory and its applications. Its target audience includes specialists in probabilistic theory, random processes, and mathematical modeling as well as engineers engaged in logical and technical design and operational management of data processing systems, communications, and computer networks.

December 2020

Alexander Dudin
Anatoly Nazarov
Alexander Moiseev

Organization

The conference was organized by the National Research Tomsk State University, International Computer Science Continues Professional Development Center, Peoples' Friendship University of Russia (RUDN University), and Trapeznikov Institute of Control Sciences of Russian Academy of Sciences.

International Program Committee Chairs

Alexander Dudin	Belarusian State University, Belarus
Anatoly Nazarov	Tomsk State University, Russia

International Program Committee

Khalid Al-Begain	Kuwait College of Science and Technology, Kuwait
Ivan Atencia	University of Málaga, Spain
Pedro Cabral	Universidade Nova de Lisboa, Portugal
Pau Fonseca i Casas	Universitat Politècnica de Catalunya, Spain
Srinivas Chakravarthy	Kettering University, USA
Bong Dae Choi	National Institute for Mathematical Sciences, South Korea
Tadeusz Czachórski	Institute of Theoretical and Applied Informatics, Polish Academy of Sciences, Poland
Rui Dinis	Universidade Nova de Lisboa, Portugal
Dmitry Efrosinin	Johannes Kepler University Linz, Austria
Mais Farhadov	Institute of Control Sciences, Russian Academy of Sciences, Russia
Yulia Gaydamaka	Peoples' Friendship University of Russia (RUDN University), Russia
Erol Gelenbe	Institute of Theoretical and Applied Informatics, Polish Academy of Sciences, Poland
Alexander Gortsev	Tomsk State University, Russia
Bara Kim	Korea University, South Korea
Che Soong Kim	Sangji University, South Korea
Udo Krieger	Universität Bamberg, Germany
B. Krishna Kumar	Anna University, Chennai, India
Achyutha Krishnamoorthy	Cochin University of Science and Technology, India
Krishna Kumar	Cochin University of Science and Technology, India
Quan-Lin Li	Yanshan University, China
Yury Malinkovsky	Francisk Skorina Gomel State University, Belarus
Agassi Melikov	National Aviation Academy of Azerbaijan, Azerbaijan
Alexander Moiseev	Tomsk State University, Russia

Svetlana Moiseeva	Tomsk State University, Russia
Paulo Montezuma-Carvalho	Universidade Nova de Lisboa, Portugal
Evsey Morozov	Institute of Applied Mathematical Research, Karelian Research Centre of Russian Academy of Sciences, Russia
Valery Naumov	Service Innovation Research Institute, Finland
Rein Nobel	Vrije Universiteit Amsterdam, The Netherlands
Michele Pagano	Pisa University, Italy
Tuan Phung-Duc	University of Tsukuba, Japan
Thomas B. Preußer	Technische Universität Dresden, Germany
Jacques Resing	Eindhoven University of Technology, The Netherlands
Vladimir Rykov	Gubkin Russian State University of Oil and Gas, Russia
Konstantin Samouylov	Peoples' Friendship University of Russia (RUDN University), Russia
Sergey Suschenko	Tomsk State University, Russia
Daniel Stamate	Goldsmiths, University of London, UK
János Sztrik	University of Debrecen, Hungary
Henk Tijms	Vrije Universiteit Amsterdam, The Netherlands
Oleg Tikhonenko	Cardinal Stefan Wyszyński University in Warsaw, Poland
Gurami Tsitsiashvili	Institute of Applied Mathematics, Far Eastern Branch of Russian Academy of Sciences, Russia
Vladimir Vishnevsky	Institute of Control Sciences, Russian Academy of Sciences, Russia
Anton Voitishek	Institute of Computational Mathematics and Mathematical Geophysics, Siberian Branch, Russian Academy of Sciences, Russia
Vladimir Zadorozhny	Omsk State Technical University, Russia
Alexander Zamyatin	Tomsk State University, Russia
Andrey Zorin	Lobachevsky State University of Nizhni Novgorod, Russia

Local Organizing Committee

Svetlana Moiseeva (Chair)
Svetlana Paul (Co-chair)
Valentina Broner
Elena Danilyuk
Ekaterina Fedorova
Anastasiya Galileyskaya
Elena Glukhova
Yana Izmailova
Irina Kochetkova
Ivan Lapatin
Ekaterina Lisovskaya
Olga Lizyura
Anna Morozova
Svetlana Rozhkova
Daria Semenova
Maria Shklennik
Alexey Shkurkin
Konstantin Voytikov

Contents

Some Special Features of Finite-Source Retrial Queues with Collisions, an Unreliable Server and Impatient Customers in the Orbit

János Sztrik(✉) and Ádám Tóth

Faculty of Informatics, University of Debrecen, Debrecen, Hungary
{sztrik.janos,toth.adam}@inf.unideb.hu

Abstract. The goal of the paper is to study a finite-source retrial queuing system with collisions and customers' impatience behavior in the orbit. The situation when an incoming customer from the orbit or from the source finds the server busy causes a collision and both requests are directed toward the orbit. It is assumed that every request in the source is eligible to generate customers whenever the server is failed but these requests immediately go into orbit. A customer after some waiting in the orbit can depart without fulfilling its service requirement these are the so-called impatient/reneging/abandoned customers. In that case it returns to the source. A customer who is under service when the server fails is also sent to the orbit. The source, service, retrial, impatience, operation and repair times are supposed to be independent of each other. The novelty of the investigation is to carry out a sensitivity analysis comparing various distributions of impatience time of customers on the performance measures such as mean number of customers in the orbit, mean waiting time of an arbitrary, successfully served and reneging customers, probability of abandonment, server utilization, etc.

Keywords: Finite-source queuing system · Retrial queues · Collisions · Server breakdowns and repairs · Impatient customers · Stochastic simulation

1 Introduction

Impatience of the customers is a natural phenomenon and an interesting topic in queueing theory. The process of reneging and balking is extensively studied by many researchers, for example in [1–10,13,14,20,24]. Whenever an arriving customer decides not to enter the system, which is called balking while in reneging a customer in the system after waiting for some time leaves the system

The research of both authors was supported by the construction EFOP-3.6.3-VEKOP-16-2017-00002. The project was supported by the European Union, co-financed by the European Social Fund.

A. Dudin et al. (Eds.): ITMM 2020, CCIS 1391, pp. 1–15, 2021.
https://doi.org/10.1007/978-3-030-72247-0_1

without being served. In our investigated model reneging customers are considered. Queuing systems with repeated calls may competently describe major telecommunication systems, such as telephone switching systems, call centers, CSMA-based wireless mesh networks in frame level. The main feature of retrial queueing system is that customers remain in the system even if it is unable to find idle service unit and after some random time it attempts to reach the service facility again. Speaking of communication systems where the available channels or other facilities are very limited thus users (sources) usually need to fight for these resources. This results a high possibility of conflict because several sources may launch uncoordinated attempts producing collisions. In these cases the loss of transmission takes place and it is necessary to ensure of the process of retransmission. So evolving efficient procedures for preventing conflict and corresponding message delay is essential. In case of a collision both calls, the one under service and the newly arriving one go to the orbit. A review of results on finite-source retrial queues with collision and unreliable server has been published in [18]. In many papers of retrial queueing literature the service unit is assumed to be available steadily. But these assumptions are quite unrealistic because in real life applications of these systems can break down, different types of problems can arise like power outage, human error or other failures. Various factors have effect on the transmission rate of the wireless channel in a wireless communication scenario and these are apt to suffer transmission failure, interruptions throughout transferring the packets. Investigating retrial queueing systems with random server breakdowns and repairs has a great importance as the operation of non-reliable systems modifies system characteristics and performance measures. In this paper, we assume that in the case of a failure of the server, the request generation from the source continues and calls go to the orbit. Moreover, a customer who is under service when the server fails is also sent to the orbit.

The novelty of this investigation is to carry out sensitivity analysis using different distributions of impatient calls on performance measures. Different Figures help to understand the special features of the system. The model is a generalization of [21] and a continuation of the works [12,22].

The aim of the present paper is to show some special features of finite-source retrial queuing systems with impatient customers in the orbit. In general we could see that the steady-state distribution of the number of customers in the service facility can be approximated by a normal distribution with given mean and variance. By the help of stochastic simulation several systems are analyzed showing directions for further analytic investigations. Tables and Figures are collected to illustrate unexpected properties of these systems.

2 System Model

A retrial queueing system of type $M/M/1//N$ is considered with a non-reliable server and impatient customers. In the finite-source N customers reside and each of them is able to generate calls towards the server with rate λ/N so the inter-request time is exponential with parameter λ/N. A customer cannot generate a new call until the previous call returns to the source. Every incoming customer has a random impatience time which determines how much time the customer spends in the orbit without getting its service requirement. Exceeding this time results that the customer no longer waits for the service unit and departs without being served properly. This random time follows gamma, hypo-exponential, hyper-exponential, Pareto and lognormal distribution with different parameters but with the same mean value. If an arriving customer either from the source or from the orbit finds the server in idle state its service starts immediately. The service times of the customers is assumed to be exponentially distributed with parameter μ. After its successful service customers return to the source. Encountering the service unit in a busy state the arriving customer causes a collision with the call under service and both enter the orbit. After an exponentially distributed time with parameter σ/N customers located in the orbit make another attempt to get into the service. The server is not reliable so from time to time can break down. The lifetime is an exponentially distributed random variable with parameter γ_0 in case of an idle server and γ_1 when the server is busy. The repair process starts immediately upon the breakdown which also follows an exponential distribution with parameter γ_2. If server failure takes place during the service of a customer it is transferred to the orbit. Furthermore, in the case of a failure, we can distinguish two options. Namely, the failure either stops entering new customers from the source or allows them to go to the orbit. Usually, we treat the system with the latter option if the other one is not stated. The source, service, retrial, impatience, operation, and repair times are supposed to be independent of each other.

3 Simulation Case Studies

The simulation approach is a very important method that helps us in performance modeling when the system is too complicated to investigate with the help of other standard methods, like analytical, numerical, or asymptotic ones. For the interested readers we list some of the most important works, such as [11,15–17,19,23].

In this section first, we deal with exponentially distributed impatience time which helps us to check our simulation results with the help of those we got using MOSEL (MOdeling, Specification, and Evaluation Language), published in [12]. As soon as we realized that the simulation program operates correctly we can investigate the effect of the distribution of impatience time on the performance measures as we will do in the second part. One of the advantages of the simulation that we can make a difference between abandoned and successfully served requests as we show in our examples. Reading the papers dealing with abandonment we noticed that mainly the distribution of the customers in the system has been investigated and then using the Little-formula the mean response time of an arbitrary customer has been obtained. There are no performance measures for different types of customers.

3.1 Exponentially Distributed Impatience Times

In these cases we would like to show the effect of the impatience rate on the distribution of the number of customers in the system. All the random variables mentioned above are exponentially distributed (Table 1).

Table 1. Different impatience rates

	N	λ/N	σ/N	γ_0	γ_1	γ_2	μ	τ
Case 1	100	0.01	0.1	0.1	0.1	1	1	1E−10
Case 2	100	0.01	0.1	0.1	0.1	1	1	0.000001
Case 3	100	0.01	0.1	0.1	0.1	1	1	0.0001
Case 4	100	0.01	0.1	0.1	0.1	1	1	0.001
Case 5	100	0.01	0.1	0.1	0.1	1	1	0.01
Case 6	100	0.01	0.1	0.1	0.1	1	1	0.1
Case 7	100	0.01	0.1	0.1	0.1	1	1	1
Case 8	100	0.01	0.1	0.1	0.1	1	1	5

The results are understandable and illustrate what we expected, namely the higher the impatience rate the less the number of customers in the system see Fig. 1. However, we are able to give other measures which show some unexpected features of the system. It should be mentioned that we need some special parameter set up so that these cases should happen. We have some experience to find this setup from our previous works.

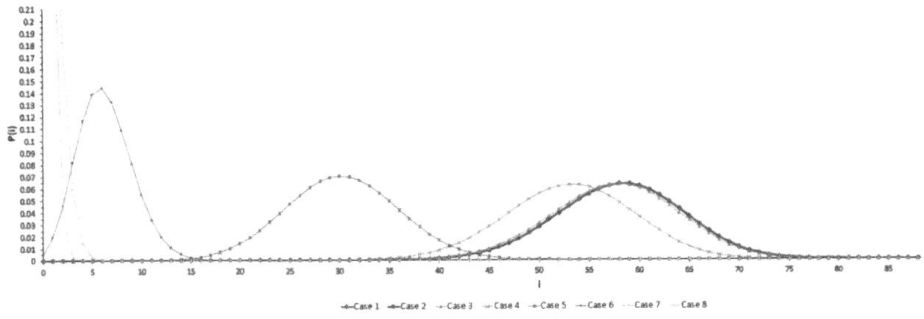

Fig. 1. Distribution of number of customers for different impatience rates

In the following we give detailed estimations for the different type of customers, namely successfully served and abandoned requests. In our opinion only the simulation can help us to receive these measures. In Tables 2, 3, 4 we can see some special features of these systems for which the notations are given in Table 5.

Table 2. Estimations 1

	E(NS)	E(O)	E(T)	E(TS)	E(TA)	E(W)	E(WS)	E(WA)	E(WAO)
Case 1	57.9829	57.5627	137.9729	137.9729	0.0000	136.9730	136.9730	0.0000	0.0000
Case 2	57.9771	57.5570	137.9197	137.9182	148.6876	136.9203	136.9188	147.7836	20.1735
Case 3	57.4366	57.0166	134.9306	134.7932	145.0371	133.9441	133.8056	144.1288	20.3340
Case 4	52.9963	52.5786	112.7086	111.5587	121.8498	111.8203	110.6612	121.0345	19.9312
Case 5	30.2477	29.8485	43.3468	39.3924	48.6341	42.7747	38.7689	48.1307	16.3978
Case 6	6.3188	5.9797	6.7437	3.6688	8.4876	6.3818	3.1664	8.2053	6.3764
Case 7	0.9897	0.6868	0.9996	0.5514	1.1972	0.6937	0.0633	0.9716	0.9432
Case 8	0.4365	0.1398	0.4384	0.4849	0.4187	0.1404	0.0030	0.1987	0.1974

Table 3. Estimations 2

E(ST)	E(STS)	E(STA)	E(STSI)	E(STSUI)	Pa	Pao	Us	UsS	UsA
0.9999	0.9999	0.0000	0.8606	0.1394	0.0000	0.0000	0.4202	0.4200	0.0000
0.9994	0.9994	0.9040	0.8603	0.1394	0.0001	0.0739	0.4201	0.4200	0.0001
0.9866	0.9876	0.9083	0.8474	0.1403	0.0134	0.0777	0.4200	0.4149	0.0052
0.8883	0.8975	0.8152	0.7489	0.1487	0.1117	0.0906	0.4177	0.3749	0.0428
0.5721	0.6235	0.5034	0.4095	0.2140	0.4279	0.1874	0.3992	0.2489	0.1504
0.3619	0.5024	0.2823	0.1118	0.3907	0.6381	0.4272	0.3391	0.1703	0.1688
0.3059	0.4881	0.2256	0.0162	0.4718	0.6940	0.5318	0.3029	0.1479	0.1553
0.2980	0.4819	0.2200	0.0039	0.4780	0.7020	0.5426	0.2968	0.1431	0.1538

It is easy to check that the mean response/waiting and total service time of an arbitrary customer can be obtained by the help of law of total expectation. Furthermore, the Little-formulas are also valid.

Let us make some comments concerning the results. If we take a closer look at the mean waiting time of an abandoned customer $E(WA)$ and the conditional mean waiting time $E(WAO)$ of those abandoned customers who never left the orbit we can see one of the unexpected features. Namely, one might think that they should be around the mean of the assumed impatience time, in this case $1/\tau$. But as we can observe the estimations are much less in the rows $2-5$ of Table 2. Our explanation is the following: since the impatience rates are small the customers abandon very rarely and in the realizations only the short durations happen. Thus the sample mean of these few durations cannot be considered as the true estimation of the hypothetical expectation $E(WAO)$.

In Table 3 the mean total service time of a successful customer $E(STS)$ decreases but the mean of the uninterrupted service time of a successful request increases. We have similar explanation, namely as more and more request abandons the system less and less customer needs service. Thus the number of collisions decreases and as a consequence the uninterrupted service times increase but their mean is less than the expectation of the hypothetical service time since only the shorter durations are considered. The behavior of the probabilities are reasonable, but again the utilization of the server with respect to the abandoned

Table 4. Estimations 3

Var(NS)	Var(T)	Var(TS)	Var(TA)	Var(W)	Var(WS)	Var(WA)	Var(ST)
40.007	21911.908	21911.908	0.000	21657.862	21657.862	0.000	1.000
40.124	21881.572	21881.345	23376.692	21627.907	21627.682	23112.600	0.999
40.251	20963.887	20960.814	21088.319	20719.418	20716.321	20844.001	0.975
40.608	14752.973	14728.385	14854.131	14571.818	14547.085	14672.623	0.805
32.106	2326.767	2264.987	2360.505	2287.902	2225.639	2321.008	0.386
7.570	65.502	44.016	69.292	63.994	41.980	67.298	0.233
0.980	0.999	0.387	1.141	0.867	0.118	0.945	0.205
0.344	0.215	0.235	0.204	0.036	0.001	0.039	0.199

Table 5. Notation for the estimations

E(NS): mean number of customers in the system
E(T): mean sojourn time of an arbitrary customer
E(TS): mean sojourn time of a successfully served customer
E(TA): mean sojourn time of a reneging customer
E(O): mean number of customers in the orbit
E(W): mean waiting time of an arbitrary customer
E(WS): mean waiting time of a successfully served customer
E(WA): mean waiting time of a abandoned customer
E(ST): mean total service time of an arbitrary customer
E(STS): mean total service time of a successfully served customer
E(STA): mean total service time of a reneging customer
Pa: probability of abandonment
Us: server utilization
Var(NS): variance of number of customers in the system
Var(T): variance of sojourn time of an arbitrary customer
Var(TS): variance of sojourn time of a successfully served customer
Var(TA): variance of sojourn time of a reneging customer
Var(W): variance of waiting time of an arbitrary customer
Var(WS): variance of waiting time of a successfully served customer
Var(WA): variance of waiting time of a reneging customer
Var(ST): variance of total service time of an arbitrary customer
Var(STS): variance of total service time of a successful customer
Var(STA): variance of total service time of an abandoned customer
Pao: conditional probability that an abandoned customer never leaves the orbit
E(WAO): mean waiting time of an abandoned customer who never leaves the orbit
E(STSI): mean total interrupted service time of a successful customer
E(STSUI): mean total uninterrupted service time of a successful customer
UsA: server utilization of an abandoned customer
UsS: server utilization of a successfully served customer

customers UsA is surprising. We cannot explain why at the beginning it increases then decreases.

3.2 Generally Distributed Impatience Times

Our aim is to examine how the different distributions of impatience of calls have an effect on the performance measure when the mean and variance are equal, respectively. The investigations are divided into two parts depending on the squared coefficient of variation.

Squared Coefficient of Variation is Greater than One

In the first part Table 6 shows the parameters of distinct distributions. The parameters are chosen in such a way that the squared coefficient of variation would be greater than one. For comparison hyper-exponential, gamma, lognormal and Pareto distributions are used besides the case when the impatience time is constant. Our simulation program is equipped with random number generators and these functions need input parameters which are different in every distribution.

Numerical values of model parameters are the following:

$$N = 100 \quad \lambda/N = 0.01 \quad \gamma_0 = 0.1 \quad \gamma_1 = 0.1 \quad \gamma_2 = 1 \quad \sigma/N = 0.1 \quad \mu = 1$$

Table 6. Parameters of impatience distributions, squared coefficient of variation is greater than one

Distribution	Gamma	Hyper-exponential	Pareto	Lognormal
Parameters	$\alpha = 0.390625$	$p = 0.33098$	$\alpha = 2.1792$	$m = 5.57973$
	$\beta = 0.0007813$	$\lambda_1 = 0.00132$	$k = 270.56302$	$\sigma = 1.12684$
		$\lambda_2 = 0.00268$		
Mean	500			
Variance	640000			
Squared coefficient of variation	2.56			

Figure 2 shows the comparison of steady-state distribution of the number of customers in the system. Taking a closer look on the results all the curves correspond to a normal distribution, the explanation can be found in paper [18]. However, this figure clearly displays the contrast among the applied distributions. Although the shape of the curves is almost the same the average number of customers in the system varies a little bit especially in case of Pareto distribution and when the impatience time of calls is constant the mean is greater compared to the others.

The mean response time of different types of customer is shown in function of arrival intensity on Figs. 3, 4, 5. Figure 4 illustrates how the mean response time of impatient customers changes. The mean waiting time in the orbit should

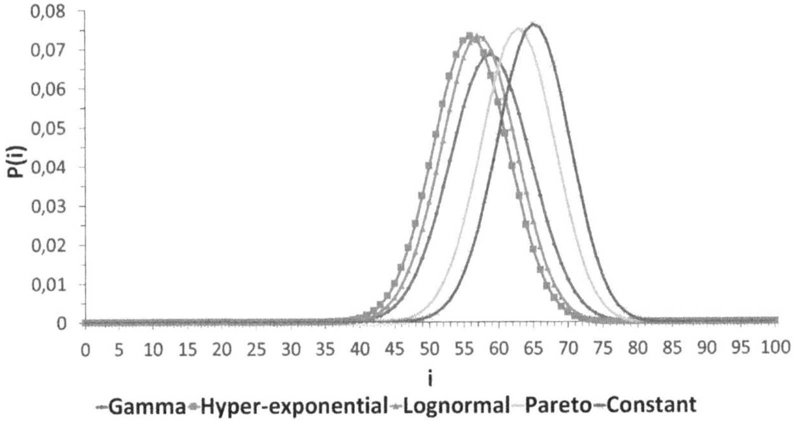

Fig. 2. Comparison of steady-state distributions

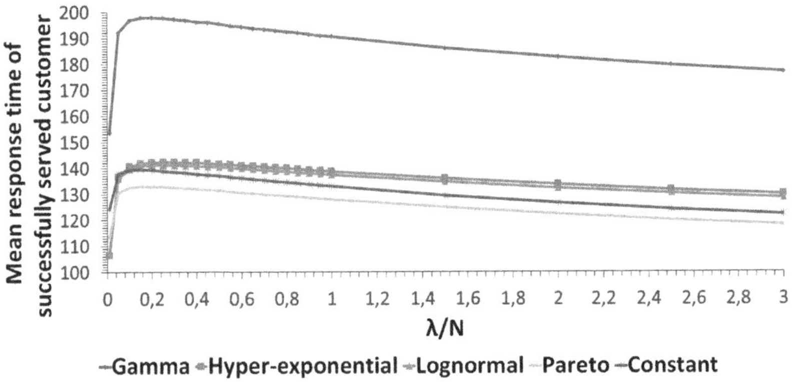

Fig. 3. Mean response time vs. arrival intensity using various distributions

be constant, due to the constant impatient time of a customer. Of course, Fig. 5 can be obtained with the help of the law of total expectation, too. Interestingly, differences can be observed even though the first two moments are equal, respectively. Results clearly illustrate the effect of various distributions. Highest values are experienced at gamma distribution in the case of successful customers, but in the case of impatient calls, constant impatience time gives the greatest values. Despite the increasing arrival intensity the maximum property characteristic of finite-source retrial queueing systems occurs under suitable parameter settings as we mentioned in [18].

Figure 6 demonstrates how the probability of abandonment of a customer changes with the increment of the arrival intensity. Under probability of abandonment we mean the probability of that a customer leaves the system without getting its full service requirement (through the orbit). After a slow increase of the value of this performance measure it stagnates which is true for every used

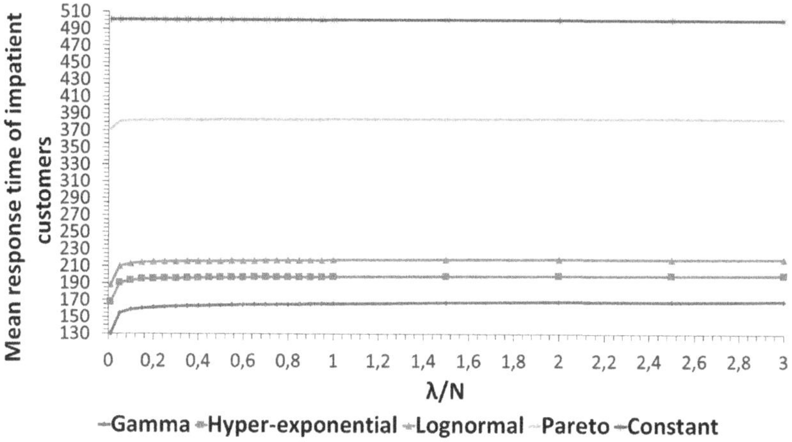

Fig. 4. Mean response time vs. arrival intensity using various distributions

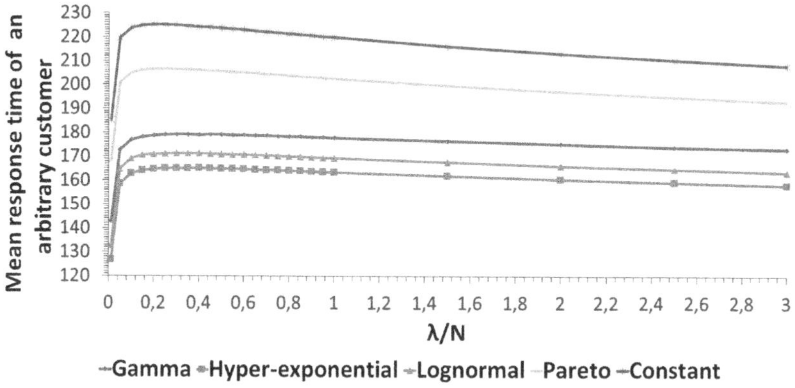

Fig. 5. Mean response time vs. arrival intensity using various distributions

distributions of impatience of calls but they differ significantly from each other. At gamma distribution the tendency of leaving the system earlier is much higher than the others especially compared to at constant mean of impatience of calls. Here the disparity is much higher among the applied distributions compared to the previous Figures. An explanation of this feature could be the following: if the squared coefficient of variation is greater than one the gamma distribution takes small values with great probability, so the customers leave the system quite early and thus the probability of abandonment is high.

Figure 7 is related to the total utilization of server versus arrival intensity. Total utilization contains every service time including the interrupted ones no matter whether a call departed from the service unit or from the orbit. By examining closely the Figure we find prominent results when gamma distribution is applied and regarding the others the received values are almost identical. With

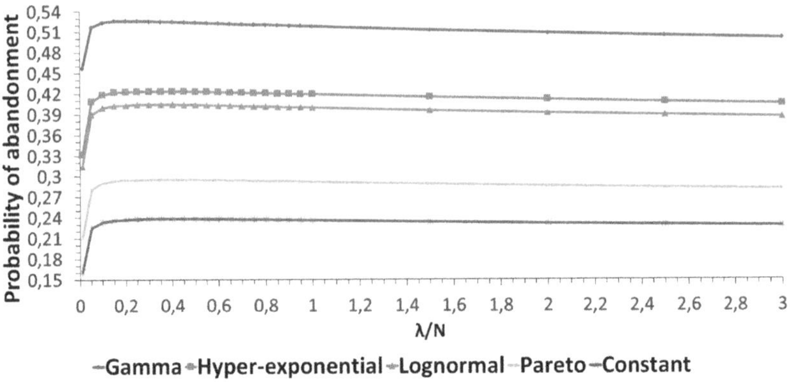

Fig. 6. Probability of abandonment of a customer vs. arrival intensity using various distributions

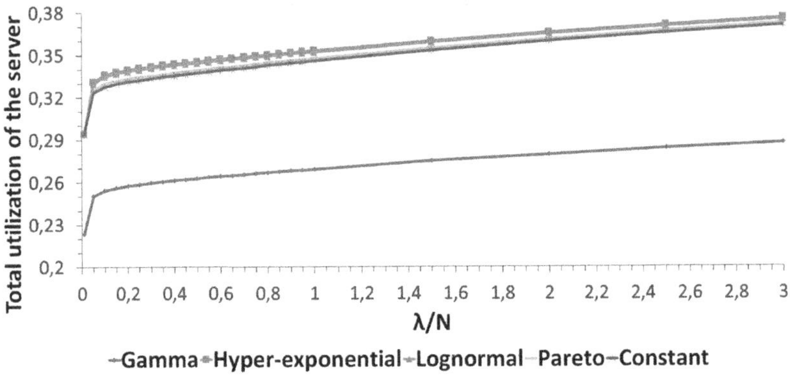

Fig. 7. Total utilization of server vs. arrival intensity using various distributions

the increment of arrival intensity the total utilization of the service unit increases as well. Here the explanation is the same as in the previous Figure, that is there are less customers in the system and hence the utilization is the smallest.

Squared Coefficient of Variation is Less than One
We carried the simulation in the case when the squared coefficient of variation is less than one, the mean is the same as before. The differences are noticeable but not as big as in the previous cases.

After viewing the above outcomes and figures we are intrigued to know how the operation of the system changes if another parameter setting is used. To do so we modify the parameters in order the squared coefficient of variation to be less than one so hyper-exponential is exchanged for hypo-exponential distribution. Table 7 contains the modified parameter setting of distribution of impatience of calls. Other parameters remain unchanged (see Table 6).

Table 7. Parameters of impatience distributions, squared coefficient of variation is less than one

Distribution	Gamma	Hypo-exponential	Pareto	Lognormal
Parameters	$\alpha = 1.47059$	$\mu_1 = 0.01$	$\alpha = 2.5718$	$m = 5.9552$
	$\beta = 0.002941$	$\mu_2 = 0.0025$	$k = 305.5844$	$\sigma = 0.72027$
Mean	500			
Variance	170000			
Squared coefficient of variation	0.68			

First, Fig. 8 represents the steady state distribution of the number of customers in the system. Analyzing the curves in more detail they are much closer to each other as on Fig. 2. Differences appear among the applied distributions with this parameter setting, too. As regard to the values with these parameters the mean number of customer is higher in case of every distribution. We think that this feature is due to the smaller variance of impatience time and customers stay in the orbit for a longer time since in the realization of the simulation there are less early abandonment.

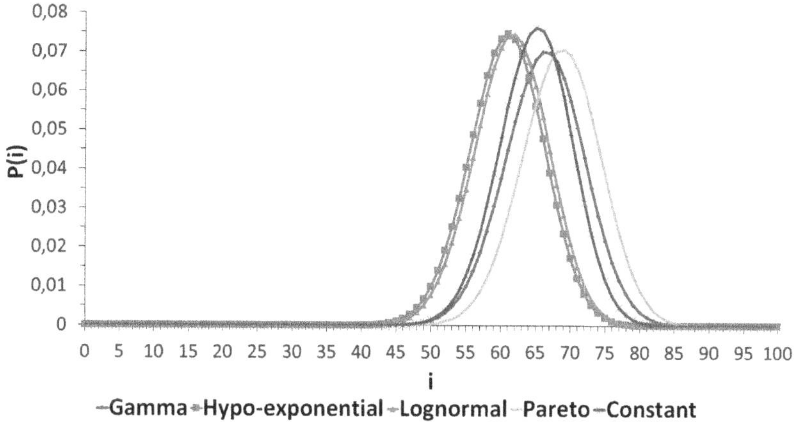

Fig. 8. Comparison of steady-state distributions

The next Figures show the mean response time of different types of customers in function of arrival intensity. Examining Figures the same tendency can be seen as on the previous corresponding Figures but differences can still be discovered especially in case of gamma distribution. These Figures also reveal that successful customers on the average spend less time in the system compared to the previous parameter setting but the impatient and arbitrary ones spend longer time in the system (Figs. 9, 10 and 11).

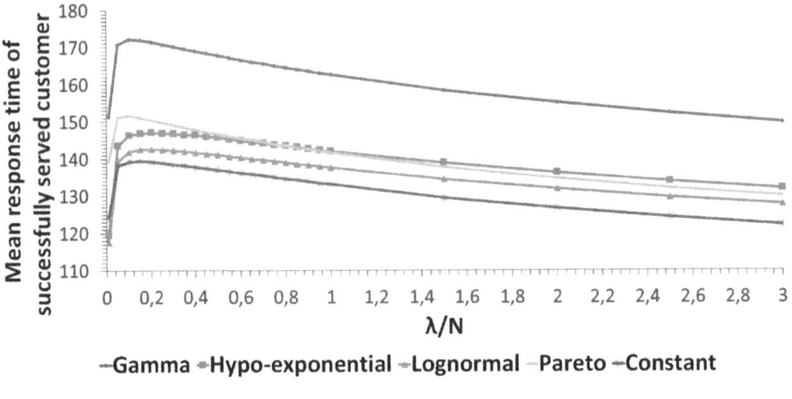

Fig. 9. Mean response time vs. arrival intensity using various distributions

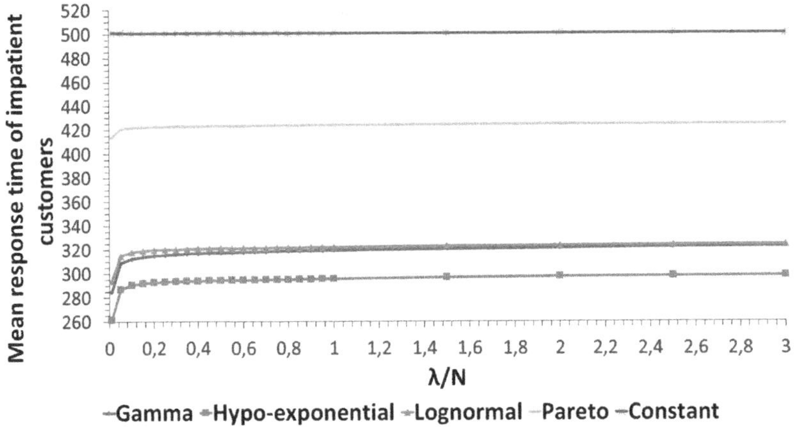

Fig. 10. Mean response time vs. arrival intensity using various distributions

Figure 12 demonstrates the probability of abandonment of a customer versus arrival intensity. Not surprisingly after seeing the previous two Figures the difference of achieved values are relatively far from each other, disparity is still present among the applied distributions. We can observe that the probability of abandonment is less than in the case of previous case, see Fig. 6. Our explanation is the same as in the case of number of customers in the system.

Lastly, on Fig. 13 the running parameter (value of x-axis) is the arrival intensity and value of y-axis is the total utilization of the server. Among the lines there are not so significant differences, they coincide with each other meaning that the utilization is almost the same except in case of Pareto and gamma distribution where the utilization of service unit is significantly less.

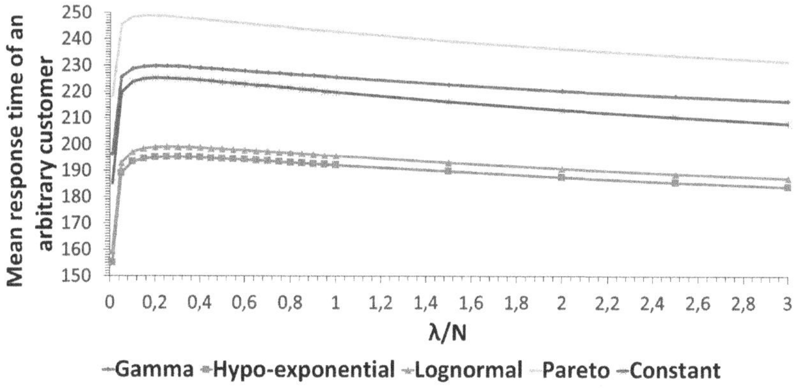

Fig. 11. Mean response time vs. arrival intensity using various distributions

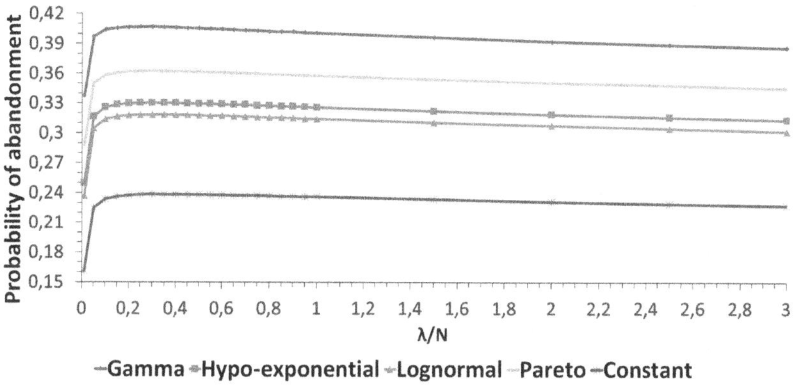

Fig. 12. Probability of abandonment of a customer vs. arrival intensity using various distributions

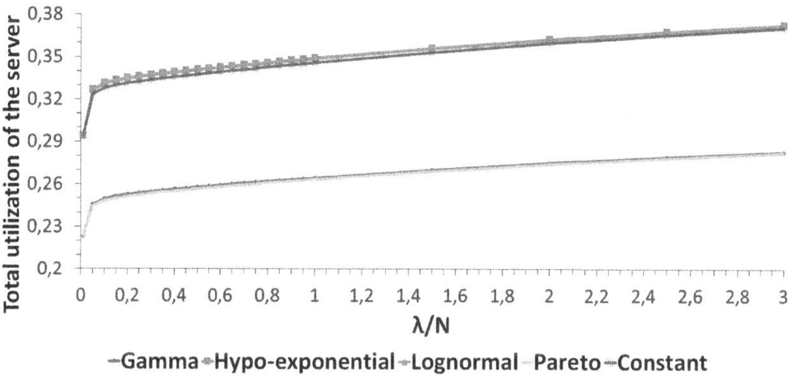

Fig. 13. Total utilization of server vs. arrival intensity using various distributions

4 Conclusion

In this paper a finite-source retrial queueing system is presented with a non-reliable server, collisions and impatient customers. The obtained results fully demonstrated how essential is the distribution of impatience of calls because it has a great influence on the system characteristics despite the fact that the mean and the variance are the same, respectively. Figures in connection several performance measures, for example the probability of abandonment clearly assure this phenomenon. Results evidently indicated the distinction is noticeable and significant among the performance measures when the squared coefficient of variation is greater than one and moderate when it is less than one. In the future we would like to deal with more distributions to expand our investigation and examine the performance measures when the distribution of service time is not exponential. We also would like to analyze systems of two-way communications with impatient customers.

References

1. Adan, I., Hathaway, B., Kulkarni, V.: On first-come, first-served queues with two classes of impatient customers. Queueing Syst. **91**(1–2), 113–142 (2019)
2. Aissani, A., Lounis, F., Hamadouche, D., Taleb, S.: Analysis of customers' impatience in a repairable retrial queue under postponed preventive actions. Am. J. Math. Manag. Sci. **38**(2), 125–150 (2019)
3. Artalejo, J.R., Pla, V.: On the impact of customer balking, impatience and retrials in telecommunication systems. Comput. Math. Appl. **57**(2), 217–229 (2009)
4. Danilyuk, E., Moiseeva, S., Nazarov, A.: Asymptotic analysis of retrial queueing system $M/GI/1$ with collisions and impatient calls. In: Dudin, A., Nazarov, A., Moiseev, A. (eds.) ITMM 2019. CCIS, vol. 1109, pp. 230–242. Springer, Cham (2019). https://doi.org/10.1007/978-3-030-33388-1_19
5. Danilyuk, E., Moiseeva, S., Sztrik, J.: Asymptotic analysis of retrial queueing system $M/M/1$ with impatient customers, collisions and unreliable server. J. Siberian Federal Univ. Math. Phys. **13**(2), 218–230 (2020)
6. Dudin, A.: Operations research perspectives. Oper. Res. **5**, 245–255 (2018)
7. Haight, F.A.: Queueing with reneging. Metrika **2**(1), 186–197 (1959)
8. Jain, M., Rani, S.: Markovian model of unreliable server retrial queue with discouragement. Proc. Natl. Acad. Sci. India Section A: Phys. Sci. 1–8 (2020)
9. Kim, C., Dudin, A., Dudina, O., Klimenok, V.: Analysis of queueing system with non-preemptive time limited service and impatient customers. Methodol. Comput. Appl. Probab. 1–32 (2019)
10. Kim, J.S.: Retrial queueing system with collision and impatience. Commun. Korean Math. Soc. **25**(4), 647–653 (2010)
11. Kobayashi, H., Mark, B.L.: System Modeling and Analysis: Foundations of System Performance Evaluation. Pearson Education India (2009)
12. Kuki, A., Bérczes, T., Tóth, Á., Sztrik, J.: Numerical analysis of finite source Markov retrial system with non-reliable server, collision, and impatient customers. In: Annales Mathematicae et Informaticae, vol. 51, pp. 53–63. Liceum University Press (2020)

13. Kumar, R., Som, B.K.: A multi-server queue with reverse balking and impatient customers. Pak. J. Stat. **36**(2), 91–101 (2020)
14. Lakaour, L., Aissani, D., Aissanou, K., Barkaoui, K.: $M/M/1$ retrial queue with collisions and transmission errors. Methodol. Comput. Appl. Probab. 1–12 (2018)
15. Law, A.M.: Statistical analysis of simulation output data: the practical state of the art. In: Proceedings of the 2015 Winter Simulation Conference, pp. 1810–1819. IEEE Press (2015)
16. Law, A.M., Kelton, W.D.: Simulation Modeling and Analysis. McGraw-Hill, New York (1991)
17. Lowndes, V., Berry, S.: Introduction to the use of queueing theory and simulation. In: Berry, S., Lowndes, V., Trovati, M. (eds.) Guide to Computational Modelling for Decision Processes. SFMA, pp. 145–171. Springer, Cham (2017). https://doi.org/10.1007/978-3-319-55417-4_5
18. Nazarov, A., Sztrik, J., Kvach, A.: A survey of recent results in finite-source retrial queues with collisions. In: Dudin, A., Nazarov, A., Moiseev, A. (eds.) ITMM/WRQ -2018. CCIS, vol. 912, pp. 1–15. Springer, Cham (2018). https://doi.org/10.1007/978-3-319-97595-5_1
19. Rubinstein, R.Y., Kroese, D.P.: Simulation and the Monte Carlo Method. Wiley, Hoboken (2016)
20. Satin, Y., Zeifman, A., Sipin, A., Ammar, S., Sztrik, J.: On probability characteristics for a class of queueing models with impatient customers. Mathematics **8**(4), 594 (2020)
21. Tóth, Á., Bérczes, T., Sztrik, J., Kvach, A.: Simulation of finite-source retrial queueing systems with collisions and non-reliable server. In: Vishnevskiy, V.M., Samouylov, K.E., Kozyrev, D.V. (eds.) DCCN 2017. CCIS, vol. 700, pp. 146–158. Springer, Cham (2017). https://doi.org/10.1007/978-3-319-66836-9_13
22. Tóth, Á., Sztrik, J.: Simulation of finite-source retrial queuing systems with collisions, non-reliable server and impatient customers in the orbit. In: Proceedings of 11th International Conference on Applied Informatics, Eger, Hungary. CEUR Workshop Proceedings (CEUR-WS.org), vol. 2650, pp. 408–419 (2020). http://ceur-ws.org/Vol-2650/
23. Wehrle, K., Günes, M., Gross, J.: Modeling and Tools for Network Simulation. Springer, Heidelberg (2010). https://doi.org/10.1007/978-3-642-12331-3
24. Wu, H., He, Q.M.: Double-sided queues with marked Markovian arrival processes and abandonment. Stoch. Models 1–36 (2020)

Improved Priority Scheme for Unreliable Queueing System

Alexander Dudin[1,2(✉)], Olga Dudina[1], and Sergey Dudin[1]

[1] Belarusian State University, 4 Nezavisimosti Avenue, 220030 Minsk, Belarus
{dudin,dudina,dudins}@bsu.by

[2] Peoples' Friendship University of Russia (RUDN University),
6 Miklukho-Maklaya Street, Moscow 117198, Russian Federation

Abstract. A novel flexible discipline for providing priority in a single-server queue is applied to the system where the service can be unreliable what results in the loss of a customer or repetition of its service. According to this discipline, arriving customers are stored in the finite buffers dedicated to the customers of the corresponding type if the buffer is not full. After the staying in the corresponding buffer during the exponentially distributed time, the customers try to enter the main buffer of a finite capacity which is common for both types of customers. If this main buffer is full, the customer returns to the dedicated buffer and repeats the attempts to enter the main buffer later. Customers staying in the dedicated buffers are impatient and can go away from the system after a certain patience period duration of which has an exponential distribution. Customers of both types, which succeed to enter the main buffer, are picked up for the service in the order of their admission to this buffer. Providing the preference to priority customers is managed via the corresponding choice of capacities of the dedicated buffers and the rate of trials to transfer to the main buffer. Performance measures of this system are obtained under the assumption that the arrival flows of two types of customers are defined by the Markov arrival processes and the service time has the distribution of phase-type with failures type. Some aspects relating to an optimal choice of the parameters of the system are discussed via numerical experiments.

Keywords: Fair priority · Markov arrival process · Impatience · Phase-type distribution with failures

1 Introduction

Queueing theory is a powerful tool for performance evaluation, resource planning, and scheduling in various areas including various telecommunication systems and networks, information and health care systems, transportation, etc. Despite the

This paper has been supported by the RUDN University Strategic Academic Leadership Program.

A. Dudin et al. (Eds.): ITMM 2020, CCIS 1391, pp. 16–30, 2021.
https://doi.org/10.1007/978-3-030-72247-0_2

very quick development of new equipment and technologies in these networks and systems, the demand for service also quickly grows and the problem of the effective use of the available resources remains to be very important. Since the users are usually quite heterogeneous in respect to the most critical indicators of their service, there is an opportunity to provide some priority to the classes of users having more strict requirements, e.g., to the latency, jitter, loss probability, etc., and the corresponding financial resources to pay for better quality of service. Therefore, although the theory of priority queues is already quite old and well developed, see, e.g., classical books [6,8], new models and new challenges arise in various real-world systems. This stimulates the continuation of research in the theory of priority queues.

In priority queues, all arriving customers are divided into several classes and customers from the different classes have a different treatment in the system. For simplicity, we will speak about two classes. Users from one class are called the priority customers while the users from the second class are considered non-priority users. The priority can work at the stage of the user arrival to the system and decision making whether to admit this customer to the system or drop it and (or) at the moment of service completion when it is necessary to decide user of which type will be picked up for the next service first. More flexible priority schemes, so-called dynamic priorities, take into account at the decision moment both the established in advance priorities of classes and lengths of queues of two types. However, this flexibility (and overall quality of service) is achieved at the expense of permanent monitoring the system states which can be costly. In contrast to them, the so-called static priorities, take into account only the established in advance priorities.

There are two kinds of static priorities: preemptive and non-preemptive priorities. The preemptive priority assumes that service of a non-priority customer is interrupted when a priority customer arrives at the system. E.g., the arrival of a primary user having a license for service in the radio system interrupts the service of a cognitive user that occupied a temporarily available server. The non-preemptive priority suggests that service of a non-priority customer is not interrupted when a priority customer arrives at the system. The priority plays a role only at the moment of starting a new service. E.g., the arrival of a handover user into the cell of a mobile communication network does not interrupt service of an ordinary user. But the handover user has the preference in access to the server that becomes idle.

The preemptive priority creates ideal conditions for service of the priority users that may practically ignore the existence of non-priority customers. However, this type of priority is rather unfair to non-priority customers whose waiting time can be very long. Therefore, various disciplines that are more benevolent concerning non-priority customers are invented. E.g., these can be the disciplines when one non-priority customer can receive service after service of the certain number of priority customers in turn or in the case when all priority customers finish the service, not only one but several non-priority customers can receive service in turn. Sometimes, the possibility of an increase of the priority of a customer is allowed if its waiting time becomes too long, see, e.g., [9,11,13,14]. Another way to provide more fair priority was offered in [5]. There, before acceptance to the main buffer, from which the customers are picked up for service, customers of both types attend the dedicated finite auxiliary buffers (storages). Each customer from these buffers can transit to the main buffer after a random amount of time having an exponential distribution with the rate depending on the type of the customer. The degree of the preference given to the priority customers can be smoothly varied via the proper choice of the capacities of the dedicated buffers and the rates of transitions from them into the main buffer. Analysis of this discipline as implemented in [5] under assumptions that the arrival flow of customers is defined by the Marked Markov arrival process, the main buffer has an infinite capacity and the service time of all customers has phase-type (PH) distribution.

In this paper, we suggest that arrivals are defined by two independent Markov arrival processes, the main buffer has a finite capacity and the service times have so-called phase-type with failures (PHF) distribution (see, [3]) that supposes possibilities of incorrect service of a customer with options to lose the customer or repeat its service from the early beginning or from a certain phase of the underlying process of service at which the breakdown occurred.

The outline of the presentation of the results in this paper is the following. In Sect. 2, the mathematical model is completely described. The process of system states is formally defined in Sect. 3. This process is a multi-dimensional continuous-time Markov chain. The generator of this Markov chain is derived. Expressions for the key performance measures of the system are given in Sect. 4. The results of numerical experiments are presented in Sect. 5. Section 6 concludes the paper.

2 Mathematical Model

We consider a queueing system the structure of which is presented in Fig. 1.

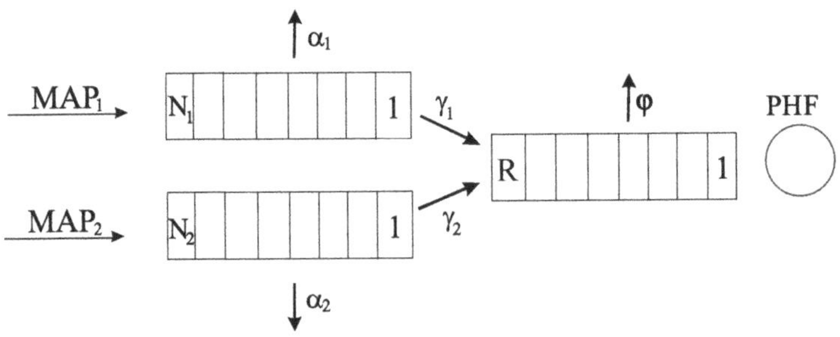

Fig. 1. Queueing system under study

The system has one server and a finite main buffer of capacity R. Customers of two types arrive at the system. The arrival flow of type-1 customers is defined by the MAP (Markov Arrival Process), see, e.g., [2,12,15]. This process is coded as MAP_1 and is defined by the irreducible continuous-time Markov chain ν_t, $t \geq 0$, having a finite state space $\{1, 2, ..., W_1\}$ and the matrices $D_0^{(1)}$ and $D_1^{(1)}$. The matrix $D_1^{(1)}$ consists of the intensities of transitions of the chain ν_t that are accompanied by the arrival of a customer. The non-diagonal entries of the matrix $D_0^{(1)}$ define the intensity of the corresponding transition of the chain ν_t without the generation of customers, and the modules of the negative diagonal entries define the rates of the exit of the process ν_t from the corresponding states. The matrix $D^{(1)}(1) = D_0^{(1)} + D_1^{(1)}$ is the infinitesimal generator of the Markov chain ν_t.

The arrival flow of type-2 customers is also defined by the Markov arrival process. It coded as MAP_2 and is defined by the irreducible continuous-time Markov chain v_t, $t \geq 0$, having a finite state space $\{1, 2, ..., W_2\}$ and the generator $D^{(2)}(1) = D_0^{(2)} + D_1^{(2)}$.

The average intensity of type-l, $l = 1, 2$, customers (fundamental rate) λ_l is defined by the formula

$$\lambda_l = \boldsymbol{\theta}_l D_1^{(l)} \mathbf{e}$$

where $\boldsymbol{\theta}_1$ and $\boldsymbol{\theta}_2$ are the row vectors of the stationary probabilities of the Markov chains ν_t and v_t correspondingly. The vector $\boldsymbol{\theta}_l$, $l = 1, 2$, is the unique solution to the system

$$\boldsymbol{\theta}_l D^{(l)}(1) = \mathbf{0}, \ \boldsymbol{\theta}_l \mathbf{e} = 1.$$

Here and throughout this paper, $\mathbf{e}$ is a column vector of appropriate size consisting of ones, and $\mathbf{0}$ is a row vector of appropriate size consisting of zeroes.

We assume that each type customers are placed upon arrival into the dedicated finite buffer (storage). The storage for type-l, $l = 1, 2$, customers has the finite capacity N_l. If the storage is full at a customer arrival epoch, the customer leaves the system permanently (is lost). Customers of the first (second) flow transfer from the storage, independently of all other customers staying in the storage, to the finite main buffer after a random amount of time that is exponentially distributed with the parameter $\gamma_1(\gamma_2)$. We assume that $\gamma_1 > \gamma_2$. It means that customers from the first flow more quickly transfer to the main buffer what creates for them a priority over customers from the second flow. If the main buffer is full at the moment when a customer of any type from the storage tries to enter into it, this customer is returned into the corresponding storage.

The customers staying in each storage are assumed to be impatient. A type-l customer leaves the storage (is lost) independently of all other customers staying in the storage after an exponentially distributed with the parameter α_l, $\alpha_l \geq 0$, $l = 1, 2$, amount of time.

After entering the main buffer, the customers of both types are assumed to become identical. The service time of an arbitrary customer has a PHF distribution, see [3]. The PHF type distribution of the service time is defined by a continuous-time Markov chain m_t, $t \geq 0$, with M transient states $\{1, \ldots, M\}$ and two absorbing states $M+1$ and $M+2$. The initial state of the process m_t is chosen among the transient states in accordance with a stochastic row vector $\boldsymbol{\beta} = (\beta_1, \ldots, \beta_M)$. The intensities of the transition of the process m_t between transient states are defined by the matrix S.

The intensities of the transition to the first absorbing state $M+1$ are defined by the entries of the column vector $\mathbf{S}_1$. We assume that the transition to the first absorbing state corresponds to successful service completion. The transition to the second absorbing state $M+2$ means that a failure occurs. The intensities of such transition are given by the entries of the column vector $\mathbf{S}_2$. Here $\mathbf{S}_1 + \mathbf{S}_2 = -S\mathbf{e}$.

We suppose, that if during the service of a customer failure occurs, then with probability q_1 the customer leaves the system forever (is lost); with probability q_2, the service of the customer starts from the early beginning, and with probability $1 - q_1 - q_2$ the service of the customer resumes from the state of the process m_t where the failure occurred.

The customers staying in the main buffer are assumed to be impatient. Each customer, which is not picked up for service, leaves the buffer independently of all other customers after an exponentially distributed with the parameter φ amount of time.

To avoid starvation of the server and improve the performance of the system, we assume that if at a service completion moment, the buffer is empty, but the first storage is not empty, a type-1 customer is immediately picked up from the storage and starts service. If there is no type-1 customer in the storage, but there are type-2 customers in the second storage, the type-2 customer is picked up from this storage and immediately starts service. If both storages are empty, the server stays idle until the first arrival of any type customer, who immediately starts service without visiting storage.

It is worth noting that the considered system can be also interpreted as an unreliable single-server system with a finite buffer, customer retrials, two separate finite orbits for retrying customers and instantaneous search of a customer in the orbits in the case when the server becomes idle and the main buffer is empty.

3 Process of System States

Let, during the epoch t, $t \geq 0$,

- i_t, $i_t = \overline{0, R+1}$, be the number of customers in the main buffer and on the server,
- $n_t^{(l)}$, $n_t^{(l)} = \overline{0, N_l}$, be the number of customers in the lth storage, $l = 1, 2$,
- ν_t, $\nu_t = \overline{1, W_1}$, be the state of the underlying process of the MAP_1,
- v_t, $v_t = \overline{1, W_2}$, be the state of the underlying process of the MAP_2,
- m_t, $m_t = \overline{1, M}$, be the state of PHF service process.

The Markov chain $\xi_t = \{i_t, n_t^{(1)}, n_t^{(2)}, \nu_t, v_t, m_t\}$, $t \geq 0$, is a regular irreducible continuous-time Markov chain. It has the following finite state space:

$$\Big(\{0,0,0,\nu,v\}\Big)\bigcup\Big(\{i,n^{(1)},n^{(2)},\nu,v,m\}, i=\overline{1,R+1},\ n^{(1)}=\overline{0,N_1},\ n^{(2)}=\overline{0,N_2},$$

$$m=\overline{1,M}\Big),\ \nu=\overline{1,W_1},\ v=\overline{1,W_2}.$$

Let us enumerate the states of the Markov chain ξ_t in the lexicographic order and refer to the set of states of the chain having value i of the first component of the Markov chain as level i, $i=\overline{0,R+1}$.

Let Q be the generator of the Markov chain ξ_t, $t \geq 0$.

Lemma 1. The generator Q has the following block-tridiagonal structure:

$$Q = \begin{pmatrix} Q_{0,0} & Q_{0,1} & O & O & O & \dots & O & O \\ Q_{1,0} & Q_{1,1} & Q^{+} & O & O & \dots & O & O \\ O & Q_{2,1} & Q_{2,2} & Q^{+} & O & \dots & O & O \\ O & O & Q_{3,2} & Q_{3,3} & Q^{+} & \dots & O & O \\ \vdots & \vdots & \vdots & \vdots & \vdots & \ddots & \vdots & \vdots \\ O & O & O & O & O & \dots & Q_{R,R} & Q^{+} \\ O & O & O & O & O & \dots & Q_{R+1,R} & Q_{R+1,R+1} \end{pmatrix}.$$

The non-zero blocks $Q_{i,j}$, $i,j=\overline{0,R+1}$, containing the intensities of the transitions from level i to level j have the following form:

$$Q_{0,0} = D_0^{(1)} \oplus D_0^{(2)},$$

$$\begin{aligned} Q_{1,1} &= I_{(N_1+1)(N_2+1)} \otimes (D_0^{(1)} \oplus D_0^{(2)} \oplus S) \\ &+ Z_1 \otimes I_{N_2+1} \otimes D_1^{(1)} \otimes I_{W_2 M} + I_{N_1+1} \otimes Z_2 \otimes I_{W_1} \otimes D_1^{(2)} \otimes I_M \\ &- (\alpha_1+\gamma_1)A_1 \otimes I_{(N_2+1)W_1W_2M} - (\alpha_2+\gamma_2)I_{N_1+1} \otimes A_2 \otimes I_{W_1W_2M} \\ &+ \alpha_1 A_1 E_1 \otimes I_{(N_2+1)W_1W_2M} + \alpha_2 I_{N_1+1} \otimes A_2E_2 \otimes I_{W_1W_2M} \\ &+ (E_1 \otimes I_{N_2+1} + F \otimes E_2) \otimes I_{W_1W_2} \otimes (\boldsymbol{S}_1 + q_1\boldsymbol{S}_2)\boldsymbol{\beta} \\ &+ I_{(N_1+1)(N_2+1)W_1W_2} \otimes (q_2\boldsymbol{S}_2\boldsymbol{\beta} + (1-q_1-q_2)\mathrm{diag}\{(\boldsymbol{S}_2)_l,\ l=\overline{1,M}\}), \end{aligned}$$

$$\begin{aligned} Q_{i,i} &= I_{(N_1+1)(N_2+1)} \otimes (D_0^{(1)} \oplus D_0^{(2)} \oplus S) \\ &+ Z_1 \otimes I_{N_2+1} \otimes D_1^{(1)} \otimes I_{W_2 M} + I_{N_1+1} \otimes Z_2 \otimes I_{W_1} \otimes D_1^{(2)} \otimes I_M \\ &- (i-1)\varphi I_{(N_1+1)(N_2+1)W_1W_2M} \\ &- (\alpha_1+\gamma_1)A_1 \otimes I_{(N_2+1)W_1W_2M} - (\alpha_2+\gamma_2)I_{N_1+1} \otimes A_2 \otimes I_{W_1W_2M} \\ &+ \alpha_1 A_1 E_1 \otimes I_{(N_2+1)W_1W_2M} + \alpha_2 I_{N_1+1} \otimes A_2E_2 \otimes I_{W_1W_2M} \end{aligned}$$

$$+I_{(N_1+1)(N_2+1)W_1W_2} \otimes (q_2\boldsymbol{S}_2\boldsymbol{\beta} + (1-q_1-q_2)\text{diag}\{(\boldsymbol{S}_2)_l,\ l=\overline{1,M}\}),\ i=\overline{2,R},$$

$$Q_{R+1,R+1} = I_{(N_1+1)(N_2+1)} \otimes (D_0^{(1)} \oplus D_0^{(2)} \oplus S)$$

$$+Z_1 \otimes I_{N_2+1} \otimes D_1^{(1)} \otimes I_{W_2M} + I_{N_1+1} \otimes Z_2 \otimes I_{W_1} \otimes D_1^{(2)} \otimes I_M$$

$$-(i-1)\varphi I_{(N_1+1)(N_2+1)W_1W_2M}$$

$$-\alpha_1 A_1 \otimes I_{(N_2+1)W_1W_2M} - \alpha_2 I_{N_1+1} \otimes A_2 \otimes I_{W_1W_2M}$$

$$+\alpha_1 A_1 E_1 \otimes I_{(N_2+1)W_1W_2M} + \alpha_2 I_{N_1+1} \otimes A_2 E_2 \otimes I_{W_1W_2M}$$

$$+I_{(N_1+1)(N_2+1)W_1W_2} \otimes (q_2\boldsymbol{S}_2\boldsymbol{\beta} + (1-q_1-q_2)\text{diag}\{(\boldsymbol{S}_2)_l,\ l=\overline{1,M}\}),$$

$$Q_{0,1} = \boldsymbol{c}_1 \otimes \boldsymbol{c}_2 \otimes (D_1^{(1)} \oplus D_1^{(2)}) \otimes \boldsymbol{\beta},$$

$$Q_{i,i+1} = Q^+ = \gamma_1 A_1 E_1 \otimes I_{(N_2+1)W_1W_2M}$$

$$+\gamma_2 I_{N_1+1} \otimes A_2 E_2 \otimes I_{W_1W_2M},\ i=\overline{1,R},$$

$$Q_{1,0} = (\boldsymbol{c}_1)^T \otimes (\boldsymbol{c}_2)^T \otimes I_{W_1W_2} \otimes (\boldsymbol{S}_1 + q_1\boldsymbol{S}_2),$$

$$Q_{i,i-1} = I_{(N_1+1)(N_2+1)W_1W_2} \otimes (\boldsymbol{S}_1 + q_1\boldsymbol{S}_2)\boldsymbol{\beta}$$

$$+(i-1)\varphi I_{(N_1+1)(N_2+1)W_1W_2M},\ i=\overline{2,R+1},$$

where

I is the identity matrix, and O is a zero matrix. If the dimension of the matrix is not clear from the context, it can be indicated as subscript;

$A_1 = \text{diag}\{0,1,\ldots,N_1\}$, where $\text{diag}\{0,1,\ldots,a\}$ is the diagonal matrix with the diagonal entries $\{0,1,\ldots,a\}$;

$A_2 = \text{diag}\{0,1,\ldots,N_2\}$;

E_l, $l=1,2$, is the square matrix of size N_l+1 with all zero entries except the entries $(E_l)_{k,k-1}$, $k=\overline{1,N_l}$, which are equal to one;

Z_l, $l=1,2$, is the square matrix of size N_l+1 with all zero entries except the entries $(Z_l)_{k,k+1}$, $k=\overline{0,N_l-1}$, and $(Z_l)_{N_l,N_l}$, which are equal to one;

F is the square matrix of size N_1+1 with all zero entries except the entry $(F)_{0,0}$, which is equal to one;

$\boldsymbol{c}_l$, $l=1,2$, is the row vector of size N_l+1 with all zero entries except the entry $(\boldsymbol{c}_l)_0$, which is equal to one;

$\otimes$ and $\oplus$ are the symbols of the Kronecker product and the sum of matrices; see, e.g., [7].

The Markov chain ξ_t, $t \geq 0$, is irreducible and has a finite state space. Thus, the following limits (stationary probabilities) exist:

$$\pi(0,0,0,\nu,v) = \lim_{t\to\infty} P\{i_t=0, n_t^{(1)}=0, n_t^{(2)}=0, \nu_t=\nu, v_t=v\},$$

$$\pi(i, n_t^{(1)}, n_t^{(2)}, \nu, v, m)$$

$$= \lim_{t\to\infty} P\{i_t=i, n_t^{(1)}=n^{(1)}, n_t^{(2)}=n^{(2)}, \nu_t=\nu, v_t=v,\ m_t=m\},$$

$i = \overline{1, R+1}$, $n^{(1)} = \overline{0, N_1}$, $n^{(2)} = \overline{0, N_2}$, $\nu = \overline{1, W_1}$, $v = \overline{1, W_2}, m = \overline{1, M}$.

Let us form the row vectors $\boldsymbol{\pi}_i$, $i = \overline{0, R+1}$, of these probabilities enumerated in the direct lexicographical order of components $n^{(1)}$, $n^{(2)}$, ν, v, m.

It is well known that the probability vectors $\boldsymbol{\pi}_i$, $i \geq 0$, satisfy the following system of linear algebraic equations:

$$(\boldsymbol{\pi}_0, \boldsymbol{\pi}_1, \ldots, \boldsymbol{\pi}_{R+1})Q = \mathbf{0}, \quad (\boldsymbol{\pi}_0, \boldsymbol{\pi}_1, \ldots, \boldsymbol{\pi}_{R+1})\mathbf{e} = 1$$

called equilibrium or Chapman–Kolmogorov equations.

This system is the finite one and there are several numerically stable methods for its solving that effectively use the sparse structure of the generator, see, e.g., [1,4,10].

4 Performance Measures

Having computed the vectors of the stationary probabilities $\boldsymbol{\pi}_i$, $i = \overline{0, R+1}$, it is possible to compute a variety of the performance measures of the system.

The average number of customers in the system is computed by:

$$L = \sum_{i=1}^{R+1} \sum_{n_1=0}^{N_1} \sum_{n_2=0}^{N_2} (i + n_1 + n_2)\boldsymbol{\pi}(i, n_1, n_2)\mathbf{e}.$$

The average number of customers in the main buffer is computed by:

$$N^{buf} = \sum_{i=2}^{R+1} (i-1)\boldsymbol{\pi}_i\mathbf{e}.$$

The average number of customers in the first storage is computed by:

$$N_1^{stor} = \sum_{i=1}^{R+1} \sum_{n_1=1}^{N_1} n_1\boldsymbol{\pi}(i, n_1)\mathbf{e}.$$

The average number of customers in the second storage is computed by:

$$N_2^{stor} = \sum_{i=1}^{R+1} \sum_{n_1=0}^{N_1} \sum_{n_2=1}^{N_2} n_2\boldsymbol{\pi}(i, n_1, n_2)\mathbf{e}.$$

The loss probability of an arbitrary type-1 customer upon arrival due to the first storage overflow is computed by:

$$P_1^{ent-loss} = \frac{1}{\lambda_1} \sum_{i=1}^{R+1} \sum_{n_2=0}^{N_2} \boldsymbol{\pi}(i, N_1, n_2)(D_1^{(1)} \otimes I_{W_2 M})\mathbf{e}.$$

The loss probability of an arbitrary type-2 customer upon arrival due to the second storage overflow is computed by:

$$P_2^{ent-loss} = \frac{1}{\lambda_2} \sum_{i=1}^{R+1} \sum_{n_1=0}^{N_1} \boldsymbol{\pi}(i, n_1, N_2)(I_{W_1} \otimes D_1^{(2)} \otimes I_M)\mathbf{e}.$$

The loss probability of an arbitrary type-1 customer due to impatience in the first storage is computed by:

$$P_1^{imp-loss} = \frac{1}{\lambda_1} \sum_{i=1}^{R+1} \sum_{n_1=1}^{N_1} n_1 \alpha_1 \boldsymbol{\pi}(i, n_1)\mathbf{e} = \frac{\alpha_1}{\lambda_1} N_1^{stor}.$$

The loss probability of an arbitrary type-2 customer due to impatience in the second storage is computed by:

$$P_2^{imp-loss} = \frac{1}{\lambda_2} \sum_{i=1}^{R+1} \sum_{n_1=0}^{N_1} \sum_{n_2=1}^{N_2} n_2 \alpha_2 \boldsymbol{\pi}(i, n_1, n_2)\mathbf{e} = \frac{\alpha_2}{\lambda_2} N_2^{stor}.$$

The intensity of the output flow of successfully served customers is computed by:

$$\lambda_{out} = \sum_{i=1}^{R+1} \boldsymbol{\pi}_i(\mathbf{e}_{(N_1+1)(N_2+1)W_1W_2} \otimes \mathbf{S}_1).$$

The intensity of the output flow of customers lost due to failures is computed by:

$$\lambda_{fail} = \sum_{i=1}^{R+1} \boldsymbol{\pi}_i(\mathbf{e}_{(N_1+1)(N_2+1)W_1W_2} \otimes q_1\mathbf{S}_2).$$

The probability of an arbitrary customer loss is computed by:

$$P_{loss} = 1 - \frac{\lambda_{out}}{\lambda_1 + \lambda_2}. \quad (1)$$

The probability of an arbitrary type-1 customer loss from the first storage is computed by:

$$P_1^{loss} = P_1^{ent-loss} + P_1^{imp-loss}.$$

The probability of an arbitrary type-2 customer loss from the second storage is computed by:

$$P_2^{loss} = P_2^{ent-loss} + P_2^{imp-loss}.$$

The intensity of the input flow of customers into the system (to the main buffer or directly to the server) is computed by:

$$\lambda_{in} = (1 - P_1^{loss})\lambda_1 + (1 - P_2^{loss})\lambda_2.$$

The probability that an arbitrary customer leaves the main buffer due to impatience is computed by:

$$P^{imp-loss} = \frac{1}{\lambda_{in}}\varphi\sum_{i=2}^{R+1}(i-1)\boldsymbol{\pi}_i\mathbf{e}.$$

The probability of an arbitrary customer who enters the system loss due to failure is computed by:

$$P^{fail-loss} = \frac{\lambda_{fail}}{\lambda_{in}}.$$

Instead of using formula (1) for the computation of the probability P_{loss} of an arbitrary customer loss, the following formula can be used:

$$P_{loss} = \frac{\lambda_1 P_1^{loss} + \lambda_2 P_2^{loss} + \lambda_{in}(P^{imp-loss} + P^{fail-loss})}{\lambda_1 + \lambda_2}.$$

The existence of two different formulas for the probability P_{loss} can help in the validation of the correctness of calculation of the stationary distribution of the Markov chain ξ_t.

5 Numerical Examples

Let us assume that the first arrival flow of customers MAP_1 is defined by the following matrices:

$$D_0^{(1)} = \begin{pmatrix} -10 & 0 \\ 0 & -2 \end{pmatrix},\ D_1^{(1)} = \begin{pmatrix} 9 & 1 \\ 0.1 & 1.9 \end{pmatrix}.$$

The average rate of customers in MAP_1 is $\lambda_1 = 2.72727$. The coefficient of correlation of successive inter-arrival times in this arrival process is 0.147027, and the squared coefficient of variation is 1.52893.

The second arrival process MAP_2 is defined as follows:

$$D_0^{(2)} = \begin{pmatrix} -0.35 & 0 \\ 0.07 & -5.6 \end{pmatrix},\ D_1^{(2)} = \begin{pmatrix} 0.28 & 0.07 \\ 0.07 & 5.46 \end{pmatrix}.$$

The average intensity of customers is $\lambda_2 = 2.07667$. The coefficient of correlation is 0.334815, and the squared coefficient of variation is 7.20778.

We assume that the capacity of the first storage is $N_1 = 10$ and the capacity of the second storage is $N_2 = 15$. The intensities of impatience in the first and second storages are equal to $\alpha_1 = 0.06$ and $\alpha_2 = 0.02$, and the intensity φ of impatience in the main buffer is equal to 0.02.

The PHF service process is defined by the row vector $\boldsymbol{\beta} = (0.4, 0.4, 0.2)$, column vectors $\boldsymbol{S}_1 = (5.5, 11, 4)^T$, $\boldsymbol{S}_2 = (0.5, 1, 1)^T$, the sub-generator $S = \begin{pmatrix} -9 & 2 & 1 \\ 1 & -15 & 2 \\ 1 & 2 & -8 \end{pmatrix}$, and the probabilities $q_1 = 0.2$, $q_2 = 0.4$.

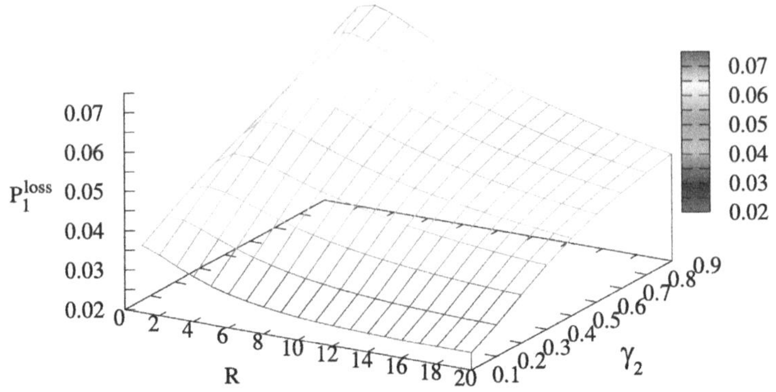

Fig. 2. Dependence of the probability P_1^{loss} on γ_2 and R

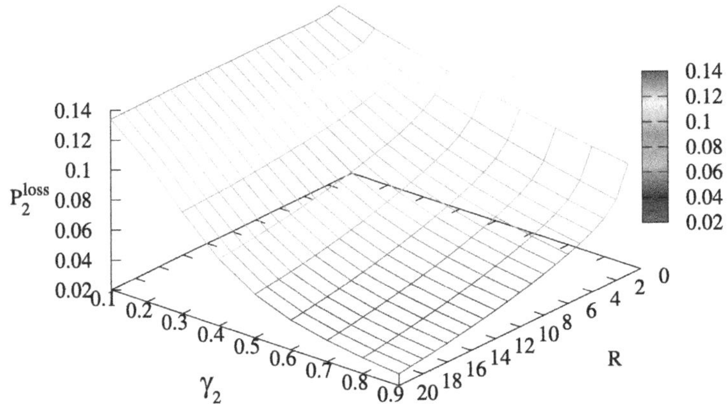

Fig. 3. Dependence of the probability P_2^{loss} on γ_2 and R

The probability of successful service of an arbitrary customer is equal to 0.892453, and the probability that a failure occurs during the service is 0.107547.

Let us assume that the intensity γ_1 is equal to 1 and vary the intensity γ_2 over the interval $[0.1, \gamma_1)$ with a step of 0.1. We also vary the capacity of the buffer R over the interval $[1, 20]$ with a step of 1.

Figure 2 and 3 illustrate the dependence of the probability P_1^{loss} of an arbitrary type-1 customer loss from the first storage and loss probability P_2^{loss} of a type-2 customer from the second storage on the parameters γ_2 and R.

As it is seen from Fig. 2, the loss probability P_1^{loss} increases with grows of γ_2. This is because as γ_2 grows, more type-2 customers enter the buffer, and the buffer becomes full more often. If the buffer is full, customers cannot enter it. This implies the increase in the number of type-1 customers in Storage-1, which increases the probability $P_1^{ent-loss}$ of losing a type-1 customer at the entrance of the system and the loss probability $P_1^{imp-loss}$ of an arbitrary type-1 customer due to impatience from the first storage. The dependence of the loss probability P_1^{loss} on R is not monotonic for larger values of γ_2. As one can see, for example, for $\gamma_2 = 0.9$, at first, the probability P_1^{loss} increases with R and then decreases. This fact can be explained as follows. For small values of R, for example, for $R = 1$, a situation when at the end of service the buffer turns out to be empty (for example, because the only customer in it left because of impatience) occurs rather often. In this case, customers of the first type (if any) are always selected for service. As R grows, the probability that the buffer will be empty at the service completion epoch decreases significantly, which, at the initial stages, leads to an increase in the probability P_1^{loss}. With a further increase of the parameter R, the probability of P_1^{loss} decreases because the buffer is less often full, and customers leave the first storage faster.

As one can see from Fig. 3, the loss probability P_2^{loss} decreases with an increase of γ_2 and R. The increase of the parameters γ_2 and R obviously increases the rate of the transfer of type-2 customers to the main buffer and improves the chances for type-2 customer to enter the storage at an arrival moment.

Figures 4 and 5 illustrate the dependence of the probability $P^{imp-loss}$ of loss of an arbitrary customer from the buffer due to impatience and the probability $P^{fail-loss}$ of a customer loss due to failure on the parameters γ_2 and R.

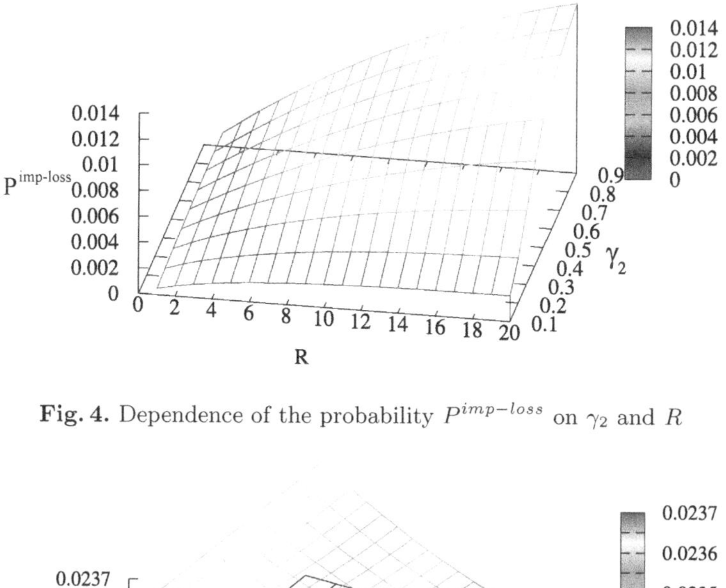

Fig. 4. Dependence of the probability $P^{imp-loss}$ on γ_2 and R

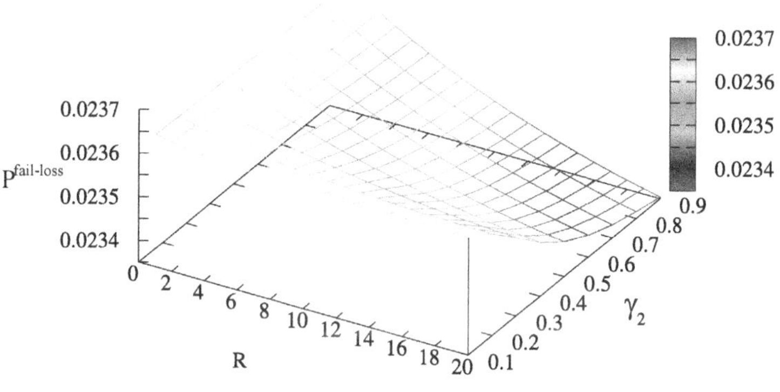

Fig. 5. Dependence of the probability $P^{fail-loss}$ on γ_2 and R

As it is seen from Fig. 4, the loss probability $P^{imp-loss}$ increases with an increase of γ_2 and R. This is because the growth of γ_2 and R implies an increase in the number of customers in the buffer. Thus, more customers leave the buffer due to impatience. The probability $P^{fail-loss}$ slightly decreases with increase of γ_2 and R. It can be explained by the fact that with growth in γ_2 and R the part of customers who leave the buffer due to impatience grows.

Figure 6 illustrates the dependence of the loss probability P_{loss} on γ_2 and R.

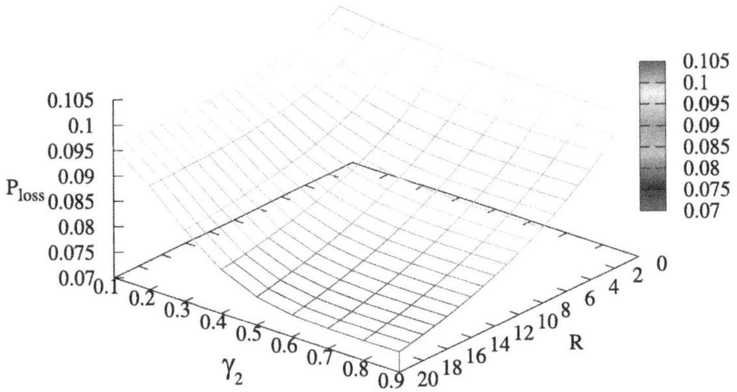

Fig. 6. Dependence of the probability P_{loss} on γ_2 and R

In the considered example, the probability P_{loss} decreases with increase in γ_2 and R. Our results allow us to quantify this dependence.

Let us assume that the quality of system operation is described by the following economical cost criterion:

$$J(\gamma_2, R) = a\lambda_{out} - b\lambda_1 P_1^{loss} - c\lambda_2 P_2^{loss} - d\lambda_{in} P^{imp-loss} - e\lambda_{in} P^{fail-loss} - fR$$

where a is the profit obtained by the system for one successful service, b is a charge paid by the system for the loss of a priority customer from the first storage, c is a charge paid for the loss of a non-priority customer from the second storage, d is a charge paid for the loss of a customer from buffer due to impatience, e is a charge paid for the loss of a customer due to failure occurrence, and f is a cost of maintenance of one place in the main buffer per unit of time.

We fix the following values of cost coefficients in the cost criterion: $a = 4$, $b = 12$, $c = 6$, $d = 15$, $e = 1$, $f = 0.05$. The following Fig. 7 illustrates the dependence of the cost criterion $J(\gamma_2, R)$ on the parameters γ_2 and R. As it is seen from this figure, the maximal value of the economic criterion is $J^*(\gamma_2, R) = 14.282$. It is achieved when $R = 8$ and $\gamma_2 = 0.3$.

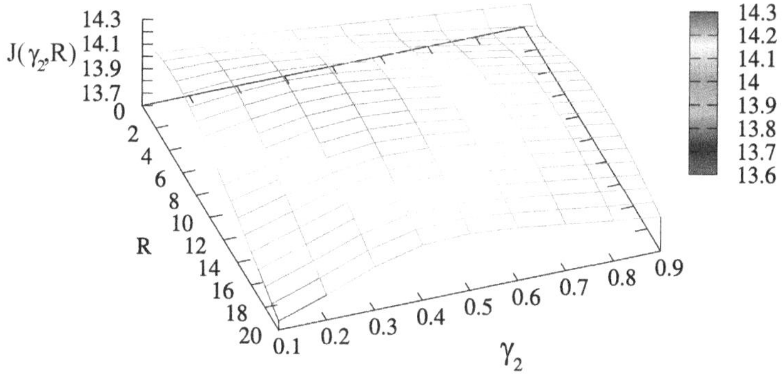

Fig. 7. Dependence of the cost criterion $J(\gamma_2, R)$ on γ_2 and R.

6 Conclusion

We considered a discipline of service of two competitive flows of customers at the single-server device with a finite buffer. The first flow is assumed to be a priority flow. To avoid shortcomings of the standard priority schemes (preemptive priority can cause high monopolization of the server and discriminate low priority customers while the non-preemptive priority can provide not enough privilege to high priority customers), we propose the new scheme. This scheme does not use the information about the lengths of queues and suggests some kind of randomization in decision making. The scheme is rather flexible because the degree of the privilege of high-priority customers depends on many parameters that are subject to control. These parameters include the capacities of the buffers for different types of customers, the rate of transition from these buffers to the main buffer, the rates of customers dropping from the buffers (due to impatience, obsolescence, or departure of the moving user from the coverage area, etc.), the capacity of the main buffer and the rate of dropping customers from this buffer. The existence of so many factors having an impact on performance measures of the system makes challenging various optimization problems.

References

1. Baumann, H., Sandmann, W.: Multi-server tandem queue with Markovian arrival process, phase-type service times, and finite buffers. Eur. J. Oper. Res. **256**(1), 187–195 (2017). https://doi.org/10.1016/j.ejor.2016.07.035
2. Chakravarthy, S.: The batch Markovian arrival process: a review and future work. Adv. Probab. Theory Stoch. Process. **1**, 21–39 (2001)
3. Dudin, A., Dudin, S.: Analysis of a priority queue with phase-type service and failures. Int. J. Stoch. Anal. **2016**, 1–11 (2016). https://doi.org/10.1155/2016/9152701. Article ID 9152701
4. Dudin, S.A., Dudina, O.S.: Call center operation model as a $MAP/PH/N/R-N$ system with impatient customers. Prob. Inf. Transm. **47**, 364–377 (2011). https://doi.org/10.1134/S0032946011040053

5. Dudin, S., Dudina, O., Samouylov, K., Dudin, A.: Improvement of the fairness of non-preemptive priorities in the transmission of heterogeneous traffic. Mathematics **8**(6), 929 (2020). https://doi.org/10.3390/math8060929
6. Gnedenko, B.V., Dannelin, E.A., Dimitrov, B.N., et al.: Priority Queueing Systems. MSU, Moscow (1973). (in Russian)
7. Graham, A.: Kronecker Products and Matrix Calculus with Applications. Ellis Horwood, Cichester (1981)
8. Jaiswal, N.K.: Priority Queues. Elsevier, Amsterdam (1968)
9. Klimenok, V., Dudin, A., Dudina, O., Gudkova, I.: Queuing system with two types of customers and dynamic change of a priority. Mathematics **8**(5), 824 (2020). https://doi.org/10.3390/math8050824
10. Klimenok, V., Kim, C.S., Orlovsky, D., Dudin, A.: Lack of invariant property of the erlang loss model in case of MAP input. Queueing Syst. **49**, 187–213 (2005). https://doi.org/10.1007/s11134-005-6481-z
11. Lee, S., Dudin, S., Dudina, O., Kim, C., Klimenok, V.: A priority queue with many customer types, correlated arrivals and changing priorities. Mathematics **8**(8) (2020). https://doi.org/10.3390/math8081292
12. Lucantoni, D.M.: New results on the single server queue with a batch Markovian arrival process. Commun. Stat. Stoch. Models **7**(1), 1–46 (1991). https://doi.org/10.1080/15326349108807174
13. Maertens, T., Walraevens, J., Bruneel, H.: On priority queues with priority jumps. Perform. Eval. **63**(12), 1235–1252 (2006). https://doi.org/10.1016/j.peva.2005.12.003
14. Stanford, D.A., Taylor, P., Ziedins, I.: Waiting time distributions in the accumulating priority queue. Queueing Syst. **77**(3), 297–330 (2013). https://doi.org/10.1007/s11134-013-9382-6
15. Vishnevskii, V.M., Dudin, A.N.: Queueing systems with correlated arrival flows and their applications to modeling telecommunication networks. Autom. Remote. Control. **78**(8), 1361–1403 (2017). https://doi.org/10.1134/S000511791708001X

Simulation Analysis in Cognitive Radio Networks with Unreliability and Abandonment

Hamza Nemouchi, Mohamed Hedi Zaghouani(✉), and János Sztrik

Doctoral School of Informatics, Faculty of Informatics, University of Debrecen, Debrecen, Hungary
{nemouchi.hamza,zaghouani.hedi,sztrik.janos}@inf.unideb.hu

Abstract. The current paper presents a Cognitive Radio Network with impatient customers and unreliable servers by the help of a finite-source retrial queueing system. We consider two types of customers (primary and secondary) assigned to two interconnected frequency bands. A first frequency band with a priority queue and a second frequency band with an orbit are reserved for Primary Users (PUs) and Secondary Users (SUs), respectively. If the servers are busy, both customers (licensed and unlicensed) enter either the queue or orbit. Before they enter orbit, the secondary customers receive a random retrial time according to the exponential distribution, i.e. the waiting time before the next retry. Unlicensed users (impatient) are obliged to leave the system as soon as their total waiting time exceeds a random maximum waiting time. It should be noted, that the secondary service unit of our system is subject to random breakdowns and repairs. The novelty of this work consists in the investigation of the abandonment and secondary server unreliability impact on various performance measures of the system (Cognitive Radio Network), such as the mean response and waiting time of users, the probability of abandonment of SU, etc. Several figures illustrate the problem in question through simulation.

Keywords: Finite source queuing systems · Simulation · Cognitive radio networks · Performance and reliability measures · Non-reliable servers · Impatient customers

1 Introduction

Cognitive Radio (CR) is an intelligent technology capable of overcoming the problems of spectrum under-utilization by allowing secondary customers to use the primary channel opportunistically without disrupting primary customer

The research work of János Sztrik is supported by the EFOP-3.6.1-16-2016-00022 project. The project is co-financed by the European Union and the European Social Fund. Mohamed Hedi Zaghouani is supported by the Stipendium Hungaricum Scholarship.

A. Dudin et al. (Eds.): ITMM 2020, CCIS 1391, pp. 31–45, 2021.
https://doi.org/10.1007/978-3-030-72247-0_3

communications, in order to improve network performance. This intelligent technology is able to modify its transmitter parameters in compliance with the interaction of the environment in which it operates. The main objective of the CRN is to use the unused portions of the primary frequency bands for the benefit of unlicensed customers. Further details can be found in [1,4,6,9–12,17]. Several studies and researches, such as [18,20] show that often many parts of the channels are unused in time and space by licensed users (white spaces). Secondary users in these parts of the service unit can detect this non-use and communicate freely without any harmful effects on the primary users. Today there are two types of Cognitive Radio Network. The first type is called (underground) network, where unlicensed users can use the primary channels simultaneously with the licensed users under certain predetermined conditions. The second type is referred to as (overlay) networks, in which the unlicensed users can use the primary service at any time as far as the primary unit is not occupied by licensed users, the authors of [13,15,19,21] have introduced further information. However, the present paper deals with an overlay CR technique by modelling a CRN system containing two finite source subsystems (primary and secondary).

In this queuing system, we take two elements into account. A first subsystem is intended for the jobs of Primary Users (PU) with a finite number of sources. In this part of the system, each source generates a primary call for the PUs after an exponentially distributed time. The latter requests are forwarded to a single server Primary Channel Service (PCS) with a preemptive discipline (FIFO queue) to start the service, assuming that the service time is exponentially distributed. The second component of the model is created for Secondary Users (SU) coming from a finite source and forwarded to Secondary Channel Service (SCS), knowing that the inter-arrival and service times of secondary users are exponentially distributed. The generated primary tasks aim to check the PCS for accessibility. If this service unit is not occupied, the service starts immediately. However, if the PCS is busy with another primary task, this last task will join a First In First Out (FIFO) queue. However, if a second job is being handled in the primary unit, this job is immediately disconnected and should be routed back to the secondary Channel Service. Per the secondary channel's status, the aborted task either restarts the service on its original server (SCS) or joins the retrial queue (Orbit). Besides, the secondary channel also receives low priority requests. If the targeted unit is idle, the service may start immediately. Otherwise, these secondary requests will attempt to join the primary unit. If the primary unit is idle, the secondary requests will have the opportunity to start. If not, they will automatically enter the orbit. From orbit, the postponed requests will retry to receive service after an exponentially distributed random interval. Further details are given in [6,13,17,21]. In this study, we assume that impatient customers in orbit whose total waiting time exceeds a random abandonment time which is generally distributed have to leave the system and the second service unit is unreliable which are the novelty of this work. Several studies have examined the Abandonment and/or Unreliability on the basis of different scenarios and systems. At [22] as an example, the authors have presented a retrial queueing system with a single server which is subject to random breakdowns and assuming that collisions may occur when a customer arrives at a busy server which forces both jobs to join the orbit. However, to get closer to real-life situations and involving more servers to the system, the authors of [23] examined the abandonment concept on a Cognitive Radio Network by setting a constant value for the maximum waiting time (abandonment time) of secondary users. In an extended work [24], the same authors of the above-mentioned paper assumed that the abandonment time is random, using various distributions to investigate their influence on the main performance measures of such a system. Other probes analysed the abandonment in other types of networks and showed that customers can leave systems from queues, server units while receiving services and while waiting; more details are given in [7,16]. However, in the current paper, we assume that impatient users (secondary) are forced to leave the system only from the orbit while waiting. Unreliability of servers was investigated in [25,26], without taking in consideration that customers have the opportunity to leave the system. Several figures will show the effects of the abandonment and unlicensed server unreliability on the performance measures of the system using simulation.

2 System Model

Figure 1 demonstrates a finite source queuing system that models the considered cognitive radio network. Our queuing system consists of two not independent, interconnected sub-systems. The first part is allocated to primary requests, with

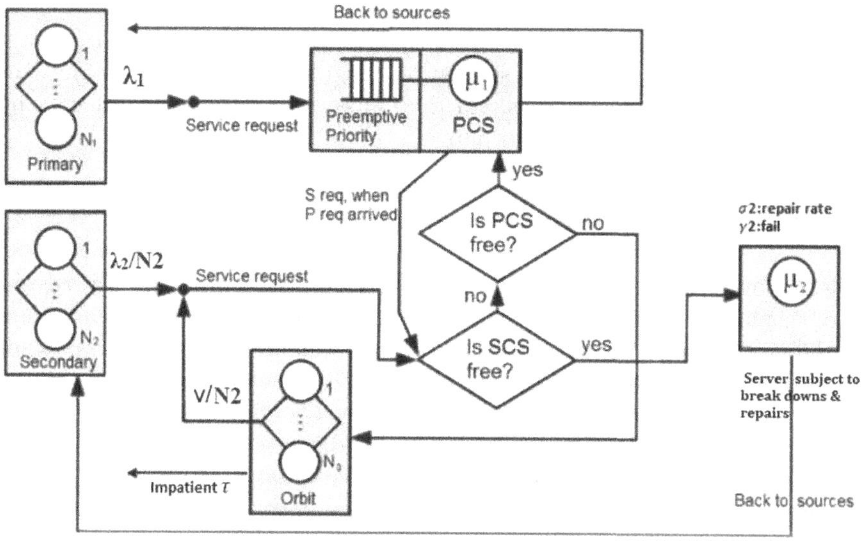

Fig. 1. Finite-source retrial queuing system: Modeling the Cognitive Radio Network with unreliability and abandonment.

N_1 the number of sources. These sources will be responsible for generating high priority requests with an inter-request and service times that are exponentially distributed, using parameters λ_1 and μ_1, respectively. All the produced requests are directed to a single server unit (PCS) with a preemptive priority over the secondary users. The second subsystem is devoted to the low-priority requests with the number of sources denoted by N_2, the inter-arrival times and service times in this subsystem are assumed to be exponentially distributed as well, with parameter λ_2/N_2 and μ_2, respectively. Based on the state of both server (idle or busy), the generated primary packet goes to the primary server (if the server is idle) or joins the FIFO queue (if it is busy with a PU). However, if an unlicensed user occupies the PCS, its service is instantly stopped and will be sent back to the Secondary unit.

Depending on the secondary unit's availability, the aborted task is addressed either to the server or the retrial queue from which reties to get served from the beginning after an exponentially distributed time with parameter ν/N_2. On the other hand, requests from SUs are directed to SCS. If it is idle, the service begins. If not, this unlicensed task will sense the PCS. In case of an idle status for PCS, this service may opportunistically join the high priority channel. If the PCS is engaged, the request goes to orbit. It should be noted that Secondary Users in orbit are obliged to leave the system once their total waiting time exceeds a random abandonment time which is generally distributed (Hyper, Hypo, Gamma, Log-normal and Pareto) a rate τ. Random breakdowns during a busy and idle state of the secondary service unit may occur after an exponentially distributed random time with parameters γ_1 and γ_2, respectively. The repair time is also exponentially distributed random variable with parameter σ.

Assuming that all random variables included in the system are exponentially distributed except the impatience time which is generally distributed random variable, we created a stochastic simulation program written in C coding language with SimPack [29] libraries. All the numerical results were collected by the validation of the simulation outputs.

3 Simulation Results

In this section, several cases are analyzed using a simulation program. The advantage of this later is to make difference between observations during a single run. This difference allows us to investigate the performance measure of two types of cognitive customers (SUs), those who leave the system with successful service and those who abandon the system without a service due to their limited waiting time. Also, the difference between secondary users who left the system with a successful service from the primary service channel and secondary users that leave the system without service after several interruptions at the primary service unit due to the preemptive priority of the primary customers over the secondary ones. In order to estimate the performance measures of these categories, the batch mean value method was used in the simulation. This method is a common confidence interval technique used for the analysis of the steady-state simulation output. See for example [3–5,8]. By dividing the cognitive users into two categories (Successful and Abandon, we could generate several results. This section of the simulation results is organized following these scenarios:

- **Scenario 1:** Impatience time of the customers is exponentially distributed.
- **Scenario 2:** Impatience time is generally distributed with $C_x^2 > 1$, using Hyper-Exponential, Gamma, Lognormal and Pareto.
- **Scenario 3:** Impatience time is generally distributed with $C_x^2 < 1$, using Hypo-Exponential, Gamma, Lognormal and Pareto.

In the above scenarios, we suppose that the interrupted secondary service from the PCS due to PUs arrival or from the SCS due to server breakdown will be repeated from the beginning (non-intelligent). Also, the service unit failure will not block the system and the free sources keep generating new calls.

3.1 Impatience Time is Exponentially Distributed

In this case, we would like to analyze the main features of the system while all the involved random inter-times are exponentially distributed random variables. The investigation is based on increasing the rate of the impatience times distribution τ.

For the sake of obtaining the following results, the set of parameters values defined in Table 1 must be used.

Table 1. Different impatience rates

	N_2	λ_2/N_2	ν/N_2	γ_1	γ_2	σ	μ_1, μ_2	τ
Case 1	100	0.01	0.1	0.1	0.1	1	1	0.000001
Case 2	100	0.01	0.1	0.1	0.1	1	1	0.0001
Case 3	100	0.01	0.1	0.1	0.1	1	1	0.001
Case 4	100	0.01	0.1	0.1	0.1	1	1	0.01
Case 5	100	0.01	0.1	0.1	0.1	1	1	0.1
Case 6	100	0.01	0.1	0.1	0.1	1	1	1

For the numerous categories of cognitive users, including successfully served and abandoned ones, we provide accurate estimates in the following. One of the advantages of the simulation is to assist us to perform these measures. We could see some characteristics of our systems in Tables 2, 3, 4, 5 for which the notations are provided in 7. These results are the estimations mean and variance of the measures based on two scenarios:

- **Scenario A:** The arrival traffic of the primary customers is low and small number of sources $\lambda_1 = 0.01$, $N_1 = 10$.
- **Scenario B:** The arrival intensity of the primary users is high and large number of sources $\lambda_1 = 0.1$, $N_1 = 100$.

Table 2. Estimation of the expectations for scenario A

	$E(TS)$	$E(WS)$	$E(T)$	$E(W)$	$E(NS)$	$E(TA)$	Pa
Case 1	14.0437	13.8001	14.04	13.8001	48.59	0.0000	0.0000
Case 2	14.0525	13.8165	14.05	13.8284	44.64	15.0001	0.001
Case 3	13.8333	13.5979	13.59	13.0235	38.17	15.226	0.012
Case 4	12.3461	12.1107	12.48	12.27	28.33	13.5472	0.15
Case 5	5.5598	5.3241	6.0853	5.9801	12.21	6.4914	0.56
Case 6	0.8258	0.5908	0.9654	0.3491	5.2454	0.9772	0.9217

Table 3. Estimation of the variances for scenario A

	$Var(TS)$	$Var(WS)$	$Var(T)$	$Var(W)$	$Var(TA)$
Case 1	197.227	190.66	197.227	197.227	0.0000
Case 2	197.473	190.897	197.47	190.87	185.249
Case 3	191.36	184.902	191.35	181.35	185.652
Case 4	152.42	146.67	152.48	146.66	183.52
Case 5	30.9119	28.3471	30.91	28.43	42.13
Case 6	0.6818	0.3491	0.6820	0.3491	0.955

Table 2 and 3 determine the values of the expectations and variances of different types of cognitive users, respectively. These results are the outputs of the simulation while $\lambda_1 = 0.01$ and $N_1 = 10$. The rows of the tables define the cases where the impatience rate τ is increasing. It is clearly seen that the mean and variance values of the response, waiting and arbitrary users are decreasing while the probability of abandonment is increasing. This later increases while an elevation of impatience rate. Unexpectedly, the mean and variance values of the impatient customers are increasing then decreasing during the growth of the abandonment rate. The interpretation of this phenomenon can be as follows: when the impatience rate is very small, the waiting time of the customers is very long, as a result, they rarely leave the system. Therefore, the confidence interval of the expectation for a small set of observations can be very large, thus, the estimation is not accurate.

In Table 4 and 5, the same features as above were treated but in this case the primary traffic more intense supposing $\lambda_1 = 0.1$ and $N1 = 100$. The first thing we notice comparing with Tables 2 and 3 is the efficiency of cognitive technology. It

Table 4. Estimation of the expectations for scenario B

	$E(TS)$	$E(WS)$	$E(T)$	$E(W)$	$E(NS)$	$E(TA)$	Pa
Case 1	25.0657	24.8343	25.0651	24.8338	57.28	29.5803	0.000023
Case 2	24.9055	24.66	24.8423	24.6119	57.28	26.5792	0.002
Case 3	24.2940	24.0632	24.3561	24.1307	52.26	26.9554	0.02
Case 4	20.0967	19.8643	20.4913	20.3042	30.15	22.118	0.194
Case 5	6.5178	6.2909	7.2363	7.1664	6.07	7.5563	0.61
Case 6	0.8242	0.6068	0.9743	0.3682	0.9598	0.9834	0.943

Table 5. Estimation of the variances for scenario B

	$Var(TS)$	$Var(WS)$	$Var(T)$	$Var(W)$	$Var(TA)$
Case 1	628.037	616.75	628.29	616.57	726.24
Case 2	620.037	608.519	608.59	606.84	724.40
Case 3	590.201	579.04	590.2013	579.03	726.65
Case 4	403.8815	394.591	403.8813	394.5913	489.2139
Case 5	42.4828	39.5766	42.4828	39.4741	57.0989
Case 6	0.6771	0.3682	0.6794	0.3682	0.9671

is shown that when a very small impatience rate ($\tau = 0.000001$), the probability of abandonment is zero in Table 2, row 1 since the licensed service channel in scenario A is most of the time idle from primary customers. However, besides the larger values of the mean and variance in scenario B. Same explanation as previously for the expectation and variance waiting time of the impatient customers (Table 6).

3.2 Impatience Time is Generally Distributed with $C_x^2 > 1$

This subsection is Scenario 2 of our investigation, we have analyzed the impact of abandonment time distributions on the main characteristics of the system.

Table 6. Estimation of the variances for scenario A

Notation	Definition
E(TS), Var(TS)	Mean and variance response time of successful cognitive users
E(WS), Var(WS)	Mean and variance waiting time of successful cognitive users
E(T), Var(T)	Mean and variance response time of arbitrary cognitive users
E(W), Var(W)	Mean and variance waiting time of arbitrary cognitive users
E(NS)	Mean number of secondary customers in the system
E(TA), Var(TA)	Mean and variance waiting time of impatient customers
Pa	Probability of abandonment

In this scenario, we consider that the impatience time is hyper-exponentially, gamma, Pareto and, lognormally distributed random variable with the same mean and same variances. In order to calculate the shape, rate, and scale of these distributions, see [28]. Table 7 defines the numerical values of distribution parameters.

The inter-events times are generated using several methods of random numbers generator. These methods require input parameters which in our case are the rates of the distributions. The numerical values of these parameters are defined in Table 8.

Table 7. Parameters of the distributions

Distribution	Gamma	Hyper-exponential	Pareto	Lognormal
Parameters	$\alpha = 0.390625$	$p = 0.33098$	$\alpha = 2.1892$	$m = 5.5797$
	$\beta = 0.0007813$	$\lambda_1 = 0.00132$	$k = 270.5630$	$\sigma = 1.12684$
		$\lambda_2 = 0.00268$		
Mean	500			
Variance	640000			
C_x^2	2.56			

Table 8. Numerical values of the parameters

N_1	N_2	λ_1	λ_2/N_2	μ_1	μ_2	ν/N_2	τ	γ_1	γ_2	σ_2
10	100	0.01	x - axis	4	4	0.01	0.002	0.01	0.01	1

Comments

In all the obtained results, it should be noticed that beside the abandonment time, all the inter-event times in the system are supposed to be exponentially distributed random variables.

Figure 2 illustrates the impact of the impatience time distributions on the mean sojourn time of the cognitive users that leave the system after a successful service while increasing the secondary arrival intensity. The Pareto distribution gives the smallest value of the mean response while the gamma distribution gives the greatest value. This relative difference could be explained by the help of the density function of each distribution. Also, if we read the papers that investigated a single server finite-source retrial queueing system with the abandonment of customers, we notice that the relative difference of the mean values between the lognormal and the hyper-exponential distributions is smaller then the one

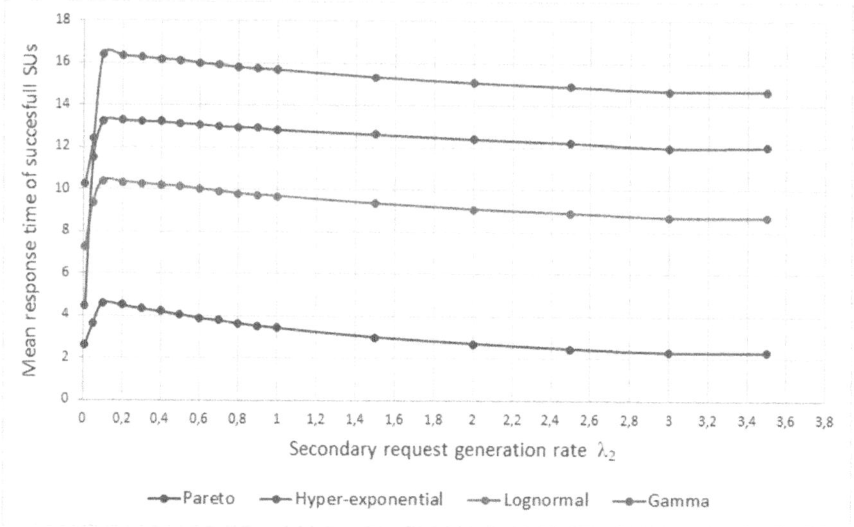

Fig. 2. The impact of the impatience time distributions on the mean sojourn time of successful cognitive users vs secondary request generation rate

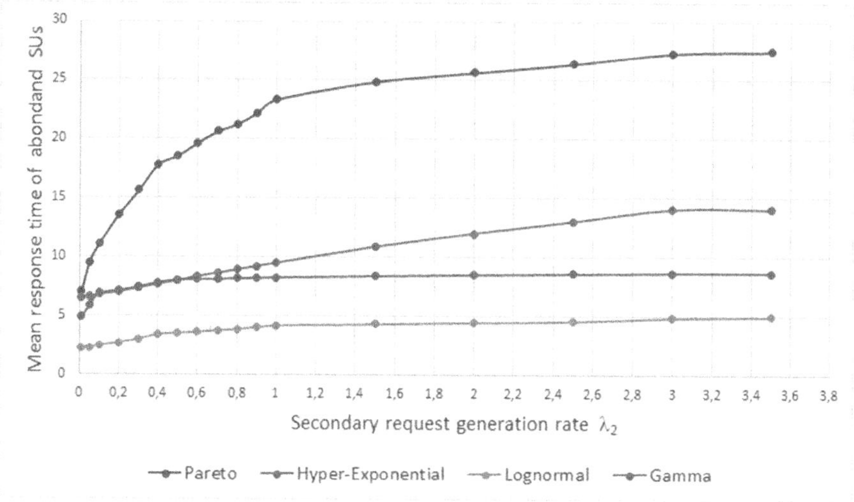

Fig. 3. The impact of the impatience time distributions on the mean sojourn time of impatient cognitive users vs secondary request generation rate

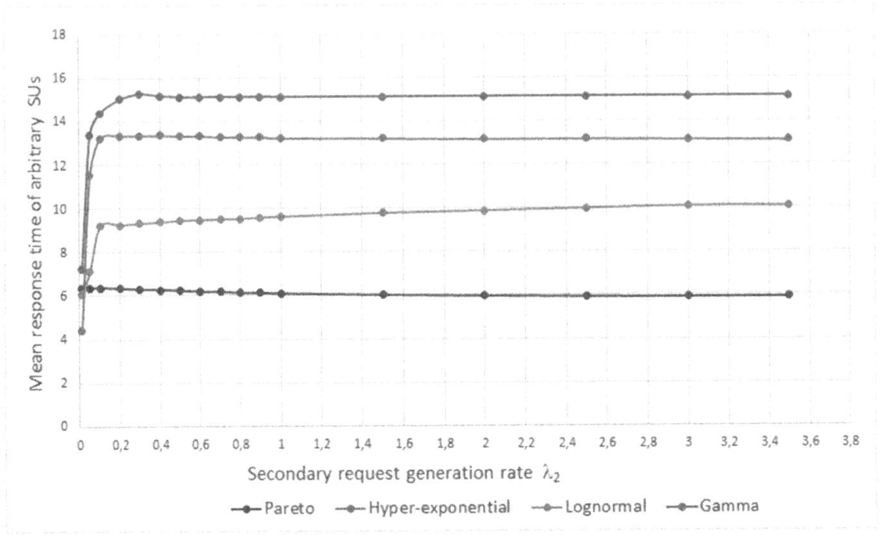

Fig. 4. The impact of the impatience time distributions on the mean sojourn time of arbitrary cognitive users vs secondary request generation rate

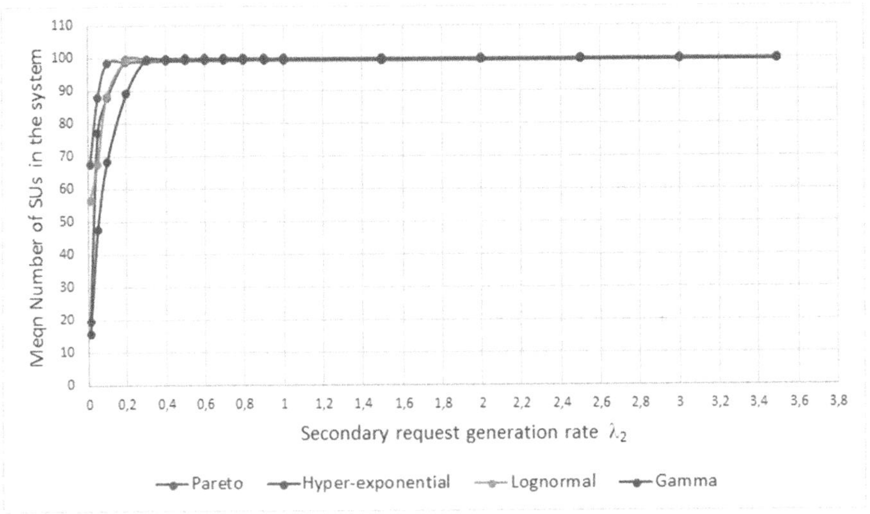

Fig. 5. The mean number of cognitive users in the system vs secondary arrival intensity

obtained in Fig. 1. This difference is obviously due to the introduction of a second server and a cognitive characteristic to the secondary customers which make the system more complex. However, the maximum property of the mean response time that was noticed in [27] is obtained.

Figure 3 shows the effect of the abandonment time distributions on the mean response time of the secondary customers that leave the system without getting served. The result shows that while increasing the request generation intensity, the value of the mean increases. The aim of making a difference between successful and impatient customer is obtained in this figure, we see that the Pareto distribution gives the greatest value of the mean, while for this feature, the lognormal distribution shows the smallest value.

With the help of the law of total expectation, it is easy to calculate the mean of arbitrary users. Figure 4 illustrates the effect of the impatience time distribution on the mean response time of an arbitrary user while the secondary request generation rate is increasing. The mean value of arbitrary users depends on the probability of success and the probability of abandonment. As the gamma distribution gives a small value of the impatience time, it involves a high probability of abandonment, thus, the value of the mean response time of arbitrary users is the greatest as shown in the figure.

Figure 5 illustrates the mean number of secondary customers in the function of the second generation request rate while the impatience time is generally distributed. The effect of the distributions can be seen when the system is low loaded. When the arrival intensity increases, the mean number of cognitive customers increases, and the distributions have no more impact on its value.

3.3 Impatience Time is Generally Distributed with $C_x^2 < 1$

In Scenario 3 of our investigation, we set the parameters of the used distribution in a way their squared coefficient of variation becomes less than one. For this case, we use the two phases of hypo-exponential distribution. The aim is to analyze their effects on the main performance measures of the system. The new set of the distribution parameters and the numerical values of the simulation input parameters are shown in Table 9 and Table 10, respectively.

Table 9. Parameters of the distributions

Distribution	Gamma	Hypo-exponential	Pareto	Lognormal
Parameters	$\alpha = 1.47059$	$\lambda_1 = 0.01$	$\alpha = 2.5718$	$m = 5.9552$
	$\beta = 0.002941$	$\lambda_2 = 0.0025$	$k = 305.5844$	$\sigma = 0.72027$
Mean	500			
Variance	170000			
C_x^2	0.68			

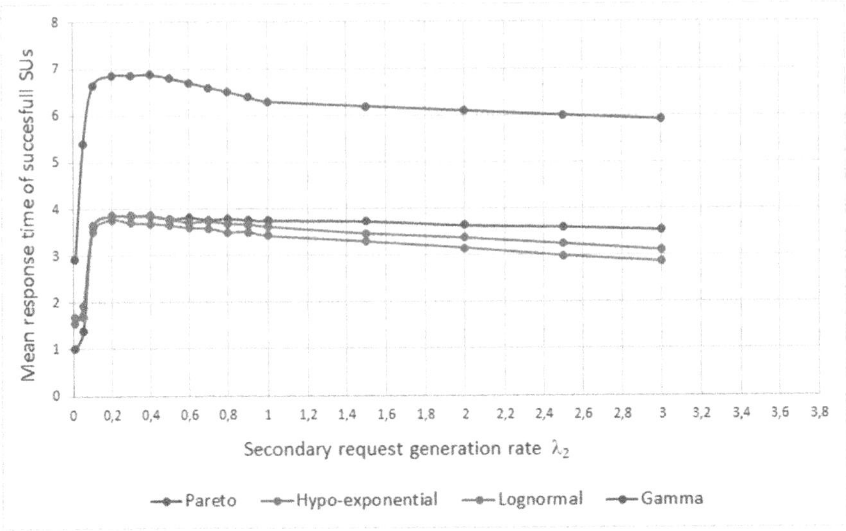

Fig. 6. The effect of the impatience time distributions on the mean sojourn time of successful cognitive users vs secondary request generation rate

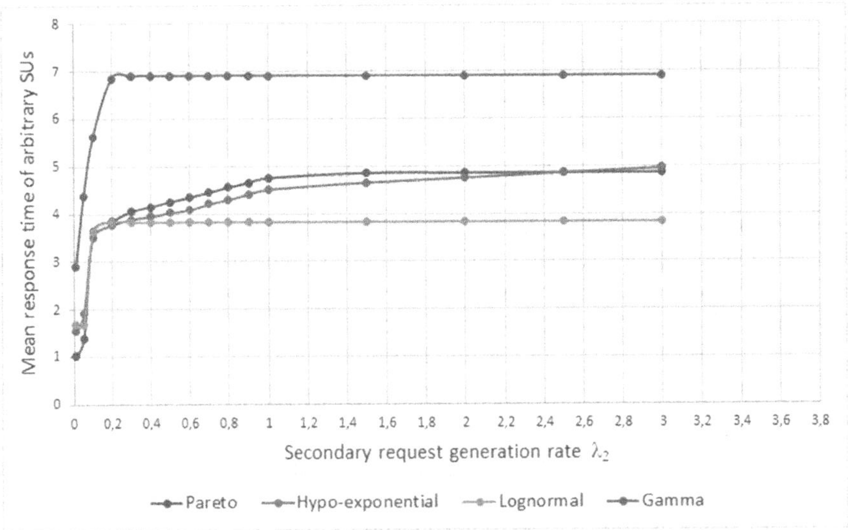

Fig. 7. The effect of the impatience time distributions on the mean sojourn time of arbitrary cognitive users vs secondary request generation rate

Table 10. Numerical values of the parameters

N_1	N_2	λ_1	λ_2/N_2	μ_1	μ_2	ν/N_2	τ	γ_1	γ_2	σ_2
10	100	0.01	x - axis	4	4	0.01	0.002	0.01	0.01	1

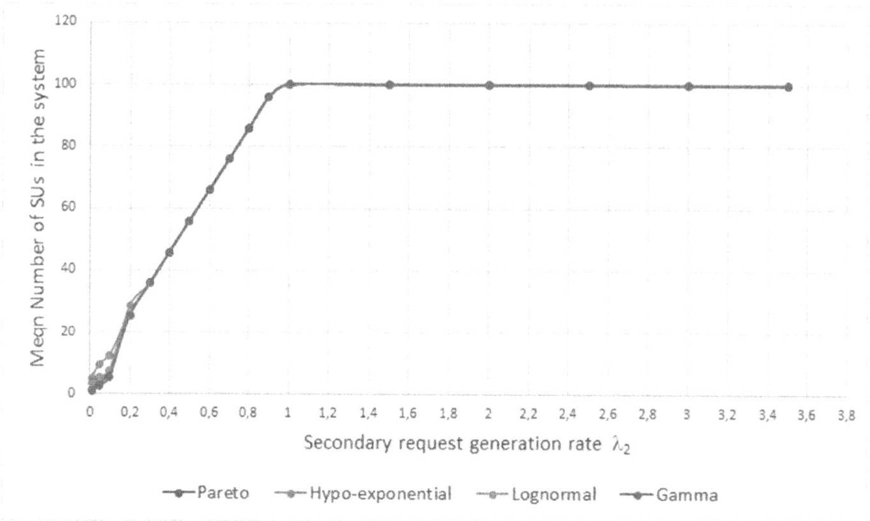

Fig. 8. The mean number of cognitive users in the system vs secondary arrival intensity

Comments

In terms of arrival intensity, Fig. 6 and Fig. 7 display the mean residence time of successful and arbitrary customers, respectively. The same pattern can be seen in the analysis of the outputs as in the previous corresponding figures, but variations can also be seen, especially in case of the gamma distribution. These statistics also indicate that, relative to the previous parameter setting, successful and arbitrary users spend less time on the system on average.

Lastly, Fig. 8 illustrates the mean number of secondary customers in the system in the function of the secondary arrival intensity. Among a highly loaded system, the figure shows no impact of the distribution on the mean number of customers, but as a low loaded system, a slight difference can be investigated. Also, relative to the previous set of parameters with the corresponding figure, there are fewer customers in the system when the squared coefficient of variation of the distributions is less than one.

4 Conclusion

In this paper, a finite-source cognitive radio network was modelled with the help of a retrial queueing system with impatient customers and a secondary server non-reliable. The results have demonstrated the impact of the abandonment time distribution on the mean and variance of the main characteristic of such a complex system. The efficiency of the primary service channel and of the cognitive property at the secondary users was also demonstrated. Using simulation, we succeeded to separate the secondary customers into three categories (impatient/successful/arbitrary) and analyze their performances separately. Based on this work, our perspective for the future is to introduce intelligent cognitive users which mean that a secondary user will continue the interrupted (due to primary arrival or server breakdown) service and he will not repeat it from the beginning and demonstrate the distribution of the customers (primary and secondary) in such a system while both subsystems are non-independent. Also, with the help of simulation, we can investigate separately the customers that leave the system successfully from the primary service channel and their mean interrupted service time.

References

1. Akyildiz, I.F., Lee, W.Y., Vuran, M.C., Mohanty, S.: Next generation/dynamic spectrum access/cognitive radio wireless networks: a survey. Comput. Netw. **50**(13), 2127–2159 (2006)
2. Almási, B., Bérczes, T., Kuki, A., Sztrik, J., Wang, J.: Performance modeling of finite-source cognitive radio networks. Acta Cybernetica **22**(3), 617–631 (2016)
3. Carlstein, E., et al.: The use of subseries values for estimating the variance of a general statistic from a stationary sequence. Ann. Stat. **14**(3), 1171–1179 (1986)
4. Chen, E.J., Kelton, W.D.: A procedure for generating batch-means confidence intervals for simulation. Checking independence and normality. Simulation **83**(10), 683–694 (2007)
5. Fishman, G.S., Yarberry, L.S.: An implementation of the batch means method. FORMS J. Comput. **9**(3), 296–310 (1997)
6. Gunawardena, S., Zhuang, W.: Modeling and Analysis of Voice and Data in Cognitive Radio Networks. Springer, Heidelberg (2014)
7. He, Q.M., Zhang, H., Ye, Q.: An M/PH/K queue with constant impatient time. Math. Methods Oper. Res. **87**(1), 139–168 (2018)
8. Law, A.M., Kelton, W.D.: Simulation Modeling and Analysis, vol. 3. McGraw-Hill, New York (2000)
9. Nemouchi, H., Sztrik, J.: Performance simulation of finite-source cognitive radio networks with servers subjects to breakdowns and repairs. J. Math. Sci. **237**(5), 702–711 (2019)
10. Nemouchi, H., Sztrik, J.: Performance evaluation of finite-source cognitive radio networks with collision using simulation. In: 8th IEEE International Conference on Cognitive Infocommunications (CogInfoCom), pp. 000127–000131. IEEE (2017)
11. Nemouchi, H., Sztrik, J.: Performance evaluation of finite-source cognitive radio networks with non-reliable services using simulation. Annales Mathematicae et Informaticae, vol. 49, pp. 109–122. Eszterházy Károly University Institute of Mathematics and Informatics (2018)
12. Nemouchi, H., Sztrik, J.: Performance simulation of non-reliable servers in finite-source cognitive radio networks with collision. In: Dudin, A., Nazarov, A., Kirpichnikov, A. (eds.) ITMM 2017. CCIS, vol. 800, pp. 194–203. Springer, Cham (2017). https://doi.org/10.1007/978-3-319-68069-9_16
13. Paluncic, F., Alfa, A.S., Maharaj, B.T., Tsimba, H.M.: Queueing models for cognitive radio networks: a survey. IEEE Access **6**, 50801–50823 (2018)
14. Sztrik, J., Almási, B., Roszik, J.: Heterogeneous finite-source retrial queues with server subject to breakdowns and repairs. J. Math. Sci. **132**(5), 677–685 (2006)
15. Van Do, T., Do, N.H., Horváth, Á., Wang, J.: Modelling opportunistic spectrum renting in mobile cellular networks. J. Netw. Comput. Appl. **52**, 129–138 (2015)

16. Wang, J., Abouee Mehrizi, H., Baron, O., Berman, O.: Staffing Tandem Queues with Impatient Customers-Application in Financial Service Operations. Rotman School of Management Working Paper 3116815 (2018)
17. Wang, L., Wang, C., Adachi, F.: Load-balancing spectrum decision for cognitive radio networks. IEEE J. Sel. Areas Commun. **29**(4), 757–769 (2011)
18. Weiss, T.A., Jondral, F.K.: Spectrum pooling: an innovative strategy for the enhancement of spectrum efficiency. IEEE Commun. Mag. **42**(3), S8–14 (2004)
19. Wong, E.W., Foh, C.H.: Analysis of cognitive radio spectrum access with finite user population. IEEE Commun. Lett. **13**(5), 294–296 (2009)
20. Zaghouani, M.H., Sztrik, J., Uka, A.: Simulation of the performance of Cognitive Radio Networks with unreliable servers. Annales Mathematicae et Informaticae (2020)
21. Zekavat, S.A., Li, X.: User-central wireless system: ultimate dynamic channel allocation. In: First IEEE International Symposium on New Frontiers in Dynamic Spectrum Access Networks. DySPAN, pp. 82–87. IEEE (2005)
22. Kuki, A., Bérczes, T., Tóth, Á., Sztrik, J.: Numerical analysis of finite source Markov retrial system with non-reliable server, collision, and impatient customers. Annales Mathematicae et Informaticae **51**, 53–63 (2020)
23. Zaghouani, M.H., Sztrik, J.: Performance evaluation of finite-source Cognitive Radio Networks with impatient customers. Annales Mathematicae et Informaticae **51**, 89–99 (2020)
24. Zaghouani, M.H., Sztrik, J.: Performance simulation of finite-source Cognitive Radio Networks with impatient calls in the orbit, ISSPSM (2020)
25. Sztik, J., Zaghouani, M.H., Uka, A.: Reliability analysis of cognitive radio networks. In: 18th International Conference named after A.F. Terpugov. Information Technologies and Mathematical Modelling ITMM-2019, Saratov, Russia (2019)
26. Nemouchi, H., Zaghouani, M.H., Sztrik, J.: The impact of servers reliability on the characteristics of cognitive radio systems. In: The 1st Conference on Information Technology and Data Science CITDS (2020)
27. Nazarov, A., Sztrik, J., Kvach, A.: A survey of recent results in finite-source retrial queues with collisions. In: Dudin, A., Nazarov, A., Moiseev, A. (eds.) ITMM/WRQ -2018. CCIS, vol. 912, pp. 1–15. Springer, Cham (2018). https://doi.org/10.1007/978-3-319-97595-5_1
28. Kuki, A., Berczes, T., Sztrik, J., Toth A.: Reliability analysis of a two-way communication system with searching for customers. In: Proceedings of The International Conference on Information and Digital Technologies, pp. 260–265. IEEE (2019)
29. Fishwick, P.A.: Getting started with simulation programming in C and C++. In: Winter Simulation Conference, pp. 154–162 (1992)

A Retrial Queueing System with Processor Sharing

Valentina Klimenok and Alexander Dudin(✉)

Department of Applied Mathematics and Computer Science,
Belarusian State University, 4 Nezavisimosti Avenue, 220030 Minsk, Belarus
dudin@bsu.by

Abstract. We consider a retrial queueing system with limited processor sharing which can be used for modeling the operation of a cell of fixed capacity in a wireless cellular network with two types of customers (handover and new customers). Customers of two types arrive at the system according to the Marked Markovian Arrival Process (MMAP). Arriving customers of each type follow a bandwidth sharing policy. In period when the number of customers of definite type in the system exceeds a threshold (different between new and handover customers) newly arriving customers of one type (handover customers) are considered to be lost while the customers of another type (new customers) go to orbit of infinite size. From the orbit, they try their attempts to reach a server in exponentially distributed time.

We describe the system operation by multi-dimensional Markov chain, calculate the steady state distribution and main performance measures of the system. Illustrative numerical examples are presented.

Keywords: Queueing system · Two type of customers · Limited processor sharing · Ergodicity condition · Steady state distribution · Performance measures

1 Introduction

The discipline of processor sharing is characterized by the fact that several (possibly all) users of a service resource can receive service simultaneously. This discipline is often used in computer systems and telecommunication networks for various purposes. In particular, this discipline is used in scheduling problems in multiprogrammed computer systems, in mobile cellular communication networks, in caching popular multimedia content, etc. There are a large number of works devoted to the study of the functioning of real systems, where the processor sharing discipline is used. For examples and links see articles [1–5].

The models considered in the listed works do not take into account the complex nature of traffic in modern telecommunication networks and systems. They assume that the flows of customers are stationary Poisson, and the service times have an exponential distribution. These restrictions are removed in works where

A. Dudin et al. (Eds.): ITMM 2020, CCIS 1391, pp. 46–60, 2021.
https://doi.org/10.1007/978-3-030-72247-0_4

Markovian arrival process (MAP) is considered as an input flow, and service times are distributed according to the phase type (PH) distribution, see the papers [6,7].

At the same time, there remains a problem associated with the heterogeneity of the input flow, indicated, in particular, in [2]. With such a flow, customers of different types can have different distributions of service time and share the processor resource in different proportions. In addition, we take into account the retrial phenomenon which is typical for computer systems and telecommunications networks.

Queuing systems with repeated customers (retrials) differ from classical systems with buffers and systems with losses in the following respect. An arriving customer, which meets all service facilities busy, does not queue and does not leave the system forever, but goes to the so-called "orbit" that is a virtual place for such customers, from where it attempts to get service at random times. Note that retrial queueing systems theory is much less developed than loss or buffer systems theory. This is due to the fact that the random processes describing the operation of such systems are more complex in structure due to spatial inhomogeneity what greatly complicates their analytical study.

A good mathematical model for heterogeneous correlated traffic is a Marked Markovian arrival process ($MMAP$). In this paper we investigate a retrial queuing system with $MMAP$ of customers of two types, various schemes for dividing the processor between customers of different types and restrictions on the number of customers of each of types on the server. We describe the system operation by multi-dimensional Markov chain, calculate the steady state distribution and the main performance characteristics of the system.

2 Mathematical Model

We consider a single-server queueing system without a buffer. Customers of two different types arrive at the system in the Marked Markovian arrival process ($MMAP$). For the reader's convenience, we give below the brief description of the $MMAP$.

In general case, $MMAP$ can model the arrival process of customers of K different types. In the $MMAP$, customers arrive under control of the regular irreducible Markov chain $\nu_t, t \geq 0$, which takes values in the set $\{0, 1, 2, \dots, W\}$. This chain is called as an underlying process of the $MMAP$. The underlying process stays in the state ν during an exponentially distributed time interval with parameter λ_ν, $\nu = \overline{0, W}$. After that with probability $p_k(\nu, \nu')$ the underlying process enters the state ν' with generation of a customer of kth type, $k \in \{1, 2, \dots, K\}$, or, with probability $p_0(\nu, \nu')$, it goes to the state ν' without generating a customer. For the indicated probabilities, natural constraints are satisfied: $p_0(\nu, \nu) = 0$, $\sum\limits_{k=1}^{K} \sum\limits_{\nu'=0}^{W} p_k(\nu, \nu') = 1$, $\nu = \overline{0, W}$.

Thus, the $MMAP$ is given by the parameters $W + 1$; K; λ_ν, $\nu = \overline{0, W}$; $p_k(\nu, \nu'), k = \overline{1, K}, \nu, \nu' = \overline{0, W}$. It is convenient to storage all information about the $MMAP$ as a set of matrices D_k, $k = \overline{1,K}$, of order $(W+1) \times (W+1)$ with entries

$$(D_k)_{\nu,\nu'} = \lambda_\nu p_k(\nu,\nu'),\ \nu,\nu' = \overline{0,W}, k = \overline{1,K},$$
$$(D_0)_{\nu,\nu'} = \begin{cases} \lambda_\nu p_0(\nu,\nu'), \ \nu \neq \nu', \nu,\nu' = \overline{0,W}, \\ -\lambda_\nu, \qquad\qquad \nu = \nu' = \overline{0,W}. \end{cases}$$

It is easy to see that the entries of the matrices D_k, $k = \overline{1,K}$, are the rates of transitions of the process ν_t accompanied by generating a customer of the k-th type. The off-diagonal entries of the matrix D_0 define the rates of transitions of the process ν_t without generation of customers and the moduli of diagonal entries of this matrix give the rates of exit of the process ν_t from the corresponding states. A natural requirement for the matrices D_k, $k = \overline{1,K}$, is that not all of them are zero. When this requirement is met, the matrix D_0 is irreducible and, moreover, stable. The matrices D_k, $k = \overline{1,K}$, can be specified by their matrix generating function $D(z) = \sum\limits_{k=0}^{K} D_k z^k$, $|z| < 1$. Note that the value of this function at the point $z = 1$ is an infinitesimal generator of the underlying process ν_t, $t \geq 0$. Stationary distribution of this process, the row vector $\boldsymbol{\theta}$, is defined as the unique solution of the system of linear algebraic equations $\boldsymbol{\theta} D(1) = \mathbf{0}, \boldsymbol{\theta}\mathbf{e} = 1$. Hereafter $\mathbf{0}$ is a row vector consisting of zeros, $\mathbf{e}$ is a column vector consisting of ones.

The arrival rate of customers of type k in the $MMAP$ is given by the formula

$$\lambda_k = \boldsymbol{\theta} D_k \mathbf{e}, k = \overline{1,K}.$$

The variance of inter-arrival times of customers of type k is calculated by the formula

$$v_k = \frac{2\boldsymbol{\theta}(-D_0 - \sum\limits_{l=1, l\neq k}^{K} D_l)^{-1}\mathbf{e}}{\lambda_k} - \left(\frac{1}{\lambda_k}\right)^2, \ k = \overline{1,K}.$$

The coefficient of correlation of the lengths of two adjacent intervals between the arrivals of customers of type k is calculated by

$$c_{cor}^{(k)} = \left[\frac{\boldsymbol{\theta}(D_0 + \sum\limits_{l=1, l\neq k}^{K} D_l)^{-1}}{\lambda_k} D_k (D_0 + \sum_{l=1, l\neq k}^{K} D_l)^{-1}\mathbf{e} - \left(\frac{1}{\lambda_k}\right)^2\right] v_k^{-1}, \ k = \overline{1,K}.$$

More information about the $MMAP$ can be found in [8,9].

In this paper, we assume that the system receives customers of two types in the $MMAP$, i.e., $K = 2$.

The server can simultaneously serve up to N customers of type 1 and up to R customers of type 2. If only one customer of type k is serviced on the server, $k = 1, 2$, then its service time has the PH distribution given by the irreducible representation $(\boldsymbol{\beta}_k, S_k)$ and the underlying process $m_t^{(k)}$, $t \geq 0$, with the state space $\{1,\ldots,M_k,M_k+1\}$, where the state M_k+1 is the absorbing one. The rates of transitions to the absorbing state are determined by the column vector $\boldsymbol{S}_0^{(k)} = -S_k\mathbf{e}$. The mean service time of type k customer is calculated as $b_1^{(k)} = \boldsymbol{\beta}_k(-S_k)^{-1}\mathbf{e}$ and the service rate is equal to $\mu_k = (b_1^{(k)})^{-1}$. The squared coefficient of the variation of the service time of the k type customer is given as $(c_{var}^{serv,k})^2 = 2\frac{\boldsymbol{\beta}_k(-S_k)^{-2}\mathbf{e}}{(\boldsymbol{\beta}_k(-S_k)^{-1}\mathbf{e})^2} - 1$. More detail about PH distribution can be found in [9,10].

Customers of each type divide the service bandwidth assigned for them equally. If the server simultaneously serves n_k customers of type k, then the service time of any of these customers has the PH distribution given by the irreducible representation $(\boldsymbol{\beta}_k, \frac{1}{n_k}S_k)$ and the underlying process $m_t^{(k)}$, $t \geq 0$, with state space $\{1,\ldots,M_k,M_k+1\}$ where the state M_k+1 is absorbing. The rates of transitions to the absorbing state are determined by the column vector $\frac{1}{n_k}\boldsymbol{S}_0^{(k)}$.

If an arriving customer of type 1 sees that the system already has N customers of type 1, then it goes to the orbit of infinite size, from where it attempts to get service at exponentially distributed times with the parameter γ. If an arriving customer of type 2 finds that there are already R customers of type 2 in the system, then it leaves the system forever (is lost).

3 Process of the System States

We describe the operation of the system by a regular irreducible continuous time Markov chain

$$\xi_t = \{i_t, n_t, \eta_t^{(1)}, \eta_t^{(2)}, \ldots, \eta_t^{(M_1)}, r_t, \tau_t^{(1)}, \tau_t^{(2)}, \ldots, \tau_t^{(M_2)}, \nu_t\},$$

where, at time instant t,

- i_t is the number of type 1 customers in the orbit $i \geq 0$;
- n_t is the number of type 1 customers in the system, $n_t = \overline{0,N}$;
- $\eta_t^{(m^{(1)})}$ is the number of customers of type 1 that are served in the phase $m^{(1)}$, $\eta_t^{(m^{(1)})} = \overline{0,n_t}$, $m^{(1)} = \overline{1,M_1}$;
- r_t is the number of type 2 customers in the system, $r_t = \overline{0,R}$;
- $\tau_t^{(m^{(2)})}$is the number of customers of type 2 that are served in the phase $m^{(2)}$, $\tau_t^{(m^{(2)})} = \overline{0,r_t}$, $m^{(2)} = \overline{1,M_2}$;
- ν_t is the state of underlying process of the $MMAP$, $\nu_t = \overline{0,W}$.

In the sequel we will use the following notation:

- $\bar{W} = W + 1$;
- $\otimes(\oplus)$ is the symbol of the Kronecker product (sum) of matrices, see, e.g., [11];
- $\mathcal{R} = \sum\limits_{r=0}^{R} C_{r+M_2-1}^{M_2-1}$;

- $diag\{a_1, a_2, ..., a_n\}$ is a block diagonal matrix in which the diagonal blocks are equal to the elements listed in brackets, and the other blocks are zero;
- $diag^+\{a_1, a_2, ..., a_n\}$ ($diag^-\{a_1, a_2, ..., a_n\}$) is a square block matrix in which the off-diagonal (below-diagonal) blocks are equal to the elements listed in brackets, and the other blocks are zero;

  $C_n^m = \begin{pmatrix} n \\ m \end{pmatrix} = \frac{n!}{m!(n-m)!}$;
- $\mathbf{u}_t^{(1)} = \{\eta_t^{(1)}, \eta_t^{(2)}, \dots, \eta_t^{(M_1)}\}$;
- $\mathbf{u}_t^{(2)} = \{\tau_t^{(1)}, \tau_t^{(2)}, \dots, \tau_t^{(M_2)}\}$.

Using the last two notation, we can represent the Markov chain $\xi_t, t \geq 0$, in the form $\xi_t = \{i_t, n_t, \mathbf{u}_t^{(1)}, r_t, \mathbf{u}_t^{(2)}, \nu_t\}$. We assume that the states of the chain are enumerated as follows: the components i_t, n_t, r_t, ν_t are enumerated in the direct lexicographic order and the states of the processes $\mathbf{u}_t^{(1)}$ and $\mathbf{u}_t^{(2)}$ are enumerated in the reverse lexicographic order. Reverse lexicographic ordering is required to describe the transition rates of the processes $\mathbf{u}_t^{(1)}$ and $\mathbf{u}_t^{(2)}$ using matrices $P_i(\cdot)$, $A_i(\cdot, \cdot)$, and $L_i(\cdot, \cdot)$ introduced in the papers [12,13]. Below we give a brief explanation of the probabilistic meaning of these matrices.

Let us introduce the matrices $\tilde{S}_l = \begin{pmatrix} 0 & O \\ \mathbf{S}_0^{(l)} & S_l \end{pmatrix}$, $l = 1, 2$. Then

- the matrix $L_k(n, \tilde{S}_l)$ contains the transition rates of the process $\mathbf{u}_t^{(l)}$, leading to the service completion of one of $n - k$ customers of the lth type (k is the number of free channels for customers of type l, n is the total number of free channels for customers of type l and customers of this type that are being serviced);
- the matrix $P_n(\boldsymbol{\beta}_l)$ contains the transition probabilities of the process $\mathbf{u}_t^{(l)}$ leading to an increase in the number of customers of the lth type on the server from n to $n + 1$;
- the matrix $A_n(k, S_l)$ contains the transition rates of the process $\mathbf{u}_t^{(l)}$ in its state space without increasing or decreasing the number of customers of the lth type on the server (n is the number of customers of the lth type, k is the total number customers of the lth type and free channels for customers of this type).

Algorithm for calculating matrices $P_i(\cdot)$, $A_i(\cdot, \cdot)$, and $L_i(\cdot, \cdot)$ follows from the results by V. Ramaswami and D. Lucantoni and is described clearly step by step in [14].

Let us introduce the notation $Q_{i,j}$ for the rates of the chain transitions from states corresponding to the value i of the first component to states corresponding to the value j of this component, $i, j \geq 0$. Then the infinitesimal generator of the chain is defined by the following theorem.

Lemma 1. *The infinitesimal generator Q of a Markov chain $\xi_t, t \geq 0$, has the block three-diagonal structure*

$$Q = \begin{pmatrix} Q_{0,0} & Q_{0,1} & O & O & O & \dots \\ Q_{1,0} & Q_{1,1} & Q_{1,2} & O & O & \dots \\ O & Q_{2,1} & Q_{2,2} & Q_{2,3} & O & \dots \\ O & O & Q_{3,2} & Q_{3,3} & Q_{3,4} & \dots \\ \vdots & \vdots & \vdots & \vdots & \vdots & \ddots \end{pmatrix}$$

where

$$Q_{i,i-1} = i\gamma\, diag^{+}\{P_n(\boldsymbol{\beta}_1), n = \overline{0, N-1}\} \otimes I_{\mathcal{R}} \otimes I_{\bar{W}},\ i \geq 1,$$

$$Q_{i,i+1} = diag\{O_{C^{M_1-1}_{n+M_1-1}}, n = \overline{0, N-1}, I_{C^{M_1-1}_{N+M_1-1}}\} \otimes I_{\mathcal{R}} \otimes D_1,\ i \geq 0,$$

$$(Q_{i,i})_{n,n-1} = \frac{1}{n} L_{N-n}(N, \tilde{S}_1) \otimes I_{\mathcal{R}} \otimes I_{\bar{W}}, n = \overline{1, N},\ i \geq 0,$$

$$(Q_{i,i})_{n,n+1} = P_n(\boldsymbol{\beta}_1) \otimes I_{\mathcal{R}} \otimes D_1, n = \overline{0, N-1},\ i \geq 0,$$

$$(Q_{i,i})_{n,n} = \Psi_n - i\gamma I_{C^{M_1-1}_{n+M_1-1}\mathcal{R}\bar{W}} + \Delta_{i,n}, n = \overline{0, N-1}, i \geq 0,$$

$$(Q_{i,i})_{N,N} = \Psi_N + \Delta_N, i \geq 0,$$

where

$$\Psi_n = I_{C^{M_1-1}_{n+M_1-1}} \otimes diag^{-}\{\tfrac{1}{r} L_{R-r}(R, \tilde{S}_2), r = \overline{1, R}\} \otimes I_{\bar{W}}$$

$$+ \tfrac{1}{n} A_n(N, S_1) \oplus diag\{0, \tfrac{1}{r} A_r(R, S_2), r = \overline{1, R}\} \oplus D_0$$

$$+ I_{C^{M_1-1}_{n+M_1-1}} \otimes diag\{O_{\sum\limits_{r=0}^{R-1} C^{M_2-1}_{r+M_2-1}}, I_{C^{M_2-1}_{R+M_2-1}}\} \otimes D_2$$

$$+ I_{C^{M_1-1}_{n+M_1-1}} \otimes diag^{+}\{P_r(\boldsymbol{\beta}_2), r = \overline{0, R-1}\} \otimes D_2.$$

Here $\Delta_{i,n}, i \geq 0, n = \overline{0, N-1}$, Δ_N are diagonal matrices that ensure that the equality $Q\mathbf{e} = \mathbf{0}$ holds (the sums over the rows of the generator are equal to zero).

Proof. Proof of the lemma is carried out by analyzing the behavior of the Markov chain $\xi_t, t \geq 0$, during an infinitely small time interval. Let's describe briefly the meaning of non-zero blocks of the generator.

The block $Q_{i,i-1}$, $i \geq 1$, consists of the rates of transitions of the chain accompanied by a "successful" retry from the orbit of a type 1 customer. This customer occupies one of the free channels reserved for customers of this type. In this case, the number of customers of type 1 on the server increases by one, and the number of customers in the orbit decreases from i to $i-1$.

The block $Q_{i,i+1}$, $i \geq 0$, consists of the rates of transitions caused by the arrival of a primary customer of type 1, which meets all the channels allocated for this type of customers are busy and goes to the orbit. In this case, the number of customers in the orbit increases from i to $i+1$.

The block $(Q_{i,i})_{n,n-1}, n = \overline{1,N}$, $i \geq 0$, consists of the rates of transitions caused by the end of service of one of n customers of type 1. In this case, the number of customers of type 1 on the server decreases from n to $n-1$.

The block $(Q_{i,i})_{n,n+1}, n = \overline{0, N-1}$, $i \geq 0$, consists of the transition rates caused by the arrival of a primary customer of type 1, which found empty channels reserved for customers of this type. In this case, the number of customers of type 1 on the server increases from n to $n+1$.

The off-diagonal entries of the block $(Q_{i,i})_{n,n}, n = \overline{0,N}, i \geq 0$, are the rates of transitions that do not cause a change in the number of customers in the orbit or a change in the number of type 1 customers on the server. The corresponding transitions can be caused either by the end of servicing one of the customers of type 2 (matrix $I_{C^{M_1-1}_{n+M_1-1}} \otimes diag^-\{\frac{1}{r}L_{R-r}(R, \tilde{S}_2), r = \overline{1,R}\} \otimes I_{\bar{W}})$, or by redistributing the number of servers serving customers at different phases and idle transitions of the underlying process of the $MMAP$ (the matrix $\frac{1}{n}A_n(N, S_1) \oplus diag\{0, \frac{1}{r}A_r(R, S_2), r = \overline{1,R}\} \oplus D_0$), or the arrival of a type 2 customer that found all the channels occupied (the matrix $I_{C^{M_1-1}_{n+M_1-1}} \otimes diag\{O_{\sum_{r=0}^{R-1} C^{M_2-1}_{r+M_2-1}}, I_{C^{M_2-1}_{R+M_2-1}}\} \otimes D_2$, or the arrival of a type 2 customer that finds free channel for customers of this type and goes to the service (the matrix $I_{C^{M_1-1}_{n+M_1-1}} \otimes diag^+\{P_r\boldsymbol{\beta}_2), r = \overline{0, R-1}\} \otimes D_2)$. The diagonal entries of the block under consideration are taken with the opposite sign the rates of the chain exit from the states corresponding to i customers in the orbit and n customers of type 1 on the server.

For further investigation of the process ξ_t we will use the fact that this process is a multidimensional asymptotically quasi-toeplitz Markov chains (AQTMC). To prove this fact, we follow the definition of AQTMC presented in [15]. According the definition, a Markov chain belongs to the class of AQTMC if there exist the following limits:

$$Y_0 = \lim_{i\to\infty} T_i^{-1}Q_{i,i-1},\ Y_1 = \lim_{i\to\infty} T_i^{-1}Q_{i,i} + I,\ Y_2 = \lim_{i\to\infty} T_i^{-1}Q_{i,i+1}, \tag{1}$$

and the matrix $Y_0 + Y_1 + Y_2$ is stochastic.

Here T_i is a diagonal matrix with diagonal entries given by modules of the diagonal entries of the matrix $Q_{i,i}, i \geq 0$. Note that the last $C^{M_1-1}_{N+M_1-1}\mathcal{R}\bar{W}$ diagonal entries of the matrices T_i, $i \geq 1$, do not depend on i. Denote by T the diagonal matrix formed by these entries.

After some algebra we obtain the following expressions for the matrices $Y_l, l = 0, 1, 2:$

$$Y_0 = diag^+\{P_n(\boldsymbol{\beta}_1), n = \overline{0, N-1}\} \otimes I_{\mathcal{R}} \otimes I_{\bar{W}},\ i \geq 1,$$

$$Y_1 = \begin{pmatrix} O & O & O \\ O & T^{-1}\frac{1}{N}L_0(N, \tilde{S}_1) \otimes I_{\mathcal{R}\bar{W}} & T^{-1}(\Psi_N + \Delta_N) + I \end{pmatrix},$$

$$Y_2 = \begin{pmatrix} O & O \\ O & T^{-1}(I_{C_{N+M_1}^{M_1-1}\mathcal{R}} \otimes D_1) \end{pmatrix}.$$

Thus, limits (1) exist. The sum $Y_0 + Y_1 + Y_2$ is a stochastic matrix. Indeed, it is obvious that the sums of the entries of the first N block rows of this matrix are equal to one since $P_n(\boldsymbol{\beta}_1)\mathbf{e} = \mathbf{e}$. The sums over the rows of the entries of the $(N+1)th$ block row of this matrix are determined by the column vector

$$[T^{-1}\tfrac{1}{N}L_0(N, \tilde{S}_1) \otimes I + T^{-1}(\Psi_N + \Delta_N) + I + T^{-1}(I \otimes D_1)]\mathbf{e}$$

$$= T^{-1}[\tfrac{1}{N}L_0(N, \tilde{S}_1) \otimes I + \Psi_N + \Delta_N + (I \otimes D_1)]\mathbf{e} + \mathbf{e} = \mathbf{0} + \mathbf{e}$$

From the above it follows that the Markov chain ξ_t belongs to the class of *AQTMC*.

4 Ergodicity Condition

According to [15], the condition for ergodicity of the *AQTMC* $\xi_t, t \geq 0$, is formulated in terms of matrices Y_0, Y_1, Y_2. Using the corresponding results from [15], and a number of algebraic transformations, we got the following statement.

Theorem 1. *(i) A sufficient condition for the existence of the ergodic distribution of the Markov chain* ξ_t *is the fulfillment of the inequality*

$$\rho = \lambda/\bar{\mu} < 1, \tag{2}$$

where λ *is the input rate,*

$$\bar{\mu} = \mathbf{y}\frac{1}{N}L_0(N, \tilde{S}_1)\,\mathbf{e}, \tag{3}$$

$\mathbf{y}$ *is the unique solution of the system of linear algebraic equations*

$$\mathbf{y}[A_N(N, S_1) + \bar{\Delta}_N + L_0(N, \tilde{S}_1)P_{N-1}(\boldsymbol{\beta}_1)] = \mathbf{0},\ \ \mathbf{y}\,\mathbf{e} = 1, \tag{4}$$

where $\bar{\Delta}_N$, *is a diagonal matrix with diagonal entries defined by the vector* $-[L_0(N, \tilde{S}_1)\mathbf{e} + A_N(N, S_1)\mathbf{e}]$.

(ii) If $\rho > 1$, *then the ergodic distribution of the Markov chain* ξ_t *does not exist.*

Proof. (*i*) Carrying the matrix $Y(z) = Y_0 + Y_1 z + Y_2 z^2$ into the normal form, it is easy to see that this matrix at the point $z = 1$ is reducible with a single irreducible stochastic diagonal block corresponding to the matrix

$$\tilde{Y}(z) = \begin{pmatrix} zT^{-1}[\Psi_N + \Delta_N + zI_{C_{N+M_1-1}^{M_1-1}\mathcal{R}} \otimes D_1] + zI & zT^{-1}\frac{1}{N}L_0(N, \tilde{S}_1) \otimes I_{\mathcal{R}\bar{W}} \\ P_{N-1}(\boldsymbol{\beta}_1) \otimes I_{\mathcal{R}\bar{W}} & O \end{pmatrix}.$$

From [15], Lemma 2, it follows that the sufficient condition for the ergodicity of the Markov chain under consideration is formulated in terms of the matrix $\tilde{Y}(z)$ and has the form of the inequality

$$\left[\det(zI - \tilde{Y}(z))\right]'_{z=1} > 0. \tag{5}$$

Using the block structure of the matrix $\tilde{Y}(z)$, we can reduce the determinant in (5) to the following form:

$$\det(zI - \tilde{Y}(z)) = \det(zT^{-1})\det(-z[\Psi_N + \Delta_N + zI_{C_{N+M_1-1}^{M_1-1}\mathcal{R}} \otimes D_1]$$
$$-(\frac{1}{N}L_0(N, \tilde{S}_1) \otimes I_{\mathcal{R}\bar{W}})(P_{N-1}(N, \boldsymbol{\beta}_1) \otimes I_{\mathcal{R}\bar{W}}). \tag{6}$$

Using in (6) the property of the generator $Q\mathbf{e} = \mathbf{0}$, we reduce inequality (5) to the form

$$[\det(-z[\Psi_N + \Delta_N + zI_{C_{N+M_1-1}^{M_1-1}\mathcal{R}} \otimes D_1]$$
$$-(\frac{1}{N}L_0(N, \tilde{S}_1) \otimes I_{\mathcal{R}\bar{W}})(P_{N-1}(N, \boldsymbol{\beta}_1) \otimes I_{\mathcal{R}\bar{W}}))]'_{z=1} > 0. \tag{7}$$

which in turn is equivalent to

$$\mathbf{x}[(\Psi_N + \Delta_N + 2I_{C_{N+M_1-1}^{M_1-1}\mathcal{R}} \otimes D_1)]\mathbf{e} < 0, \tag{8}$$

where $\mathbf{x}$ is the unique solution of the system of linear algebraic equations

$$\mathbf{x}[\Psi_N + \Delta_N + I_{C_{N+M_1-1}^{M_1-1}\mathcal{R}} \otimes D_1$$
$$+(\frac{1}{N}L_0(N, \tilde{S}_1) \otimes I_{\mathcal{R}\bar{W}})(P_{N-1}(N, \boldsymbol{\beta}_1) \otimes I_{\mathcal{R}\bar{W}}))] = \mathbf{0}, \tag{9}$$

$$\mathbf{x}\mathbf{e} = 1. \tag{10}$$

Let us reduce inequality (8) to form (2), and the system of Eqs. (9), (10) to form (4). First we consider system (9)–(10).

For further proof, it is important for us to know the entries of the matrix Δ_N. Diagonal entries of this matrix are equal to corresponding entries of the column vector

$$-[\Psi_N + (Q_{i,i})_{N,N-1} + (Q_{i,i+1})_{N,N}]\mathbf{e} =$$

$$-\Bigg[\mathbf{e}_{C^{M_1-1}_{N+M_1-1}} \otimes [diag^-\{\frac{1}{r}L_{R-r}(R,\tilde{S}_2), r=\overline{1,R}\}$$

$$+ diag\{0, \frac{1}{r}A_r(R,S_2)\mathbf{e}, r=\overline{1,R}\}]\mathbf{e}\otimes\mathbf{e}_{\bar{W}}$$

$$+\frac{1}{N}[A_N(N,S_1)+L_0(N,\tilde{S}_1)]\mathbf{e}\otimes\mathbf{e}_{\mathcal{R}\bar{W}}\Bigg], \quad (11)$$

from which it follows that the diagonal matrix Δ_N has the form

$$\Delta_N = \Bigg[I_{C^{M_1-1}_{N+M_1-1}} \otimes \bar{\Delta}_N \otimes I_{\bar{W}} + \tilde{\Delta}_N \otimes I_{\mathcal{R}} \otimes I_{\bar{W}}\Bigg], \quad (12)$$

where $\bar{\Delta}_N$ is a diagonal matrix, the diagonal entries of which are formed by the corresponding entries of the column vector

$$-\frac{1}{N}[A_N(N,S_1)+L_0(N,\tilde{S}_1)]\mathbf{e},$$

$\tilde{\Delta}_N$ is a diagonal matrix, the diagonal entries of which are formed by the corresponding entries of the column vector

$$-[diag^-\{\frac{1}{r}L_{R-r}(R,\tilde{S}_2), r=\overline{1,R}\} + diag\{0, \frac{1}{r}A_r(R,S_2), r=\overline{1,R}\}]\mathbf{e}.$$

Substitute in (9) the expression for Ψ_N given in the statement of the theorem and expression (12) for Δ_N. Multiplying the resulting equation on the right by $I_{C^{M_1-1}_{N+M_1-1}} \times \mathbf{e}_{\mathcal{R}} \otimes I_{\bar{W}}$ and taking into account that

$$\Bigg[I_{C^{M_1-1}_{n+M_1-1}} \otimes diag^-\{\tfrac{1}{r}L_{R-r}(R,\tilde{S}_2), r=\overline{1,R}\}\otimes I_{\bar{W}}$$

$$+ I_{C^{M_1-1}_{n+M_1-1}} \otimes diag\{0, \tfrac{1}{r}A_r(R,S_2), r=\overline{1,R}\}\otimes I_{\bar{W}} + I_{C^{M_1-1}_{N+M_1-1}} \otimes \bar{\Delta}_N \otimes I_{\bar{W}}\Bigg]$$

$$\times (I_{C^{M_1-1}_{N+M_1-1}} \otimes \mathbf{e}_{\mathcal{R}} \otimes I_{\bar{W}}) = O,$$

we reduce (9) to the form

$$\mathbf{x}[\frac{1}{N}A_N(N,S_1)\otimes\mathbf{e}_{\mathcal{R}}\otimes I_{\bar{W}} + \frac{1}{N}L_0(N,\tilde{S}_1)P_{N-1}(N,\boldsymbol{\beta}_1)\otimes\mathbf{e}_{\mathcal{R}}\otimes I_{\bar{W}}$$

$$+\bar{\Delta}_N\otimes\mathbf{e}_{\mathcal{R}}\otimes I_{\bar{W}} + I_{C^{M_1-1}_{N+M_1-1}}\otimes\mathbf{e}_{\mathcal{R}}\otimes(D_0+D_1+D_2)] = \mathbf{0}. \quad (13)$$

Now we represent the vector $\mathbf{x}$ as

$$\mathbf{x} = \mathbf{y}\otimes\mathbf{z}\otimes\boldsymbol{\theta}, \quad (14)$$

where $\mathbf{y}$ is a stochastic vector of size $C_{N+M_1-1}^{M_1-1}$ and $\mathbf{z}$ is a stochastic vector of size $\mathcal{R}$. Then (13) takes the form

$$\mathbf{y}[A_N(N,S_1) + L_0(N,\tilde{S}_1)P_{N-1}(N,\boldsymbol{\beta}_1) + \bar{\Delta}_N)]\frac{1}{N} \otimes \boldsymbol{\theta} = \mathbf{0},$$

whence formula (4) of the theorem follows.

Now we consider inequality (8). In this inequality

$$(\Psi_N + \Delta_N)\mathbf{e} = -\frac{1}{N}L_0(N,\tilde{S}_1)\mathbf{e} \otimes \mathbf{e}_{\mathcal{R}\bar{W}} - \mathbf{e}_{C_{N+M_1}^{M_1-1}\mathcal{R}} \otimes D_1\mathbf{e}. \tag{15}$$

Substituting (15) into (8), we obtain the following inequality:

$$\mathbf{x}[-\frac{1}{N}L_0(N,\tilde{S}_1)\mathbf{e} \otimes \mathbf{e}_{\mathcal{R}\bar{W}} + \mathbf{e}_{C_{N+M_1}^{M_1-1}\mathcal{R}} \otimes D_1\mathbf{e}] < 0. \tag{16}$$

Substituting the vector $\mathbf{x}$ of form (14) into (16), and taking into account that $\mathbf{y}$ and $\mathbf{z}$ are stochastic vectors, and $\boldsymbol{\theta}D_1\mathbf{e} = \lambda_1$, we get

$$[-\mathbf{y}\frac{1}{N}L_0(N,\tilde{S}_1)\mathbf{e} \otimes \mathbf{e}_{\mathcal{R}} + \lambda_1 \otimes \mathbf{e}_{\mathcal{R}}] = [-\mathbf{y}\frac{1}{N}L_0(N,\tilde{S}_1)\mathbf{e} + \lambda_1] \otimes \mathbf{e}_{\mathcal{R}} < 0,$$

whence the required inequality (2) in the statement of the theorem follows.

The statement (ii) of the theorem follows from [15], Lemma 2.

In what follows, we will assume that inequality (2) is satisfied. Then the ergodic distribution of the Markov chain ξ_t exists and coincides with the stationary distribution.

5 Stationary Distribution. Performance Measures

Let us introduce the notation for the stationary probabilities of the chain:

$$p(i,n,\eta^{(1)},\eta^{(2)},\ldots,\eta^{(M_1)},r,\tau^{(1)},\tau^{(2)},\ldots,\tau^{(M_2)},\nu)$$

$$= \lim_{t\to\infty} P\{i_t, n_t = n, \eta_t^{(1)} = \eta^{(1)}, \eta_t^{(2)} = \eta^{(2)}, \ldots, \eta_t^{(n_t)} = \eta^{(n)},$$

$$r_t = r, \tau_t^{(1)} = \tau^{(1)}, \tau_t^{(2)} = \tau^{(2)}, \ldots, \tau_t^{(r_t)} = \tau^{(r)}, \nu_t = \nu\},\ i \geq 0, n = \overline{0,N},$$

$$\eta^{(m^{(1)})} = \overline{1,n}, m^{(1)} = \overline{1,M_1}, r = \overline{0,R}, \tau_t^{(m^{(2)})} = \overline{1,r}, m^{(2)} = \overline{1,M_2}, \nu = \overline{0,W}.$$

Let us arrange these probabilities according to the order adopted above for the states of the considered chain ξ_t and form the row vectors $\mathbf{p}_0, \mathbf{p}_1, \ldots$ corresponding the values i_t of the first component of the chain.

To find these vector, we use a special algorithm for calculating the stationary distribution of asymptotically quasi-Toeplitz Markov chains, proposed in [15].

Once the vectors $\mathbf{p}_i, i \geq 0$, of the steady state probabilities are calculated, we can find a number of stationary performance measures of the system. Below we give expressions for most important of them.

1. Distribution of the number of type 1 customers in the orbit $p_i = \mathbf{p}_i\mathbf{e}, i \geq 0$.
2. Mean number of type 1 customers in the orbit $Z_{orbit} = \sum_{i=1}^{\infty} ip_i$.
3. Joint distribution of the number of type 1 customers on the server and states of the $MMAP$

$$\boldsymbol{\pi}_n^* = \sum_{i=0}^{\infty} \mathbf{p}_i \left[\begin{pmatrix} \mathbf{0}^T_{\sum_{m=0}^{n-1} C_{m+M_1-1}^{M_1-1}\mathcal{R}} \\ \mathbf{e}_{C_{n+M_1-1}^{M_1-1}\mathcal{R}} \\ \mathbf{0}^T_{\sum_{m=n+1}^{N} C_{m+M_1-1}^{M_1-1}\mathcal{R}} \end{pmatrix} \otimes I_{\bar{W}} \right], \; n = \overline{0, N}.$$

4. Distribution of the number of type 1 customers on the server

$$\pi_n = \boldsymbol{\pi}_n^* \mathbf{e}_{\bar{W}}, \; n = \overline{0, N}.$$

5. Mean number of type 1 customers on the server $\bar{N} = \sum_{n=1}^{N} n\pi_n$.
6. Probability that a primary customer of type 1 will go to service immediately without visiting orbit

$$P_{imm} = 1 - \frac{1}{\lambda_1} \boldsymbol{\pi}_N^* D_1 \mathbf{e}.$$

Explanation. In the formula the subtrahend is the ratio of the rate of primary type 1 customers that meets all the channels allocated for this type of customers are busy and goes into the orbit and the rate of input flow of type 1 customers. Thereat, the subtrahend gives a probability that any primary customer will go into the orbit. The probability P_{imm} is calculated as an additional probability.
7. Joint distribution of the number of type 2 customers on the server and states of the $MMAP$

$$\mathbf{q}_r^* = \sum_{i=0}^{\infty} \mathbf{p}_i \sum_{n=0}^{N} \begin{pmatrix} \mathbf{0}^T_{\sum_{m=0}^{n-1} C_{m+M_1-1}^{M_1-1}} \\ \mathbf{e}_{C_{n+M_1-1}^{M_1-1}} \\ \mathbf{0}^T_{\sum_{m=n+1}^{N} C_{m+M_1-1}^{M_1-1}} \end{pmatrix} \otimes \begin{pmatrix} \mathbf{0}^T_{\sum_{m=0}^{r-1} C_{m+M_2-1}^{M_2-1}} \\ \mathbf{e}_{C_{r+M_2-1}^{M_2-1}} \\ \mathbf{0}^T_{\sum_{m=r+1}^{R} C_{m+M_2-1}^{M_2-1}} \end{pmatrix} \otimes I_{\bar{W}}, \; r = \overline{0, R}.$$

8. Distribution of the number of type 2 customers on the server

$$q_r = \mathbf{q}_r^* \mathbf{e}, \; r = \overline{0, R}.$$

9. Mean number of type 2 customers on the server $\bar{R} = \sum_{r=1}^{R} rq_r$.
10. Probability that type 2 customer will be lost $P_{loss,2} = \frac{1}{\lambda_2} \mathbf{q}_R^* D_2 \mathbf{e}$.

6 Numerical Results

In this section, we consider the optimization problem. From the above it follows that the service bandwidth is divided between customers of types 1 and 2 in the proportion $\mu_1 : \mu_2$. Then the optimization problem for the system under consideration can be formulated as follows: what should this proportion be in order the economic criterion of the quality of the system operation to reach minimum? Let us consider the cost criterion of the form

$$E = c_1 Z_{orbit} + c_2 \lambda_2 P_{loss,2}, \tag{17}$$

where c_1 is the penalty charged per unit of time for type 1 customer staying in the orbit, and c_2 is the penalty charged for the loss of type 2 customer.

Let the general bandwidth μ of the server be constant, $\mu = \mu_1 + \mu_2 = 50$. In this example we find the value of μ_1 in this sum that minimizes the quality criterion (17).

We consider the following input data. $N = 4, R = 3, \gamma = 5$.

The input $MMAP$ is defined by matrices D_0 and D of the form

$$D_0 = \begin{pmatrix} -86 & 0.01 \\ 0.02 & -2.76 \end{pmatrix}, \quad D = \begin{pmatrix} 85 & 0.99 \\ 0.2 & 2.54 \end{pmatrix}.$$

Based on the matrix D, we define the matrices D_1 and D_2 as $D_1 = 0.7D$ and $D_2 = 0.3D$. For this $MMAP$ $\lambda_1 = 12.4266$, $\lambda_2 = 5.3257$ $v_1 = 0.0602$, $v_2 = 0.2859$, $c_{cor}^{(1)} = 0.3902, c_{cor}^{(2)} = 0.3284$.

PH distribution of the service time of a single customer of type 1 has Erlang distribution of order 2 with parameter 20. It is defined by the vector $\boldsymbol{\beta}^{(1)} = (1, 0)$ and the matrix $S_1 = \begin{pmatrix} -20 & 20 \\ 0 & -20 \end{pmatrix}$.

PH distribution of the service time of a single customer of type 2 has Erlang distribution of order 2 with parameter 80. It is defined by the vector $\boldsymbol{\beta}^{(2)} = (1, 0)$ and the matrix $S_2 = \begin{pmatrix} -80 & 80 \\ 0 & -80 \end{pmatrix}$.

Table 1. The values of μ_1^*, E^* E_1, E_2 for different values of c_2

c_2	μ_1^*	E^*	E_1	E_2
20	27.6147	44.5519	343.2395	103.7683
40	24.8532	69.0413	355.4567	204.4983
50	24.8532	80.2572	361.5652	254.8634

The cost criterion E as a function of bandwidth μ_1 under condition $\mu_1 + \mu_2 = 50$ is depicted in Fig. 1. This figure shows the values of the criterion under $c_1 = 1$ and three different values of $c_2 = 20, 40, 50$. In Table 1 we present the minimum points μ_1^*, minimum values E^* of the criterion and values E_1, E_2 of the criterion on the boundaries of the interval $(13.8, 49.7)$ where we vary μ_1. Note that in the region $\mu_1 < 12.4266$ the ergodicity condition is violated. This explains our choice of the left boundary of the interval in which we vary μ_1.

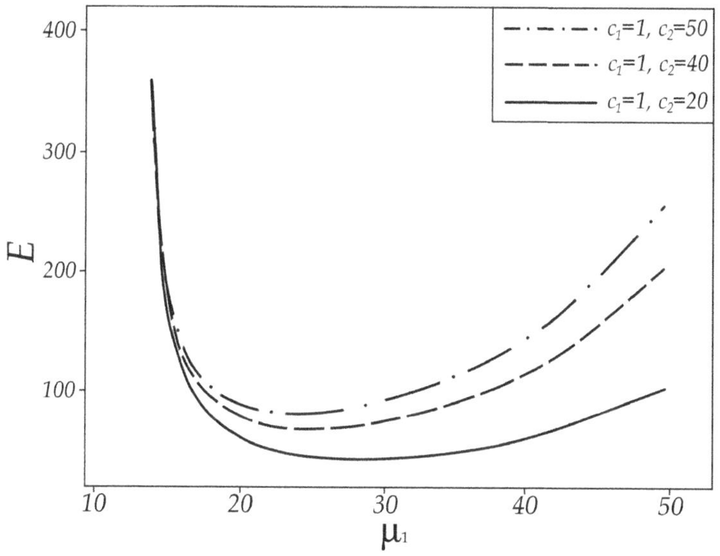

Fig. 1. Cost criterion E as a function of μ_1 for different cost coefficients under condition $\mu_1 + \mu_2 = 50$

It is seen from Fig. 1 and from Table 1 that it is most advantageous to divide the bandwidth approximately equally, i.e. $\mu_1 \approx \mu_2 \approx 25$. One can also see that the relative gains from applying optimal bandwidth sharing between two types of customers can be very large. So, in the case $c_2 = 20$ the relative gains δ_1, δ_2 from the optimal value of the cost criterion E^* in comparison with the values E_1, E_2 of the criterion on the boundaries of the region $\mu_1 \in (13.5, 49.5)$ are equal to $\delta_1 = \frac{E_1 - E^*}{E*} 100\% = 680\%$, $\delta_2 = \frac{E_2 - E^*}{E^*} 100\% = 133\%$.

7 Conclusion

In this paper, we have investigated the retrial queueing system with processor sharing and two types of customers which arrive to the system according the $MMAP$. The service times of customers have PH distribution different for customer of different types. The operation of the system is described in terms of multidimensional Markov chain. We derive the ergodicity condition for this chain and calculate its stationary distribution. Based on the stationary probabilities we derive formulas for a number of performance measures of the system. In order to demonstrate the use of the results obtained for optimisation of the quality of service in the system, we introduced the cost criterion, defined as average penalty per unit of time, and presented the example of numerical optimisation in which it is solved in what proportion should the channel capacity be divided in order to minimize the value of the criterion.

The results of the research can be used for modeling the operation of a cell of fixed capacity in a wireless cellular network and others real world systems operating in processor sharing mode.

References

1. Ghosh, A., Banik, A.D.: An algorithmic analysis of the $BMAP/MSP/1$ generalized processor-sharing queue. Comput. Oper. Res. **79**, 1–11 (2017)
2. Telek, M., van Houdt, B.: Response time distribution of a class of limited processor sharing queues. ACM SIGMETRICS Perform. Eval. Rev. **45**, 143–155 (2018). https://doi.org/10.1145/3199524.3199548
3. Yashkov, S., Yashkova, A.: Processor sharing: a survey of the mathematical theory. Autom. Remote. Control. **68**, 662–731 (2007)
4. Zhen, Q., Knessl, C.: On sojourn times in the finite capacity $M/M/1$ queue with processor sharing. Oper. Res. Lett. **37**, 447–450 (2009)
5. Masuyama, H., Takine, T.: Sojourn time distribution in a $MAP/M/1$ processor-sharing queue. Oper. Res. Lett. **31**, 406–412 (2003)
6. Dudin, S., Dudin, A., Dudina, O., Samouylov, K.: Analysis of a retrial queue with limited processor sharing operating in the random environment. In: Koucheryavy, Y., Mamatas, L., Matta, I., Ometov, A., Papadimitriou, P. (eds.) WWIC 2017. LNCS, vol. 10372, pp. 38–49. Springer, Cham (2017). https://doi.org/10.1007/978-3-319-61382-6_4
7. Dudin, A., Dudin, S., Dudina, O., Samouylov, K.: Analysis of queuing model with limited processor sharing discipline and customers impatience. Oper. Res. Perspect. **5**, 245–255 (2018)
8. He, Q.M.: Queues with marked customers. Adv. Appl. Probab. **28**, 567–587 (1996)
9. Dudin, A.N., Klimenok, V.I., Vishnevsky, V.M.: The Theory of Queuing Systems with Correlated Flows. Springer, Heidelberg (2020). ISBN 978-3-030-32072-0
10. Neuts, M.F.: Matrix-Geometric Solutions in Stochastic Models. The Johns Hopkins University Press, Baltimore (1981)
11. Graham, A.: Kronecker Products and Matrix Calculus with Applications. Ellis Horwood, Cichester (1981)
12. Ramaswami, V.: Independent Markov processes in parallel. Commun. Stat. Stoch. Models **1**, 419–432 (1985)
13. Ramaswami, V., Lucantoni, D.M.: Algorithms for the multi-server queue with phase-type service. Commun. Stat. Stoch. Models **1**, 393–417 (1985)
14. Dudina, O., Kim, C.S., Dudin, S.: Retrial queueing system with Markovian arrival flow and phase type service time distribution. Comput. Ind. Eng. **66**, 360–373 (2013)
15. Klimenok, V.I., Dudin, A.N.: Multi-dimensional asymptotically quasi-Toeplitz Markov chains and their application in queueing theory. Queueing Syst. **54**, 245–259 (2006)

Multi-level MMPP as a Model of Fractal Traffic

Anatoly Nazarov[1], Alexander Moiseev[1(✉)], Ivan Lapatin[1], Svetlana Paul[1], Olga Lizyura[1], Pavel Pristupa[1], Xi Peng[2], Li Chen[2], and Bo Bai[2]

[1] Institute of Applied Mathematics and Computer Science, National Research Tomsk State University, 36 Lenina Avenue, Tomsk 634050, Russia
nazarov.tsu@gmail.com, moiseev.tsu@gmail.com, ilapatin@mail.ru, paulsv82@mail.ru, oliztsu@mail.ru, pristupa@gmail.com

[2] Theory Lab, Central Research Center, 2012Labs, Huawei Tech. Investment Co., Ltd., 8/F, Bio-informatics Center, No. 2 Science Park West Avenue, Hong Kong Science Park, Pak Shek Kok, Shatin, N.T., Hong Kong
{pancy.pengxi,chen.li7}@huawei.com, ee.bobbai@gmail.com

Abstract. In this paper, we propose the multi-level Markov modulated Poisson process with arbitrary distribution of the packet length as a model of fractal traffic. For the total amount of information received in multi-level MMPP, we investigate the probability distribution and present the algorithm of calculating the first and the second moments. Using asymptotic analysis method, we build Gaussian approximation of aforementioned distribution. We show that the convergence time of the probability distribution to the Gaussian distribution forms the period where the Hurst parameter is stable and reflects the self-similarity of the multi-level MMPP.

Keywords: Traffic modeling · Self-similarity · Fractality · Burstiness · Markov modulated Poisson process · Asymptotic analysis · Hurst parameter

1 Introduction

Various statistical studies of real telecommunication flows show the fractality, self-similarity and burstiness of the traffic. The will to build the model that takes into account all of such properties leads to the emergence of the models that are complicated to investigate. Moreover, considering queueing systems with such flows does not make any sense since the analysis will be impossible. Our goal is to build the model with clear structure that captures mentioned properties of the traffic and allows analytical investigation.

The property of burstiness is characterized by the existence of intervals where traffic rate is rather high and intervals with low or zero intensity. Such behavior of telecommunication flows led to the creation of On-Off models [1,8,13]. On-Off models are doubly stochastic point processes having two states: "on" and "off". Doubly stochastic point processes are widely described in paper [5]. The

A. Dudin et al. (Eds.): ITMM 2020, CCIS 1391, pp. 61–77, 2021.
https://doi.org/10.1007/978-3-030-72247-0_5

class of On-Off models contains models with self-transitions and alternating models. Such models are simple for simulation and have clear interpretation. However, models with general distributed interarrival times may be complicated for investigation.

Since the interarrival times in self-similar flows usually follow heavy-tailed distribution, another commonly used models are renewal processes using heavy-tailed distributions, e.g. Pareto, Weibull, Lognormal, hyperexponential distributions with a special choice of parameter values to describe the interarrival times [2,3,14]. Renewal processes allow to capture long-range dependence and self-similarity of the traffic. On the other hand, heavy-tailed distributions often have infinite moments, which does not allow using method of moments to evaluate the parameters. In renewal processes, interarrival times are represented as independent identically distributed events. Due to the lack of intercorrelation between arrivals, these models are in general inadequate to capture traffic burstiness. Paper [4] is devoted to the heavy-tailed distributions arise in fractal traffic.

Another class of traffic models that reflects burstiness and self-similarity are Markovian arrival processes, which are analiticaly tractable. The intensity of the flow in such models depends on the state of underlying Markov chain. Due to this, setting high intensity for one state and low intensity for another state yields the burstiness of the flow [9,12,17].

Statistical studies of the fractal traffic are devoted to the deriving measures that can capture the burstiness of the traffic [7,10] and the order of self-similarity [11,16]. In our research, we use the Hurst parameter (Hurst exponent), calculated using algorithm described in [15].

This paper is dedicated to the multi-level Markov modulated Poisson process (MMPP), which is a special case of the MAP. MMPP has a more demonstrative interpretation than MAP. The set of parameters in MAP model is wider, however such model does not invest more in burstiness of the simulated traffic than MMPP. Moreover, the parameters of MMPP are easier to evaluate using statistical measures of the real traffic. Thus, we use simpler MMPP model. We construct the structure of the MMPP in certain form to reflect the fractality, self-similarity and burstiness of the traffic.

The rest of the paper is organized as follows. In Sect. 2, we describe the mathematical model and the structure of the MMPP. Section 3 contains the study of the total amount of information received in the flow. Section 4 is devoted to the asymptotic analysis of the model under the limit condition of growing observation time. Section 5 comprises the algorithm of calculation the moments of the total amount of information. In Sect. 6, we present the numerical implementation of the asymptotic results and the estimation of approximation accuracy. Section 7 is dedicated to the estimation of the values of Hurst exponent for the total amount of information received in multi-level MMPP. Finally, Sect. 8 is devoted to concluding remarks.

2 Mathematical Model of Multi-level MMPP

In bursty flows the intervals of high intensity alternate with intervals of low or even zero intensity.

The property of self-similarity is characterized by the similarity of flow behavior on different time scales. If we consider the first level interval of high intensity and decrease the time scale by two or more orders, then the flow structure coincides with the cyclic structure of the first level flow. On the second level of the flow, we also have alternating between high and low intensity intervals.

If we decrease the time unit again, then we obtain the same structure on the third level. Figure 1 shows the structure of the flow with four levels. Intervals $H1$, $H2$ and $H3$ are intervals with high intensity on the different time scales. On the other hand, $L1$, $L2$ and $L3$ denote low rate intervals.

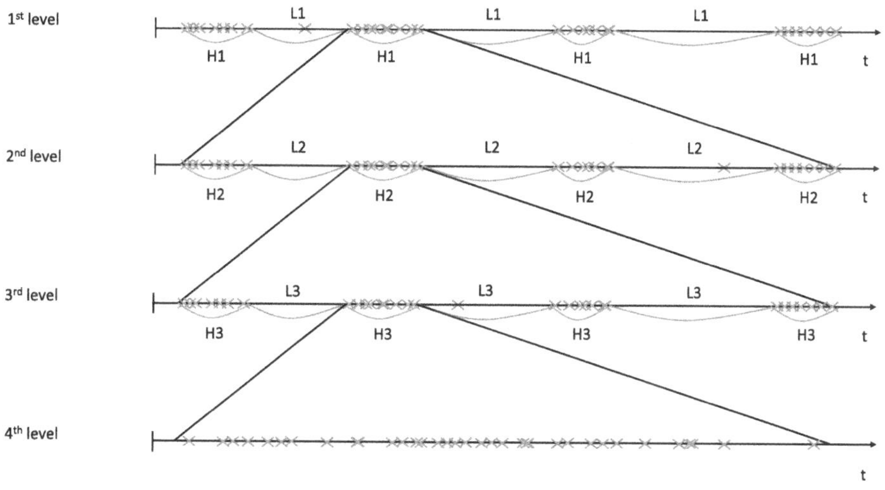

Fig. 1. Structure of fractal traffic

Of course, a real-life traffic can not be true fractal because, due to the technological constraints, there exists some level of the time scale where we can observe only regular arrivals or absence of arrivals. So, in the example, we can suppose that at the last level of time scaling (inside the $H3$ intervals) we have Poisson process with rather big rate. We can vary the number of levels of the model and, so, we will vary the number of modeled self-similarity levels.

Consider the model of traffic with N levels of self-similarity, zero intensity inside the L_i intervals and Poisson arrivals inside the last $(N+1)$-th level. For modeling such behavior, we consider MMPP given by infinitesimal generator $\mathbf{Q}$ of underlying Markov chain $n(t) \in \{0, \ldots, N\}$, and diagonal matrix $\mathbf{\Lambda}$ with conditional intensities λ_n on the main diagonal which are the rates of arrivals corresponding to the n-th state of the flow. In order to reflect the properties of burstiness and self-similarity of the flow, we must set matrices $\mathbf{Q}$ and $\mathbf{\Lambda}$ in certain form.

We will put corresponding parameters into matrices $\mathbf{Q}$ and $\mathbf{\Lambda}$ in reverse order numbering them starting from zero. The zeroth state of the MMPP will correspond to the $(N+1)$-th level where we have a regular (Poisson) flow with high intensity $\lambda_0 = \lambda$. Other states of the MMPP will model only zero-intensity intervals L_i between periods H_i at different levels of self-similarity (see Fig. 1). So, all rest conditional intensities will be equal to zero: $\lambda_1 = \cdots = \lambda_N = 0$.

We suppose that if the MMPP is in the state with zero intensity (states of the underlying Markov chain $1, \ldots, N$) then, when this state ends, the MMPP may move only in state 0 with non-zero intensity. In such a way, we can model changing of periods with zero intensity by periods with non-zero intensity at different time scales (levels of self-similarity). Due to this, we have transitions from states $1, \ldots, N$ only to state 0. Denote intensities of these transitions by $q_1, \ldots, q_N$. Also, denote the intensity of leaving state 0 by q_0 and probabilities of transitions in states $1, \ldots, N$ after state 0 ends by $v_1, \ldots, v_N$, respectively. So, we obtain the following structures of matrices $\mathbf{Q}$ and $\mathbf{\Lambda}$ that determine the proposed model of multi-level MMPP:

$$\mathbf{Q} = \begin{bmatrix} -q_0 & q_0 v_1 & q_0 v_2 & \ldots & q_0 v_N \\ q_1 & -q_1 & 0 & \ldots & 0 \\ q_2 & 0 & -q_2 & \ldots & 0 \\ \ldots & \ldots & \ldots & \ldots & \ldots \\ q_N & 0 & 0 & \ldots & -q_N \end{bmatrix}, \quad \mathbf{\Lambda} = \begin{bmatrix} \lambda & 0 & 0 & \ldots & 0 \\ 0 & 0 & 0 & \ldots & 0 \\ 0 & 0 & 0 & \ldots & 0 \\ \ldots & \ldots & \ldots & \ldots & \ldots \\ 0 & 0 & 0 & \ldots & 0 \end{bmatrix}, \tag{1}$$

where $\sum_{n=1}^{N} v_n = 1$.

In practice, the following conditions should be satisfied to the proposed model be close to real-life traffic as possible:

$$\lambda >> q_0 >> q_1 >> q_2 >> \cdots >> q_N,$$

where sign ">>" means "much more than" ("greater than in several orders").

Each request arrived in the MMPP contains some amount of information which we will model as continuous random variable with given distribution function $B(x)$. Let $S(t)$ denote the total amount of information arrived in the flow during a time period with length t. The goal problem of the paper is to obtain the probability distribution of process $S(t)$.

3 Derivation of Equations for Probability Distribution of the Total Amount of Information Received in Multi-level MMPP for a Certain Time

Denote probability distribution

$$P_n(s,t) = P\{n(t) = n,\ S(t) < s\},\ n = 0,\ \ldots,\ N. \tag{2}$$

Two-dimensional process $\{n(t),\ S(t)\}$ is Markovian. Thus, we can derive equations

$$P_0(s,t+\Delta t) = P_0(s,t)(1-\lambda\Delta t)(1-q_0\Delta t)$$

$$+\lambda\Delta t\int\limits_0^s P_0(s-x,t)dB(x) + \sum_{n=1}^{N} P_n(s,t)q_n\Delta t + o(\Delta t),$$

$$P_n(s,t+\Delta t) = P_n(s,t)(1-q_n\Delta t) + P_0(s,t)q_0 v_n\Delta t + o(\Delta t),\ n=1,\ \dots,\ N,$$

which allow us to obtain system of differential equations

$$\frac{\partial P_0(s,t)}{\partial t} = -(\lambda+q_0)P_0(s,t) + \lambda\int\limits_0^s P_0(s-x,t)dB(x) + \sum_{n=1}^{N} P_n(s,t)q_n,$$

$$\frac{\partial P_n(s,t)}{\partial t} = -q_n P_n(s,t) + P_0(s,t)q_0 v_n,\ n=1,\ \dots,\ N. \tag{3}$$

We introduce Fourier transform

$$H_n(u,t) = \int\limits_0^\infty e^{jus} d_s P_n(s,t),\ n=0,\ \dots,\ N.$$

Since the sum of functions $H_n(u,t)$ by $n=0,\dots,N$ gives the characteristic function of process $S(t)$, we will call functions $H_n(u,t)$ as partial characteristic functions. Using such notation, we can rewrite system of Eqs. (3) as follows:

$$\frac{\partial H_0(u,t)}{\partial t} = \lambda(B^*(u)-1)H_0(u,t) - q_0 H_0(u,t) + \sum_{n=1}^{N} H_n(u,t)q_n,$$

$$\frac{\partial H_n(u,t)}{\partial t} = -q_n H_n(u,t) + H_0(u,t)q_0 v_n. \tag{4}$$

Here $B^*(u) = \int\limits_0^\infty e^{jux}dB(x)$ is a characteristic function of the amount of information in one request of MMPP.

We denote row vector

$$\mathbf{H}(u,t) = \{H_1(u,t),\ H_2(u,t),\ \dots,\ H_N(u,t)\}$$

and represent system (4) as a matrix equation

$$\frac{\partial \mathbf{H}(u,t)}{\partial t} = \mathbf{H}(u,t)\{\mathbf{Q} + (B^*(u)-1)\mathbf{\Lambda}\}. \tag{5}$$

Denoting

$$\mathbf{H}(0,t) = \mathbf{r},$$

we obtain the system of equations for vector $\mathbf{r}$, which is the stationary distribution of the states of the underlying Markov chain

$$\mathbf{rQ} = 0, \ \mathbf{re} = 1. \tag{6}$$

Here $\mathbf{e}$ is a vector of ones with $(N+1) \times 1$ dimension.

Using matrix exponential method, we can write the following solution of the problem (5)–(6):

$$\mathbf{H}(u,t) = \mathbf{r} \exp\{[\mathbf{Q} + (B^*(u) - 1)\mathbf{\Lambda}]t\}.$$

From two-dimensional distribution defined by vector characteristic function $\mathbf{H}(u,t)$, we can derive one-dimensional distribution in the form of scalar characteristic function

$$\mathbf{H}(u,t)\mathbf{e} = \mathbb{E}e^{juS(t)} = \mathbf{r} \exp\{[\mathbf{Q} + (B^*(u) - 1)\mathbf{\Lambda}]t\}\mathbf{e}. \tag{7}$$

Similar results were derived in [6] for discrete distribution of the number of arrivals in MAP.

Inverse Fourier transform

$$P(s,t) = \frac{1}{2\pi} \int_{-\infty}^{\infty} \frac{1 - e^{-jus}}{ju} \mathbf{H}(u,t)\mathbf{e} \, du$$

uniquely defines distribution function $P(s,t) = P\{S(t) < s\}$ of the total amount of received information $S(t)$ per time t in multi-level MMPP.

Multiplying matrix Eq. (5) by vector $\mathbf{e}$, we obtain system

$$\frac{\partial \mathbf{H}(u,t)}{\partial t} = \mathbf{H}(u,t)\{\mathbf{Q} + (B^*(u) - 1)\mathbf{\Lambda}\},$$

$$\frac{\partial \mathbf{H}(u,t)}{\partial t}\mathbf{e} = \mathbf{H}(u,t)(B^*(u) - 1)\mathbf{\Lambda e}, \tag{8}$$

which is the main system for our analysis.

4 Asymptotic Analysis of the Amount of Information Received in MMPP Under Growing Time Limit Condition

We consider system (8) under the limit condition of growing time of the flow observation, which is characterized by $t = \tau T$, where T is infinite parameter.

Theorem 1. *The limit of characteristic function*

$$\mathbf{H}(u,t)\mathbf{e} = \mathbb{E}e^{juS(t)}$$

of the total amount of information received per time t in multi-level MMPP satisfies the equality

$$\lim_{t\to\infty}\left[\mathbb{E}e^{juS(t)} - \exp\left\{ju\kappa_1 t + \frac{(ju)^2}{2}\kappa_2 t\right\}\right] = 0, \tag{9}$$

where

$$\kappa_1 = b_1\mathbf{r}\mathbf{\Lambda}\mathbf{e}, \tag{10}$$

$$\kappa_2 = 2b_1\mathbf{g}\mathbf{\Lambda}\mathbf{e} + b_2\mathbf{r}\mathbf{\Lambda}\mathbf{e}. \tag{11}$$

Here b_1 and b_2 are the first and the second raw moments of distribution $B(x)$, $\mathbf{g}$ is the solution of system

$$\mathbf{g}\mathbf{Q} = \mathbf{r}(\kappa_1\mathbf{I} - b_1\mathbf{\Lambda}),$$

$$\mathbf{g}\mathbf{e} = 0, \tag{12}$$

where $\mathbf{I}$ is the identity matrix of dimension $(N+1)\times(N+1)$. Vector $\mathbf{r}$ is the solution of system (6).

Proof. In system (8) we denote $\frac{1}{T} = \varepsilon$ and make substitutions

$$\tau = t\varepsilon,\ u = w\varepsilon,\ \mathbf{H}(u,t) = \mathbf{F}(w,\tau,\varepsilon), \tag{13}$$

then we derive system

$$\varepsilon\frac{\partial\mathbf{F}(w,\tau,\varepsilon)}{\partial\tau} = \mathbf{F}(w,\tau,\varepsilon)\{\mathbf{Q} + (B^*(w\varepsilon)-1)\mathbf{\Lambda}\},$$

$$\varepsilon\frac{\partial\mathbf{F}(w,\tau,\varepsilon)}{\partial\tau}\mathbf{e} = \mathbf{F}(w,\tau,\varepsilon)(B^*(w\varepsilon)-1)\mathbf{\Lambda}\mathbf{e}. \tag{14}$$

Taking the limit by $\varepsilon\to 0$ in the first equation of system (14) and taking into account that $B^*(0) = 1$ and $\mathbf{F}(w,0) = \mathbf{r}$ yields to Cauchy problem

$$\mathbf{F}(w,\tau)\mathbf{Q} = \mathbf{0},$$

$$\mathbf{F}(w,0) = \mathbf{r}. \tag{15}$$

Since the determinant of matrix $\mathbf{Q}$ is zero, then the homogeneous solution of problem (15) has the following form:

$$\mathbf{F}(w,\tau) = \Phi(w,\tau)\mathbf{r}, \tag{16}$$

where vector $\mathbf{r}$ is the stationary distribution of MMPP states given by (6) and function $\Phi(w,\tau)$ is the scalar multiplier defining the space of particular solutions of problem (15).

We derive the equation for function $\Phi(w,\tau)$, considering the second equation of system (14). We use decomposition of function $B^*(w\varepsilon)$ up to $O(\varepsilon^2)$ and substitute solution (16) into the second equation of system (14) taking the limit by $\varepsilon\to 0$

$$\frac{\partial\Phi(w,\tau)}{\partial\tau} = \Phi(w,\tau)jwb_1\mathbf{r}\mathbf{\Lambda}\mathbf{e},$$

where b_1 is the first raw moment of distribution $B(x)$. The solution of obtained equation is as follows:

$$\Phi(w,\tau) = e^{jw\kappa_1\tau},$$

where

$$\kappa_1 = b_1\mathbf{r}\mathbf{\Lambda}\mathbf{e},$$

which coincides with (10).

Making backward substitutions $w = \frac{u}{\varepsilon}$ and $\tau = t\varepsilon$, we can write expression

$$e^{jw\kappa_1\tau} = e^{ju\kappa_1 t}.$$

For more detailed analysis, we make substitutions

$$\mathbf{H}(u,t) = e^{ju\kappa_1 t}\mathbf{H}^{(1)}(u,t) \tag{17}$$

and obtain system

$$\frac{\partial \mathbf{H}^{(1)}(u,t)}{\partial t} + ju\kappa_1\mathbf{H}^{(1)}(u,t) = \mathbf{H}^{(1)}(u,t)\{\mathbf{Q} + (B^*(u)-1)\mathbf{\Lambda}\},$$

$$\frac{\partial \mathbf{H}^{(1)}(u,t)}{\partial t}\mathbf{e} + ju\kappa_1\mathbf{H}^{(1)}(u,t)\mathbf{e} = \mathbf{H}^{(1)}(u,t)(B^*(u)-1)\mathbf{\Lambda}\mathbf{e}. \tag{18}$$

Denoting $\frac{1}{T} = \varepsilon^2$, we introducee notations

$$\tau = t\varepsilon^2,\ u = w\varepsilon,\ \mathbf{H}^{(1)}(u,t) = \mathbf{F}^{(1)}(w,\tau,\varepsilon) \tag{19}$$

in system (18) and get

$$\varepsilon^2\frac{\partial \mathbf{F}^{(1)}(w,\tau,\varepsilon)}{\partial \tau} + jw\varepsilon\kappa_1\mathbf{F}^{(1)}(w,\tau,\varepsilon) = \mathbf{F}^{(1)}(w,\tau,\varepsilon)\{\mathbf{Q} + (B^*(w\varepsilon)-1)\mathbf{\Lambda}\},$$

$$\varepsilon^2\frac{\partial \mathbf{F}^{(1)}(w,\tau,\varepsilon)}{\partial \tau}\mathbf{e} + jw\varepsilon\kappa_1\mathbf{F}^{(1)}(w,\tau,\varepsilon)\mathbf{e} = \mathbf{F}^{(1)}(w,\tau,\varepsilon)(B^*(w\varepsilon)-1)\mathbf{\Lambda}\mathbf{e}. \tag{20}$$

Solution $\mathbf{F}^{(1)}(w,\tau,\varepsilon)$ of the first equation of system (20) can be written as decomposition

$$\mathbf{F}^{(1)}(w,\tau,\varepsilon) = \Phi_2(w,\tau)\{\mathbf{r} + jw\varepsilon\mathbf{f}\} + O(\varepsilon^2), \tag{21}$$

then from the first equation of system (20), we have

$$jw\varepsilon\kappa_1\mathbf{r} = \mathbf{r}\{\mathbf{Q} + jw\varepsilon b_1\mathbf{\Lambda}\} + jw\varepsilon\mathbf{f}\mathbf{Q} + O(\varepsilon^2).$$

From the last equation, we obtain

$$\mathbf{f}\mathbf{Q} = \mathbf{r}(\kappa_1\mathbf{I} - b_1\mathbf{\Lambda}). \tag{22}$$

Using superposition principle for inhomogeneous systems, we write the solution as sum

$$\mathbf{f} = C\mathbf{r} + \mathbf{g}, \tag{23}$$

which we substitute into (22) and obtain equation

$$\mathbf{gQ} = \mathbf{r}(\kappa_1 \mathbf{I} - b_1 \mathbf{\Lambda}). \tag{24}$$

Due to (23), vector $\mathbf{g}$ is a particular solution of an inhomogeneous system. Hence, it satisfies an additional condition, which we choose as $\mathbf{ge} = 0$. Thus, vector $\mathbf{g}$ is the solution of system

$$\mathbf{gQ} = \mathbf{r}(\kappa_1 \mathbf{I} - b_1 \mathbf{\Lambda}),$$

$$\mathbf{ge} = 0. \tag{25}$$

Further, we consider the second equation of system (20) substituting decomposition (21) and derive equation

$$\varepsilon^2 \frac{\partial \Phi_2(w,\tau)}{\partial \tau} + jw\varepsilon\kappa_1 \Phi_2(w,\tau)\{1 + jw\varepsilon \mathbf{fe}\} =$$

$$= \Phi_2(w,\tau)\{\mathbf{r} + jw\varepsilon \mathbf{f}\}\left(jw\varepsilon b_1 + \frac{(jw\varepsilon)^2}{2} b_2\right) \mathbf{\Lambda e} + O(\varepsilon^3),$$

where b_2 is the second raw moment of distribution $B(x)$. Then we divide the equation by ε^2 taking (10) into account and obtain

$$\frac{\partial \Phi_2(w,\tau)}{\partial \tau} + (jw)^2 \kappa_1 \Phi_2(w,\tau)\mathbf{fe} = \frac{(jw)^2}{2} \Phi_2(w,\tau)(2b_1 \mathbf{f\Lambda e} + b_2 \mathbf{r\Lambda e}). \tag{26}$$

We substitute solution (23) into equation (26) and get

$$\frac{\partial \Phi_2(w,\tau)}{\partial \tau} + (jw)^2 \kappa_1 \Phi_2(w,\tau)(C\mathbf{r} + \mathbf{g}) =$$

$$= \frac{(jw)^2}{2} \Phi_2(w,\tau)(2b_1(C\mathbf{r} + \mathbf{g})\mathbf{\Lambda e} + b_2 \mathbf{r\Lambda e}).$$

From the last equation, taking equalities $\mathbf{re} = 1$, $\mathbf{ge} = 0$ into account, we can write

$$\frac{\partial \Phi_2(w,\tau)}{\partial \tau} = \frac{(jw)^2}{2} \Phi_2(w,\tau)(2b_1 \mathbf{g\Lambda e} + b_2 \mathbf{r\Lambda e}).$$

We denote

$$\kappa_2 = 2b_1 \mathbf{g\Lambda e} + b_2 \mathbf{r\Lambda e}, \tag{27}$$

which coincides with (11). Then function $\Phi_2(w,\tau)$ has the following form:

$$\Phi_2(w,\tau) = \exp\left\{\frac{(jw)^2}{2}\kappa_2 \tau\right\}.$$

By substitutions (19), we can write

$$w = \frac{u}{\varepsilon},\ \tau = t\varepsilon^2.$$

Then for function $\Phi_2(w,\tau)$, we can write expression

$$\Phi_2(w,\tau) = \exp\left\{\frac{(jw)^2}{2}\kappa_2\tau\right\} = \exp\left\{\frac{(ju)^2}{2\varepsilon^2}\kappa_2\varepsilon^2 t\right\} = \exp\left\{\frac{(ju)^2}{2}\kappa_2 t\right\}.$$

Hence, we obtain scalar asymptotic characteristic function

$$\mathbf{H}(u,t)\mathbf{e} = \exp\left\{ju\kappa_1 t + \frac{(ju)^2}{2}\kappa_2 t\right\}.$$

Therefore, equality (9) holds. **Theorem is proved.**

5 Calculation Algorithm for the Moments of the Total Amount of Information Received in Multi-level MMPP

Vector characteristic function $\mathbf{H}(u,t)$ is the solution of Cauchy problem

$$\frac{\partial \mathbf{H}(u,t)}{\partial t} = \mathbf{H}(u,t)\{\mathbf{Q} + (B^*(u) - 1)\mathbf{\Lambda}\},$$

$$\mathbf{H}(u,0) = \mathbf{r}. \tag{28}$$

We differentiate the equations of system (28) by u at zero point and obtain the Cauchy problems for vector moments $\mathbf{m}_1(t)$ and $\mathbf{m}_2(t)$

$$\mathbf{m}_1'(t) = \mathbf{m}(t)\mathbf{Q} + b_1\mathbf{r}\mathbf{\Lambda},$$

$$\mathbf{m}_1(0) = \mathbf{0}. \tag{29}$$

$$\mathbf{m}_2'(t) = \mathbf{m}_2(t)\mathbf{Q} + 2b_1\mathbf{m}_1(t)\mathbf{\Lambda} + b_2\mathbf{r}\mathbf{\Lambda},$$

$$\mathbf{m}_2(0) = \mathbf{0}. \tag{30}$$

We denote $m_1(t) = \mathbf{m}_1(t)\mathbf{e}$ as a scalar first raw moment. Multiplying the equations of system (29) by vector $\mathbf{e}$, we obtain system

$$m_1'(t) = b_1\mathbf{r}\mathbf{\Lambda}\mathbf{e},$$

$$m_1(0) = 0,$$

the solution of which is

$$m_1(t) = \mathbf{m}_1(t)\mathbf{e} = b_1\mathbf{r}\mathbf{\Lambda}\mathbf{e}t. \tag{31}$$

Likewise, for the second raw moment we have

$$(\mathbf{m}_2(t)\mathbf{e})' = 2b_1\mathbf{m}_1(t)\mathbf{\Lambda}\mathbf{e} + b_2\mathbf{r}\mathbf{\Lambda}\mathbf{e},$$

$$\mathbf{m}_2(0)\mathbf{e} = 0.$$

Thus, the second raw moment is given by

$$m_2(t) = \mathbf{m}_2(t)\mathbf{e} = 2b_1 \int_0^t \mathbf{m}_1(z)\, dz\, \mathbf{\Lambda e} + b_2\mathbf{r\Lambda e}t. \tag{32}$$

Here we have to derive the value of vector integral $\int_0^t \mathbf{m}_1(z)dz$, where $\mathbf{m}_1(t)$ is a solution of problem (29).

Despite the fact that problem (29) is rather simple to solve, we give the algorithm of obtaining its explicit solution. First, we rewrite the problem in canonical form denoting

$$\mathbf{m}_1^T(t) = \mathbf{x}(t),\ \mathbf{Q}^T = \mathbf{A},\ b_1\mathbf{\Lambda r}^T = \mathbf{f},$$

then the problem has the following form:

$$\mathbf{x}'(t) = \mathbf{Ax}(t) + \mathbf{f},$$

$$\mathbf{x}(0) = \mathbf{0}. \tag{33}$$

We denote k_i and $\mathbf{v}_i$ as eigenvalues and eigenvectors of matrix $\mathbf{A} = \mathbf{Q}^T$. Since the determinant of matrix $\mathbf{Q}$ is zero, we always have zero value among eigenvalues. We assume that $k_0 = 0$ and all eigenvalues are simple. Solution $\mathbf{x}(t)$ of problem (33) is given as follows

$$\mathbf{x}(t) = \sum_{i=1}^{N} C_i(t)e^{k_i t}\mathbf{v}_i.$$

Let $\mathbf{V}$ denote the matrix of eigenvectors and $\mathbf{c}(t)$ is a column vector with components $C_i(t)$, then we rewrite solution $\mathbf{x}(t)$ as

$$\mathbf{x}(t) = \mathbf{V}\mathrm{diag}(e^{k_i t})\mathbf{c}(t). \tag{34}$$

Substituting this expression into (33), we obtain equation

$$\mathbf{V}\mathrm{diag}(e^{k_i t})\mathbf{c}'(t) + \mathbf{V}\mathrm{diag}(k_i e^{k_i t})\mathbf{c}(t) = \mathbf{AV}\mathrm{diag}(e^{k_i t})\mathbf{c}(t) + \mathbf{f},$$

from which we derive

$$\mathbf{c}'(t) = \mathrm{diag}(e^{-k_i t})\mathbf{V}^{-1}\mathbf{f}.$$

Since $k_0 = 0$, we rewrite diagonal matrix $\mathrm{diag}(e^{-k_i t})$ as $\mathrm{diag}(1, e^{-k_i t})$, where the first element of the matrix is equal to one. Thus, we obtain the formula for $\mathbf{c}(t)$

$$\mathbf{c}(t) = \mathrm{diag}\left(t, \frac{1 - e^{-k_i t}}{k_i}\right)\mathbf{V}^{-1}\mathbf{f}.$$

From (34) we obtain

$$\mathbf{x}(t) = \mathbf{V}\mathrm{diag}(1, e^{-k_i t})\mathrm{diag}\left(t, \frac{1 - e^{-k_i t}}{k_i}\right)\mathbf{V}^{-1}\mathbf{f}$$

$$= \mathbf{V}\mathrm{diag}\left(t, \frac{1-e^{-k_i t}}{k_i}\right)\mathbf{V}^{-1}\mathbf{f}.$$

Since $\mathbf{x}(t) = \mathbf{m}_1^T(t)$, $\mathbf{f} = b_1\mathbf{\Lambda}\mathbf{r}^T$, we can write the expression for the first vector moment

$$\mathbf{m}_1(t) = b_1\mathbf{r}\mathbf{\Lambda}\mathbf{V}\mathrm{diag}\left(t, \frac{1-e^{-k_i t}}{k_i}\right)\mathbf{V}^{-1},$$

and obtain the value of integral

$$\int_0^t \mathbf{m}_1(z)\,dz = b_1\mathbf{r}\mathbf{\Lambda}\mathbf{V}\mathrm{diag}\left(\frac{t^2}{2}, \frac{e^{k_i t}-1}{k_i^2} - \frac{t}{k_i}\right)\mathbf{V}^{-1}.$$

Further, we substitute the value of integral into (32) and obtain the value of the second raw moment of process $S(t)$

$$m_2(t) = \mathbf{m}_2(t)\mathbf{e} = 2b_1^2\mathbf{r}\mathbf{\Lambda}\mathbf{V}\mathrm{diag}\left(\frac{t^2}{2}, \frac{e^{k_i t}-1}{k_i^2} - \frac{t}{k_i}\right)\mathbf{V}^{-1}\mathbf{\Lambda}\mathbf{e} + b_2\mathbf{r}\mathbf{\Lambda}\mathbf{e}t. \quad (35)$$

The variance $d(t)$ can be derived as follows

$$d(t) = m_2(t) + m_1(t)^2,$$

where $m_1(t)$ is given by (31).

6 Numerical Implementation of the Analytical Results for a Three-Level MMPP

In this section, we demonstrate the obtained results for the case of three-level MMPP. We set matrices $\mathbf{Q}$ and $\mathbf{\Lambda}$ as follows:

$$\mathbf{Q} = \begin{bmatrix} -100 & 99 & 0.99 & 0.01 \\ 50 & -50 & 0 & 0 \\ 0.2 & 0 & -0.2 & 0 \\ 0.001 & 0 & 0 & -0.001 \end{bmatrix}, \quad \mathbf{\Lambda} = \begin{bmatrix} 100 & 0 & 0 & 0 \\ 0 & 0 & 0 & 0 \\ 0 & 0 & 0 & 0 \\ 0 & 0 & 0 & 0 \end{bmatrix}.$$

The amount of information in one request follows gamma distribution with shape parameter $\alpha = 0.5$ and rate $\beta = \alpha$. Solving system $\mathbf{rQ} = 0$, $\mathbf{re} = 1$, we obtain the vector of steady-state distribution of the states of multi-level MMPP

$$\mathbf{r} = \{0.056,\ 0.110,\ 0.276,\ 0.558\}$$

and intensity $\mathbf{r}\mathbf{\Lambda}\mathbf{e} = 5.577$.

6.1 Gaussian Approximation Accuracy

We denote matrix $\mathbf{A}(u) = \mathbf{Q}+(B^*(u)-1)\mathbf{\Lambda}$. Let $\gamma(u)$ and $x(u)$ denote the vector of eigenvalues and the matrix of eigenvectors of matrix $\mathbf{A}(u)$, respectively.

The value of matrix exponential is given by

$$e^{\mathbf{A}(u)t} = x(u)\mathrm{diag}(e^{\gamma(u)t})x(u)^{-1}.$$

Using expression (7), we write scalar characteristic function

$$\mathbf{H}(u)\mathbf{e} = \mathbf{r}e^{\mathbf{A}(u)t}\mathbf{e}.$$

Numerical implementation of inverse Fourier transform

$$P(s,t) = \frac{1}{2\pi}\int_{-\infty}^{\infty} \frac{1-e^{-jus}}{ju}\mathbf{H}(u)\mathbf{e}\;du$$

causes computational problems due to the presence of component $W(t)$, which is constant by u, in characteristic function $\mathbf{H}(u)\mathbf{e}$

$$W(t) = \lim_{u\to\infty} \mathbf{H}(u)\mathbf{e} = \mathbf{r}\exp\{[\mathbf{Q}-\mathbf{\Lambda}]t\}\mathbf{e}.$$

Hence, we represent characteristic function $\mathbf{H}(u)\mathbf{e}$ in the following form:

$$\mathbf{H}(u)\mathbf{e} = W(t) + (1-W(t))h(u,t),$$

$$h(u,t) = \frac{\mathbf{H}(u)\mathbf{e} - W(t)}{1-W(t)}.$$

Numerical implementation of the inverse Fourier transform of function $h(u,t)$ does not make any problems. Thereby, we first calculate the values of function

$$p(s,t) = \frac{1}{2\pi}\int_{-\infty}^{\infty} \frac{1-e^{-jus}}{ju}h(u,t)\;du$$

and then obtain the values of distribution function $P(s,t)$ using formula

$$P(s,t) = W(t) + (1-W(t))p(s,t).$$

It is interesting to note that for small values of t and $s=0$, distribution function $P(s,t)$ has a gap equal to $W(t)$ and the gap decreases to zero as the value of t increases. Having the values of κ_1 and κ_2 given by formulas (10) and (11), we can write Gaussian distribution function $N(s,t)$ with parameters $a=\kappa_1 t$ and $\sigma=\sqrt{\kappa_2 t}$. In our case, the values of κ_1 and κ_2 are

$$\kappa_1 = 5.577,\;\; \kappa_2 = 3.48\cdot 10^4.$$

We also note that the value of κ_2 is four orders greater than κ_1 which can be interpreted as a heavy tail of Gaussian approximation of the probability distribution of process $S(t)$.

The error of Gaussian approximation $N(s,t)$ against prelimit distribution $P(s,t)$ we will evaluate using the Kolmogorov distance

$$\Delta(t) = \max_{0 \leqslant s < \infty} \left| P(s,t) - N(s,t) \right|.$$

Table 1 shows values of $\Delta(t)$ depending on t. We note that the error of the approximation decreases while t increasing, which confirms the obtained analytical results.

Table 1. Kolmogorov distance between distribution function $P(s,t)$ and Gaussian approximation $N(s,t)$.

t	100	1000	1250	1500	2500	5000	10000	100000
$\Delta(t)$	0.236	0.065	0.050	0.039	0.022	0.015	0.008	0.002

We assume that an approximation is enough accurate when the Kolmogorov distance is less or equal to 0.05. In our case, the approximation is acceptable when $t \geqslant 1250$.

7 Hurst Parameter Evaluation for Multi-level MMPP

Having expected value $m_1(t)$ given by (31) and the second raw moment $m_2(t)$ given by (35) for the total amount of information received in multi-level MMPP per time t, we evaluate Fano factor $F(t)$

$$F(t) = \frac{d(t)}{m_1(t)},$$

where $d(t) = m_2(t) + m_1^2(t)$ is the variance of considered process. For Hurst parameter $h(t)$ of fractality of the process the following equality holds

$$\ln(F(t) - 1) = (2h(t) - 1)\ln t + y. \tag{36}$$

We derive the expression for parameter y from equality (36) setting $t = 1$

$$y = \ln(F(1) - 1) = 3.308.$$

Let us resolve equality (36) for $h(t)$

$$h(t) = \frac{1}{2}\left\{1 + \frac{1}{\ln t}\ln\frac{F(t) - 1}{F(1) - 1}\right\}. \tag{37}$$

Table 2. Hurst parameter of the total amount of information received in multi-level MMPP.

t	100	200	400	600	800	1000	1200	1250
$h(t)$	0.856	0.863	0.865	0.864	0.860	0.857	0.854	0.853

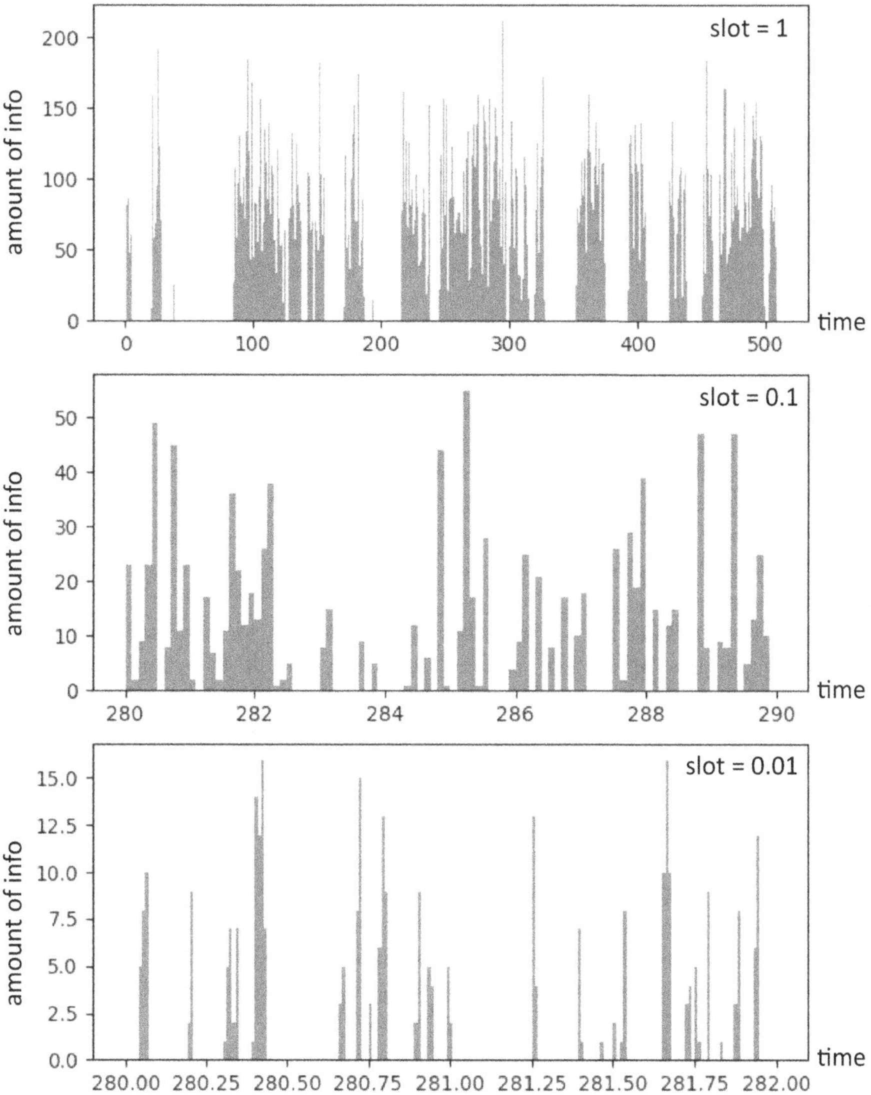

Fig. 2. Traffic trace simulated using multi-level MMPP presented using various lengths of time slot and scales.

We calculate Hurst parameter for multi-level MMPP with parameters, which are presented in the Sect. 6. Table 2 depicts the values of $h(t)$ for different t.

The mean of the Hurst parameter values in the table is equal to 0.86. The range of t was selected from the obtained in Table 1 condition $\Delta(t) > 0.05$ for which the Gaussian approximation is not recommended.

We note that increasing time t leads to decreasing of the Hurst parameter. This is due to the fact that with increasing time t the mean $m(t)$ and variance $d(t)$ converge to their limit values $\kappa_1 t$ and $\kappa_2 t$ while Fano factor $F(t)$ converges to the constant κ_2/κ_1, which does not depend on t. Therefore, naturally, by virtue of (37), the Hurst parameter decreases, and process $S(t)$ becomes closer to Gaussian one.

In Fig. 2, we show the traffic simulated using multi-level MMPP at various levels of time scaling (similar to Fig. 1). As we can see, the multi-level MMPP model gives bursts at different time scales. Numerical experiments and comparison with real-life traffic data show a good quality of the proposed multi-level MMPP in modeling network flows.

8 Conclusion

In this paper, we have provided the research of the multi-level MMPP as a model that captures the fractality, self-similarity and burstiness of the traffic. We have derived the characterictic function of the total amount of information received in the flow during time t. Applying the inverse Fourier transform, we have obtained the probability distribution of the considered process.

Moreover, we have presented the calculation algorithm and explicit formulas for the first and the second raw moments of the total amount of information received in the flow.

The next contribution is a Gaussian approximation for the probability distribution of the total amount of received information built under limit condition of growing observation time. We have shown that the convergence time of considered process to the Gaussian process is rather large and the Hurst parameter of the process is stably high during this interval.

References

1. Adas, A.: Traffic models in broadband networks. IEEE Commun. Mag. **35**(7), 82–89 (1997)
2. Arfeen, M.A., Pawlikowski, K., Willig, A., McNickle, D.: Fractal renewal process based analysis of emerging network traffic in access networks. In: 2016 26th International Telecommunication Networks and Applications Conference (ITNAC), pp. 265–270. IEEE (2016)
3. Becchi, M.: From poisson processes to self-similarity: a survey of network traffic models. Washington University in St. Louis, Technical report (2008)
4. Cappé, O., Moulines, E., Pesquet, J.C., Petropulu, A.P., Yang, X.: Long-range dependence and heavy-tail modeling for teletraffic data. IEEE Signal Process. Mag. **19**(3), 14–27 (2002)

5. Cox, D.R.: The analysis of non-Markovian stochastic processes by the inclusion of supplementary variables. In: Mathematical Proceedings of the Cambridge Philosophical Society, vol. 51, pp. 433–441. Cambridge University Press (1955)
6. Dudin, A., Klimenok, V., Vishnevsky, V.: The Theory of Queuing Systems with Correlated Flows. Springer, Heidelberg (2020)
7. Goh, K.I., Barabási, A.L.: Burstiness and memory in complex systems. EPL (Europhys. Lett.) **81**(4), 48002 (2008)
8. Han, X., Schormans, J.: Cross-layer queueing analysis for aggregated on-off arrivals with adaptive modulation and coding. IET Commun. **10**(17), 2336–2343 (2016)
9. Hryn, G., Jerzak, Z., Chydzinski, A.: MMPP-based HTTP traffic generation with multiple emulated sources. Archiwum Informatyki Teoretycznej i Stosowanej **16**(4), 321–335 (2004)
10. Johnson, M.A., Narayana, S.: Descriptors of arrival-process burstiness with application to the discrete Markovian arrival process. Queueing Syst. **23**(1–4), 107–130 (1996)
11. Kettani, H., Gubner, J.A.: A novel approach to the estimation of the hurst parameter in self-similar traffic. In: Proceedings of the 27th Annual IEEE Conference on Local Computer Networks, LCN 2002, pp. 160–165. IEEE (2002)
12. Klemm, A., Lindemann, C., Lohmann, M.: Modeling IP traffic using the batch Markovian arrival process. Perform. Eval. **54**(2), 149–173 (2003)
13. Marvi, M., Aijaz, A., Khurram, M.: On the use of on/off traffic models for spatio-temporal analysis of wireless networks. IEEE Commun. Lett. **23**(7), 1219–1222 (2019)
14. Ryu, B.K., Lowen, S.B.: Point process approaches to the modeling and analysis of self-similar traffic. I. Model construction. In: Proceedings of IEEE INFOCOM 1996. Conference on Computer Communications, vol. 3, pp. 1468–1475. IEEE (1996)
15. Trenogin, N., Sokolov, D.: Fractal properties of events flow of application level in information systems. Vestnik SibGUTI **4**, 97–103 (2017)
16. Xue, F.: Performance analysis of a wavelet-based hurst parameter estimator for self-similar traffic. In: 2000 SCS Symposium on Performance Evaluation of Computer and Telecommunication Systems-SPECTS'2K. Citeseer (2000)
17. Yoshihara, T., Kasahara, S., Takahashi, Y.: Practical time-scale fitting of self-similar traffic with Markov-modulated poisson process. Telecommun. Syst. **17**(1–2), 185–211 (2001)

Asymptotic Analysis of Intensity of Poisson Flows Assembly

Gurami Tsitsiashvili[1], Alexander Moiseev[2(✉)], and Konstantin Voytikov[3]

[1] Institute for Applied Mathematics, Far Eastern Branch of RAS, Vladivostok, Russia
guram@iam.dvo.ru
[2] Tomsk State University, Tomsk, Russia
moiseev.tsu@gmail.com
[3] Moscow Institute of Physics and Technology, Moscow, Russia
voytikovk@gmail.com

Abstract. The paper is devoted to obtaining estimations of the rate of convergence of the intensity of an assembly of Poisson flows to the intensity of a stationary Poisson flow. Analysis of the results shows that this problem should combine analytical and numerical studies. An important role is played by the Central limit theorem for both random variables and stochastic processes which is understood in the sense of C-convergence. Exact asymptotic formulas are derived for intensity of the assembly flow of identical Poisson flows, and estimations of the convergence rate are build for the case of non-identical original flows.

Keywords: Assembly of flows · Asymptotic analysis · Central limit theorem

1 Introduction

In the paper, we analyze the assembly of independent Poisson flows which is interpreted as connection of customers with the same order numbers taken from different flows. The assembly processes may be found in computer networks [1], in conveyor systems for the manufacture of products [2–4], in open queueing networks with a single input flow, division and merging of customers and with sufficiently general configuration of network [5,6], in closed queueing networks with discrete time transitions of batches of customers and dynamic control of service rates [7]. However, the study of the flow of customers coming out after the assembly is a very complicated problem.

It is shown in [8] that the average intensity of the assembled flow tends to the lower of the original Poisson flow intensities while time tends to infinity. However, computational experiments performed by approximating the Poisson distribution with a large parameter by a normal distribution showed that it is possible to improve the obtained estimations of the convergence rate. This paper is devoted to obtaining, in a certain sense, unimproved estimations of the convergence rate of the intensity of the assembly flow to the intensity of a stationary Poisson flow. Analysis of the results shows that this problem should

A. Dudin et al. (Eds.): ITMM 2020, CCIS 1391, pp. 78–94, 2021.
https://doi.org/10.1007/978-3-030-72247-0_6

combine analytical and numerical methods comparing their results with each other. Moreover, an important role is played by the Central limit theorem for both random variables and stochastic processes which is understood in the sense of C-convergence [9].

The authors are grateful to Professor Anatoly Nazarov from Tomsk State University for his ideas and advice regarding the research presented in the article, which made it possible to significantly improve quality of the paper and move further in the studies.

2 Mathematical Model

Assume that there are r independent stationary Poisson processes (we will call them as "original flows") with intensities $\lambda_1, \dots, \lambda_r$. Let us denote an instants of arrivals in the flows by $t_{k,i}$, where k is the number of original flow and i is the order number of the arrival in this flow. The original flows we denote as $T_k = \{0 \le t_{k,1} \le t_{k,2} \le \dots\}$, where $k = 1, 2, \dots$, $k = 1, \dots, r$. We will call the flow $A_r = \{0 \le \max(t_{1,1}, \dots, t_{r,1}) \le \max(t_{1,2}, \dots, t_{r,2}) \le \dots\}$ as an *assembly of flows* $T_1, \dots, T_r$ or as an *assembly flow*.

Denote the number of points in k-th flow in interval $[0, t)$ by $n_k(t)$. Then the number of points in the assembly flow $N_r(t)$ in the interval may be expressed as

$$N_r(t) = \min_{k=1,\dots,r} n_k(t). \tag{1}$$

Flow A_r is not Poisson, because its increments are not independent due to formula (1).

3 Central Limit Theorem for the Assembly Flow

Suppose that several original flows have minimal intensities: $\lambda = \lambda_1 = \dots = \lambda_s < \lambda_{s+1} \le \dots \le \lambda_r$, $s \le r$. Then the following statement can be proved.

Theorem 1. *For any $v \in (-\infty, \infty)$, the following limit relation is true:*

$$\mathrm{P}\left\{\frac{N_r(t) - \lambda t}{\sqrt{\lambda t}} > v\right\} \to \left[\int_v^\infty \frac{1}{\sqrt{2\pi}} \exp(-u^2/2) du\right]^s, \ t \to \infty. \tag{2}$$

Proof. From formula (1) and the independence of flows $T_1, \dots, T_r$, the equality follows

$$\mathrm{P}\left\{N_r(t) > i\right\} = \prod_{k=1}^{r} \mathrm{P}\left\{n_k(t) > i\right\}. \tag{3}$$

Then due to the Central limit theorem, we derive

$$\mathrm{P}\left\{\frac{n_k(t) - \lambda t}{\sqrt{\lambda t}} > v\right\} \to \int_v^\infty \frac{1}{\sqrt{2\pi}} \exp(-u^2/2) du, \quad \text{for } k = 1, \dots, s. \tag{4}$$

and

$$\mathrm{P}\left\{\frac{n_k(t)-\lambda t}{\sqrt{\lambda t}}>v\right\}=\mathrm{P}\left\{\frac{n_k(t)-\lambda_k t}{\sqrt{\lambda_k t}}>v\sqrt{\frac{\lambda}{\lambda_k}}-(\lambda_k-\lambda)t\right\}\geq$$
$$\mathrm{P}\left\{\frac{n_k(t)-\lambda_k t}{\sqrt{\lambda_k t}}>-(\lambda_k-\lambda)t\right\}\to 1,\ t\to\infty,\ \text{for } k=s+1,\dots,r.$$

So, we obtain

$$\mathrm{P}\left\{\frac{n_k(t)-\lambda t}{\sqrt{\lambda t}}>v\right\}\to 1,\ t\to\infty,\quad \text{for } k=s+1,\dots,r. \tag{5}$$

Using formulas (1), (3)–(5), we derive

$$\mathrm{P}\left\{\frac{N_r(t)-\lambda t}{\sqrt{\lambda t}}>v\right\}=\mathrm{P}\left\{\frac{\min\limits_{k=1,\dots,r} n_k(t)-\lambda t}{\sqrt{\lambda t}}>v\right\}=$$
$$\mathrm{P}\left\{\min_{k=1,\dots,r}\frac{n_k(t)-\lambda t}{\sqrt{\lambda t}}>v\right\}=$$
$$\prod_{k=1}^{r}\mathrm{P}\left\{\frac{n_k(t)-\lambda t}{\sqrt{\lambda t}}>v\right\}\to\left[\int_v^\infty\frac{1}{\sqrt{2\pi}}\exp(-u^2/2)du\right]^s,\ t\to\infty. \tag{6}$$

So, the theorem is proved.

Remark 1. The stochastic process $\frac{n_k(tu)-\lambda tu}{\sqrt{\lambda t}}$ as a function of variable $u\geq 0$ tends to Wiener process $\xi_k(u),\ k=1,\dots,s$ while $t\to\infty$. Since process $n_k(t)$ is a process with independent increments, then the stochastic process $\frac{n_k(tu)-\lambda tu}{\sqrt{\lambda t}}$ is also a process with independent increments. Moreover, while $t\to\infty$, due to the Central limit theorem, the increment of process $\frac{n_k(tu)-\lambda tu}{\sqrt{\lambda t}}$ in interval $[u_1,u_2],\ u_1<u_2$, tends to Gaussian random variable with zero mean and variance equal to u_2-u_1. Therefore, if $t\to\infty$, process $\frac{n_k(tu)-\lambda tu}{\sqrt{\lambda t}}$ converges to Wiener process $w_k(u)$ in the sense of C-convergence [9, Chapter 4, § 3, Theorem 10].

Remark 2. Let $r=s$, then stochastic process $\frac{N_r(tu)-\lambda tu}{\sqrt{\lambda t}}$ for $t\to\infty$ converges to process $\min\limits_{k=1,\dots,r} w_k(u)$ in the sense of C-convergence, where $w_1(u),\dots,w_r(u)$ are independent Wiener processes ($u\geq 0$). This statement follows from formula (9) and Remark 1.

4 Limit Relations for Intensity of Assembly of Independent and Identically Distributed Poisson Flows

Consider Markov process $\{n_1(t), \ldots, n_r(t)\}$. A jump of this process in instant t from state $(n_1, \ldots, n_i, \ldots, n_m)$, where $n_i < \min\limits_{k \neq i} n_k$, to state $(n_1, \ldots, n_i + 1, \ldots, n_r)$ causes a new point to appear in flow A_r in time moment t. Therefore, instant intensity $\overline{\lambda}(t)$ of assembly flow in this moment satisfies the equality

$$\overline{\lambda}(t) = \lambda \sum_{i=1}^{r} \mathrm{P} \left\{ n_i(t) < \min_{k \neq i} n_k(t) \right\}. \tag{7}$$

Lemma 1. *The following equality is true:*

$$\overline{\lambda}(t) = \lambda(1 - \mathrm{P}\{n_1(t) = \ldots = n_r(t)\}). \tag{8}$$

Proof. Let us denote the following sets of indices:

$$J = \{1, \ldots, r\}, \ J_i = J \setminus i, \ i = 1, \ldots, r.$$

Then equality (7) can be transformed as follows:

$$\overline{\lambda}(t) = \lambda \mathrm{P} \left\{ \bigcup_{i=1}^{r} (n_i(t) < \min_{k \in J_i} n_k(t)) \right\} = \lambda \left(1 - \mathrm{P} \left\{ \bigcap_{i=1}^{r} (n_i(t) \geq \min_{k \in J_i} n_k(t)) \right\} \right)$$

$$= \lambda \left(1 - \mathrm{P} \left\{ \bigcap_{i=1}^{r} (n_i(t) \geq \min_{k \in J} n_k(t)) \right\} \right) = \lambda(1 - \mathrm{P}\{n_1(t) = \ldots = n_r(t)\}).$$

The lemma is proved.

Let us denote $a = \lambda t$ and

$$p(k, a) = \frac{e^{-a} a^k}{k!}, \ k = 0, 1, \ldots,$$

$$f(a) = P(n_1(t) = \ldots = n_r(t)) = \sum_{k=0}^{\infty} p^r(k, a).$$

We will search for approximation $g(a)$ of function $f(a)$ in the form

$$g(a) = \int_{-\infty}^{\infty} \left[\frac{1}{\sqrt{2\pi a}} \exp\left(-\frac{(x-a)^2}{2a} \right) \right]^r dx =$$

$$(2\pi a)^{-r/2} \int_{-\infty}^{\infty} \exp\left(-\frac{(x-a)^2}{2a/r} \right) dx =$$

$$(2\pi a)^{-r/2} \sqrt{2\pi a/r} \int_{-\infty}^{\infty} \frac{1}{\sqrt{2\pi a/r}} \exp\left(-\frac{(x-a)^2}{2a/r} \right) dx = \frac{1}{\sqrt{r}} (2\pi a)^{(1-r)/2}. \tag{9}$$

Theorem 2. *For $\lambda > 0$, $r = 2$, the following limit ratio is true:*

$$\mathrm{P}\{n_1(t) = n_2(t)\} \sim \left(2\sqrt{\pi a}\right)^{-1} \to 0, \; a \to \infty, \tag{10}$$

and therefore, $\lambda(t) \to \lambda$, $\lambda(t) - \lambda \sim \lambda(2\sqrt{\pi\lambda t})^{-1}$, $t \to \infty$.

Proof. Indeed, the following equalities are fulfilled:

$$\mathrm{P}\{n_1(t) = n_2(t)\} = \sum_{k=0}^{\infty} \exp(-2a)\frac{a^{2k}}{(k!)^2} = \exp(-2a)B, \; B = \sum_{k=0}^{\infty} \frac{a^{2k}}{(k!)^2}.$$

Here $B = B(a)$ is the Infeld function [10, Chapter 4, Sect. 11] satisfying the asymptotic relation

$$B(a) = \frac{\exp(2a)}{2\sqrt{\pi a}}\left(1 + O\left(\frac{1}{a}\right)\right). \tag{11}$$

Replacing here a by λt, we derive relation (10). The theorem is proved.

Theorem 3. *If $\lambda > 0$, $r > 2$, $\frac{1}{2} < \gamma < \frac{2}{3}$, then the following limit relation takes place:*

$$f(a) = g(a)(1 + O(a^{3\gamma-2})) \sim g(a), \; a \to \infty, \tag{12}$$

and therefore, $\lambda(t) \to \lambda$, $\lambda(t) - \lambda \sim \lambda\dfrac{(2\pi\lambda t)^{(1-r)/2}}{\sqrt{r}}$, $t \to \infty$.

Proof. Consider the following integrals:

$$g_1(a) = \int_{-\infty}^{a-a^\gamma} \left[\frac{1}{\sqrt{2\pi a}} \exp\left(-\frac{(x-a)^2}{2a}\right)\right]^r dx,$$
$$g_2(a) = \int_{a+a^\gamma}^{\infty} \left[\frac{1}{\sqrt{2\pi a}} \exp\left(-\frac{(x-a)^2}{2a}\right)\right]^r dx,$$
$$g_3(a) = \int_{a-a^\gamma}^{a+a^\gamma} \left[\frac{1}{\sqrt{2\pi a}} \exp\left(-\frac{(x-a)^2}{2a}\right)\right]^r dx,$$
$$g(a) = g_1(a) + g_2(a) + g_3(a). \tag{13}$$

Let us prove the following supplementary statement.

Lemma 2. *The following limit relations are true:*

$$g_1(a) = g_2(a) = o(g(a)), \; g_3(a) = g(a)(1 + o(g(a))), \; a \to \infty. \tag{14}$$

Proof. By replacing variable $t = x - a$, we obtain the equalities

$$g_1(a) = \int_{-\infty}^{-a^\gamma} \left[\frac{1}{\sqrt{2\pi a}} \exp\left(-\frac{t^2}{2a}\right)\right]^r dt, \; g_2(a) = \int_{a^\gamma}^{\infty} \left[\frac{1}{\sqrt{2\pi a}} \exp\left(-\frac{t^2}{2a}\right)\right]^r dt.$$

From here, we get the following relations (for $a \to \infty$):

$$g_1(a) = g_2(a) = \int_{a^\gamma}^{\infty} \frac{(2\pi a)^{-r/2}}{a^{-1}rt} \exp\left(-\frac{rt^2}{2a}\right) d\left(\frac{rt^2}{2a}\right) \leq$$

$$\frac{(2\pi a)^{-r/2}}{ra^{\gamma-1}} \int_{a^\gamma}^{\infty} \exp\left(-\frac{rt^2}{2a}\right) d\frac{rt^2}{2a} \leq \frac{(2\pi a)^{-r/2}}{ra^{\gamma-1}} \exp\left(-\frac{ra^{2\gamma-1}}{2}\right) = o(g(a)).$$

From these relations and formulas (9), (13), limit relations (14) follow. The lemma is proved.

Let us now consider the sums

$$f_1(a) = \sum_{0 \leq k < a - a^\gamma} \left(\frac{e^{-a}a^k}{k!}\right)^r, \tag{15}$$

$$f_2(a) = \sum_{a + a^\gamma < k \leq \infty} \left(\frac{e^{-a}a^k}{k!}\right)^r, \quad f_3(a) = \sum_{a - a^\gamma \leq k \leq a + a^\gamma} \left(\frac{e^{-a}a^k}{k!}\right)^r. \tag{16}$$

Further, we denote an integer part of some real number x by $[x]$.
Let us prove an additional supplementary statement.

Lemma 3. *The following limit relations are true:*

$$f_1(a) = O\left(\frac{a}{(2\pi a)^{r/2}} \exp\left(-\frac{ra^{2\gamma-1}}{2}\right)\right) = o(g(a)), \quad a \to \infty, \tag{17}$$

$$f_2(a) = O\left(\frac{a}{(2\pi a)^{r/2}} \exp\left(-\frac{ra^{2\gamma-1}}{2}\right)\right) = o(g(a)), \quad a \to \infty. \tag{18}$$

Proof. We construct an estimation of $f_1(a)$, assuming $c = [a - a^\gamma] \sim a, \ a \to \infty$:

$$f_1(a) \leq c\left(\frac{e^{-a}a^c}{c!}\right)^r \sim a\left(\frac{e^{-a}a^c}{c^c e^{-c}\sqrt{2\pi a}}\right)^r \leq$$

$$\frac{a}{(2\pi a)^{r/2}} \left(\frac{e^{-a}a^{a-a^\gamma}}{(a - a^\gamma - 1)^{a-a^\gamma-1}e^{-a+a^\gamma}}\right)^r = \frac{a}{(2\pi a)^{r/2}} e^{rF_1(a)}, \tag{19}$$

where

$$F_1(a) = -a^\gamma + (a - a^\gamma)\ln a - (a - a^\gamma - 1)\ln(a - a^\gamma - 1) = -\frac{a^{2\gamma-1}}{2}(1 + o(1)). \tag{20}$$

Thus, from the condition $\frac{1}{2} < \gamma$, definition of function $g(a)$ and formulas (15), (19), (20), it leads us to (17).

We construct an estimation of $f_2(a)$, assuming in the proof of Lemma 3 that $d = [a + a^\gamma] \sim a, \ a \to \infty$:

$$f_2(a) \leq \sum_{d \leq k} \left(e^{-a}\frac{a^k}{k!}\right)^r \leq \left(\frac{e^{-a}a^d}{d!}\right)^r \sum_{k \geq 0} \left(\frac{a}{d}\right)^{kr} \sim \left(\frac{e^{-a}a^d}{d!}\right)^r \frac{a^{1-\gamma}}{r}. \tag{21}$$

So, for $a \to \infty$ we derive

$$\left(\frac{e^{-a}a^d}{d!}\right)^r \sim \left(\frac{e^{-a}a^d}{d^d e^{-d}\sqrt{2\pi a}}\right)^r \leq$$

$$\left(\frac{e^{-a}a^{a+a^\gamma}}{(a+a^\gamma-1)^{a+a^\gamma-1}e^{-a-a^\gamma}\sqrt{2\pi a}}\right)^r = \frac{1}{(2\pi a)^{r/2}}e^{rF_2(a)}, \tag{22}$$

where

$$F_2(a) = a^\gamma + (a+a^\gamma)\ln a - (a+a^\gamma-1)\ln(a+a^\gamma-1) = -\frac{a^{2\gamma-1}}{2}(1+o(1)). \tag{23}$$

From (16), (21)–(23) and the condition $\frac{1}{2} < \gamma$, we obtain (18). The lemma is proved.

Let us denote

$$\varphi_3(a) = \sum_{a-a^\gamma \leq k \leq a+a^\gamma} (2\pi a)^{-r/2} \exp\left(-\frac{r(k-a)^2}{2}\right)$$

and prove the following supplementary statements.

Lemma 4. *The following limit relation is true:*

$$f_3(a) = \varphi_3(a)(1+O(a^{3\gamma-2})),\ a \to \infty. \tag{24}$$

Proof. We analyze the expression $e^{-ra}\dfrac{a^{rk}}{(k!)^r}$ using the Stirling formula in the form

$$k! = k^k e^{-k}\sqrt{2\pi k}\exp\left(\frac{\theta(k)}{12k}\right),\ 0 \leq \theta(k) \leq 1.$$

From this formula, it follows that

$$(k!)^{-r} = k^{-rk}e^{rk}(2\pi k)^{-r/2}\exp\left(-\frac{r\theta(k)}{12k}\right),\ 0 \leq \theta(k) \leq 1. \tag{25}$$

It is obvious that the following relations are true:

$$\sup_{k:\ |k-a|\leq a^\gamma}\left|\exp\left(-\frac{r\theta(k)}{12k}\right) - 1\right| = O(a^{-1}), \tag{26}$$

$$\sup_{k:\ |k-a|\leq a^\gamma}\left|\frac{(2\pi k)^{-r/2}}{(2\pi a)^{-r/2}} - 1\right| = O(a^{\gamma-1}). \tag{27}$$

From formulas (25)–(27), we obtain the following relation:

$$\sup_{k:\ |k-a|\leq a^\gamma}\left|e^{-ra}\frac{a^{rk}}{(k!)^r}\cdot e^{ra}\left(\frac{ea}{k}\right)^{-rk}(2\pi a)^{r/2} - 1\right| = O(a^{\gamma-1}).$$

The notation

$$p(k,a) = q(k,a)(1 + O(a^t)), \ |k-a| \le a^\gamma, \ a \to \infty$$

means that the following relation is satisfied:

$$\sup_{k:\ |k-a|\le a^\gamma} \left| \frac{p(k,a)}{q(k,a)} - 1 \right| = O(a^t), \ a \to \infty.$$

Therefore, the following relation is true:

$$e^{-ra}\frac{a^{rk}}{(k!)^r} = e^{-ra}\left(\frac{ea}{k}\right)^{rk}(2\pi a)^{-r/2}(1 + O(a^{\gamma-1})), \ |k-a| \le a^\gamma, \ a \to \infty. \quad (28)$$

Using the Taylor series expansion of the function $\ln(1+u) = u - \frac{u^2}{2} + O(u^3)$, $|u| < 1$, we evaluate $\ln\left[e^{-a}\left(\frac{ea}{k}\right)^k\right]$. To do this, we set $k = a + v$, $|v| \le a^\gamma$ and evaluate the ratio

$$\ln\left[e^{-a}\left(\frac{ea}{k}\right)^k\right] = -a + (a+v)(1 + \ln a - \ln a - \ln(1 + v/a)) = -\frac{v^2}{2a} + O\left(\frac{v^3}{a^2}\right),$$

from which it follows that

$$e^{-a}\left(\frac{ea}{k}\right)^k = \exp\left(-\frac{(k-a)^2}{2a}\right)\left(1 + O\left(a^{3\gamma-2}\right)\right), \ |k-a| \le a^\gamma, \ a \to \infty. \quad (29)$$

From expressions (28) and (29), we obtain the asymptotic relation

$$e^{-ra}\frac{a^{rk}}{(k!)^r} = (2\pi a)^{-r/2}\exp\left(-\frac{r(k-a)^2}{2a}\right)(1 + O(a^{3\gamma-2})), \ |k-a| \le a^\gamma, \ a \to \infty.$$

By combining this relation with formula (16), we obtain limit relation (24). The lemma is proved.

Corollary 1. *It follows from Lemma 4 that the following asymptotic formula holds uniformly for all* $k: \ |k-a| \le a^\gamma$*:*

$$p(k,a) \sim \exp\left(-\frac{(k-a)^2}{2a}\right)\frac{1}{\sqrt{2\pi a}}, \ a \to \infty.$$

Lemma 5. *The following limit ratio is true:*

$$\varphi_3(a) = g_3(a)(1 + O(a^{\gamma-1})), \ a \to \infty. \quad (30)$$

Proof. Without a significant generality constraint (to simplify the proof), we assume that a, a^γ are integer. Then the following equality is true:

$$g_3(a) = \sum_{a-a^\gamma \le k < a+a^\gamma} (2\pi a)^{-r/2}\int_k^{k+1} \exp\left(-\frac{r(x-a)^2}{2a}\right)dx.$$

For $k < a$, the function $\exp\left(-\frac{r(x-a)^2}{2a}\right)$ is monotonically increasing, and for $k \geq a$, it decreases monotonically on the interval $[k, k+1]$. So, the following relation is true:

$$\exp\left(-\frac{r(k-a)^2}{2a}\right) = \exp\left(-\frac{r(k+1-a)^2}{2a}\right)(1+O(a^{\gamma-1})),\ |k-a| \leq a^{\gamma},\ a \to \infty.$$

It follows that the following relation is satisfied:

$$\exp\left(-\frac{r(k-a)^2}{2a}\right) = \int_k^{k+1} \exp\left(-\frac{r(x-a)^2}{2a}\right) dx(1+O(a^{\gamma-1})),\ |k-a| \leq a^{\gamma},$$

when $a \to \infty$, and in addition, we have $\varphi_3(a+a^{\gamma}, a) = (2\pi a)^{-r/2}\exp(-ra^{2\gamma-1})$. From these relations and formulas (12), (14), we derive (30). The lemma is proved.

From formulas (24), (30), we obtain the relation

$$f_3(a) = g_3(a)(1+O(a^{3\gamma-2}))(1+O(a^{\gamma-1})) = g_3(a)(1+O(a^{3\gamma-2})). \tag{31}$$

Combining Formulas (9), (14), (17), (18), and (31), we obtain (12). So, Theorem 3 is proved.

Remark 3. The series $\sum_{k\geq 0}\left(\frac{a^k}{k!}\right)^r$ considered in Theorem 3 is a generalized hypergeometric series. However, it is impossible to use a well-known asymptotic formulas [11, Chapter 16] for it.

Remark 4. We present results of a computational experiment illustrating the accuracy of the obtained approximations. Denote error of the approximation by $\Delta(a) = \left|\frac{f(a)-g(a)}{f(a)}\right|$. Values of $\Delta(a)$ are presented in Table 1. We may notice that the error is decreasing while a grows.

Table 1. Values of $\Delta(a)$ for $r = 2, 5, 20$; $a = 10^k$, $k = 1, \ldots, 6$.

r	a					
	10	10^2	10^3	10^4	10^5	10^6
2	6.4×10^{-3}	6.3×10^{-4}	6.3×10^{-5}	6.3×10^{-6}	6.2×10^{-7}	6.2×10^{-8}
5	2.0×10^{-2}	2.0×10^{-3}	2.0×10^{-4}	2.0×10^{-5}	2.0×10^{-6}	2.0×10^{-7}
20	8.3×10^{-2}	8.3×10^{-3}	8.3×10^{-4}	8.3×10^{-5}	8.3×10^{-6}	8.2×10^{-7}

5 Assembly of Independent Flows with Different Intensities

Consider now the case when there are two Poisson flows T_1, T_2 with intensities $\lambda_1, \lambda_2 : \lambda_1 < \lambda_2$, and denote $d = \lambda_2 t$, $cd = \lambda_1 t$, so,

$$0 < c = \frac{\lambda_1}{\lambda_2} < 1.$$

For instant intensity of assembly flow $\overline{\lambda}(t)$ in this case, we have

$$\overline{\lambda}(t) = \lambda_1 P(n_2(t) > n_1(t)) + \lambda_2 P(n_1(t) > n_2(t)) = \lambda_1 (P(n_1(t) > n_2(t)) \\ + P(n_2(t) > n_1(t))) + (\lambda_2 - \lambda_1) P(n_1(t) > n_2(t)) = \lambda_1 (1 - P(n_1(t) = n_2(t))) \\ + (\lambda_2 - \lambda_1) P(n_1(t) > n_2(t)) = \lambda_1 - \lambda_1 P(n_1(t) \geq n_2(t)) + \lambda_2 P(n_1(t) > n_2(t)),$$

therefore,

$$|\overline{\lambda}(t) - \lambda_1| \leq \lambda_2 P(n_1(t) \geq n_2(t)), \tag{32}$$

where

$$P(n_1(t) \geq n_2(t)) = \sum_{k=0}^{\infty} e^{-d} \frac{d^k}{k!} \sum_{i=k}^{\infty} e^{-cd} \frac{(cd)^i}{i!} = G(d). \tag{33}$$

Consider function $G(d)$ in a form of the sum $G(d) = G_1(d) + G_2(d)$, where

$$G_1(d) = \sum_{k>d} e^{-d} \frac{d^k}{k!} \sum_{i=k}^{\infty} e^{-cd} \frac{(cd)^i}{i!}, \quad G_2(d) = \sum_{k \leq d} e^{-d} \frac{d^k}{k!} \sum_{i=k}^{\infty} e^{-cd} \frac{(cd)^i}{i!}.$$

For a fixed $c: \ 0 < c < 1$, we define the function $\psi(c) = c - 1 - \ln c$. Function $\psi(c)$ satisfies the relations $\psi(1) = 0$, $\psi'(c) = 1 - 1/c < 0$, so, $\psi(c)$ is positive and monotonically decreasing for $0 < c < 1$.

We will call that positive functions $p(d)$ and $q(d)$ satisfy the relation

$$p(d) \preceq q(d), \ d \to \infty, \ \text{ if } \limsup_{d \to \infty} \frac{p(d)}{q(d)} < \infty.$$

Lemma 6. *For any $c: \ 0 < c < 1$, the following formula holds:*

$$d^{-1/2} \exp(-d\psi(c)) \preceq G_1(d) \preceq d^{1/2} \exp(-d\psi(c)), \ d \to \infty. \tag{34}$$

Proof. Really, we have:

$$G_1(d) \leq \sum_{k>d} e^{-d} \frac{d^k}{k!} \sum_{i=[d]}^{\infty} e^{-cd} \frac{(cd)^i}{i!} \leq \sum_{i=[d]}^{\infty} e^{-cd} \frac{(cd)^i}{i!} \leq \\ \frac{e^{-cd}(cd)^d}{[d]!} \sum_{i=[d]}^{\infty} \left(\frac{cd}{[d]}\right)^{i-[d]} = \frac{e^{-cd}(cd)^d}{[d]!} \left(1 - \frac{cd}{[d]}\right)^{-1} \sim \frac{e^{-cd}(cd)^d}{(1-c)[d]!}.$$

Due to the Stirling formula, we derive

$$\frac{e^{-cd}(cd)^d}{[d]!} \leq \frac{e^{-cd}(cd)^d}{[d]^{[d]}e^{-[d]}\sqrt{2\pi[d]}} \leq \frac{e^{d-cd}(cd)^d}{[d]^d\sqrt{2\pi[d]}} \leq$$

$$\frac{e^{d-cd}(cd)^d}{(d-1)^{d-1}\sqrt{2\pi(d-1)}} \sim e\sqrt{\frac{d}{2\pi}}\exp(-d\psi(c)).$$

As a result, we come to the right relation in formula (34).

We now construct the lower bound of the function $G_1(d)$, assuming $j = [d]+1,\ d \to \infty$:

$$e^{-d}\frac{d^j}{j!} \geq e^{-1/12d}e^{-d+j}\frac{[d]^j}{j^j\sqrt{2\pi j}} \geq \frac{1}{\sqrt{2\pi j}}\left(1+\frac{1}{[d]}\right)^{-j} \sim \frac{1}{e\sqrt{2\pi d}}, \tag{35}$$

$$e^{-cd}\frac{(cd)^j}{j!} \sim e^{-cd}\frac{(cd)^j}{j^je^{-j}\sqrt{2\pi j}} \geq e^{-cd+j}\frac{c^{d+1}d^j}{j^j\sqrt{2\pi j}} \sim \frac{c\exp(-d\psi(c))}{e\sqrt{2\pi d}}. \tag{36}$$

From formulas (35) and (36), the left relation in formula (34) follows. The lemma is proved.

Let us fix $s:\ 0<c<s<1$, let $l=[sd]$, and estimate $G_2(d) = G_2'(d)+G_2''(d)$, where

$$G_2'(d) = \sum_{0\leq k<sd} e^{-d}\frac{d^k}{k!}\sum_{i=k}^{\infty} e^{-cd}\frac{(cd)^i}{i!},\ G_2''(d) = \sum_{sd\leq k\leq d} e^{-d}\frac{d^k}{k!}\sum_{i=k}^{\infty} e^{-cd}\frac{(cd)^i}{i!}.$$

Denote $\mu(c,s) = c - s(1+\ln c - \ln s)$, $q(s) = 1-s(1-\ln s)$, $0<c<s<1$. For any $c,s:\ 0<c<s<1$, the relations

$$\mu(c,1) = \psi(c) > 0,\ \frac{\partial\mu(c,s)}{\partial s} = -\ln\frac{c}{s} > 0$$

take place for fixed c. Function $\mu(c,s)$ increases on argument $s:\ c<s<1$ and

$$\mu(c,c) = 0,\ \mu(c,s) > 0,\ c<s<1.$$

Function $q(s)$ satisfies the relations

$$q(1) = 0,\ q'(s) = \ln s < 0,\ q(s) > 0,\ 0<s<1,$$

and hence, it is positive and monotonically decreasing for $0<s<1$.

Lemma 7. *For any $c,\ s:\ 0<c<s<1$, the following formula holds:*

$$d^{-1/2}\exp(-dq(s)) \preceq G_2'(d) \preceq d^{1/2}\exp(-dq(s)),\ d\to\infty. \tag{37}$$

Proof. Function $G_2'(d)$ satisfies the following relations for $cd > 2$:

$$G_2'(d) \leq l\frac{e^{-d}d^l}{l!} \sim e^{-d+l}\sqrt{\frac{l}{2\pi}}\left(\frac{d}{l}\right)^l \leq e^{-d+sd}\sqrt{\frac{l}{2\pi}}\left(\frac{d}{sd-1}\right)^{sd} \sim$$

$$\sqrt{\frac{sd}{2\pi}}\frac{l \cdot d^l}{l^l\sqrt{2\pi sd}} \sim \frac{e}{\sqrt{2\pi sd}}\exp(-dq(s)).$$

Thus, the right relation in formula (37) is true.

On the other hand $G_2'(d) \geq \frac{e^{-d}d^l}{l!}\left(1 - \sum_{0\leq i\leq l} e^{-cd}\frac{(cd)^i}{i!}\right)$, where

$$\frac{e^{-d}d^l}{l!} \sim \frac{e^{-d+l}d^l}{l^l\sqrt{2\pi l}} \geq \frac{e^{-d+l}}{\sqrt{2\pi l}}\left(\frac{1}{s}\right)^l \sim \frac{\exp(-dq(s))}{e\sqrt{2\pi sd}}, \tag{38}$$

and

$$\sum_{0\leq i<l}\frac{e^{-cd}(cd)^i}{i!} \leq sde^{-cd}\frac{(cd)^l}{l!} \sim \sqrt{\frac{1}{2\pi sd}}e\exp(-d\mu(s,c)) \to 0,\ d\to\infty. \tag{39}$$

From (38), (39), we derive the left relation of (37). The lemma is proved.

Lemma 8. *For any $c,\ s:\ 0 < c < s < 1$, the following formula holds:*

$$d^{-1}\exp(-d(\mu(c,s)+q(s)) \preceq G_2''(d) \preceq d^{1/2}\exp(-d\mu(c,s))),\ d\to\infty. \tag{40}$$

Proof. Function $G_2''(d)$ satisfies the following relations for $cd > 2$:

$$G_2''(d) \leq \sum_{i\geq sd} e^{-cd}\frac{(cd)^i}{i!} \leq e^{-cd}\frac{(cd)^l}{l!}\sum_{i\geq 0}\frac{(cd)^i}{l^i} =$$

$$e^{-cd}\frac{(cd)^l}{l!}\left(1-\frac{cd}{l}\right)^{-1} \sim e^{-cd}\frac{(cd)^l}{l!}\left(1-\frac{c}{s}\right)^{-1}.$$

We derive

$$e^{-cd}\frac{(cd)^l}{l!} \sim e^{-cd+l}\frac{(cd)^{sd}}{l^l\sqrt{2\pi sd}} \leq e^{-cd+sd}\frac{(cd)^{sd}}{(sd-1)^{sd-1}\sqrt{2\pi sd}} \leq$$

$$sde^{-cd+sd}\frac{(cd)^{sd}}{(sd-1)^{sd}\sqrt{2\pi sd}} = \sqrt{\frac{sd}{2\pi}}e^{-cd+sd}\left(\frac{c}{s}\right)^{sd}\left(1-\frac{1}{sd}\right)^{-sd} \sim$$

$$e\sqrt{\frac{sd}{2\pi}}\exp(-d\mu(c,s)). \tag{41}$$

Hence, the right relation in formula (40) is true.

At the same time, using (38) in a similar way to formula (41), we obtain

$$G_2''(d) \geq \frac{e^{-d}d^l}{l!}\cdot\frac{e^{-cd}(cd)^l}{l!} \sim \frac{\exp(-dq(s))}{e\sqrt{2\pi sd}}\cdot\frac{\exp(-d\mu(c,s))}{e\sqrt{2\pi sd}}. \tag{42}$$

From (47), the left relation of formula (40) follows. The lemma is proved.

For fixed $c, s:\ 0 < c < s < 1$, we define the functions

$$\vartheta(c,s) = \min(\mu(c,s), q(s));\ \nu(c,s) = \min(\vartheta(c,s), \psi(c)),$$

$$\alpha(c) = \sup_{s:\ c<s<1} \nu(c,s),\ s^*(c) = -\tfrac{1-c}{\ln c}.$$

Lemma 9. *For any $c:\ 0 < c < 1$, the following formula holds:*

$$\alpha(c) = q(s^*(c)). \tag{43}$$

Proof. Since function $\mu(c,s)$ is increasing, the function $q(s)$ is decreasing for $s:\ c \le s \le 1$, and $\mu(c,c) = 0,\ q(1) = 0,\ s = 1$, then there exists the unique point $s^*(c) = -\dfrac{1-c}{\ln c} > c$ that satisfies the equality $\mu(c, s^*(c)) = q(s^*(c))$. Therefore, the following equality is fulfilled:

$$\sup_{s:\ c<s<1} \vartheta(c,s) = q(s^*(c)). \tag{44}$$

Now we prove that the inequality $q(s^*(c)) < \psi(c)$ holds. Indeed, since $c < s^*(c)$ and function $q(s)$ is decreasing, then $q(s^*(c)) < q(c)$. Therefore, the function

$$\omega(c) = q(c) - \psi(c) = 2 - 2c + (1+c)\ln c$$

satisfies the equality $\omega(1) = 0$, and its derivative satisfies $\omega'(c) = \dfrac{q(c)}{c} > 0$. This means that $\omega(c) < 0$ for $0 < c < 1$. So, the following inequalities are true:

$$q(s^*(c)) < q(c) < \psi(c),\ 0 < c < s^*(c) < 1. \tag{45}$$

From relations (44), (45), we obtain formula (43) for a fixed $c:\ 0 < c < 1$. The lemma is proved.

Theorem 4. *For any $c:\ 0 < c < 1$, the following formula holds:*

$$d^{-1}\exp(-d\alpha(c)) \preceq G(d) \le d^{1/2}\exp(-d\alpha(c)), \tag{46}$$

and therefore, $\lambda(t) \to \lambda,\ \lambda(t) - \lambda = G(\lambda_2 t)$ while $t \to \infty$.

Proof. For any $c, s:\ 0 < c < s < 1$, it follows from Lemmas 6–9 that

$$G(d) \preceq d^{1/2}\exp(-d\min(\psi(c), q(s), \mu(c,s)).$$

So, for any $c:\ 0 < c < 1$, we derive

$$G(d) \preceq d^{1/2}\exp(-dq(s^*(c)) = d^{1/2}\exp(-d\alpha(c)).$$

The right relation in (46) is proved.

For $c, s:\ 0 < c < s < 1$, it follows from Lemmas 6–9 that

$$G(d) \succeq d^{-1}\exp(-d\min(\psi(c), q(s), \mu(c,s) + q(s)).$$

Assuming $s = s^*(c)$ in the last inequality, we obtain

$$G(d) \succeq d^{-1}\exp(-d\min(\psi(c), q(s^*(c)), \mu(c, s^*(c)) + q(s^*(c)))$$

$$= d^{-1}\exp(-d\min(\psi(c), q(s^*(c))) = d^{-1}\exp(-d\alpha(c).$$

The left relation in formula (46) is proved. So, Theorem 4 is proved.

Remark 5. In Remark 4, the estimation of probability $\mathrm{P}\{n_1(t) = \ldots = n_r(t)\}$ uses a Gaussian approximation of a Poisson distribution with a large parameter. It is shown that this approximation gives results similar to the results of the analytical study. Consider how this approximation works when estimating the probability $\mathrm{P}\{n_1(t) \geq n_2(t)\}$.

To do this, we write the following approximations of random variables $n_1(t)$ and $n_2(t)$:

$$n_1(t) \approx \sqrt{cd}\xi_1 + cd, \quad n_2(t) \approx \sqrt{d}\xi_2 + d,$$

where ξ_1 and ξ_2 are independent random variables having a standard normal distribution (with zero mean and variance equal to one). Then by analogy with the proof of Lemma 5, we can construct a Gaussian approximation of the probability

$$\mathrm{P}\{n_1(t) \geq n_2(t)\} \approx \mathrm{P}\{\sqrt{cd}\,\xi_1 + cd \geq \sqrt{d}\,\xi_2 + d\} =$$

$$\mathrm{P}\{\xi_2 \leq \sqrt{c}\,\xi_1 + \sqrt{d}(c-1)\} = S(d),\ d \to \infty.$$

Denote a random variable with a standard normal distribution by η and put $h = (c-1)\sqrt{\dfrac{d}{c+1}}$. Since random vector (ξ_1, ξ_2) has a two-dimensional normal distribution with zero mean and with an identity covariance matrix, then using well-known asymptotic formula

$$\mathrm{P}\{\eta > R\} \sim \frac{1}{R\sqrt{2\pi}}\exp\left(-\frac{R^2}{2}\right),\ R \to \infty,$$

it is possible to obtain the following ratio based on the Gaussian approximation:

$$\mathrm{P}\{n_1(t) \geq n_2(t)\} \approx \frac{1}{h\sqrt{2\pi}}\exp\left(-\frac{h^2}{2}\right) = \frac{\sqrt{c+1}}{\sqrt{2\pi d}(c-1)}\exp\left(-d \cdot \frac{(c-1)^2}{2(c+1)}\right)$$

$$= \frac{\sqrt{c+1}}{\sqrt{2\pi d}(c-1)}\exp(-dA(c)) = S(d),\ A(c) = \frac{(c-1)^2}{2(c+1)},\ d \to \infty. \quad (47)$$

Now compare factors $\alpha(c)$ and $A(c)$ in the exponents of (46) and (47). When $c = 5/6$, we have $\alpha(c) \approx 0,0038,$, $A(c) \approx 0,0076$. If $c = 2/3$, then $\alpha(c) \approx 0,0168$, $A(c) \approx 0,0333$. Thus, factor $A(c)$ calculated by the Gaussian approximation is greater than factor $\alpha(c)$ calculated analytically.

We denote $\delta(d) = \left|\dfrac{G(d) - S(d)}{G(d)}\right|$ and numerically evaluate an accuracy of the Gaussian approximation for $c = 5/6$ and $c = 2/3$. The results are presented in Tables 2 and 3. We may notice that the rate of decreasing of values $\delta(d)$ decreases while d grows for the case $c = 5/6$ (Table 2). On other hand, after some decreasing, value of $\delta(d)$ starts to grow while d grows for the case $c = 2/3$ (Table 3). Thus, the results given in Tables 2 and 3 indicate a much worse quality of the Gaussian approximation than the results given in Table 1.

Table 2. Values of $\delta(d)$ for $c = 5/6$.

d	100	200	500	1000	2000
$\delta(d)$	0.267	0.143	0.051	0.021	0.018

Table 3. Values of $\delta(d)$ for $c = 2/3$.

d	10	50	100	200	500
$\delta(d)$	0.321	0.059	0.029	0.047	0.192

Remark 6. Using the proof of Lemma 1, it is easy to consider the case of assembling r independent Poisson flows with intensities $\lambda_1 = \lambda_2 = \ldots = \lambda_s < \lambda_{s+1} \leq \ldots \leq \lambda_r$, to get the inequality

$$|\overline{\lambda}(t) - \lambda_1| \leq \sum_{i=s+1}^{r} \lambda_i P(n_1(t) \geq n_i(t))$$

and to use Theorem 4 for estimating the probabilities $\mathrm{P}\{n_1(1) \geq n_i(t)\}$, $i = s+1, \ldots, r$. The results of performed numerical experiments are very sensitive to the correct or incorrect choice of the corresponding asymptotic formulas.

6 Convergence of Assembly Flow A_2 to Poisson Flow

Consider the union T^2 of independent Poisson flows T_1 and T_2 with equal intensities λ. It is well-known that T^2 is a Poisson flow with intensity 2λ. Denote its points as $T^2 = \{0 = t(0) < t(1) < \ldots\}$.

Also, consider assembly A_2 of the flows T_1 and T_2. Its points $\{0 = t_0 < t_1 < t_2 < \ldots\}$ are defined by the expressions

$$t_k = \inf\{t > t_{k-1} : n_1(t) = n_2(t)\}. \tag{48}$$

Define the Markov process $\nu(t) = n_2(t) - n_1(t)$ with state space $\{0, \pm 1, \pm 2, \ldots\}$ and transient intensities $\lambda_{k,k+1} = \lambda_{k,k-1} = \lambda$, $k = 0, \pm 1, \pm 2, \ldots$

Process $\nu(t)$ has jumps ± 1 in the points of flow T^2, and its zeroing points coincide with points $0 = t_0 < t_1 < t_2 < \ldots$ of assembly flow A_2. The random sequence $\{\nu(t(j)),\ j = 0, 1, 2, \ldots\}$ is a symmetric random walk on the set $\{0, \pm 1, \pm 2, \ldots\}$, therefore, accordingly to [12, Chapter III, § 3, Lemma 1], we can write

$$\mathrm{P}\{\nu(t(2j)) = 0\} = C_{2j}^{j} 2^{-2j} = p_{2j} \leq p_{2(j+1)},\ j = 1, 2, \ldots,\ p_{2j} \sim \frac{1}{\sqrt{\pi j}},\ j \to \infty. \tag{49}$$

Define the random event

$$\bigcup_{j=k}^{k+K} \{\nu(t(2j)) = 0\} = \bigcup_{j=2k}^{2(k+K)} \{\nu(t(j)) = 0\}.$$

Using (49) for the given ε and K, we can derive the following expression:

$$k(\varepsilon, K) = \left[\frac{K^2}{\pi \varepsilon^2}\right], \tag{50}$$

and for any $k > k(\varepsilon, K)$ we obtain

$$P\left\{\bigcup_{j=2k}^{2(k+K)} \{\nu(t(j)) = 0\}\right\} \leq \frac{K}{\sqrt{\pi k(\varepsilon, K)}} \leq \varepsilon. \tag{51}$$

Then due to (51), the equality $n_1(t) = n_2(t)$ does not hold in any $2K$ points following the moment $t(2k(\varepsilon, K))$. Therefore, at this time interval, the assembly flow is Poisson with parameter λ with probability not greater than ε. Note that in this case, due to formula (50), value of $k(\varepsilon, K)$ increases quite rapidly while ε decreases.

7 Conclusion

Despite the apparent simplicity of the considered model of the assembly flow of independent Poisson flows, the study have shown that the model is quite complex for the analysis. In the paper, we have obtained various versions of the Central limit theorem for the assembly flow both in terms of random variables and in terms of stochastic processes. Exact asymptotic formulas are derived for intensity of the assembly flow of identical Poisson flows, and estimations of the convergence rate are build for the case of non-identical original flows. Estimations of the convergence rate of the assembly flow of identical Poisson flows to a Poisson flow are derived.

References

1. Prabhakar, B., Bambos, N., Mountford, T.S.: The synchronization of Poisson processes and queueing networks with service and synchronization nodes. Adv. Appl. Probab. **32**(3), 824–843 (2000)

2. Cohen, Y., Faccio, M., Pilati, F., Yao, X.: Design and management of digital manufacturing and assembly systems in the Industry 4.0 era. Int. J. Adv. Manuf. Technol. **105**(9), 3565–3577 (2019)
3. Oestreich, H., Wrede, S., Wrede, B.: Learning and performing assembly processes: an overview of learning and adaptivity in digital assistance systems for manufacturing. In: PETRA 2020: Proceedings of the 13th ACM International Conference on PErvasive Technologies Related to Assistive Environments, Article No. 42, pp. 1–8 (2020)
4. De Cuypere, E., De Turck, K., Fiems, D.: Performance analysis of a kitting process as a paired queue. Math. Prob. Eng. Article ID 843184 (2013)
5. Narahari, Y., Sundarrajan, P.: Performability analysis of Fork-Join queueing systems. J. Oper. Res. Soc. **46**(10), 1237–1249 (1995)
6. Thomasian, A.: Analysis of Fork/Join and related queueing systems. ACM Comput. Surv. **47**(2), 1–71 (2014)
7. Mitrophanov, Y.I., Rogachko, E.S., Stankevich, E.P.: Analysis of queueing networks with batch movements of customers and control of flows among clusters. Autom. Control. Comput. Sci. **49**(4), 221–230 (2015). https://doi.org/10.3103/S0146411615040094
8. Tsitsiashvili, G.S., Osipova, M.A.: The study of the assembly of Poisson flows. Tomsk State Univ. J. Control Comput. Sci. **48**, 51–56 (2019). (In Russian)
9. Borovkov, A.A., Mogulsky, A.A., Sakhanenko A.I.: Limit theorems for random processes. Itogi Nauki i Tekhniki. Ser. Sovrem. Probl. Mat. Fundam. Napravleniia, **82**, 5–194 (1995). (in Russian)
10. Sveshnikov, A.G., Bogolubov, A.N., Kravcov, V.V.: Lectures on Mathematical Physics: Textbook. Moscow State University (1993). (in Russian)
11. Askey, R.A., Olde Daalhuis, A.B.: Generalized Hypergeometric Functions and Meijer G-function. In: NIST Handbook of Mathematical Functions, U.S. Dept. Commerce, Washington, DC, pp. 403–418 (2010)
12. Feller, V.: Introduction to Probability Theory and its Applications, vol. 1. Wiley, Hoboken (1968)

On Multiserver State-Dependent Retrial Queues Operating in Stationary Regime

Eugene A. Lebedev[1], Vadym D. Ponomarov[2](✉), and Oksana V. Pryshchepa[3]

[1] Taras Shevchenko National University of Kyiv, Kyiv, Ukraine
leb@unicyb.kiev.ua
[2] Taras Shevchenko National University of Kyiv, Kyiv, Ukraine
[3] National University of Water and Environmental Engineering, Rivne, Ukraine
o.v.pryshchepa@nuwm.edu.ua

Abstract. In this paper, we consider Markov models for two types of multiserver retrial queues with an input flow rate that depends on a number of calls in an orbit: classical state-dependent models and state-dependent queues with limited number of retrials. For both systems, the conditions for the existence of the stationary regime are defined and formulas for steady-state probabilities are presented. The investigative technique is based on approximation of the input system by the system with truncated state space.

Keywords: Retrial queue · Steady-state probabilities · Service process · Quasi-birth-and-death process

1 Introduction

Retrial queues form a special class of stochastical models that take into account an important property of the service process. The call that finds all servers busy becomes a source of repeated calls. It is usually assumed, that the calls can try to get service an infinite number of times. However, in some cases, the number of repeated attempts can be limited. This can be caused by the impatience of the calls, limited resources, etc. Such type of models is widely used for modeling and analysis of computer and telecommunication systems, call centers, airport control systems.

From a practical standpoint, the systems with varying input flow rate are of special interest. The input flow rate in such systems depends on the number of repeated calls in the current moment of time (see, e.g. [4]). This allows to control the input flow in order to maximize the quality of service, eliminate the possibility of overflow, maximize the income from the system, etc.

Despite the widespread of the retrial queueing systems analytical representation of the steady state probabilities were obtained only for the simplest cases [1,3,9]. The mathematical analysis of such systems faces a number of difficulties. The phenomenon of repeated calls leads to a multi-dimensional service process,

A. Dudin et al. (Eds.): ITMM 2020, CCIS 1391, pp. 95–107, 2021.
https://doi.org/10.1007/978-3-030-72247-0_7

which is typical for stochastic networks and as a consequence complicates the theory. The transitions matrix does not have properties, that allow its transformations for obtaining an explicit solution. Moreover, local characteristics of the service process in controlled retrial queueing systems depend on the phase point, which does not allow to use classical approaches of stochastic systems analysis (e.g. method of generating functions).

In this paper, we develop approach for the analysis of retrial queues with controlled input flow rate by the systems with limited phase space. For classical systems and systems with a limited number of retrials, this approach allows finding explicit formulas of vector-matrix type for probability characteristics of the truncated system.

2 Mathematical Model of the Classical Retrial Queue with Controlled Input

Consider a continuous time Markov chain $X(t) = (X_1(t); X_2(t))$, $X_1(t) \in \{0, 1, \dots, c\}$, $X_2(t) \in \{0, 1, \dots\}$, that is defined by infinitesimal rates $q_{(i,j)(i',j')}$, $(i, j), (i', j') \in S(X) = \{0, 1, \dots, c\} \times \{0, 1, \dots\}$:

1. For $i = \{0, 1, \dots, c-1\}$

$$q_{(i,j)(i',j')} = \begin{cases} \lambda_j, & when \quad (i', j') = (i+1, j); \\ i\nu, & when \quad (i', j') = (i+1, j-1); \\ i\mu, & when \quad (i', j') = (i-1, j); \\ -(\lambda_j + j\nu + i\mu), & when \quad (i', j') = (i, j); \\ 0, & otherwise. \end{cases}$$

2. For $i = c$

$$q_{(c,j)(i',j')} = \begin{cases} \lambda_j, & when \quad (i', j') = (c, j+1); \\ c\mu, & when \quad (i', j') = (c-1, j); \\ -(\lambda_j + c\mu), & when \quad (i', j') = (c, j); \\ 0, & otherwise. \end{cases}$$

Two-dimensional Markov chain $X(t)$ describes a service process in the following system. The service facility contains c identical servers. Service rate of each server is $\mu > 0$, $\nu > 0$ - rate of retrial calls flow, $\lambda_j > 0$ - input flow rate when there are j sources of retrial calls (j calls in the orbit). The first component $X_1(t) \in \{0, 1, \dots, c\}$ indicates the number of busy servers at the instant $t \geq 0$ and the second one $X_2(t) \in \{0, 1, \dots\}$ is the number of retrial sources.

Let's write up the ergodicity conditions for $X(t), t \geq 0$.

Lemma 1. *Let* $\lambda = \overline{\lim_{j \to \infty}} \lambda_j < \infty$. *Then under* $\lambda < c\mu$ *the chain* $X(t)$ *is ergodic and its limit distribution is the same as the single stationary one.*

Proof. We will consider the following functions as Lyapunov test functions

$$\phi(i,j) = \alpha i + j, (i,j) \in S(X),$$

where the α parameter will be defined later.

For the given test functions the average transfer

$$y_{ij} = \sum_{(i',j')\neq(i,j)} q_{(i,j)(i',j')} \left(\phi(i',j') - \phi(i,j)\right)$$

will be

$$y_{ij} = \begin{cases} \lambda_j\alpha - i\mu + j\nu(\alpha-1), 0 \le i \le c-1, \\ \lambda_j\alpha - c\mu, \qquad i = c. \end{cases}$$

Under condition $\lambda < c\mu$ for any $\alpha \in (\lambda/c\mu, 1)$ there exists such $\varepsilon > 0$, so that $y_{ij} < -\varepsilon$ for all $(i,j) \in S(X)$ except of finite number of states (i,j). So, the conditions of Tweedy theorem ([1], p. 97) are held for test functions $\phi(i,j) = \alpha i + j$, $\alpha \in \left(\frac{\lambda}{n\mu}, 1\right)$.

Lemma is proved.

For the construction of calculating schemes and explicit formulas, we will use a system with a truncated state space. Such a model operates similarly to the original queue but has a restriction on the size of an orbit: all new calls are lost when all servers are occupied and there are N calls in the orbit already. Formally, the service process in such queue is described by the Markov chain $X(t,N) = (X_1(t,N), X_2(t,N))$, $X_1(t,N) \in \{0,1,\ldots,c\}$, $X_2(t,N) \in \{0,1,\ldots,N\}$. Its infinitesimal transition rates $q^{(N)}_{(i,j)(i',j')}$, $(i,j),(i',j') \in S(X,N) = \{0,\ldots,c\} \times \{0,\ldots,N\}$ are equal to $q_{(i,j)(i',j')}$ of the chain $X(t)$ in all phase points except the boundary case $i = c$, $j = N$, where

$$q^{(N)}_{(c,N)(i',j')} = \begin{cases} c\mu, \ \textit{when} \ (i',j') = (c-1,N); \\ -c\mu, \ \textit{when} \ (i',j') = (c,N); \\ 0, \ \textit{otherwise}. \end{cases}$$

A state space $S(N)$ of the Markov chain $X(t,N)$ is finite. Therefore for $X(t,N)$ there always exists a stationary regime, and via $\pi_{ij}(N)$, $(i,j) \in S(X,N)$ we will designate its stationary probabilities.

Next, we introduce matrices which are given by the model parameters:

$A(j) = \| a_{ik}(j)\|_{i,k=1}^{c}$ is a matrix with entries $a_{ii-1}(j) = (j+1)\nu$, $i = 1,\ldots, c-1$; $a_{ck}(j) = \frac{(j+1)\nu c\mu}{\lambda_j}$, $k \neq c-1$; $a_{cc-1}(j) = \frac{(j+1)\nu(\lambda_j + c\mu)}{\lambda_j}$ and all other entries are equal to 0;

$B(j) = \| b_{ik}(j)\|_{i,k=1}^{c}$ is a three-diagonal matrix with entries $b_{ii-1}(j) = -\lambda_j$, $i = 2,\ldots,c$; $b_{ii}(j) = \lambda_j + j\nu + (i-1)\mu$, $i = 1,\ldots,c$; $b_{ii+1}(j) = -i\mu$, $i = 1,\ldots,c-1$;

$C(N) = \| c_{ik}(N)\|_{i,k=1}^{c}$, where $(c_{11}(N), c_{12}(N),\ldots,c_{1c}(N)) = e_1^T = (1, 0, \ldots, 0)$; and for $i = 2,\ldots,c$: $(c_{i1}(N), c_{i2}(N), \ldots, c_{ic}(N)) = (b_{i-11}(N), b_{i-12}(N), \ldots, b_{i-1c}(N))$.

Lemma 2. *Matrices $B(j)$, $j = 0, 1, \ldots, N$ and $C(N)$ are nonsingular.*

Proof. Let us check the Adamar's condition for columns of the matrix $B(j)$, $j = 0, 1, \ldots, N$ (see [11]):

$$G_i^j \equiv |b_{ii}(j)| - \sum_{\substack{k=0\\k \neq i}}^{c-1} |b_{ki}(j)| > 0, i = 0, \ldots, c-1.$$

$$G_i^j = \lambda_j + j\nu + i\mu - i\mu - \lambda_j = j\nu, i = 0, \ldots, c-2,$$

$$G_{c-1}^j = \lambda_j + j\nu + (c-1)\mu - (c-1)\mu = \lambda_j + j\nu.$$

The above conditions are satisfied for all $j = 1, \ldots, N$, which means nonsingularity of $B(j)$, $j = 0, 1, \ldots, N$. For $B(0)$ the weakened Adamar's condition holds true. Taking into account the fact that $B(0)$ is indecomposable matrix, we obtain nonsingularity of $B(0)$.

Since $C(N)$ is a triangular matrix, therefore

$$|C(N)| = \sum_{i=1}^{c-1} b_{ii+1}(N) = (-1)^{c-1}(c-1)!\mu^{c-1} \neq 0.$$

So nonsingularity of *C(N)* takes place and lemma is proved.

For $\pi_{ij}(N)$ Kolmogorov equations may be solved in an explicit vector-matrix form.

Theorem 1. *Let*

$$\pi_j^T(N) = (\pi_{0j}(N), \pi_{1j}(N), ..., \pi_{c-1j}(N)),$$

$$\Delta_j(N) = \frac{\left(\prod_{i=j}^{N-1} B^{-1}(i)A(i)\right) C^{-1}(N)e_1}{e_1^T \left(\prod_{i=0}^{N-1} B^{-1}(i)A(i)\right) C^{-1}(N)e_1}, j = 0, 1, ..., N,$$

where we use the commonly accepted agreement $\prod_{i=N}^{N-1} B^{-1}(i)A(i) = E$.

Then

$$\pi_j(N) = \pi_{00}(N)\Delta_j(N), j = 0, 1, \ldots, N,$$

$$\pi_{cj}(N) = \pi_{00}(N)\tfrac{(j+1)\nu}{\lambda_j}\bar{1}^T(c)\Delta_{j+1}(N), j = 0, 1, \ldots, N-1,$$

$$\pi_{cN}(N) = \pi_{00}(N)\tfrac{1}{c\mu}(\lambda_N e_c^T + N\nu\bar{1}^T(n))\Delta_N(N),$$

$$\pi_{00}(N) = \left\{\bar{1}^T(c)\Delta_0(N) + \sum_{j=1}^{N}\left(1 + \frac{j\nu}{\lambda_{j-1}}\right)\bar{1}^T(c)\Delta_j(N) + \frac{1}{n\mu}\left(\lambda_N e_n^T + N\nu\bar{1}^T(n)\right)\Delta_N(N)\right\}^{-1}.$$

Proof. First, let us consider the Kolmogorov equations corresponding to the phase points $(i, j) \in \{0, \ldots, c-1\} \times \{0, \ldots, N-1\}$:

$$-\lambda_j \pi_{i-1j}(N) + (\lambda_j + i\mu + j\nu)\pi_{ij}(N) - (i+1)\mu\pi_{i+1j}(N) = (j+1)\nu\pi_{i-1j+1}(N) \quad (1)$$

By an agreement we set $\pi_{-1k}(N) = 0$.
Next, we supplement the following equations into (1)

$$\lambda_j \pi_{cj}(N) = (j+1)\nu \sum_{i=0}^{c-1} \pi_{ij+1}(N), j = 0, \ldots, N-1. \quad (2)$$

The above equations represent an equality of probability flows in stationary regime through a separation boundary of the phase set $S(X, N) = E(j, N) \cup \bar{E}(j, N)$, where $E(j, N) = \{(\alpha, \beta) \in S(X, N) : \beta \leq j\}$ (see [12], Section II).

Taking into account the introduces notations, the Eqs. (1), (2) take the following vector-matrix form

$$B(j)\pi_j(N) = A(j)\pi_{j+1}(N), j = 0, \ldots, N-1. \quad (3)$$

For the level $j = N$ we observe

$$-\lambda_N \pi_{i-2N}(N) + (\lambda_N + (i-1)\mu + N\nu)\pi_{i-1N}(N) - i\mu\pi_{iN}(N) = 0, i = 1, \ldots, c-1.$$

If we supplement the above equations with the identity $\pi_{0N}(N) = \pi_{0N}(N)$, then as a result we will get the system of the following vector-matrix form

$$C(N)\pi_N(N) = \pi_{0N}(N)e_1. \quad (4)$$

From (3), (4) we find

$$\pi_j(N) = \pi_{0N}(N) \left(\prod_{i=j}^{N-1} B^{-1}(i)A(i) \right) C^{-1}(N)e_1, j = 0, 1, \ldots, N.$$

The last formula for $j = 0$ can be deduced

$$\pi_{0N}(N) = \pi_{00}(N) \left\{ e_1^T \left(\prod_{i=0}^{N-1} B^{-1}(i)A(i) \right) C^{-1}(N)e_1 \right\}^{-1}.$$

The probability $\pi_{00}(N)$ can be found from the normalization condition.
The theorem is proved.

Steady state probabilities $\pi_{ij}(N)$ converge to the probabilities π_{ij} when the truncation level N increases. This fact is a direct corollary of the stochastic ordering of the migration processes ([1], pp. 111–116).

In case of $c = 1, 2$ the results of Theorem 1 turn into the explicit formulas of a scalar type.

Corollary 1. *Let for the state-dependent M|M|1-queue the conditions of Lemma 1 hold. Then a stationary regime exists and the stationary probabilities take the form:*

$$\pi_{0j} = \pi_{00} \prod_{i=1}^{j} \rho_{i-1} \left[1 + \frac{1}{i} \left(\lambda_{i-1}/\nu - 1 \right) \right],$$

$$\pi_{1j} = \pi_{00} \frac{j+1}{\lambda_j/\nu} \prod_{i=1}^{j+1} \rho_{i-1} \left[1 + \frac{1}{i} \left(\lambda_{i-1}/\nu - 1 \right) \right], j = 0, 1, \ldots,$$

$$\pi_{00} = \left\{ 1 + \sum_{j=1}^{\infty} \left(1 + \frac{j}{\lambda_{j-1}/\nu} \right) \prod_{i=1}^{j+1} \rho_{i-1} \left[1 + \frac{1}{i} \left(\lambda_{i-1}/\nu - 1 \right) \right] \right\}^{-1},$$

where $\rho_i = \lambda_i/\mu$ is a queue load provided that the number of calls in an orbit is equal to i.

Corollary 2. *Let for the state-dependent M|M|2-queue the conditions of Lemma 1 hold. Then a stationary regime exists and the stationary probabilities take the form:*

$$\pi_{0j} = \pi_{00} \cdot \frac{\lambda_0^2}{\left(\lambda_j + j\nu \right)^2 + j\nu\mu} \cdot \delta_j,$$

$$\pi_{1j} = \pi_{00} \cdot \frac{\lambda_0^2}{\mu} \cdot \frac{\lambda_0^2}{\left(\lambda_j + j\nu \right)^2 + j\nu\mu} \cdot \delta_j,$$

$$\pi_{2j} = \pi_{00} \cdot \frac{\lambda_0^2}{\mu} \cdot \frac{(j+1)\nu}{\lambda_j} \cdot \frac{\mu + \lambda_{j+1} + (j+1)\nu}{\left(\lambda_{j+1} + (j+1)\nu \right)^2 + (j+1)\nu\mu} \cdot \delta_{j+1}, j = 0, 1, \ldots,$$

where $\pi_{00} = \left\{ 1 + \frac{\lambda_0}{\mu} + \frac{\lambda_0^2}{\mu} \sum_{j=1}^{\infty} \left(1 + \frac{j\nu}{\lambda_{j-1}} \right) \frac{\mu + \lambda_j + j\nu}{(\lambda_j + j\nu)^2 + j\nu\mu} \delta_j \right\}^{-1}$, $\delta_j = \prod_{i=1}^{j} \rho_{i-1} \cdot \frac{(\lambda_i + i\nu)^2 + i\nu\mu}{i\nu(\lambda_i + i\nu + \mu(1 + \rho_{i-1}))}$, $j = 0, 1, \ldots,$ $\rho_i = \lambda_i/2\mu$ *is a load for the M|M|2-queue.*

Results of Theorem 1 and its corollaries give explicit representations of stationary probabilities of the multi-channel retrial queue with controlled input flow. They allow to estimate various integral characteristics of the system, to solve optimal control and optimization problems.

Next, we present results for queues with limited number of retrials.

3 Steady State Analysis of Systems with Limited Number of Retrials

Consider an $(m+1)$-dimensional Markov chain $Q^m(t) = (Q_0^m(t), Q_1^m(t), \ldots, Q_m^m(t))$, $t \geq 0$ in the phase space $S(Q^m) = \{0, 1, \ldots, c\} \times Z_+^m$, where $Z_+ = \{0, 1, \ldots\}$ that is defined by its infinitesimal characteristics $q_{\beta\beta'}$, $\beta = (i, j_1, \ldots, j_m)$, $\beta' = (i', j'_1, \ldots, j'_m) \in S(Q^m)$:

1. For $i = 0, \dots, c-1$, $j_k \in Z_+, k = 1, 2, \dots, m$

$$q_{\beta\beta'} = \begin{cases} \lambda_{j_1,\dots,j_m}, & when\ \beta' = (i+1, j_1, \dots, j_k, \dots, j_m), \\ i\mu, & when\ \beta' = (i-1, j_1, \dots, j_k, \dots, j_m), \\ j_1\nu_1, & when\ \beta' = (i+1, j_1 - 1, \dots, j_k, \dots, j_m), \\ j_2\nu_2, & when\ \beta' = (i+1, j_1, j_2 - 1, \dots, j_k, \dots, j_m), \\ \dots & \\ j_k\nu_k, & when\ \beta' = (i+1, j_1, \dots, j_k - 1, \dots, j_m), \\ \dots & \\ j_m\nu_m, & when\ \beta' = (i+1, j_1, \dots, j_k, \dots, j_m - 1), \\ -(\lambda_{j_1,\dots,j_m} + i\mu + \sum_{k=1}^{m} j_k\nu_k), & when\ \beta' = (i, j_1, \dots, j_k, \dots, j_m), \\ 0, & otherwise \end{cases}$$

2. For $i = c$, $j_k \in Z_+$, $k = 1, 2, \dots, m$

$$q_{\beta\beta'} = \begin{cases} \lambda_{j_1,\dots,j_m}, & when\ \beta' = (c, j_1 + 1, \dots, j_k, \dots, j_m), \\ c\mu, & when\ \beta' = (c-1, j_1, \dots, j_k, \dots, j_m), \\ j_1\nu_1, & when\ \beta' = (c, j_1 - 1, j_2 + 1, \dots, j_m), \\ j_2\nu_2, & when\ \beta' = (c, j_1, j_2 - 1, j_3 + 1, \dots, j_m), \\ \dots & \\ j_k\nu_k, & when\ \beta' = (c, j_1, \dots, j_k - 1, j_{k+1} + 1, \dots, j_m), \\ \dots & \\ j_{m-1}\nu_{m-1}, & when\ \beta' = (c, j_1, \dots, j_k, \dots, j_{m-1} - 1, j_m + 1), \\ j_m\nu_m, & when\ \beta' = (c, j_1, \dots, j_k, \dots, j_m - 1), \\ -(\lambda_{j_1,\dots,j_m} + c\mu + \sum_{k=1}^{m} j_k\nu_k), & when\ \beta' = (c, j_1, \dots, j_k, \dots, j_m), \\ 0, & otherwise. \end{cases}$$

The component $Q_0^m(t)$ indicates the number of busy servers at the instant $t \geq 0$ and $Q_k^m(t)$, $k = 1, \dots, m$ – is equal to the number of retrial calls that have made k unsuccessful attempts to get a service. In the queue the number of retrials is limited by m.

Markov chain $Q^m(t)$ models the service process in the following state-dependent queue. Service facility of the system consists of c identical servers. If there is at least one free server on the call arrival, it immediately gets service and then leaves the system. Service time is a exponentially distributed random variable with parameter μ. If all the servers are busy, then the call creates a source of repeated call and tries to get service in a random period of time. Each call is allowed to make m repeated attempts. If there are no free server at the time of the last repeated attempt, the call abandons the system and does not get service. The rate of $k-th$ repeated attempt is ν_k, $k = 1, 2, \dots, m$. The rate of input flow is $\lambda_{j_1,\dots,j_k,\dots,j_m}$, $j_k \in Z_+$, $k = 1, 2, \dots, m$ depends on the number of sources of repeated calls is the system. The similar process with $\lambda_{j_1,\dots,j_k,\dots,j_m} = \lambda = const$ was analyzed in [7,8,13] using numerical methods.

Ergodicity conditions for the process $Q^m(t)$ are presented by the following statement.

Lemma 3. *If* $\overline{\lim\limits_{j_m\to\infty}} j_m^{-1}\lambda_{j_1\dots j_k\dots j_m} < \nu_m$ *and* $\lambda_{j_1\dots,j_k\dots j_m}$, μ, $\nu_k > 0$, $j_k \in Z_+$, $k = 1, 2, \dots, m$, *then* $Q^m(t)$ *is ergodic and its ergodic distribution* $\pi_{ij_1\dots j_k\dots j_m}$, $(i, j_1, \dots, j_m) \in S(Q^m)$ *is the same as the single stationary one.*

This result is analog to the result of Lemma 1 and also follows from the Tweedie theorem [1], p. 97.

The service process for $m \geq 2$ becomes complicated and explicit formulas of the steady state probabilities have not been found so far. However, for $m = 1$ we give representation of $\pi_{ij_1\dots j_k\dots j_m}$, $(i, j_1, \dots, j_m) \in S(Q^m)$ in terms of system parameters in an explicit form for the case $c = 1, 2$.

The service process of the retrial queue with one repeated attempt is a two-dimensional continuous time Markov chain $Q^1(t) = (Q_0^1(t), Q_1^1(t)) \in S(Q^1)$, where $Q_0^1(t)$ is the number of busy servers at the instant $t \geq 0$ and $Q_1^1(t)$, $k = 1, \dots, m$ is equal to the number of retrial call with a single retrial attempt to get a service, $S(Q^1) = \{0, \dots, c\} \times \{0, 1, \dots\}$. The input flow rate in this case is denoted by $\lambda_j > 0$. It means that λ_j depends on the number of sources for single repeated attempt.

Our next goal is to obtain results similar to Corollary 1 for stationary probabilities π_{ij}, $(i, j) \in S(Q^1)$ of the process $Q^1(t)$.

Theorem 2. *Let for the one-channel state-dependent queue with one retrial* ($m = 1$) *the conditions of Lemma 3 hold. Then for the service process* $Q^1(t) = (Q_0^1(t), Q_1^1(t))$ *there exists a stationary regime and stationary probabilities take the form:*

$$\pi_{0j} = \frac{1}{j!\nu_1^j}\prod_{i=1}^{j}\frac{\lambda_{i-1}(\lambda_{i-1} + (i-1)\nu_1)}{\lambda_i + i\nu_1 + \mu}\pi_{00}, j = 1, 2, \dots,$$

$$\pi_{1j} = \frac{\lambda_0}{\mu}\frac{1}{j!\nu_1^j}\prod_{i=1}^{j}\frac{\lambda_{i-1}(\lambda_i + i\nu_1)}{\lambda_i + i\nu_1 + \mu}\pi_{00}, j = 0, 1, \dots,$$

where

$$\pi_{00}^{-1} = \sum_{j=0}^{\infty}\left\{\frac{1}{j!\nu_1^j}\prod_{i=1}^{j}\frac{\lambda_{i-1}(\lambda_{i-1} + (i-1)\nu_1)}{\lambda_i + i\nu_1 + \mu} + \frac{\lambda_0}{\mu}\frac{1}{j!\nu_1^j}\prod_{i=1}^{j}\frac{\lambda_{i-1}(\lambda_i + i\nu_1)}{\lambda_i + i\nu_1 + \mu}\right\}.$$

Proof. Let us write the Kolmogorov set of equations

$$(\lambda_j + j\nu_1)\pi_{0j} = \mu\pi_{1j}, j = 0, 1, \dots, \tag{5}$$

$$(\lambda_j + \mu + j\nu_1)\pi_{1j} = \lambda_j\pi_{0j} + \lambda_{j-1}\pi_{1j-1} + (j+1)\nu_1\pi_{0j+1} + (j+1)\nu_1\pi_{1j+1}, j = 0, 1, \dots. \tag{6}$$

Using an equality of probability flows in stationary regime through a separation boundary of the phase set $S(Q^1) = S_j(Q^1) \cup \overline{S_j(Q^1)}$, where $S_j(Q^1) = \{(n,m) \in S(Q^1) : m \le j\}$, we get

$$\lambda_j \pi_{cj} = (j+1)\nu_1(\pi_{0j+1} + \pi_{1j+1}), j = 0, 1, \dots.$$

Taking into account the last equation, we can write Eq. (6) in the following form:

$$(\mu + j\nu_1)\pi_{1j} = \lambda_j \pi_{0j} + \lambda_{j-1}\pi_{1j-1}, j = 0, 1, \dots.$$

The expression from (5)

$$\pi_{1j} = \frac{1}{\mu}(\lambda_j + j\nu_1)\pi_{0j}, j = 0, 1, \dots.$$

is substituted into last equation and leads to the recurrence relation for π_{0j}:

$$j\nu_1(\lambda_j + j\nu_1 + \mu)\pi_{0j} = \lambda_{j-1}(\lambda_{j-1} + (j-1)\nu_1)\pi_{0j-1}, j = 0, 1, \dots.$$

Then we can find probability π_{0j}, $j = 0, 1, \dots$:

$$\pi_{0j} = \frac{1}{j!\nu_1^j}\prod_{i=1}^{j}\frac{\lambda_{i-1}(\lambda_{i-1} + (i-1)\nu_1)}{\lambda_i + i\nu_1 + \mu}\pi_{00}, j = 0, 1, \dots$$

and can write π_{1j}:

$$\pi_{1j} = \frac{1}{\mu}(\lambda_j + j\nu_1)\pi_{0j} = \frac{\lambda_0}{\mu}\frac{1}{j!\nu_1^j}\prod_{i=1}^{j}\frac{\lambda_{i-1}(\lambda_i + i\nu_1)}{\lambda_i + i\nu_1 + \mu}\pi_{00}, j = 0, 1, \dots.$$

The stationary probability π_{00} we obtain from the normalization condition $\sum_{j=0}^{\infty}(\pi_{0j} + \pi_{1j}) = 1$.
Theorem is proved.

In order to formulate an analog of Corollary 2, we again use approximation with truncated system technique. This system has a fixed number of N places in the orbit. This means that the repeated call leaves the system if all servers are busy and there are N calls in the orbit already.

We can define the service process for truncated model as two-dimensional Markov chain $Q^1(t,N) = (Q_0^1(t,N), Q_1^1(t,N))$ with continuous time in the phase space $S(Q^1, N) = \{0, \dots, c\} \times \{0, 1, \dots, N\}$.

The infinitesimal transitions rates of $Q^1(t,N)$ are the same as the infinitesimal transitions rates of the chain $Q^1(t)$, except the case $i = c, j = N$ where:

$$q_{(c,N)(i',j')} = \begin{cases} c\mu, & \textit{when} \quad (i',j') = (c-1,N), \\ N\nu, & \textit{when} \quad (i',j') = (c,N-1), \\ -(c\mu + N\nu), & \textit{when} \quad (i',j') = (c,N), \\ 0, & \textit{otherwise}; \end{cases}$$

We introduce the following notations. Let $x_j = \sum_{n=1}^{\infty} \frac{\alpha_{j+n}|}{|\beta_{j+n}}$ be the continued fraction where for $j = 1, 2, \ldots,$

$$\alpha_j = -\frac{\lambda_{j-1}((\lambda_{j-1}+(j-1)\nu_1)^2+(j-1)\nu_1\mu)}{j(j+1)\nu_1^2\mu},$$

$$\beta_j = -\frac{j\nu_1((\lambda_j+\mu+j\nu_1)^2+\mu(\lambda_{j-1}+\mu+j\nu_1))}{j(j+1)\nu_1^2\mu}.$$

Theorem 3. *Let for the two-channel state-dependent queue with one retrial the conditions of Lemma 3 hold. Then for the service process there exists a stationary regime and stationary probabilities take the form:*

$$\pi_{0j} = \left(\prod_{k=0}^{j-1} x_k\right) \pi_{00},$$

$$\pi_{1j} = \frac{\lambda_j + j\nu_1}{\mu} \left(\prod_{k=0}^{j-1} x_k\right) \pi_{00},$$

$$\pi_{2j} = \frac{1}{2\mu^2}\left((\lambda_j + j\nu_1)^2 + j\nu_1\mu - (j+1)\nu_1\mu x_j\right)\left(\prod_{k=0}^{j-1} x_k\right)\pi_{00}, j = 0, 1, \ldots,$$

where $\pi_{00}^{-1} = \frac{1}{2\mu^2}\sum_{j=0}^{\infty}\left((\lambda_j + \mu + j\nu_1)^2 + \mu(\mu + j\nu_1 - (j+1)\nu_1 x_j)\right)\prod_{k=0}^{j-1} x_k.$

Proof. Let us denote the stationary probabilities of the truncated state-dependent queue as $\pi_{ij}(N)$, $(i, j) \in S(Q^1, N)$.

We build Kolmogorov set of equations for $\pi_{ij}(N)$, $(i, j) \in S(Q^1, N)$:

$$(\lambda_j + j\nu_1)\pi_{0j}(N) = \mu\pi_{1j}(N), j = 0, \ldots, N, \tag{7}$$

$$(\lambda_j+\mu+j\nu_1)\pi_{1j}(N) = \lambda_j\pi_{0j}(N)+2\mu\pi_{2j}(N)+(j+1)\nu_1\pi_{0j+1}(N), j = 0, \ldots, N-1, \tag{8}$$

$$\begin{aligned}(\lambda_j + 2\mu + j\nu_1)\pi_{2j}(N) = \\ \lambda_j\pi_{1j}(N) + \lambda_{j-1}\pi_{2j-1}(N) + (j+1)\nu_1\pi_{1j+1}(N) + (j+1)\nu_1\pi_{2j+1}(N),\end{aligned} \tag{9}$$

$$j = 0, \ldots, N-1,$$

$$\lambda_j\pi_{cj}(N) = (j+1)\nu_1\left(\pi_{0j+1}(N) + \pi_{1j+1}(N) + \pi_{2j+1}(N)\right), j = 0, \ldots, N-1, \tag{10}$$

$$(\lambda_N + \mu + N\nu_1)\pi_{1N}(N) = \lambda_N\pi_{0N}(N) + 2\mu\pi_{2N}(N), \tag{11}$$

$$(2\mu + N\nu_1)\pi_{2N}(N) = \lambda_N\pi_{1N}(N) + \lambda_{N-1}\pi_{2N-1}(N). \tag{12}$$

Taking (7), we write expression $\pi_{1j}(N) = \frac{\lambda_j + j\nu_1}{\mu}\pi_{0j}(N), j = 0, 1, \ldots, N$ that is substituted into (8) and we have

$$2\mu^2\pi_{2j}(N) = ((\lambda_j + j\nu_1)^2 + j\nu_1\mu)\pi_{0j}(N) - (j+1)\nu_1\mu\pi_{0j+1}(N), j = 0, \ldots, N-1. \tag{13}$$

Using similar substitution, we can write Eq. (11) in following form:

$$2\mu^2\pi_{2N}(N) = ((\lambda_N + N\nu_1)^2 + N\nu_1\mu)\pi_{0N}(N). \tag{14}$$

Using formulas (12)–(14), we get

$$\lambda_{N-1}((\lambda_{N-1} + (N-1)\nu_1)^2 + (N-1)\nu_1\mu)\pi_{0N-1}(N) = \\ N\nu_1((\lambda_N + \mu + N\nu_1)^2 + \mu(\lambda_{N-1} + \mu + N\nu_1))\pi_{0N}(N). \tag{15}$$

We can write the last result in new notation

$$\alpha_N\pi_{0N-1}(N) = \beta_N\pi_{0N}(N), \tag{16}$$

where

$$\alpha_N = \lambda_{N-1}((\lambda_{N-1} + (N-1)\nu_1)^2 + (N-1)\nu_1\mu),$$

$$\beta_N = N\nu_1((\lambda_N + \mu + N\nu_1)^2 + \mu(\lambda_{N-1} + \mu + N\nu_1)).$$

Using Eq, (10) and expression for $\pi_{1j}(N)$, we can write (9) as

$$(2\mu + j\nu)\pi_{2j}(N) = \frac{\lambda_j(\lambda_j + j\nu_1)}{\mu}\pi_{0j}(N) + \lambda_{j-1}\pi_{2j-1}(N) - (j+1)\nu_1\pi_{0j+1}(N),$$

$$j = 0, \ldots, N-1.$$

Then we use (13) for the last equation. Thus, according to notation for α_j and β_j we get

$$\alpha_j\pi_{0j-1}(N) = \beta_j\pi_{0j}(N) + \pi_{0j+1}(N), j = 0, \ldots, N-1. \tag{17}$$

Let us divide the left- and right-hand sides by $\pi_{0N-1}(N)$ in (16) and by $\pi_{0j}(N)$ in (17). Passing to the new variable $x_j(N) = \sum_{n=1}^{N} \frac{\alpha_{j+n}|}{|\beta_{j+n}} = \frac{\pi_{0j+1}(N)}{\pi_{0j}(N)}$ we get:

$$\frac{\alpha_j}{x_{j-1}(N)} = \beta_j + x_j(N), j = 1, 2, \ldots, N-1,$$

$$x_{N-1}(N) = \frac{\alpha_N}{\beta_N}.$$

These formulas allow getting the stationary probabilities $\pi_{0j}(N)$, $j = 0, 1, \ldots, N-1$ in terms of continued fractions $x_j(N)$, $j = 0, 1, \ldots, N-1$:

$$\pi_{0j}(N) = \left(\prod_{k=j}^{N-1} x_k(N)\right)^{-1} \pi_{0N}(N), j = 0, 1, ..., N-1.$$

Using the last expression for $j = 0$, we get $\pi_{0N}(N)$ via $\pi_{00}(N)$:

$$\pi_{0N}(N) = \left(\prod_{k=j}^{N-1} x_k(N)\right) \pi_{00}(N).$$

Then we obtain formulas for $\pi_{0j}(N)$, $j = 1, 2, \ldots, N$, $\pi_{1j}(N)$, $j = 0, 1, \ldots, N$, that only depend on $\pi_{00}(N)$:

$$\pi_{0j}(N) = \left(\prod_{k=0}^{j-1} x_k(N)\right) \pi_{00}(N), j = 1, \ldots, N$$

$$\pi_{1j}(N) = \frac{\lambda_j + j\nu_1}{\mu} \left(\prod_{k=0}^{j-1} x_k(N)\right) \pi_{00}(N), j = 0, 1, \ldots, N.$$

Formulas (13), (14) allow finding the stationary probabilities $\pi_{2j}(N)$, $j = 0, 1, \ldots, N$:

$$\pi_{2j}(N) = \frac{1}{2\mu^2}((\lambda_j + j\nu_1)^2 + j\nu_1\mu - (j+1)\nu_1\mu x_j(N)) \left(\prod_{k=0}^{j-1} x_k(N)\right) \pi_{00}(N),$$

$$j = 0, 1, \ldots, N,$$

$$\pi_{2N}(N) = \frac{(\lambda_N + N\nu_1)^2 + N\nu_1\mu}{2\mu^2} \left(\prod_{k=0}^{N-1} x_k(N)\right) \pi_{00}(N).$$

According to the normalization condition $\sum_{j=0}^{N} (\pi_{0j}(N) + \pi_{1j}(N) + \pi_{2j}(N)) = 1$ we get

$$(\pi_{00}(N))^{-1} = \frac{1}{2\mu^2} \left(\sum_{j=0}^{N} \left((\lambda_j + \mu + j\nu_1)^2 + \mu(\mu + j\nu_1)\right) \left(\prod_{k=0}^{j-1} x_k(N)\right) \right. \\ \left. - \sum_{j=0}^{N-1} (j+1)\nu_1\mu x_j(N) \left(\prod_{k=0}^{j-1} x_k(N)\right)\right).$$

Service processes $Q^1(t)$ and $Q^1(t, N)$ are migration processes. Therefore, on the basis of the results from [1], Sect. 2, Theorems 2.3 and 2.4 as $N \to \infty$ the stationary probabilities $\pi_{ij}(N)$ of the truncated model approximate the stationary probabilities π_{ij} of the input model. Thus, we pass to the limit in formulas for $\pi_{ij}(N)$ as $N \to \infty$ we get the expressions for stationary probabilities that we need to prove.

We note that in contrast to the classical models now the explicit formulas contain the continued fractions as components. For $\pi_{ij}(N)$, $(i,j) \in S(Q^1, N)$ Kolmogorov equations can be solved in an explicit vector-matrix form (see [6]). The obtained results can be used to solve optimization problems in the class of threshold strategies (see, for example, [2–5, 10]).

4 Conclusion

In this paper, we have presented the research of multiserver state-dependent retrial queues of two types: classical systems and systems with a limited number of repeated attempts. The rate of input calls flow in the systems under investigation depends on the current number of repeated calls. For both types of systems, we have found ergodicity conditions and obtained explicit representations of steady-state probabilities. These results allow to carry out further analysis of the service process and compute different integral characteristics. They also give an efficient algorithm for solving optimization problems.

References

1. Falin, G.I., Templeton, J.G.C.: Retrial Queues. Chapman and Hall, London (1997)
2. Dudin, A.N., Chakravarthy, S.R.: Multi-threshold control of the BMAP|SM|1|K queue with group services. J. Appl. Math. Stoch. Anal. **16**(4), 327–347 (2003)
3. Kim, C.S., Klimenok, V.I., Birukov, A., Dudin, A.N.: Optimal multi-threshold control by the BMAP|SM|1 retrial system. Ann. Oper. Res. **141**(1), 193–210 (2006). https://doi.org/10.1007/s10479-006-5299-3
4. Lebedev, E.A., Ponomarov, V.D.: Retrial queues with variable service rate. Cybern. Syst. Anal. **47**(3), 434–441 (2011). https://doi.org/10.1007/s10559-011-9325-3
5. Atencia, I., Lebedev, E., Ponomarov, V., Livinska, H.: Special retrial queues with state-dependent input rate. In: Dudin, A., Nazarov, A., Moiseev, A. (eds.) ITMM 2019. CCIS, vol. 1109, pp. 73–85. Springer, Cham (2019). https://doi.org/10.1007/978-3-030-33388-1_7
6. Pryshchepa, O.V., Lebedev, E.O.: On a multi-channel retrial queueing system. Cybern. Syst. Anal. **53**(3), 441–449 (2017). https://doi.org/10.1007/s10559-017-9945-3
7. Shin, Y.W., Moon, D.H.: Approximations of retrial queue with limited number of retrials. Comput. Oper. Res. **37**(7), 1262–1270 (2010)
8. Shin, Y.W., Moon, D.H.: Retrial queues with limited number of retrials: numerical investigations. In: The Seventh International Symposium on Operations Research and Its Applications (ISORA 2008), Lijiang, China, pp. 237–247 (2008)
9. Artalejo, J.R., Gomez-Corral, A.: Retrial Queueing Systems: A Computational Approach. Springer, Berlin (2008). https://doi.org/10.1007/978-3-540-78725-9
10. Klimenok, V., Dudina, O.: Retrial tandem queue with controllable strategy of repeated attempts. Qual. Technol. Quant. Manag. **14**(1), 74–93 (2017)
11. Roger, A.H., Charles, R.J.: Matrix Analysis. Cambridge University Press, New York (1985)
12. Walrand, J.: An Introduction to Queueing Networks. Prentice Hall, Englewood Cliffs (1988)
13. Fiems, D., Phung-Duc, T.: Light-traffic analysis of random access systems without collisions. Ann. Oper. Res. **277**(2), 311–327 (2019). https://doi.org/10.1007/s10479-017-2636-7

Reducing of Service Process Dimension for a General-Type Multichannel Network in Heavy Traffic

Eugene A. Lebedev and Hanna Livinska(✉)

Applied Statistics Department, National Taras Shevchenko University of Kyiv, Volodymyrska Street, 64, Kyiv 01601, Ukraine
leb@unicyb.kiev.ua

Abstract. In the work a multichannel queueing network with a general input flow is considered [3]. There are no restrictions on the structure of the input flow. Heavy traffic conditions on the network parameters are introduced. A functional limit theorem for the service process of the network is proved provided that the conditions are satisfied. Approximative Gaussian process is constructed. An additional splittability condition for the switching matrix of the network yields an opportunity to merge network nodes and to reduce dimension of the limit process at the Gaussian approximation scheme. Convergence is proved in the uniform topology, which enables solving optimization problems for correspondent functionals.

Keywords: Multichannel queueing network · Heavy traffic regime · Gaussian approximation · Asymptotic merging

1 Introduction

In the analysis of nowadays real-life networks such as data transmission networks, computer networks, queueing models of stochastic systems and networks are efficiently used [1–4]. Their structure is determined by the probabilistic characteristics of the input information flows, data processing algorithms. Typically, the information processing in stochastic networks are high-dimensional vectors with interconnected components and complex system of stochastic relations defining the process. Thus, one of significant problems in simulation and studying of stochastic networks is connected with the large dimension of descriptive processes for the networks and complexity of the phase space of a stochastic model. To overcome such problems, an approach of asymptotic merging of nodes set is proposed for investigation of a general-type multichannel queueing network.

The merging approach of phase space for stochastic systems was pioneered by works of V.S. Korolyuk [5,6] and V.V. Anisimov [7,9], who developed different methodologies in this direction. Nowadays, many works are dedicated to the solution of the phase merging problems (see, for example, [8, 10]). According to this methodology, phase space can be merged and models under consideration can be simplified.

A. Dudin et al. (Eds.): ITMM 2020, CCIS 1391, pp. 108–119, 2021.
https://doi.org/10.1007/978-3-030-72247-0_8

In contrast to these works, we use the approach of asymptotic merging in relation to the set of nodes for multichannel queueing networks operating under heavy traffic conditions [11]. For such networks we consider a multidimensional service process that is a stochastic process indicating the number of customers being processed at the nodes of the network. Considering the network processing under assumptions of heavy traffic regime yields the opportunity to approximate the service process by a Gaussian process (see, for example, [12–14]). Since for multichannel stochastic networks the service process can be a vector-process with high dimension, the merging approach allows us to reduce dimension of considered stochastic processes, and therefore it simplifies notably network analysis. An additional condition for the switching matrix of the network provides an effect of consolidation for some network nodes to a similar-type node, and the set of the network nodes can be merged. As a result, for the service process in the merged network, we have a form of limit processes, simpler than for the original one.

Note that we prove the convergence of the corresponding service processes in the uniform topology. This can be used to the calculating of the quality functionals of network operation and to the solving optimization problems [15].

The rest of the paper is organized as follows. In Sect. 2, the main mathematical model is described in details. In Sect. 3 conditions on network parameters are formulated. We need them for proving our main result (a functional limit theorem). Section 4 presents the main result on Gaussian approximation for the merged network. For such a network operating in heavy traffic regime, an approximating Gaussian process is given, and its characteristics are specified via the network parameters. Approach with merging of the nodes set reduces the dimension of the limit Gaussian process. It should be noted that the limit process in general case is a non-Markov process. In Sect. 5 some auxiliary results are proved. Section 6 provides the proof of the main result. In Sect. 7 a particular case is described. In this case the limit process in a diffusion process. Section 8 concludes the work.

2 Model Description

In the work we consider a multichannel network of a $[G|GI|\infty]^r$ type. Such a network consists of r service nodes. At instants $\tau_k^{(i)}$, $k = 1, 2, \ldots$, the i-th node receives calls from the outside for service at the network, we denote the number of calls arrived in the period of time $[0,t]$ by $\nu_i(t)$. There are no restrictions on the structure of the input flow $\nu(t) = (\nu_1(t), \ldots, \nu_r(t))'$.

Each of the r nodes is a multi-channel queueing system. If a call arrives in such a system, its service begins immediately. For the i-th node distribution function of service time will be denoted by $G_i(t)$, and its Laplace-Stieltjes transformation by

$$G_i(s) = \int_0^\infty e^{-st} dG_i(t), \;\; i = 1, \ldots, r .$$

The direction of movement of a call within the network is controlled by the switching matrix $P = \|p_{ij}\|_1^r$. For any $i = 1, 2, \ldots, r$, if a call has its service finished in the i-th node, $p_{ir+1} = 1 - \sum_{j=1}^{r} p_{ij}$ represents a probability of the call's exit from the network.

The service process in the $[G|GI|\infty]^r$-network is defined as an r- dimensional process $Q(t) = (Q_1(t),...,Q_r(t))'$, where $Q_i(t)$, $i = 1,2,...,r$, is the number of calls in the i-th node at an instant t, $t \geq 0$.

Let a call be in the node numbered m at the initial instant $t = 0$, and let its service do not begin before $t = 0$. We connect the service process $Q(t)$ with a semi-Markov process $x^{(m)}(t) \in \{1,...,r,r+1\}$ which describes the service path of the call within the $[G|GI|\infty]^r$-network. Based on the algorithm of call service at the $[G|GI|\infty]^r$-network, we conclude that $x^{(m)}(t)$ has a semi-Markov matrix $\left\|G_{ij}(t)\right\|_1^{r+1}$ of the form

$$G_{ij}(t) = \begin{cases} p_{ij}G_i(t), & i = 1,2,...,r, \quad j = 1,2,...,r,r+1, \\ \delta_{r+1\,j}G_{r+1}(t), & i = r+1, \qquad j = 1,2,...,r,r+1, \end{cases}$$

$$G_{r+1}(t) = \begin{cases} 0, & t < 1, \\ 1, & t \geq 1, \end{cases}$$

where δ_{ij} is the Kronecker delta. At the initial instant $t = 0$ we have $x^{(m)}(0) = m$ and distribution function of the sojourn time in the initial state coincides with $G_m(t)$.

The state $r+1$ for the process $x^{(m)}(t)$ is absorbing. Absorption in $r+1$ is interpreted as an exit from the network. We denote the transient probabilities of the semi-Markov process $x^{(m)}(t)$ by

$$p_i^m(t) = \mathbf{P}\left\{x^{(m)}(t) = i\right\},$$

$$p_{ij}^m(s,t) = \mathbf{P}\left\{x^{(m)}(s) = i,\ x^{(m)}(t) = j\right\}, \quad s < t,$$

$$P(t) = \left\|p_j^i(t)\right\|_1^r, \quad P^{(m)}(s,t) = \left\|p_{ij}^m(s,t)\right\|_{i,j=1}^r$$

3 Underlying Conditions

Now let us introduce four conditions that we need for proving our limit theorems. The first condition is connected with the starting load of the network, the second and the third are the heavy traffic conditions, and the fourth is a condition for the switching matrix to be splittable.

So, firstly, during this work it is assumed that at the initial time $t = 0$ the network is empty:

1) $Q_i(0) = 0, \quad i = 1,...,r$.
   Since the main purpose of this work is to study the service process $Q(t)$ in the heavy traffic regime, we specify now correspondent conditions as conditions on the network parameters. So, the heavy traffic regime means that the network parameters depend on the series number n ($Q(t) = Q^{(n)}(t)$, $\nu(t) = \nu^{(n)}(t)$, $\tau_k^{(i)} = \tau_k^{(i,n)}$, $P = P^{(n)}$, $P(t) = P^{(n)}(t)$, $P^{(m)}(s,t) = P^{(m,n)}(s,t)$, $G_i(t) = G_i^{(n)}(t)$, $i = 1,...,r$) so that the following two conditions for the input flow and service time at the network nodes are met.

2) There exist constants $\lambda_i \geq 0,\ i = 1,...,r,\ \lambda_1 + ... + \lambda_r \neq 0$, such that

$$n^{-1/2}\left(\nu_1^{(n)}(nt) - \lambda_1 nt, ..., \nu_r^{(n)}(nt) - \lambda_r nt\right) \underset{n\to\infty}{\overset{U}{\Rightarrow}} W(t) = (W_1(t), ..., W_r(t))',$$

where $W(t)$ is an r-dimensional Brownian motion process with a zero vector of mean values $EW(1) = (0,...,0)'$ and with a correlation matrix $EW(1)W'(1) = \sigma^2 = \left\|\sigma_{ij}^2\right\|_1^r$, the symbol $\overset{U}{\Rightarrow}$ means weak convergence in the uniform topology.

3) For some numerical sequence $\{g_n\}_1^n$ such that $g_n \to \infty$ when $n \to \infty$, there exist limits for Laplace-Stieltjes transformations $G_i(s) = G_i^{(n)}(s)$:

$$\lim_{n\to\infty} g_n(1 - G_i^{(n)}(s/n)) = \mu_i(s),\ \ i = 1,...,r.$$

Note, that if in each node of the $[G|GI|\infty]^r$-network there exists a mean service time of calls and distribution function of service time does not depend on the series number n, then condition 3) is fulfilled for $g_n = n,\ n \geq 1$.

And, at least, in order to the set $I = \{1,...,r\}$ of $[G|GI|\infty]^r$-network nodes be able to be asymptotically merged, we require the following condition:

4) The set of serving nodes I can be divided into classes $I_1,...,I_{r_0}$ $(I_i \cap I_j = \varnothing, i \neq j, i,j = 1,...,r_0)$ in such a way that
   i) $P^{(n)} = P_0 + g_n^{-1}B_0 + o(g_n^{-1})$,
   $p_{ir+1}^{(n)} = g_n^{-1}b_{ir+1} + o(g_n^{-1})$;
   ii) $a_{\alpha\alpha} = -\sum\limits_{\beta=1,\beta\neq\alpha}^{r_0+1} a_{\alpha\beta} \neq 0$;
   iii) The spectral radius of the matrix $\hat{P} = \left\|\hat{p}_{\alpha\beta}\right\|_1^{r_0} = \left\|(1 - \delta_{\alpha\beta}a_{\alpha\beta}/(-a_{\alpha\alpha})\right\|_1^{r_0}$ is strictly less than 1,
   where
   $P_0 = \left\|\delta_{\alpha\beta}P(\alpha)\right\|_1^{r_0},\ P(\alpha) = \left\|p_{ij}(\alpha)\right\|_{i,j\in I_\alpha}$, is an indecomposable stochastic matrix with stationary distribution $\rho_i(\alpha),\ i \in I_\alpha$;
   $B_0 = \left\|b_{ij}\right\|_1^r = \|B(\alpha,\beta)\|_1^{r_0},\ B(\alpha,\beta) = \left\|b_{ij}\right\|_{i\in I_\alpha, j\in I_\beta}$ are rectangular matrices of size $|I_\alpha| \times |I_\beta|,\ \alpha,\beta = 1,...,r_0$;
   $a_{\alpha\beta} = \rho'(\alpha)\cdot B_0 \cdot 1(\beta),\ \alpha,\beta = 1,...,r_0,\ \alpha \neq \beta,\ a_{\alpha r_0+1} = \rho'(\alpha)\cdot b_{r+1},\ \alpha = 1,...,r_0,\ b'_{r+1} = (b_{1r+1},...,b_{rr+1})$;
   $\rho(\alpha)$ is an r- dimensional vector with its i-th entry equal to $\rho_i(\alpha)$ if $i \in I_\alpha$, and equal to zero otherwise, $\alpha = 1,...,r_0$;
   $1(\beta)$ is an r- dimensional vector with its i-th entry equal to 1 if $i \in I_\beta$, and equal to zero otherwise, $\beta = 1,...,r_0$.

4 Main Result

Let $\hat{x}_0^{(\alpha)}(t) \in \{1,...,r_0,r_0+1\}$ be a semi-Markov process, for which the state r_0+1 is an absorbing one, $\hat{x}_0^{(\alpha)}(0) = \alpha,\ \alpha = 1,...,r_0$, the matrix of transient probabilities between states $\{1,...,r_0\}$ for the embedded chain is equal to $\hat{P}$, and the sojourn time in the state α has the following Laplace-Stiltjes transformation:

$$\hat{G}_\alpha(s) = \frac{(-a_{\alpha\alpha})}{(-a_{\alpha\alpha}) + \hat{\mu}_\alpha(s)}, \quad \hat{\mu}_\alpha(s) = \sum_{i \in I_\alpha} \rho_i(\alpha)\mu_i(s), \quad \alpha = 1,2,...,r_0.$$

The corresponding probability distributions of $\hat{x}_0^{(\alpha)}(t)$ are denoted by

$$\hat{p}^\alpha_\beta(t) = \mathbf{P}\left\{\hat{x}_0^{(\alpha)}(t) = \beta\right\}, \quad \hat{P}(t) = \left\|\hat{p}^\alpha_\beta(t)\right\|^{r_0}_{\alpha,\beta=1},$$

$$\hat{p}^\alpha_{\beta\gamma}(s,t) = \mathbf{P}\left\{\hat{x}_0^{(\alpha)}(s) = \beta, \hat{x}_0^{(\alpha)}(t) = \gamma\right\}, \quad \hat{P}^{(\alpha)}(s,t) = \left\|\hat{p}^\alpha_{\beta\gamma}(s,t)\right\|^{r_0}_{\beta,\gamma=1}, \quad s < t.$$

In order to describe a limit of the sequence of stochastic processes

$$\hat{\xi}^{(n)'}(t) = n^{-1/2}\left(Q^{(n)'}(nt) - n\lambda'\left(\int_0^t P^{(n)}(nu)du\right)\right)\hat{E}, \quad n > 2,$$

where $\lambda' = (\lambda_1,...,\lambda_r)$, $\hat{E} = \|1(1)...1(r_0)\|$ is a rectangular matrix of size $r \times r_0$, it is necessary to introduce two independent Gaussian processes $\hat{\xi}^{(1)}(t)$ and $\hat{\xi}^{(2)}(t)$ which have zero mean values and correlation matrices

$$\hat{R}^{(1)}(t) = \int_0^t \hat{P}'(u)\hat{\sigma}^2\hat{P}(u)du,$$

$$\hat{R}^{(1)}(s,t) = \int_0^s \hat{P}'(u)\hat{\sigma}^2\hat{P}(u+t-s)du, \quad s < t,$$

$$\hat{R}^{(2)}(t) = \int_0^t \left[\Delta\left(\hat{\lambda}'\hat{P}(u)\right) - \hat{P}'(u)\Delta(\hat{\lambda})\hat{P}(u)\right]du,$$

$$\hat{R}^{(2)}(s,t) = \sum_{\alpha=1}^{r_0} \hat{\lambda}_\alpha \int_0^s \left[\Delta(\hat{p}_\alpha(u)) - \hat{p}_\alpha(u)\hat{p}'_\alpha(u)\right]\hat{E}^{(\alpha)}(u,u+t-s)du, \quad s < t,$$

where

$\Delta(x) = \left\|\delta_{\alpha\beta}x_\alpha\right\|_1^{r_0}$ for any vector $x' = (x_1,...,x_{r_0})$,
$\hat{p}_\alpha(t)$ is the line of the matrix $\hat{P}(t)$ with the number α,
$\hat{\sigma}^2 = \left\|\hat{\sigma}_{\alpha\beta}\right\|_1^{r_0} = \hat{E}'\sigma^2\hat{E}$,
$\hat{\lambda}' = (\hat{\lambda}_1,...,\hat{\lambda}_{r_0}) = \lambda'\hat{E}$,
$\hat{E}^{(\alpha)}(s,t) = \left\|\hat{E}^{(\alpha)}_{\beta\gamma}(s,t)\right\|$,
$\hat{E}^{(\alpha)}_{\beta\gamma}(s,t) = \hat{p}^{(\alpha)}_{\beta\gamma}(s,t)/\hat{p}^{(\alpha)}_\beta(s)$ if $\hat{p}^{(\alpha)}_\beta(s) \neq 0$ and $\hat{E}^{(\alpha)}_{\beta\gamma}(s,t) = 0$ otherwise.

For the sequence $\hat{\xi}^{(n)}(t)$, $n > 2$, such a result is valid.

Theorem 1. *Let the queueing network of $[G|GI|\infty]^r$ type satisfy conditions 1)–4). Then, on any finite interval $[0,T]$, the sequence of stochastic processes $\hat{\xi}^{(n)}(t)$, $n > 2$, weakly converges to $\hat{\xi}^{(1)}(t) + \hat{\xi}^{(2)}(t)$ in the uniform topology.*

Note that the limit process is a sum of two independent Gaussian processes. Herewith, the first term is conditioned by fluctuations of input flow while the second is conditioned by fluctuations of service times.

Before proving the theorem, we provide some auxiliary statements.

5 Auxiliary Results

According to the partition of nodes $\{1,...,r\} = \bigcup_{\beta=1}^{r_0} I_\beta$ ($I_\alpha \bigcap I_\beta = \oslash$ when $\alpha \neq \beta$), we build enlarged process

$$\hat{x}(t) = \begin{cases} \beta, & \text{if } x^{(m)}(t) \in I_\beta, \\ r_0+1, & \text{if } x^{(m)}(t) = r+1. \end{cases}$$

If $x^{(m)}(0) = m \in I_\alpha$, then $\hat{x}(0) = \alpha$, and we denote $\hat{x}(t) = \hat{x}^{(\alpha)}(t)$.

In the case when the characteristics of the $[G|GI|\infty]^r$-network depend on the series number n and conditions 3), 4) are satisfied, then it follows from the results of Chap. 8 in [9]

$$\hat{x}_n^{(\alpha)}(nt) \overset{d}{\Rightarrow} \hat{x}_0^{(\alpha)}(t),$$

where the semi-Markov process $\hat{x}_0^{(\alpha)}(t)$ is defined above by the matrix of transitions of the embedded chain $\hat{P}$, and Laplace-Stieltjes transformations $\hat{G}_\alpha(s)$ of the sojourn time in the state α. $\overset{d}{\Rightarrow}$ denotes convergence of finite-dimensional distributions.

When we prove Theorem 1, we will need limits of the integrals of the transient functions of the process $x_n^{(m)}(t)$. In compliance with methods of the asymptotic theory of perturbed linear operators (see [16], Chap. 3), these limits can be written via the correspondent integrals of transient functions for the enlarged process $\hat{x}_0^{(\alpha)}(t)$.

Lemma 1. *If conditions 3) and 4) are satisfied, then for arbitrary t, $\Delta > 0$, we have*

$$\lim_{n\to\infty} \int_0^t p_{I_\beta}^{m(n)}(nu)du = \int_0^t \hat{p}_\beta^\alpha(u)du, \tag{1}$$

$$\lim_{n\to\infty} \int_0^t p_{I_\beta I_\gamma}^{m(n)}(nu, nu+n\Delta)du = \int_0^t \hat{p}_{\beta\gamma}^\alpha(u, u+\Delta)du, \tag{2}$$

where

$$m \in I_\alpha,\ p_{I_\beta}^{m(n)}(nu) = \sum_{i\in I_\beta} p_i^{m(n)}(nu),$$

$$p_{I_\beta I_\gamma}^{m(n)}(nu, nu+n\Delta) = \sum_{i\in I_\beta,\, j\in I_\gamma} p_{ij}^{m(n)}(nu, nu+n\Delta).$$

The limit for the service process has two independent terms $\hat{\xi}^{(1)}(t)$ and $\hat{\xi}^{(2)}(t)$. In order to distinguish the term $\hat{\xi}^{(1)}(t)$ that is connected with the fluctuations of the input flow, we need the following result.

Lemma 2. *For an r-dimensional Brownian motion* $\hat{W}'(t) = \left(\hat{W}_1(t), ..., \hat{W}_{r_0}(t)\right) = W'(t)\hat{E}$

$$\int_0^t d\hat{W}'(u)\hat{P}(t-u) \stackrel{d}{=} \hat{\xi}^{(1)'}(t),$$

where $\stackrel{d}{=}$ *means the equality of finite-dimensional distributions of stochastic processes.*

Proof of this fact follows from the properties of stochastic integrals presented, for instance, in [17], Sect. 15.

6 Proof of the Theorem

Let us check that

$$\hat{\xi}^{(n)}(t) \underset{n\to\infty}{\stackrel{d}{\Rightarrow}} \hat{\xi}^{(1)}(t) + \hat{\xi}^{(2)}(t) \tag{3}$$

For the characteristic function

$$\hat{\varphi}^{(n)}(s) = \mathbf{E}\exp\left\{i\hat{\xi}^{(n)'}(t)s\right\}, \quad s' = (s_1, ..., s_{r_0}) \in \mathbb{R}^{r_0},$$

of the one-dimensional distribution $\hat{\xi}^{(n)}(t)$ the following takes place

$$\begin{aligned}\lim_{n\to\infty} \hat{\varphi}^{(n)}(s) &= \lim_{n\to\infty} \exp\left\{-in^{1/2}\hat{q}^{(n)'}(t)s\right\} \\ &\times \mathbf{E}\exp\left\{\sum_{m=1}^{r}\sum_{k=1}^{\nu_m^{(n)}(nt)}\left[in^{-1/2}p_m^{(n)'}(nt-\tau_k^{(m,n)})\tilde{s} - \frac{1}{2}n^{-1}p_m^{(n)'}(nt-\tau_k^{(m,n)})\tilde{s}^2\right.\right. \\ &\left.\left. + \frac{1}{2}n^{-1}\tilde{s}'p_m^{(n)}(nt-\tau_k^{(m,n)})p_m^{(n)'}(nt-\tau_k^{(m,n)})\tilde{s}\right]\right\} \end{aligned} \tag{4}$$

where

$$\hat{q}^{(n)'}(t) = (\hat{q}_1^{(n)}(t), ..., \hat{q}_{r_0}^{(n)}(t)) = \lambda'\left(\int_0^t P^{(n)}(nu)du\right)\hat{E},$$

$\tilde{s} = (\tilde{s}_1, \tilde{s}_2, ..., \tilde{s}_r)'$ is an r-dimensional vector with values $\tilde{s} = \sum_{\alpha=1}^{r_0} s_\alpha 1(\alpha) = \hat{E}\cdot s$, $\tilde{s}^2 = (\tilde{s}_1^2, \tilde{s}_2^2, ..., \tilde{s}_r^2)'$, $p_m^{(n)}(t)$ is the mth row of the matrix $P^{(n)}(t)$.

Let

$$W_i^{(n)}(t) = n^{-1/2}(\nu_i^{(n)}(nt) - \lambda_i nt), \quad i = 1, 2, ..., r,$$

$$W^{(n)'}(t) = \left(W_1^{(n)}(t), ..., W_r^{(n)}(t)\right),$$

$\tilde{P}^{(n)}(t) = \left\|p_{I_\alpha}^{i(n)}(t)\right\|_{i\in I,\, \alpha\in\hat{I}}$ is a rectangular matrix of size $r \times r_0$, where $I = \{1, 2, ..., r\}$, $\hat{I} = \{1, 2, ..., r_0\}$, $\tilde{p}_m^{(n)'}(t) = \left(p_{I_1}^{m(n)}(t), ..., p_{I_{r_0}}^{m(n)}(t)\right)$ is the mth row of the matrix $\tilde{P}^{(n)}(t)$.

Then, taking into account (1), we can see that (4) yields

$$\lim_{n\to\infty}\varphi^{(n)}(s)=\lim_{n\to\infty}\exp\left\{-in^{1/2}\hat{q}^{(n)'}(t)s\right\}$$
$$\times\mathbf{E}\exp\left\{i\left(\int_0^t dW^{(n)'}(u)\tilde{P}^{(n)}(nt-nu)\right)s+in^{1/2}\hat{q}^{(n)'}(t)s\right.$$
$$-\frac{1}{2}\lambda'\int_0^t\tilde{P}^{(n)}(nu)du\,s^2-\frac{1}{2}n^{-1/2}\left(\int_0^t dW^{(n)'}(u)\tilde{P}^{(n)}(nt-nu)\right)s^2$$

$$+\frac{1}{2}s'\left[\sum_{m=1}^{r}\lambda_m\int_0^t\tilde{p}_m^{(n)}(nu)\tilde{p}_m^{(n)'}(nu)\,du\right]s$$
$$\left.+\frac{1}{2}n^{-1/2}s'\left[\sum_{m=1}^{r}\int_0^t dW_m^{(n)}(u)\tilde{p}_m^{(n)}(nt-nu)\tilde{p}_m^{(n)'}(nt-nu)\,du\right]s\right\}$$
$$=\exp\left\{-\frac{1}{2}s'\left[\Delta\left(\hat{\lambda}'\int_0^t\hat{P}(u)du\right)-\int_0^t\hat{P}'(u)\Delta(\hat{\lambda})\hat{P}(u)du\right]s\right\}$$
$$\times\mathbf{E}\exp\left\{i\int_0^t d\hat{W}'(u)\hat{P}(t-u)s\right\},$$

where $s^2=(s_1^2,\ldots,s_{r_0}^2)'$, $\hat{W}'(u)=W'(u)\hat{E}$.

The last expression is a characteristic function of $\hat{\xi}^{(1)}(t)+\hat{\xi}^{(2)}(t)$. Thus, the convergence of one-dimensional distributions is proved.

Similarly, consider the characteristic function of two-dimensional distributions

$$\varphi^{(n)}(s(1),s(2))=\mathbf{E}\exp\left\{i\hat{\xi}^{(n)'}(t_1)s(1)+i\hat{\xi}^{(n)'}(t_2)s(1)\right\},\quad t_1<t_2,$$

$$s'(1)=\left(s_1(1),\ldots,s_{r_0}(1)\right),\quad s'(2)=\left(s_1(2),\ldots,s_{r_0}(2)\right)\in\mathbb{R}^{r_0}.$$

We obtain

$$\lim_{n\to\infty}\varphi^{(n)}(s(1),s(2))=$$
$$\lim_{n\to\infty}\exp\left\{-in^{1/2}\lambda'\int_0^{t_1}P^{(n)}(nu)du\tilde{s}(1)-in^{1/2}\lambda'\int_0^{t_2}P^{(n)}(nu)du\tilde{s}(2)\right\}$$
$$\times\mathbf{E}\exp\left\{\sum_{m=1}^{r}\sum_{k=1}^{\nu_m^{(n)}(nt_1)}\left[\frac{i}{\sqrt{n}}\left(p_m^{(n)'}(nt_1-\tau_k^{(m,n)})\tilde{s}(1)+p_m^{(n)'}(nt_2-\tau_k^{(m,n)})\tilde{s}(2)\right)\right.\right.$$
$$-\frac{1}{n}\left(\frac{1}{2}p_m^{(n)'}(nt_1-\tau_k^{(m,n)})\tilde{s}^2(1)+\frac{1}{2}p_m^{(n)'}(nt_2-\tau_k^{(m,n)})\tilde{s}^2(2)\right.$$
$$\left.+\tilde{s}'(1)P^{(m,n)}(nt_1-\tau_k^{(m,n)},nt_2-\tau_k^{(m,n)})\tilde{s}(2)\right)$$
$$+\frac{1}{2}\frac{1}{n}\left(\tilde{s}'(1)p_m^{(n)}(nt_1-\tau_k^{(m,n)})p_m^{(n)'}(nt_1-\tau_k^{(m,n)})\tilde{s}(1)\right.$$
$$\left.+\tilde{s}'(2)p_m^{(n)}(nt_2-\tau_k^{(m,n)})p_m^{(n)'}(nt_2-\tau_k^{(m,n)})\tilde{s}(2)\right)$$

$$+\frac{1}{n}\tilde{s}'(1)p_m^{(n)}(nt_1-\tau_k^{(m,n)})p_m^{(n)'}(nt_2-\tau_k^{(m,n)})\tilde{s}(2)\Big]$$

$$+\sum_{m=1}^{r}\sum_{k=\nu_m^{(n)}(nt_1)+1}^{\nu_m^{(n)}(nt_2)}\Big[\frac{i}{\sqrt{n}}\Big(p_m^{(n)'}(nt_2-\tau_k^{(m,n)})\tilde{s}(2)-\frac{1}{2}\frac{1}{n}p_m^{(n)'}(nt_2-\tau_k^{(m,n)})\tilde{s}^2(2)$$

$$+\frac{1}{2}\frac{1}{n}\tilde{s}'(2)p_m^{(n)}(nt_2-\tau_k^{(m,n)})p_m^{(n)'}(nt_2-\tau_k^{(m,n)})\tilde{s}(2)\Big]\Big\},$$

where $\tilde{s}(i)=\hat{E}s(i),\ i=1,2$.

Replacing the sums by the integrals along the path of the process and using relations (1), (2), we find

$$\lim_{n\to\infty}\varphi^{(n)}(s(1),s(2))=$$

$$\lim_{n\to\infty}\mathbf{E}\exp\Big\{i\Big(\int_0^{t_1}dW^{(n)'}(u)\tilde{P}^{(n)}(nt_1-nu)\Big)s(1)+i\Big(\int_0^{t_2}dW^{(n)'}(u)\tilde{P}^{(n)}(nt_2-nu)\Big)s(2)$$
$$-\frac{1}{2}\Big(\lambda'\int_0^{t_1}\tilde{P}^{(n)}(nu)du\Big)s^2(1)-\frac{1}{2}\Big(\lambda'\int_0^{t_2}\tilde{P}^{(n)}(nt_2-nu)du\Big)s^2(2)$$
$$-s'(1)\Big(\sum_{m=1}^{r}\lambda_m\int_0^{t_1}\hat{E}'P^{(m,n)}(nu,nu+n\Delta t_2)\hat{E}du\Big)s(2)$$
$$+\frac{1}{2}s'(1)\Big(\sum_{m=1}^{r}\lambda_m\int_0^{t_1}\tilde{p}_m^{(n)}(nu)\tilde{p}_m^{(n)'}(nu)du\Big)s(1)$$
$$+\frac{1}{2}s'(2)\Big(\sum_{m=1}^{r}\lambda_m\int_0^{t_2}\tilde{p}_m^{(n)}(nt_2-nu)\tilde{p}_m^{(n)'}(nt_2-nu)du\Big)s(2)$$
$$+s'(1)\Big(\sum_{m=1}^{r}\lambda_m\int_0^{t_2}\tilde{p}_m^{(n)}(nu)\tilde{p}_m^{(n)'}(nu+n\Delta t_2)du\Big)s(2)$$
$$+i\Big(\int_{t_1}^{t_2}dW^{(n)'}(u)\tilde{P}^{(n)}(nt_2-nu)\Big)s(2)-\frac{1}{2}\Big(\lambda'\int_{t_1}^{t_2}\tilde{P}^{(n)}(nt_2-nu)du\Big)s^2(2)$$
$$+\frac{1}{2}s'(2)\Big(\sum_{m=1}^{r}\lambda_m\int_{t_1}^{t_2}\tilde{p}_m^{(n)}(nt_2-nu)\tilde{p}_m^{(n)'}(nt_2-nu)du\Big)s(2)\Big\}$$
$$=\exp\Big\{-\frac{1}{2}s'(1)\Delta\Big(\hat{\lambda}'\int_0^{t_1}\hat{P}(u)du\Big)s(1)-\frac{1}{2}s'(2)\Delta\Big(\hat{\lambda}'\int_0^{t_1}\hat{P}(t_2-u)du\Big)s(2)$$
$$-s'(1)\Big(\sum_{\alpha=1}^{r_0}\hat{\lambda}_\alpha\int_0^{t_1}\hat{P}^{(\alpha)}(u,u+\Delta t_2)du\Big)s(2)$$
$$+\frac{1}{2}s'(1)\Big(\int_0^{t_1}\hat{P}'(u)\Delta(\hat{\lambda})\hat{P}(u)du\Big)s(1)$$
$$+\frac{1}{2}s'(2)\Big(\int_0^{t_2}\hat{P}'(t_2-u)\Delta(\hat{\lambda})\hat{P}(t_2-u)du\Big)s(2)$$
$$+s'(1)\Big(\int_0^{t_1}\hat{P}'(u)\Delta(\hat{\lambda})\hat{P}(u+\Delta t_2)du\Big)s(2)$$
$$-\frac{1}{2}s'(2)\Delta\Big(\hat{\lambda}'\int_{t_1}^{t_2}\hat{P}(t_2-u)du\Big)s(2)$$

$$+\frac{1}{2}s'(2)\left(\int_{t_1}^{t_2}\hat{P}'(t_2-u)\Delta(\hat{\lambda})\hat{P}'(t_2-u)du\right)s(2)\Big\}$$

$$\times \mathbf{E}\exp\left\{i\left(\int_0^{t_1}d\hat{W}'(u)\hat{P}(t_1-u)\right)s(1)+i\left(\int_0^{t_2}d\hat{W}'(u)\hat{P}(t_2-u)\right)s(2)\right\},$$

$s^2(i)=(s_1^2(i),\ldots,s_{r_0}^2(i))'$, $i=1,2$.

After performing the obvious algebraic transformations in the last expression, we come to the following equality

$$\lim_{n\to\infty}\varphi^{(n)}(s(1),s(2))$$

$$=\exp\Big\{-\frac{1}{2}s'(1)\int_0^{t_1}\left[\Delta(\hat{\lambda}'\hat{P}(u))-\hat{P}'(u)\Delta(\hat{\lambda})\hat{P}(u)\right]du\,s(1)$$

$$-\frac{1}{2}s'(2)\int_0^{t_2}\left[\Delta(\hat{\lambda}'\hat{P}(u))-\hat{P}'(u)\Delta(\hat{\lambda})\hat{P}(u)\right]du\,s(2)$$

$$-s'(1)\left[\sum_{\alpha=1}^{r_0}\hat{\lambda}_\alpha\int_0^{t_1}\left[\Delta(\hat{p}_\alpha(u))-\hat{p}_\alpha(u)\hat{p}'_\alpha(u)\right]\hat{E}^{(\alpha)}(u,u+t_2-t_1)du\right]s(2)\Big\}$$

$$\times \mathbf{E}\exp\left\{i\left(\int_0^{t_1}d\hat{W}'(u)\hat{P}(t_1-u)\right)s(1)+i\left(\int_0^{t_2}d\hat{W}'(u)\hat{P}(t_2-u)\right)s(2)\right\}. \quad (5)$$

Equation (5) means the convergence of two-dimensional distributions. Convergence of N- dimensional distributions for $N>2$ is checked similarly.

Reinforcement (3) to the convergence in the uniform topology can be done in the same way as in [18]. The theorem is proved.

7 Approximative Process as a Diffusion

Consider one important particular case of Theorem 1 when the limit process $\hat{\xi}(t)$ is Gaussian diffusion process (see [11]).

Let the distribution of calls' service time in network nodes do not depend on the series number n and there exist means of service times. So, we introduce the following condition.

3′) $\int_0^\infty t dG_i(t)=1/\mu_i<\infty,\ i=1,\ldots,r.$

Then for such a $[G|GI|\infty]^r$-network condition 3) is satisfied for the sequence $g_n=n$, $n=1,2,\ldots$.

By 4′) we denote condition 4) with the replacement of the sequence g_n by n.

Let $\hat{\theta}'=(\hat{\theta}_1,\ldots,\hat{\theta}_{r_0})$ is a solution of the balance equation

$$\hat{\theta}_\alpha=\hat{\lambda}_\alpha+\sum_{\alpha=1}^{r_0}\hat{\theta}_\beta\hat{p}_{\beta\alpha},\quad \alpha=1,\ldots,r_0,$$

for the merged network, whose nodes are classes I_α, $\alpha=1,\ldots,r_0$;

$$\hat{\mu}(-\alpha)=(\hat{\mu}_1(-a_{11}),\ldots,\hat{\mu}'_{r_0}(-a_{r_0r_0}))',$$

$$(\hat{\theta}/\hat{\mu}(-a))=\left(\hat{\theta}_1/\hat{\mu}_1(-a_{11}),\ldots,\hat{\theta}_{r_0}/\hat{\mu}_{r_0}(-a_{r_0r_0})\right)',$$

$$\hat{\mu}_\alpha=\{\rho'(\alpha)(1/\mu)\}^{-1},\quad \alpha=1,\ldots,r_0,\quad (1/\mu)=(1/\mu_1,\ldots,1/\mu_r).$$

As a consequence of Theorem 1, we have the following result.

Theorem 2. *Let a queueing network of $[G|GI|\infty]^r$ type satisfy conditions 1), 2), 3$'$), 4$'$). Then the sequence of stochastic processes $\hat{\xi}^{(n)}(t)$, $n > 2$, weakly converges on any finite interval $[0,T]$ in the uniform topology to an r_0-dimensional diffusion process $\hat{\xi}(t)$ ($\hat{\xi}(0) = (0,...,0)'$) with the transfer vector $\hat{A}(x) = \hat{A}'x$ and the diffusion matrix*

$$\hat{B}(t) = \Delta[\hat{q}'(t)\hat{A}] - \hat{A}'\Delta[\hat{q}(t)] - \Delta[\hat{q}(t)]\hat{A} + \hat{\sigma}^2,$$

where $I = \left\|\delta_{\alpha\beta}\right\|_1^{r_0}$, $\Delta(x) = \left\|x_\alpha \delta_{\alpha\beta}\right\|_1^{r_0}$ for any r_0-dimensional vector $x = (x_1,...,x_{r_0})'$, $\hat{A} = \Delta[\hat{\mu}(-a)](\hat{P} - I)$, $\hat{q}'(t) = (\hat{\theta}/\hat{\mu}(-a))'(I - \hat{P}(t))$.

Representation of the limit process as a diffusion is attractive in the sense that the diffusion process is determined only by its local characteristics, and for the analysis of its functionals it is possible to apply the developed apparatus of Markov diffusion processes. The appearance of a Markov property of the limit Gaussian process $\hat{\xi}(t)$ is due to the fact that if conditions 3$'$) and 4$'$) of Theorem 2 are satisfied, then the semi-Markov process $\hat{x}_0^{(\alpha)}(t)$, $t \geq 0$, is a Markov chain with continuous time.

In the general case, under conditions of Theorem 1, $\hat{\xi}(t)$ can be a non-Markov Gaussian process with continuous sample functions.

8 Conclusions

In this paper, a multichannel stochastic network is considered. There are no restriction on a general input flow of calls, service times are generally distributed. It is assumed that network operates in the heavy traffic regime, defined by special conditions on the network parameters as a series number $n \to \infty$. For such a multichannel network operating in the heavy traffic, we propose an asymptotic method based on approximation of normalized jump-wise service process of calls by a multidimensional continuous path Gaussian process with its characteristics written via the network parameters. We prove convergence of the process in the uniform topology, which can be used for calculating quality functionals of network and for solving optimization problems.

An additional splittability condition for the switching matrix of the network yields an opportunity to merge network nodes and to reduce dimension of the limit process at the Gaussian approximation scheme. As a result, for the service process in the merged network, we have a form of limit process, simpler than for the original one.

We consider also a particular case, when the limit process is a diffusion process. Note, that in general case the limit can be a non-Markov process.

The main feature of the limit process is that it is decomposed into a sum of two independent Gaussian processes. The first term of the limit process is associated with fluctuations of input flow while the second is associated with fluctuations of service times.

References

1. Gnedenko, B.V., Kovalenko, I.N.: Introduction to Queueing Theory. Springer, Birkhauser, Boston Inc. (1989)
2. Théorie des files d'attente 1, 2. In: Anisimov, V. Limnios, N. (eds.) Tendances avancées. ISTE Editions Ltd. (2021)
3. Massey, W.A., Witt, W.: A stochastic model to capture space and time dynamics in wireless communication systems. Probab. Eng. Inf. Sci. **8**, 541–569 (1994)
4. Moiseev, A., Nazarov, A.: Tandem of infinite-server queues with Markovian arrival process. In: Vishnevsky, V., Kozyrev, D. (eds.) DCCN 2015. CCIS, vol. 601, pp. 323–333. Springer, Cham (2016). https://doi.org/10.1007/978-3-319-30843-2_34
5. Gusak, D.V., Korolyuk, V.S.: Asymptotic behavior of semi-Markov processes with a splittable state set. Probab. Theory Math. Stat. **5**, 43–50 (1971). (in Russian)
6. Korolyuk, V., Turbin, A.: Mathematical Foundation of the State Lumping of Large Systems. Springer, Dordrecht (1993)
7. Anisimov, V.V.: Asymptotic enlargement of the states of random processes. Cybernetics **9**, 494–504 (1973). https://doi.org/10.1007/BF01069207
8. Korolyuk, V.S., Limnios, N.: Stochastic Systems in Merging Phase Space. World Scientific, Singapore (2005)
9. Anisimov, V.V.: Switching Processes in Queueing Models. ISTE Ltd. (2008)
10. Samoilenko, I.V.: Large deviations for random evolutions with independent increments in the scheme of Lévy approximation with split and double merging. Random Oper. Stochast. Eqn. **22**(2), 137–149 (2015)
11. Lebedev, E.O., Livinska, G.V.: On the asymptotic merging of the set of nodes in stochastic networks. Theor. Probab. Math. Stat. **101**, 147–156 (2019)
12. Lebedev, E., Livinska, G.: Gaussian approximation of multi-channel networks in heavy traffic. In: Dudin, A., Klimenok, V., Tsarenkov, G., Dudin, S. (eds.) BWWQT 2013. CCIS, vol. 356, pp. 122–130. Springer, Heidelberg (2013). https://doi.org/10.1007/978-3-642-35980-4_14
13. Lebedev, E., Chechelnitsky, A., Livinska, G.: Multi-channel network with interdependent input flows in heavy traffic. Theor. Probab. Math. Stat. **97**, 109–119 (2017)
14. Livinska, H., Lebedev, E.: On transient and stationary regimes for multi-channel networks with periodic inputs. Appl. Stat. Comput. **319**, 13–23 (2018)
15. Lebedev, E., Makushenko, I.: Profit maximization and risk minimization in semi-Markovian networks. Cybern. Syst. Anal. **43**(2), 213–224 (2007). https://doi.org/10.1007/s10559-007-0040-z
16. Korolyuk, V.S., Korolyuk, V.V.: Stochastic Models of Systems. Kluwer Acad. Press, Dordrecht (1999)
17. Scorokhod, A.V.: Lectures on the Theory of Stochastic Processes. Lybid, Kyiv (1990). (in Ukrainian)
18. Lebedev, E.O.: A limit theorem for stochastic networks and its application. Theor. Probab. Math. Stat. **68**, 81–92 (2003)

Central Limit Theorem for an M/M/1/1 Retrial Queue with Unreliable Server and Two-Way Communication

Anatoly Nazarov[1], Tuan Phung-Duc[2,3], Svetlana Paul[1](✉), Olga Lizyura[1], and Kseniya Shulgina[1]

[1] Institute of Applied Mathematics and Computer Science, National Research Tomsk State University, 36 Lenina Avenue, Tomsk 634050, Russia

[2] Faculty of Engineering Information and Systems, University of Tsukuba, 1-1-1 Tennodai, Tsukuba, Ibaraki 305-8573, Japan

tuan@sk.tsukuba.ac.jp

[3] VNU Vietnam Japan University, Hanoi, Vietnam

Abstract. In this paper, we consider a single-server retrial queue with unreliable server and two-way communication. Inbound calls arrive according to a Poisson process. If the server is busy upon arrival, an incoming call joins the orbit and retries to occupy the server after some exponentially distributed time. Service durations of incoming calls follow the exponential distribution. In the idle time, the server makes outgoing calls. There are multiple types of outgoing calls in the system. We assume that durations of each type of outgoing calls follow a distinct exponential distribution. The server is subject to breakdowns with rates depending on the state of the server. If the breakdown occurs, the server undergoes a repair whose duration follows an exponential distribution. The aim of our research is to show that the scaled number of calls in the orbit follows a normal distribution under the condition that the retrial rate is low.

Keywords: Retrial queue · Two-way communication · Incoming calls · Outgoing calls · Asymptotic analysis · Central limit theorem

1 Introduction

Retrial queues with two-way communication are suitable models for blended call centers, where the operator can provide both inbound and outbound calls. Retrial queues for modeling call centers are described in [1]. The research area of retrial queues is a branch of queueing theory and literature on this topic is rich and vast [4,5]. In retrial queues, instead of waiting when the server is busy, customers join a virtual queue (orbit) and retry to access the server after a random amount of time. On the other hand, the server may make outbound

A. Dudin et al. (Eds.): ITMM 2020, CCIS 1391, pp. 120–130, 2021.
https://doi.org/10.1007/978-3-030-72247-0_9

calls in idle time. This idea of call blending is to improve the productivity of classical call centers by reducing the idle time of an operator [2,3,8,9].

Models with combination of retrials and outbound calls are studied in [6,7, 12,13]. The unreliability of the server is also of interest as a common phenomenon in communication and service systems [10]. In [13], scaling limits (including the regime with slow retrial) are presented for M/G/1/1 retrial models with two-way communication. Sakurai and Phung-Duc [12] present Markovian single server retrial models with multiple types of outgoing calls. In this paper, we extend the model of Sakurai and Phung-Duc [12] by adding the unreliable feature of the server [11]. Furthermore, instead of deriving exact expressions as in Sakurai and Phung-Duc [12], our aim is to obtain compact asymptototic expressions for the number of customers in the orbit in the stationary regime.

The rest of our paper is orgainized as follows. In Sect. 2, we present the model in details and in Sect. 3 we show the problem definition and premilinary analysis. Section 4 presents the main results of the paper where we show asymptotic formulas for the distribution of the number of customers in the orbit. In Sect. 5 we present some numerical examples. Section 6 concludes our paper and presents some remarks.

2 Mathematical Model

We consider a single-server retrial queue with two-way communication. We have two classes of calls in the system: incoming calls and outgoing calls. Primary incoming calls form a Poisson process with rate λ and upon arrival idle server starts the service immediately. The durations of service times for incoming calls are exponentially distributed with parameter μ_1. If the server is busy upon arrival, the incoming call joins the orbit and retries to occupy the server after a random delay, whose duration follows the exponential distribution with parameter σ.

In its idle time, the server can make outgoing calls. There are multiple types of outgoing calls in the system. The server makes an outgoing call of type n with rate α_n and serves it for an exponentially distributed time with parameter μ_n. We number the types of outgoing calls from 2 to N.

The unreliability of the server is defined by three parameters: γ_0 is the rate of breakdowns when the server is idle, γ_1 is the rate of breakdowns when an incoming call is in service and γ_2 is the rate of restorations. If at the instant of breakdown, the server provides an incoming call, then the call joins the orbit. We assume that there are no breakdowns during service of outgoing calls, since the server itself initiates the call.

3 Problem Definition

Let $k(t)$ denote the state of the server at time t as follows: 0, if the server is idle; 1, if the server busy with an incoming call; n, if the server is busy with an outgoing call of type $n, n = 2, ..., N; N+1$, if the server is broken. Let $i(t)$ denote the number of calls in the orbit at instant t.

The two-dimensional process $\{k(t), i(t)\}$ is Markovian. Thus, for the probability distribution of this process

$$P\{k(t) = k, i(t) = i\} = P_k(i,t),$$

we derive Kolmogorov system of differential equations

$$\frac{\partial P_0(i,t)}{\partial t} = -\left(\lambda + i\sigma + \sum_{n=2}^{N} \alpha_n + \gamma_0\right) P_0(i,t) + \sum_{k=1}^{N} \mu_k P_k(i,t) + \gamma_2 P_{N+1}(i,t),$$
$$\frac{\partial P_1(i,t)}{\partial t} = -(\lambda + \mu_1 + \gamma_1) P_1(i,t) + \lambda P_1(i-1,t)$$
$$+\lambda P_0(i,t) + (i+1)\sigma P_0(i+1,t),$$
$$\frac{\partial P_n(i,t)}{\partial t} = -(\lambda + \mu_n) P_n(i,t) + \lambda P_n(i-1,t) + \alpha_n P_0(i,t), \ n = \overline{2,N},$$
$$\frac{\partial P_{N+1}(i,t)}{\partial t} = -(\lambda + \gamma_2) P_{N+1}(i,t) + \lambda P_{N+1}(i-1,t)$$
$$+\gamma_0 P_0(i,t) + \gamma_1 P_1(i-1,t).$$

In stationary regime, we can write the system of equations for stationary probability distribution $P_n(i)$, $n = \overline{0, N+1}$ of process $\{k(t), i(t)\}$ as follows:

$$\begin{gathered}
-\left(\lambda + \sum_{n=2}^{N} \alpha_n + i\sigma + \gamma_0\right) P_0(i) + \sum_{k=1}^{N} \mu_k P_k(i) + \gamma_2 P_{N+1}(i) = 0, \\
-(\lambda + \mu_1 + \gamma_1) P_1(i) + \lambda P_0(i) + \sigma(i+1) P_0(i+1) + \lambda P_1(i-1) = 0, \\
-(\lambda + \mu_n) P_n(i) + \lambda P_n(i-1) + \alpha_n P_0(i) = 0, \ n = \overline{2,N}, \\
-(\lambda + \gamma_2) P_{N+1}(i) + \lambda P_{N+1}(i-1) + \gamma_0 P_0(i) + \gamma_1 P_1(i-1) = 0.
\end{gathered} \tag{1}$$

The problem is to obtain the probability distribution of the number of calls in the orbit. We introduce partial characteristic functions

$$H_k(u) = \sum_{i=0}^{\infty} e^{jui} P_k(i), \ k = \overline{0, N+1},$$

where $j = \sqrt{-1}$.

Rewriting system (1) for the partial characteristic functions, we obtain

$$\begin{gathered}
-\left(\lambda + \sum_{n=2}^{N} \alpha_n + \gamma_0\right) H_0(u) + j\sigma H_0'(u) + \sum_{k=1}^{N} \mu_k H_k(u) + \gamma_2 H_{N+1}(u) = 0, \\
(\lambda(e^{ju} - 1) - \mu_1 - \gamma_1) H_1(u) + \lambda H_0(u) - j\sigma e^{-ju} H_0'(u) = 0, \\
(\lambda(e^{ju} - 1) - \mu_n) H_n(u) + \alpha_n H_0(u) = 0, \ n = \overline{2,N}, \\
(\lambda(e^{ju} - 1) - \gamma_2) H_{N+1}(u) + \gamma_0 H_0(u) + \gamma_1 e^{ju} H_1(u) = 0.
\end{gathered} \tag{2}$$

Summing up equations of system (2), we write the additional equation

$$j\sigma e^{-ju} H_0'(u) + (\lambda + \gamma_1) H_1(u) + \lambda \sum_{n=2}^{N+1} H_n(u) = 0, \tag{3}$$

which we will use in further analysis. We will solve system (2) using asymptotic analysis method under low rate of retrials condition ($\sigma \to 0$).

4 Asymptotic Analysis Under Low Rate of Retrials Condition

We denote $\sigma = \varepsilon$ and make the following substitutions in system (2) and Eq. (3)

$$u = \varepsilon w, H_k(u) = F_k(w, \varepsilon), k = \overline{0, N+1}$$

in order to obtain the system

$$\begin{gathered}
-\left(\lambda + \gamma_0 + \sum_{n=2}^{N} \alpha_n\right) F_0(w,\varepsilon) + j\frac{\partial F_0(w,\varepsilon)}{\partial w} + \sum_{k=1}^{N} \mu_k F_k(w,\varepsilon) + \gamma_2 F_{N+1}(w,\varepsilon) = 0, \\
(\lambda(e^{j\varepsilon w} - 1) - \mu_1 - \gamma_1) F_1(w,\varepsilon) + \lambda F_0(w,\varepsilon) - je^{-j\varepsilon w}\frac{\partial F_0(w,\varepsilon)}{\partial w} = 0, \\
(\lambda(e^{j\varepsilon w} - 1) - \mu_n) F_n(w,\varepsilon) + \alpha_n F_0(w,\varepsilon) = 0, \ n = \overline{2, N}, \\
(\lambda(e^{j\varepsilon w} - 1) - \gamma_2) F_{N+1}(w,\varepsilon) + \gamma_0 F_0(w,\varepsilon) + \gamma_1 e^{j\varepsilon w} F_1(w,\varepsilon) = 0, \\
je^{-j\varepsilon w}\frac{\partial F_0(w,\varepsilon)}{\partial w} + (\lambda + \gamma_1) F_1(w,\varepsilon) + \lambda \sum_{n=2}^{N+1} F_n(w,\varepsilon) = 0.
\end{gathered} \tag{4}$$

Solving system (4) in the limit by $\varepsilon \to 0$, we prove Theorem 1.

Theorem 1. *Suppose that $i(t)$ is the number of calls in the orbit in the Markovian retrial queue with unreliable server and two-way communication, then the following limit equality holds*

$$\lim_{\sigma \to 0} \mathbb{E} e^{jw\sigma i(t)} = e^{jw\kappa_1},$$

where κ_1 is given by

$$\kappa_1 = \lambda \frac{\gamma_2(\lambda + \gamma_1) + \lambda\gamma_1 + \gamma_0(\mu_1 + \gamma_1) + \gamma_2(\mu_1 + \gamma_1) \sum_{n=2}^{N} \frac{\alpha_n}{\mu_n}}{\gamma_2\mu_1 - \lambda(\gamma_1 + \gamma_2)}. \tag{5}$$

Proof. In system (4), we take the limit by $\varepsilon \to 0$

$$\begin{gathered}
-\left(\lambda + \gamma_0 + \sum_{n=2}^{N} \alpha_n\right) F_0(w) + jF_0'(w) + \sum_{k=1}^{N} \mu_k F_k(w) + \gamma_2 F_{N+1}(w) = 0, \\
-(\mu_1 + \gamma_1) F_1(w) + \lambda F_0(w) - jF_0'(w) = 0, \\
-\mu_n F_n(w) + \alpha_n F_0(w) = 0, \ n = \overline{2, N}, \\
-\gamma_2 F_{N+1}(w) + \gamma_0 F_0(w) + \gamma_1 F_1(w) = 0, \\
jF_0'(w) + (\lambda + \gamma_1) F_1(w) + \lambda \sum_{n=2}^{N+1} F_n(w) = 0.
\end{gathered} \tag{6}$$

We seek the solution of system (6) in the following form

$$F_k(w) = r_k \Phi(w), \tag{7}$$

where r_k is the stationary probability distribution of the server state. Substituting (7) into system (4) and dividing the equations by $\Phi(w)$, we obtain

$$\begin{aligned}
-\left(\lambda+\gamma_0+\sum_{n=2}^{N}\alpha_n\right)r_0+jr_0\tfrac{\Phi'(w)}{\Phi(w)}+\sum_{k=1}^{N}\mu_k r_k+\gamma_2 r_{N+1}=0,\\
-(\mu_1+\gamma_1)r_1+\lambda r_0-jr_0\tfrac{\Phi'(w)}{\Phi(w)}=0,\\
-\mu_n r_n+\alpha_n r_0=0,\ n=\overline{2,N},\\
-\gamma_2 r_{N+1}+\gamma_0 r_0+\gamma_1 r_1=0,\\
jr_0\tfrac{\Phi'(w)}{\Phi(w)}+(\lambda+\gamma_1)r_1+\lambda\sum_{n=2}^{N+1}r_n=0.
\end{aligned}\tag{8}$$

Since the relation $\Phi'(w)/\Phi(w)$ does not depend on w, the function $\Phi(w)$ has the following form

$$\Phi(w)=e^{jw\kappa_1}.$$

Due to this, we rewrite system (8)

$$\begin{aligned}
-\left(\lambda+\gamma_0+\sum_{n=2}^{N}\alpha_n+\kappa_1\right)r_0+\sum_{k=1}^{N}\mu_k r_k+\gamma_2 r_{N+1}=0,\\
-(\mu_1+\gamma_1)r_1+(\lambda+\kappa_1)r_0=0,\\
-\mu_n r_n+\alpha_n r_0=0,\ n=\overline{2,N},\\
-\gamma_2 r_{N+1}+\gamma_0 r_0+\gamma_1 r_1=0,\\
-\kappa_1 r_0+(\lambda+\gamma_1)r_1+\lambda\sum_{n=2}^{N+1}r_n=0.
\end{aligned}\tag{9}$$

Taking into account the normalization condition for probability distribution r_k, we obtain the following expressions

$$r_0=\left[\frac{(\lambda+\kappa_1)(\gamma_1+\gamma_2)+(\gamma_0+\gamma_2)(\mu_1+\gamma_1)}{\gamma_2(\mu_1+\gamma_1)}+\sum_{n=2}^{N}\frac{\alpha_n}{\mu_n}\right]^{-1},$$

$$r_1=\frac{\lambda+\kappa_1}{\mu_1+\gamma_1}r_0,\ r_n=\frac{\alpha_n}{\mu_n}r_0,\ n=2,\dots,N,$$

$$r_{N+1}=\left(\frac{\gamma_0}{\gamma_2}+\frac{\gamma_1(\lambda+\kappa_1)}{\gamma_2(\mu_1+\gamma_1)}\right)r_0,$$

$$\kappa_1=\frac{(\lambda+\gamma_1)r_1+\lambda\sum\limits_{n=2}^{N+1}r_n}{r_0}.$$

Substituting the expressions for r_k into the expression for κ_1 yields the equation for κ_1, the solution of which is as follows

$$\kappa_1=\lambda\frac{\gamma_2(\lambda+\gamma_1)+\lambda\gamma_1+\gamma_0(\mu_1+\gamma_1)+\gamma_2(\mu_1+\gamma_1)\sum\limits_{n=2}^{N}\frac{\alpha_n}{\mu_n}}{\gamma_2\mu_1-\lambda(\gamma_1+\gamma_2)},$$

which coincides with (5). **Theorem is proved.**

Remark 1. The necessary stability condition is $\lambda/\mu_1 < \gamma_2/(\gamma_1+\gamma_2)$ because $\kappa_1 > 0$.

The value of κ_1 defines the asymptotic mean κ_1/σ of the number of calls in the orbit. On the next stage of analysis, we will obtain the characteristic function of process $i(t)$.

We introduce the following notations in system (2) and Eq. (3)

$$H_k(u) = \exp\left\{ju\frac{\kappa_1}{\sigma}\right\} H_k^{(2)}(u), k = \overline{0, N+1},$$

to derive the system of equations

$$\begin{aligned}
-\left(\lambda + \sum_{n=2}^{N}\alpha_n + \gamma_0 + \kappa_1\right) H_0^{(2)}(u) + j\sigma\frac{\partial H_0^{(2)}(u)}{\partial u} + \sum_{k=1}^{N}\mu_k H_k^{(2)}(u)&\\
+\gamma_2 H_{N+1}^{(2)}(u) = 0,&\\
(\lambda(e^{ju}-1) - \mu_1 - \gamma_1)H_1^{(2)}(u) + \lambda H_0^{(2)}(u) + \kappa_1 e^{-ju}H_0^{(2)}(u)&\\
-j\sigma e^{-ju}\frac{\partial H_0^{(2)}(u)}{\partial u} = 0,&\\
(\lambda(e^{ju}-1) - \mu_n)H_n^{(2)}(u) + \alpha_n H_0^{(2)}(u) = 0,\ n = \overline{2, N},&\\
(\lambda(e^{ju}-1) - \gamma_2)H_{N+1}^{(2)}(u) + \gamma_0 H_0^{(2)}(u) + \gamma_1 e^{ju}H_1^{(2)}(u) = 0,&\\
-\kappa_1 e^{-ju}H_0^{(2)}(u) + j\sigma e^{-ju}\frac{\partial H_0^{(2)}(u)}{\partial u} + (\lambda+\gamma_1)H_1^{(2)}(u) + \lambda\sum_{n=2}^{N+1}H_n^{(2)}(u) = 0.&
\end{aligned} \tag{10}$$

Then we make the substitutions

$$\sigma = \varepsilon^2, u = \varepsilon w, H_k^{(2)}(u) = F_k^{(2)}(w,\varepsilon), k = \overline{0, N+1},$$

and obtain the system of equations

$$\begin{aligned}
-\left(\lambda + \sum_{n=2}^{N}\alpha_n + \gamma_0 + \kappa_1\right) F_0^{(2)}(w,\varepsilon) + j\varepsilon\frac{\partial F_0^{(2)}(w,\varepsilon)}{\partial w} + \sum_{k=1}^{N}\mu_k F_k^{(2)}(w,\varepsilon)&\\
+\gamma_2 F_{N+1}^{(2)}(w,\varepsilon) = 0,&\\
(\lambda(e^{j\varepsilon w}-1) - \mu_1 - \gamma_1)F_1^{(2)}(w,\varepsilon) + \lambda F_0^{(2)}(w,\varepsilon) + \kappa_1 e^{-j\varepsilon w}F_0^{(2)}(w,\varepsilon)&\\
-j\varepsilon e^{-j\varepsilon w}\frac{\partial F_0^{(2)}(w,\varepsilon)}{\partial w} = 0,&\\
(\lambda(e^{j\varepsilon w}-1) - \mu_n)F_n^{(2)}(w,\varepsilon) + \alpha_n F_0^{(2)}(w,\varepsilon) = 0,\ n = \overline{2, N},&\\
(\lambda(e^{j\varepsilon w}-1) - \gamma_2)F_{N+1}^{(2)}(w,\varepsilon) + \gamma_0 F_0^{(2)}(w,\varepsilon) + \gamma_1 e^{j\varepsilon w}F_1^{(2)}(w,\varepsilon) = 0,&\\
-\kappa_1 e^{-j\varepsilon w}F_0^{(2)}(w,\varepsilon) + je^{-j\varepsilon w}\frac{\partial F_0^{(2)}(w,\varepsilon)}{\partial w} + (\lambda+\gamma_1)F_1^{(2)}(w,\varepsilon)&\\
+\lambda\sum_{n=2}^{N+1}F_n^{(2)}(w,\varepsilon) = 0.&
\end{aligned} \tag{11}$$

Considering system (6) in the limit by $\varepsilon \to 0$, we prove Theorem 2.

Theorem 2. *In the context of Theorem 1, the following limit equality is true*

$$\lim_{\sigma\to 0}\mathbb{E}\exp\left\{jw\sqrt{\sigma}\left(i(t) - \frac{\kappa_1}{\sigma}\right)\right\} = \exp\left\{\frac{jw^2}{2}\kappa_2\right\},$$

where

$$\kappa_2 = \kappa_1 + \frac{\lambda}{\gamma_2\mu_1 - \lambda(\gamma_1+\gamma_2)}$$
$$\times\left[(\gamma_2(\lambda+\gamma_1) + \gamma_1(\lambda+\mu_1+\gamma_1))\frac{\lambda+\kappa_1}{\mu_1+\gamma_1} + \lambda\gamma_2(\mu_1+\gamma_1)\sum_{n=2}^{N}\frac{\alpha_n}{\mu_n^2}\right. \tag{12}$$
$$\left. + \frac{\lambda}{\gamma_2}(\gamma_0(\mu_1+\gamma_1) + \gamma_1(\lambda+\kappa_1))\right].$$

Proof. In system (11), we substitute the following decomposition

$$F_k^{(2)}(w,\varepsilon) = \Phi_2(w)\{r_k + jw\varepsilon f_k\} + o(\varepsilon^2),\ k = \overline{0, N+1},$$

in order to obtain the system

$$\begin{gathered}
-\left(\lambda + \sum_{n=2}^{N}\alpha_n + \gamma_0 + \kappa_1\right)\Phi_2(w)\{r_0 + jw\varepsilon f_0\} + j\varepsilon r_0\Phi_2'(w) \\
+\sum_{k=1}^{N}\mu_k\Phi_2(w)\{r_k + jw\varepsilon f_k\} + \gamma_2\Phi_2(w)\{r_{N+1} + jw\varepsilon f_{N+1}\} = o(\varepsilon^2), \\
(\lambda(e^{j\varepsilon w}-1) - \mu_1 - \gamma_1)\Phi_2(w)\{r_1 + jw\varepsilon f_1\} \\
+(\lambda + \kappa_1 e^{-jw\varepsilon})\Phi_2(w)\{r_0 + jw\varepsilon f_0\} - j\varepsilon e^{-j\varepsilon w}r_0\Phi_2'(w) = 0, \\
(\lambda(e^{j\varepsilon w}-1) - \mu_n)\Phi_2(w)\{r_n + jw\varepsilon f_n\} + \alpha_n\Phi_2(w)\{r_0 + jw\varepsilon f_0\} \\
= o(\varepsilon^2),\ n = \overline{2, N}, \\
(\lambda(e^{j\varepsilon w}-1) - \gamma_2)\Phi_2(w)\{r_{N+1} + jw\varepsilon f_{N+1}\} + \gamma_0\Phi_2(w)\{r_0 + jw\varepsilon f_0\} \\
+\gamma_1 e^{j\varepsilon w}\Phi_2(w)\{r_1 + jw\varepsilon f_1\} = o(\varepsilon^2), \\
-\kappa_1 e^{-j\varepsilon w}\Phi_2(w)\{r_0 + jw\varepsilon f_0\} + je^{-j\varepsilon w}r_0\Phi_2'(w) \\
+(\lambda+\gamma_1)\Phi_2(w)\{r_1 + jw\varepsilon f_1\} + \lambda\sum_{n=2}^{N+1}\Phi_2(w)\{r_n + jw\varepsilon f_n\} = o(\varepsilon^2).
\end{gathered} \tag{13}$$

Taking (9) into account and dividing the system by $w\Phi_2(w)$, as $\varepsilon \to 0$ we obtain

$$\begin{gathered}
-\left(\lambda + \sum_{n=2}^{N}\alpha_n + \gamma_0 + \kappa_1\right)f_0 + r_0\frac{\Phi_2'(w)}{w\Phi_2(w)} + \sum_{k=1}^{N}\mu_k f_k + \gamma_2 f_{N+1} = 0, \\
-(\mu_1+\gamma_1)f_1 + (\lambda+\kappa_1)f_0 + r_0\frac{\Phi_2'(w)}{w\Phi_2(w)} + \kappa_1 r_0 - \lambda r_1 = 0, \\
\alpha_n f_0 - \mu_n f_0 + \lambda r_n = 0,\ n = \overline{2, N}, \\
\gamma_0 f_0 + \gamma_1 f_1 - \gamma_2 f_{N+1} + \lambda r_{N+1} + \gamma_1 r_1 = 0, \\
-\kappa_1 f_0 + (\lambda+\gamma_1)f_1 + \lambda\sum_{n=2}^{N+1}f_n + \kappa_1 r_0 + r_0\frac{\Phi_2'(w)}{w\Phi_2(w)} = 0.
\end{gathered} \tag{14}$$

As we can see, the relation $\Phi_2'(w)/w\Phi_2(w)$ does not depend on w, then function $\Phi_2(w)$ has following form

$$\Phi_2(w) = \exp\left\{\frac{(jw)^2}{2}\kappa_2\right\}.$$

Due to this, we rewrite system (14) as follows

$$\begin{gathered}
-\left(\lambda+\sum_{n=2}^{N}\alpha_n+\gamma_0+\kappa_1\right)f_0+\sum_{k=1}^{N}\mu_k f_k+\gamma_2 f_{N+1}=\kappa_2 r_0,\\
-(\mu_1+\gamma_1)f_1+(\lambda+\kappa_1)f_0=\kappa_1 r_0-\kappa_2 r_0-\lambda r_1,\\
\alpha_n f_0-\mu_n f_0=-\lambda r_n,\ n=\overline{2,N},\\
\gamma_0 f_0+\gamma_1 f_1-\gamma_2 f_{N+1}=-\lambda r_{N+1}-\gamma_1 r_1,\\
-\kappa_1 f_0+(\lambda+\gamma_1)f_1+\lambda\sum_{n=2}^{N+1}f_n=\kappa_2 r_0-\kappa_1 r_0.
\end{gathered}\tag{15}$$

From system (15), we obtain the expressions for f_k

$$f_1=\frac{\lambda+\kappa_1}{\mu_1+\gamma_1}f_0+\frac{\kappa_2-\kappa_1}{\mu_1+\gamma_1}r_0+\frac{\lambda}{\mu_1+\gamma_1}r_1,$$

$$f_n=\frac{\alpha_n}{\mu_n}f_0+\frac{\lambda}{\mu_n}r_n,\ n=\overline{2,N},$$

$$\begin{aligned}
f_{N+1}=&\frac{\gamma_0(\mu_1+\gamma_1)+\gamma_0(\lambda+\kappa_1)}{\gamma_2(\mu_1+\gamma_1)}f_0+\frac{\gamma_1(\kappa_2-\kappa_1)}{\gamma_2(\mu_1+\gamma_1)}r_0\\
&+\frac{\gamma_1(\lambda+\mu_1+\gamma_1)}{\gamma_2(\mu_1+\gamma_1)}r_1+\frac{\lambda}{\gamma_2}r_{N+1}.
\end{aligned}$$

Substituting the expressions into the last equation of system (15), we obtain the equation for κ_2

$$\begin{aligned}
&\left[-\kappa_1+\frac{(\lambda+\gamma_1)(\lambda+\kappa_1)}{\mu_1+\gamma_1}+\lambda\sum_{n=2}^{N}\frac{\alpha_n}{\mu_n}+\lambda\frac{\gamma_0(\mu_1+\gamma_1)+\gamma_1(\lambda+\kappa_1)}{\gamma_2(\mu_1+\gamma_1)}\right]f_0\\
&\qquad+\left[\frac{(\lambda+\gamma_1)(\kappa_2-\kappa_1)}{\mu_1+\gamma_1}+\frac{\lambda\gamma_1(\kappa_2-\kappa_1)}{\gamma_2(\mu_1+\gamma_1)}-\kappa_2+\kappa_1\right]r_0\\
&+\left[\frac{\lambda(\lambda+\gamma_1)}{\mu_1+\gamma_1}+\frac{\lambda\gamma_1(\lambda+\mu_1+\gamma_1)}{\gamma_2(\mu_1+\gamma_1)}\right]r_1+\sum_{n=2}^{N}\frac{\lambda^2}{\mu_n}r_n+\frac{\lambda^2}{\gamma_2}r_{N+1}=0.
\end{aligned}$$

Since the coefficient

$$-\kappa_1+\frac{(\lambda+\gamma_1)(\lambda+\kappa_1)}{\mu_1+\gamma_1}+\lambda\sum_{n=2}^{N}\frac{\alpha_n}{\mu_n}+\lambda\frac{\gamma_0(\mu_1+\gamma_1)+\gamma_1(\lambda+\kappa_1)}{\gamma_2(\mu_1+\gamma_1)}=0,$$

we can derive the explicit expression for κ_2

$$\begin{aligned}
\kappa_2=\kappa_1+&\frac{\lambda}{\gamma_2\mu_1-\lambda(\gamma_1+\gamma_2)}\\
\times&\left[(\gamma_2(\lambda+\gamma_1)+\gamma_1(\lambda+\mu_1+\gamma_1))\frac{\lambda+\kappa_1}{\mu_1+\gamma_1}+\lambda\gamma_2(\mu_1+\gamma_1)\sum_{n=2}^{N}\frac{\alpha_n}{\mu_n^2}\right.\\
&\left.+\frac{\lambda}{\gamma_2}(\gamma_0(\mu_1+\gamma_1)+\gamma_1(\lambda+\kappa_1))\right],
\end{aligned}$$

which coincides with (12). **Theorem is proved.**

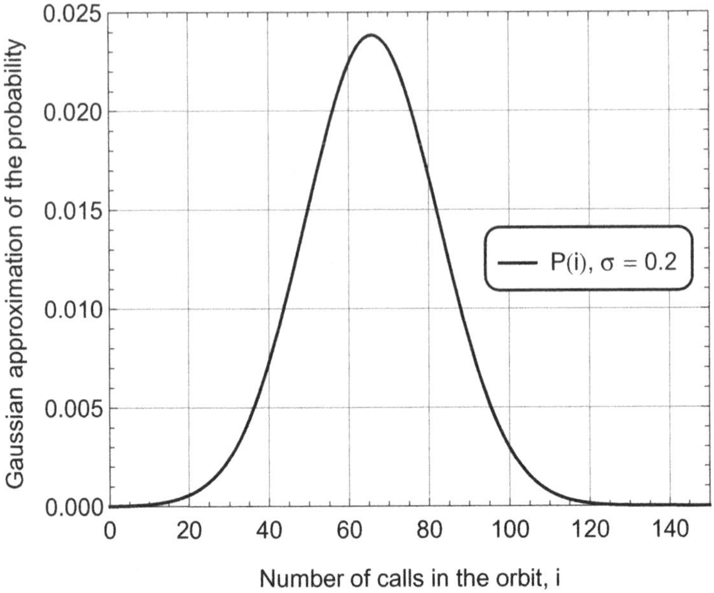

Fig. 1. Gaussian approximation of the probability distribution of the number of calls in the orbit, $\sigma = 0.2$

Theorem 2 defines the asymptotic variance κ_2/σ of the number of calls in the orbit. Thus, we can conclude that asymptotic probability distribution of the process $i(t)$ is Gaussian with mean κ_1/σ and variance κ_2/σ.

Remark 2. Theorem 2 could be interpreted as a central limit theorem for the number of customers in the orbit.

5 Numerical Examples

For example we set the parameters of the system as follows:

$$N = 4,\ \lambda = 0.5,\ \mu_1 = 1,\ \mu_2 = 2,\ \mu_3 = 3,\ \mu_4 = 4,$$

$$\alpha_2 = 1,\ \alpha_3 = 2,\ \alpha_4 = 3,\ \gamma_0 = 1,\ \gamma_1 = 1,\ \gamma_2 = 2,$$

then the value of the expression

$$\gamma_2\mu_1 - \lambda(\gamma_1 + \gamma_2) = 0.5,$$

is grower then 0, which means that the system performs in stationary regime.

Figures 1, 2 shows Gaussian approximation $P(i)$ of the probability distribution of the number of calls in the orbit given by formula

$$P(i) = \frac{G(i + 0.5) - G(i - 0.5)}{1 - G(-0.5)},$$

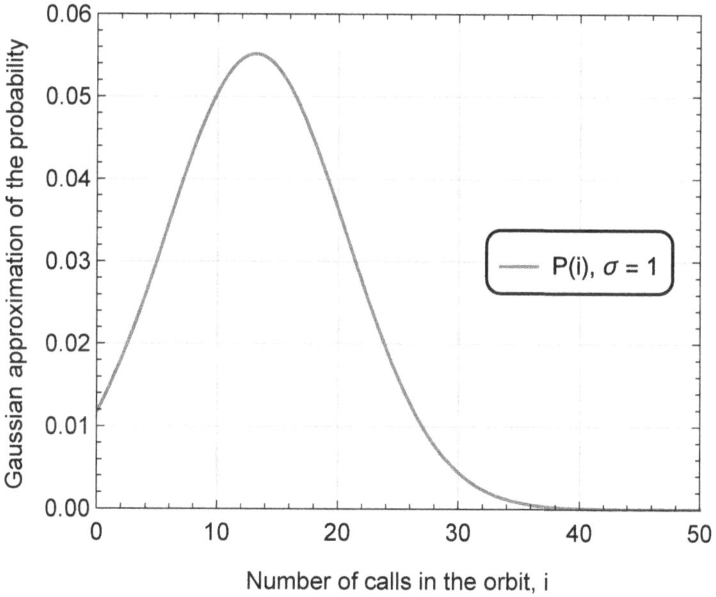

Fig. 2. Gaussian approximation of the probability distribution of the number of calls in the orbit, $\sigma = 1$

where $G(x)$ is Gaussian distribution function with mean κ_1/σ and standard deviation $\sqrt{\kappa_2/\sigma}$.

6 Conclusion

We studied the number of calls in the orbit in Markovian retrial queue with unreliable server and two-way communication. We have obtained the steady-state characteristic function of this process. Asymptotic probability distribution of the process is Gaussian with mean κ_1/σ and variance κ_2/σ. The values of κ_1 and κ_2 are given in Theorem 1 and Theorem 2, respectively. For future work, we may extend our results to the model with arbitrary service time distributions.

References

1. Aguir, S., Karaesmen, F., Akşin, O.Z., Chauvet, F.: The impact of retrials on call center performance. OR Spectr. **26**(3), 353–376 (2004). https://doi.org/10.1007/s00291-004-0165-7
2. Aissani, A., Phung-Duc, T.: Optimal analysis for M/G/1 retrial queue with two-way communication. In: Gribaudo, M., Manini, D., Remke, A. (eds.) ASMTA 2015. LNCS, vol. 9081, pp. 1–14. Springer, Cham (2015). https://doi.org/10.1007/978-3-319-18579-8_1

3. Aissani, A., Phung-Duc, T.: Profiting the idleness in single server system with orbit-queue. In: Proceedings of the 11th EAI International Conference on Performance Evaluation Methodologies and Tools, pp. 237–243. ACM (2017)
4. Artalejo, J.R.: A classified bibliography of research on retrial queues: progress in 1990–1999. Top **7**(2), 187–211 (1999). https://doi.org/10.1007/BF02564721
5. Artalejo, J.R.: Accessible bibliography on retrial queues: progress in 2000–2009. Math. Comput. Modell. **51**(9–10), 1071–1081 (2010)
6. Artalejo, J.R., Phung-Duc, T.: Markovian retrial queues with two way communication. J. Ind. Manag. Optim. **8**(4), 781–806 (2012)
7. Artalejo, J.R., Phung-Duc, T.: Single server retrial queues with two way communication. Appl. Math. Model. **37**(4), 1811–1822 (2013)
8. Bernett, H.G., Fischer, M.J., Masi, D.M.B.: Blended call center performance analysis. IT Prof. **4**(2), 33–38 (2002)
9. Bhulai, S., Koole, G.: A queueing model for call blending in call centers. IEEE Trans. Autom. Control **48**(8), 1434–1438 (2003)
10. Kumar, M.S., Dadlani, A., Kim, K.: Performance analysis of an unreliable M/G/1 retrial queue with two-way communication. Oper. Res. **20**, 1–14 (2018)
11. Paul, S., Phung-Duc, T.: Retrial queueing model with two-way communication, unreliable server and resume of interrupted call for cognitive radio networks. In: Dudin, A., Nazarov, A., Moiseev, A. (eds.) ITMM/WRQ -2018. CCIS, vol. 912, pp. 213–224. Springer, Cham (2018). https://doi.org/10.1007/978-3-319-97595-5_17
12. Sakurai, H., Phung-Duc, T.: Two-way communication retrial queues with multiple types of outgoing calls. TOP **23**(2), 466–492 (2014). https://doi.org/10.1007/s11750-014-0349-5
13. Sakurai, H., Phung-Duc, T.: Scaling limits for single server retrial queues with two-way communication. Ann. Oper. Res. **247**(1), 229–256 (2015). https://doi.org/10.1007/s10479-015-1874-9

Three-Server Queue with Consultations by Main Server with a Buffer at the Main Server

Thekkiniyedath Resmi[1(✉)] and K. Ravikumar[2]

[1] Department of Mathematics, KKTM Government College, Pullut, Trichur, India
[2] Department of Statistics, KKTM Government College, Pullut, Trichur, India

Abstract. In this paper, we analyse a three-server queueing model which is equipped with a main server and two regular servers. The main server offers consultations to the identical regular servers with a preemptive priority over customers. The service of the customers at the main server undergo interruptions during consultations. An upper bound is set for the number of interruptions during the service of a customer at the main server. We consider independent Poisson arrival processes to the main server and to the regular servers. There is a finite buffer at the main server. An arriving customer to the main server will be lost when the buffer is full. The inter occurrence time for requirement for consultation follows exponential distribution having parameter, depending upon the number of busy regular servers. When both regular severs are queued at the main server for consultation, no such fresh event can occur. The service times at the main server and the regular servers are assumed to follow mutually independent phase type distributions. We establish the stability condition and the explicit formula for mean number of interruptions to a customer at the main server. Some performance measures are studied numerically.

Keywords: Main server · Regular server · Interruption · Consultation · Buffer

1 Introduction

Queueing models with consultation by a main server are very common nowadays. In hospitals, banks, super markets, etc., one of the servers perform the role of a main server who is giving consultation to the fellow servers in addition of providing service to the customers. In some cases, the main server serves some privileged customers, while the ordinary customers are served by regular servers. For example, in a hospital, the chief physician attends patients who are suffering from serious illness or who need special attention by himself or those who are referred to him by other doctors. Thus different queues are formed at main server and regular servers. For convenience, we denote the customers at the main server and at the regular servers as Type 1 customers and as Type 2 customers, respectively.

A. Dudin et al. (Eds.): ITMM 2020, CCIS 1391, pp. 131–142, 2021.
https://doi.org/10.1007/978-3-030-72247-0_10

Queueing models with server consultation is introduced by Chakravarthy [3]. In this model, consultations are provided by the main server to the $c-1$ regular servers with a preemptive priority over customers. If more than one regular server need consultations at the same time, a queue will be formed at the main server the consultation will be provided in FIFO manner. If there is a customer being served by the main server when a request for consultation by a regular server is prompted, then the service of that customer will be stopped temporarily. At this time the service is said to be 'interrupted'. The service at the regular server is not considered to be interrupted because, consultation is a part of service at the regular service which certainly enhances the quality of service. The service times at the main server and at the identical regular servers are exponentially distributed with mean μ_1 and μ_2, respectively.

Queues with interruptions in service was first analysed by White and Christie [12]. As soon as the interruption is completed, the service will be resumed. Later on, queueing models with service interruptions are studied by Avi-Izhak and Naor [1], Boxma [2], Fiems et al. [4], Gaver [5], Ibe [6], Keilson [7] and Takine [11].

Krishnamoorthy et al. [9] considered a model where the service after interruption will be resumed or restarted according to the realisation of a threshold clock.

If a customer at the main server is interrupted infinitely many times or if the regular server receives infinite number of consultations during their service as in [3], they may get impatient and sometimes may leave the system without completing the service. So it would be better if we control the consultations and the interruptions by setting some limits to interruptions and consultations.

Krishnamoorthy et al. [8] discussed a single server queueing model in which interruptions to the server is controlled by a finite number of interruptions and the duration of a super clock.

The main differences of the present model from [3] are as follows. In this model, we consider 3 servers, one main server and two regular servers. Two different queues are maintained, one for the Type 1 customers at the main server and the other for the Type 2 customers at the regular servers. There is a finite buffer at the main server. A Type 1 customer upon arrival will not join the queue if the buffer is full. If both regular servers are free, an arriving Type 2 customer can choose either with same probability. Mutually independent phase type distributions represent the service times at the main and at the regular servers. An upper bound is set for the number of interruptions to a customer at the main server.

1.1 Notations

The following notations are used in this model.

- $L_1 = L(M+1)p+1$, $L_2 = 2LMq$, $L_3 = 2L(2M-1)q^2$
  $L_4 = 2L_1q + 2Lq + L_2p + 2Lpq$, $L_5 = L_1q^2 + 4Lq^2 + L_3p + 6Lpq^2$
- $\lambda = \lambda_1 + \lambda_2$, $\Lambda_1 = diag(\lambda I_{L-1}, \lambda_2)$, $\Lambda_2 = diag(\lambda I_L, \lambda_2)$

- $\tilde{\boldsymbol{\alpha}} = \mathbf{e}'_{M+1}(1) \otimes \boldsymbol{\alpha}$
- $\nabla = diag(I_q \otimes \boldsymbol{\beta}, \boldsymbol{\beta} \otimes I_q)$, $\nabla_1 = \begin{bmatrix} \nabla \; O \end{bmatrix}$
- $\Delta = diag(I_q \otimes V^0, V^0 \otimes I_q)$, $\Delta_1 = \begin{bmatrix} \Delta \\ O \end{bmatrix}$, $\Delta_2 = \begin{bmatrix} \Delta \otimes I_M \; O \end{bmatrix}'$,
  $\Delta_3 = diag(\Delta \otimes I_M, O)$, $\Delta_4 = diag(\Delta, O)$
- $\hat{\Delta} = diag(I_q \otimes V^0 \otimes \boldsymbol{\beta}, V^0 \otimes \boldsymbol{\beta} \otimes I_q)$, $\hat{\Delta}_1 = diag(\hat{\Delta}, O)$,
  $\Delta^* = diag(I_q \otimes V, V \otimes I_q)$
- $\ddot{I} = \begin{bmatrix} 0 & 1 \\ 1 & 0 \end{bmatrix}$, $\dot{I}_n = \begin{bmatrix} I_{n-1} & \mathbf{0} \\ \underline{\mathbf{0}} & 0 \end{bmatrix}_{n \times n}$
  $\hat{I}_n = \begin{bmatrix} \mathbf{0} & I_{n-1} \\ 0 & \underline{\mathbf{0}} \end{bmatrix}_{n \times n}$, $\tilde{I}_n = \begin{bmatrix} \underline{\mathbf{0}} & 0 \\ I_{n-1} & \mathbf{0} \end{bmatrix}_{n \times n}$,
- $I^o = I_{L_1}$, $\mathbf{e}^* = diag(1, 0)$, $\hat{\mathbf{e}}_n = \begin{bmatrix} \mathbf{e}_{n-1} \\ 0 \end{bmatrix}_{n \times 1}$.

2 Description of Model

The present model considers a 3-server queueing system having different arrival processes. The customers arrive to the main server and to the regular servers according to independent Poisson processes with rates λ_1 and λ_2, respectively. A buffer of size L is at the main server. An arriving Type 1 customer will leave the system without join the queue on seeing the buffer is full. If any one of the regular servers is free, an arriving Type 2 customer gets service immediately, else joins the queue. If all the regular servers are free, that customer can choose any one of them with equal probability.

The service times at the main and regular servers follow independent phase type distributions with representations $(\boldsymbol{\alpha}, \mathbf{U})$ and $(\boldsymbol{\beta}, \mathbf{V})$ with number of phases p and q, respectively. Note that $\mathbf{U}^0 = -\mathbf{U}\mathbf{e}$ and $\mathbf{V}^0 = -\mathbf{V}\mathbf{e}$. Let M denote the upper bound for the number of interruptions to the customer at the main server.

Requirement of consultation arises according to a Poisson process with rate θ_i, if i regular servers are busy, where $i = 1, 2$. When both the regular severs need consultation, a queue is formed for consultation and it is provided in FIFO basis. In order to distinguish the regular servers, we denote them as $\Re_1$ and $\Re_2$. Here duration of consultation follows exponential distribution with parameter ξ. After getting consultation, all the servers resume their services at the phases where they were suspended.

The regular servers can receive any number of consultations during the service a customer. At most M interruptions are allowed to a customer at the main server. Suppose that a customer is being served at the main server whose service has already interrupted M times. If a regular server needs a consultation at this time, he/she has to wait until the service of the customer at the main server is completed. Once the service is completed, the main server will attend the consultation before taking a new customer from the queue for service. Now consider another situation. The main server is giving consultation to a regular server with an interrupted customer where that customer is undergoing M^{th} interruption. If the second regular server needs a consultation by this time, he/she has to wait until the completion of the present consultation and the service of the customer at the main server after that.

The queueing model is $Z = \{Z(t), t \geq 0\}$,
where $Z(t) = \{N_2(t), S(t), B(t), J_1(t), J_2(t), N_1(t), K(t), J_3(t)\}$.
Here $N_2(t)$ and $N_1(t)$ represent the number of type 2 and type 1 customers in the system, respectively. $S(t)$ denotes the status of the servers at time t such that

$$S(t) = \begin{cases} 0, & \text{if the main together with or without regular server(s) is busy} \\ 1, & \text{if the main server is providing consultation only} \\ 2, & \text{if the main server is providing consultation with one interrupted customer at the main server} \\ 3, & \text{if the regular server is waiting for getting consultation after the present service at the main server} \\ 4, & \text{if the regular server is waiting for getting consultation after the service at the main server followed by the present interruption} \end{cases}$$

Here

$J_i(t)$ – phase of the regular server $\Re_i, i = 1, 2$
$K(t)$ – number of interruptions already befell to a customer at the main server
$J_3(t)$ – phase of the main server

For $N_2(t) = 1$, $B(t) = \{1, 2\}$ according to $\Re_1$ or $\Re_2$ is busy or under going/waiting to receive consultation.

Now consider $N_2(t) \geq 2$.
If $S(t) = \{1, 2\}$, then

$$B(t) = \begin{cases} 1 \text{ (or 2)}, & \text{if } \Re_1 \text{ is receiving consultation and } \Re_2 \text{ is idle or busy (or vice versa)} \\ 3 \text{ (or 4)}, & \text{if all the regular servers are queued for consultation with } \Re_1 \text{ is receiving consultation in the first place and } \Re_2 \text{ in the second place (or vice versa)} \end{cases}$$

If $S(t) = 3$, then $B(t)$ takes the same values $\{1, 2, 3, 4\}$ according to the above definition with 'receiving consultation' is replaced by 'waiting to receive consultation.'

If $S(t) = 4$, then

$$B(t) = \begin{cases} 1 \text{ (or 2)}, & \text{if } \Re_1 (\text{or } \Re_2) \text{is waiting to receive consultation after the present interruption followed by the service completion at the main server} \end{cases}$$

$\{Z(t), t \geq 0\}$ is a Continuous Time Markov Chain having state space

$$\Phi = \bigcup_{n=0}^{\infty} \phi(n).$$

The terms $\phi(n)$'s are defined as
$\phi(0) = \{(0,0) \cup (0,0,j,k,t_3) : 1 \leq j \leq L, 0 \leq k \leq M\}$;

$\phi(1) = \phi(1,0) \cup \phi(1,1) \cup \phi(1,2) \cup \phi(1,3)$ and
$\phi(n) = \phi(n,0) \cup \phi(n,1) \cup \phi(n,2) \cup \phi(n,3) \cup \phi(n,4)$, for $n \geq 2$,
where
$\phi(1,0) = \{(1,0,l,t_l,0) \cup (1,0,l,t_l,j,k,t_3) : 1 \leq j \leq L, 0 \leq k \leq M\}$,
$\phi(1,1) = \{(1,1,l,t_l,j) : 0 \leq j \leq L-1\}$,
$\phi(1,2) = \{(1,2,l,t_l,j,k,t_2) : 1 \leq j \leq L, 0 \leq k \leq M-1\}$ and
$\phi(1,3) = \{(1,3,l,t_l,j,t_2) : 1 \leq j \leq L\}$, for $1 \leq l \leq 2$;
and for $n \geq 2$,
$\phi(n,0) = \{(n,0,t_l,t_2,0) \cup (n,0,t_1,t_2,j,k,t_3) : 1 \leq j \leq L, 0 \leq k \leq M\}$,
$\phi(n,1) = \{(n,1,l,t_1,t_2,j) : 1 \leq l \leq 4, 0 \leq j \leq L-1\}$,
$\phi(n,2) = \{(n,2,l,t_1,t_2,j,k,t_3) : 1 \leq l \leq 2, 1 \leq j \leq L, 0 \leq k \leq M-1\} \cup \{(n,2,l,t_1,t_2,j,k,t_3) : 3 \leq l \leq 4, 1 \leq j \leq L, 0 \leq k \leq M-2\}$,
$\phi(n,3) = \{(n,3,l,t_1,t_2,j,k,t_3) : 1 \leq l \leq 4, 1 \leq j \leq L\}$ and
$\phi(n,4) = \{(n,4,l,t_1,t_2,j,t_3) : 1 \leq l \leq 2, 1 \leq j \leq L\}$;
with $1 \leq t_1, t_2 \leq q$ and $1 \leq t_3 \leq p$.

The infinitesimal generator matrix T is given by

$$T = \begin{bmatrix} Q_0 & Q_1 & & & & \\ Q_2 & Q_3 & Q_4 & & & \\ & Q_5 & P_1 & P_0 & & \\ & & P_2 & P_1 & P_0 & \\ & & & \ddots & \ddots & \ddots \\ & & & & & \end{bmatrix} \tag{1}$$

Here P_0, P_1 and P_2 are square matrices of order L_5, Q_0, Q_3 are square matrices of orders L_1 and L_4, respectively. Q_1, Q_2, Q_4 and Q_5 are matrices of orders $L_1 \times L_4$, $L_4 \times L_1$, $L_4 \times L_5$ and $L_5 \times L_4$, respectively.

$Q_0 = \begin{bmatrix} -\lambda & Q_{01} \\ Q_{02} & Q_{03} \end{bmatrix}$, $Q_1 = \frac{\lambda_2}{2} \left[\mathbf{e}_2' \otimes \boldsymbol{\beta} \otimes I^o \; O \right]$,

$Q_2 = \begin{bmatrix} \mathbf{e}_2 \otimes V^0 \otimes I^o \\ O \end{bmatrix}$, $Q_3 = \begin{bmatrix} Q_{31} & Q_{32} & Q_{33} \\ Q_{34} & Q_{35} & \\ Q_{36} & Q_{37} & Q_{38} \end{bmatrix}$,

$Q_4 = \lambda_2 \left[diag(\nabla \otimes \mathbf{e}_2 \otimes I^o, Q_{41}, Q_{42}, Q_{43}) \; O \right]$,

$Q_5 = \begin{bmatrix} diag(\mathbf{e}_2' \otimes \Delta \otimes I^o, Q_{51}, Q_{52}, Q_{53}) \\ O \end{bmatrix}$,

$$P_0 = \lambda_2 I,\ P_1 = \begin{bmatrix} P_{11} & P_{12} & P_{13} \\ P_{14} & P_{15} & \\ P_{16} & P_{17} & P_{18} \end{bmatrix},$$

$P_2 = diag(\mathbf{e}_2^{'} \otimes \hat{\Delta} \otimes \mathbf{e}_2 \otimes I^o, P_{21}, P_{22}, P_{23}, O)$.

Here

$$Q_{01} = \lambda_1 \left[\boldsymbol{\alpha}\ \underline{\mathbf{0}}\right]_{1\times L(M+1)p},\ Q_{02} = \begin{bmatrix} \mathbf{e}_{M+1} \otimes U^0 \\ \mathbf{0} \end{bmatrix}_{L(M+1)p\times 1},$$

$Q_{03} = I_{(M+1)L} \otimes U + \lambda_1 \hat{I} \otimes \mathrm{I}_{(M+1)p} + \tilde{I} \otimes \mathbf{e}_{M+1} \otimes U^0 \otimes \boldsymbol{\alpha} - \Lambda_1 \otimes I_{p(M+1)}$,
$Q_{31} = I_2 \otimes (I_q \otimes Q_0 + (V - \theta_1 I_q) \otimes I^o)$, $Q_{32} = \theta_1 I_{2q} \otimes \mathbf{e}^*$,
$Q_{33} = \theta_1 \left[I_{2q} \otimes D_1\ I_{2q} \otimes D_2\right]_{2L_1 q\times 2L(M+1)pq}$, $Q_{34} = \xi I_{2q} \otimes \Omega$,

$$Q_{35} = I_{2q} \otimes (-\xi I_{L+1} - \Lambda_2 + \lambda_1 I^*),\ Q_{36} = \begin{bmatrix} \xi I_{2q} \otimes G_1 \\ O \end{bmatrix}_{2L(M+1)pq\times 2L_1 q},$$

$$Q_{37} = \begin{bmatrix} O \\ I_{2q} \otimes U^* \end{bmatrix}_{2L(M+1)pq\times 2(L+1)q},\ Q_{38} = diag(I_{2q} \otimes F_1, I_{2q} \otimes F_2)_{2Lpq},$$

$$Q_{41} = \left[\nabla\ O\right]_{2q\times 4q^2} \otimes I_{L+1},\ Q_{42} = \left[\nabla \otimes I_{LM}\ O\right]_{L_2\times L_3} \otimes I_p,$$

$$Q_{43} = \left[\nabla\ O\right]_{2q\times 2q^2} \otimes I_{Lp},\ Q_{51} = \begin{bmatrix} \Delta \\ O \end{bmatrix}_{4q^2\times 2q} \otimes I_{L+1},$$

$$Q_{52} = \begin{bmatrix} \Delta \otimes I_{LM} \\ O \end{bmatrix}_{L_3\times L_2} \otimes I_p,\ Q_{53} = \begin{bmatrix} \Delta \\ O \end{bmatrix}_{4q^2\times 2q} \otimes I_{Lp},$$

$P_{11} = I_{q^2} \otimes Q_0 + (V \oplus V - 2\theta_2 I_{q^2}) \otimes I^o$,
$P_{12} = \theta_2 \left[\mathbf{e}_2^{'} \otimes \mathbf{e}^*\ O\right]_{L_1\times 4(L+1)} \otimes I_{q^2}$,

$$P_{13} = \theta_2 \left[M_1\ M_2\ O\right]_{L_1 q^2\times 4L(M+1)q^2} \otimes I_p,\ P_{14} = \xi \begin{bmatrix} \mathbf{e}_2 \otimes I_{q^2} \otimes \Omega \\ O \end{bmatrix}_{4(L+1)q^2\times L_1 q^2},$$

$$P_{15} = \begin{bmatrix} H_1 & \theta_1 I_{2q^2(L+1)} \\ \xi \ddot{I} \otimes I_{q^2(L+1)} & H_2 \end{bmatrix}_{4(L+1)q^2},\ P_{16} = \xi \begin{bmatrix} \mathbf{e}_2 \otimes I_{q^2} \otimes G_1 \\ O \end{bmatrix}_{L_3 p\times L_1 q^2},$$

$$P_{17} = \begin{bmatrix} O \\ I_{4q^2} \otimes U^* \\ O \end{bmatrix}_{L_3 p\times 4(L+1)q^2},\ P_{18} = \begin{bmatrix} J_1 & O & J_2 \\ & J_3 & \\ O & J_4 & J_5 \end{bmatrix}_{L_3} \otimes I_p,$$

$P_{21} = diag(\hat{\Delta}, O)_{4q^2} \otimes I_{L+1}$, $P_{22} = diag(\hat{\Delta} \otimes I_{LM}, O)_{L_3} \otimes I_p$,
$P_{23} = diag(\hat{\Delta}, O)_{4q^2} \otimes I_{Lp}$.

We describe the following terms:

$$D_1 = \begin{bmatrix} \mathbf{0} \\ I_L \otimes \dot{I}_{M+1} \otimes I_p \end{bmatrix}_{L_1\times L(M+1)p},\ D_2 = \begin{bmatrix} \mathbf{0} \\ I_L \otimes \hat{\mathbf{e}}_{M+1} \otimes I_p \end{bmatrix}_{L_1\times Lp},$$

$$\Omega = \begin{bmatrix} 1 & \mathbf{0} \\ \mathbf{0} & I_L \otimes \tilde{\boldsymbol{\alpha}} \end{bmatrix},\ G_1 = \left[O\ I_L \otimes \hat{I}_{M+1} \otimes I_p\right],\ U^* = \left[I_L \otimes U^0\ \mathbf{0}\right]_{Lp\times L+1},$$

$F_1 = (\xi I_L - \Lambda_1 + \lambda_1 \hat{I}_L) \otimes I_{Mp}$, $F_2 = I_L \otimes U + (-\Lambda_1 + \lambda_1 \hat{I}_L) \otimes I_p$,
$M_1 = \left[\mathbf{e}_2^{'} \otimes I_{q^2} \otimes D_3\ O\right]_{L_1 q^2\times L_3}$, $M_2 = \left[\mathbf{e}_2^{'} \otimes I_{q^2} \otimes D_4\ O\right]_{L_1 q^2\times Lq^2}$,
$H_1 = I_{2q^2} \otimes (\lambda_1 \hat{I}_L - (\xi + \theta_1) I_{L+1} - \Lambda_2) + I_2 \otimes \Delta^* \otimes I_{L+1}$,

$$H_2 = I_2 \otimes I_{q^2} \otimes (\lambda_1 \hat{I}_L - \xi I_{L+1} - \Lambda_2),$$

$$J_1 = \begin{bmatrix} H_3 & \theta_1 I_{2q^2L} \otimes \dot{I}_M \\ \xi \ddot{I} \otimes I_{q^2L} \otimes \hat{I}_M & H_4 \end{bmatrix}, \; J_2 = \theta_1 \begin{bmatrix} I_{2q^2L} \otimes \hat{\mathbf{e}}_M \\ O \end{bmatrix}_{L_3p \times 2Lpq^2},$$

$$J_3 = I_{4q^2} \otimes (I_L \otimes U - \Lambda_1 \otimes I_p) + \begin{bmatrix} \mathbf{e}_2' \otimes \theta_1 I_{2Lpq^2} \\ O \end{bmatrix} + \begin{bmatrix} \Delta^* \otimes I_{Lp} & O \\ O & O \end{bmatrix},$$

$$J_4 = \xi I_{2q^2} \otimes \ddot{I} \otimes I_{Lp}, \; J_5 = I_{2q^2} \otimes (\lambda_1 \hat{I}_L - \xi I_L - \Lambda_1 \otimes I_p),$$

$$D_3 = \begin{bmatrix} 0 \\ I_L \otimes \dot{I}_{M+1} \end{bmatrix}_{L(M+1)+1 \times LM}, \; D_4 = \begin{bmatrix} 0 \\ I_L \otimes \hat{\mathbf{e}}_{M+1} \end{bmatrix}_{L_1 \times L},$$

$$H_3 = I_{2q^2} \otimes (\lambda_1 \hat{I}_L - (\xi + \theta_1) I_L - \Lambda_1 \otimes I_M) + \Delta^* \otimes I_{LM},$$

$$H_4 = I_{2q^2} \otimes (\lambda_1 \hat{I}_L - \xi I_{L(M-1)} - \Lambda_1 \otimes I_{M-1}).$$

3 Steady State Analysis

The steady-state analysis of the queueing model is performed in this section. The stability condition of the queueing system is established as follows.

3.1 Stability condition

The steady-state probability vector of the generator $P_0 + P_1 + P_2$ is denoted as $\boldsymbol{\pi}$. That is, $\boldsymbol{\pi}(P_0 + P_1 + P_2) = \underline{\mathbf{0}}$; $\boldsymbol{\pi}\mathbf{e} = 1$.

Theorem 1. *The Markov chain $\{Z(t), t \geq 0\}$ is stable if and only if*

$$\boldsymbol{\pi} P_0 \boldsymbol{e} < \boldsymbol{\pi} P_2 \boldsymbol{e}. \tag{2}$$

Proof. The system is stable if and only if the rate of drift to a lower level from a given level should be greater than that to the next higher level. See Neuts [10].

The stability condition in Eq. (2) is equivalent to $\rho < 1$, where the traffic intensity ρ is given by

$$\rho = \frac{\boldsymbol{\pi} P_0 \mathbf{e}}{\boldsymbol{\pi} P_2 \mathbf{e}}. \tag{3}$$

3.2 Steady State Probability Vector

Let $\boldsymbol{z}$, partitioned as, $\boldsymbol{z} = (\boldsymbol{z}_0, \boldsymbol{z}_1, \boldsymbol{z}_2, ...)$ denotes the steady state probability vector of the Markov chain $\{Z(t), t \geq 0\}$. We know that the vector $\boldsymbol{z}$ satisfies the condition $\boldsymbol{z}T = 0$ and $\boldsymbol{z}\mathbf{e} = 1$, where $\mathbf{e}$ is a column vector of appropriate dimension. Assume that the stability condition is satisfied. Then the sub-vectors of $\boldsymbol{z}$ are obtained by the equation

$$\boldsymbol{z}_n = \boldsymbol{z}_2 R^{n-2}, n \geq 3. \tag{4}$$

Here R is the minimal non-negative solution of the matrix equation

$$R^2P_2 + RP_1 + P_0 = 0. \tag{5}$$

We can trace out the vectors $\boldsymbol{z}_0, \boldsymbol{z}_1$ and $\boldsymbol{z}_2$ by solving the equations

$$\begin{aligned} \boldsymbol{z}_0Q_0 + \boldsymbol{z}_1Q_2 &= 0 \\ \boldsymbol{z}_0Q_1 + \boldsymbol{z}_1Q_3 + \boldsymbol{z}_2Q_5 &= 0 \\ \boldsymbol{z}_1Q_4 + \boldsymbol{z}_2(P_1 + RP_2) &= 0 \end{aligned} \tag{6}$$

and the normalizing condition

$$\boldsymbol{z}_0\mathbf{e} + \boldsymbol{z}_1\mathbf{e} + \boldsymbol{z}_2(I - R)^{-1}\mathbf{e} = 1. \tag{7}$$

3.3 Mean Number of Interruptions to a Customer at the Main Server

In this section, we compute the mean number of interruptions experienced by a Type 1 customer. Consider the Markov process $Y(t) = \{(\tilde{N}(t), S^*(t), J_3(t)) : t \geq 0\}$, where $\tilde{N}(t)$ is the number of interruptions already befell to a Type 1 customer at the main server, $S^*(t) = S(t) - \{1, 3, 4\}$ and all other variables are as defined earlier. The state space of Y is$\{(i, j, t_1) : 0 \leq i \leq M, j = 0, 2, 1 \leq t_1 \leq p\} \cup \{\Gamma\}$. Γ is the absorbing state where the customer leaves the system after service completion. The infinitesimal generator $\tilde{W}$ of the process $Y(t)$ takes the form

$$\tilde{W} = \begin{bmatrix} W & W^0 \\ \underline{\mathbf{0}} & 0 \end{bmatrix}$$

where

$$W = \begin{bmatrix} W_{11} & \mathbf{O} \\ W_{12} & W_{13} \end{bmatrix}_{M+1} \text{ and } W^0 = \begin{bmatrix} \mathbf{e}_{M-1} \times C_0 \\ C_1 \\ C_2 \end{bmatrix}.$$

Here $W_{11} = I_{M-2} \otimes \begin{bmatrix} A_0 & B_0 \end{bmatrix}$, $W_{12} = \begin{bmatrix} \mathbf{O} & A_0 \\ \mathbf{O} & \mathbf{O} \end{bmatrix}$ and $W_{13} = \begin{bmatrix} B_1 & \\ A_1 & B_2 \\ & A_2 \end{bmatrix}$.

The sub matrices are given by

$$A_0 = diag(U, O)_{3p \times 3p} + \begin{bmatrix} -\theta_2 & \theta_2 & \\ & -(\theta_1 + \xi) & \theta_1 \\ & & -\xi \end{bmatrix} \otimes I_p,$$

$$A_1 = diag(U, O)_{2p \times 2p} + \begin{bmatrix} -\theta_1 & \theta_1 \\ & -\xi \end{bmatrix} \otimes I_p, \; A_2 = U;$$

$$B_0 = \xi \begin{bmatrix} \underline{\mathbf{0}} & 0 \\ I_2 & \mathbf{0} \end{bmatrix} \otimes I_p, \; B_1 = \xi \begin{bmatrix} \underline{\mathbf{0}} \\ I_2 \end{bmatrix} \otimes I_p, \; B_2 = \xi \begin{bmatrix} 0 \\ 1 \end{bmatrix} \otimes I_p;$$

$C_0 = \mathbf{e}_3(1) \otimes U^0$, $C_1 = \mathbf{e}_2(1) \otimes U^0$, $C_2 = U^0$.

If y_j is the probability that there are exactly j interruptions to a customer at the main server, then

$$y_j = \hat{\boldsymbol{\alpha}} X_{00}^j S_0 \mathbf{e},\ j = 0, 1, ..., M-2,$$

$$y_{M-1} = \hat{\boldsymbol{\alpha}} X_{00}^{M-2} X_{01} S_1 \mathbf{e} \text{ and } y_M = \hat{\boldsymbol{\alpha}} X_{00}^{M-2} X_{01} X_{12} S_2 \mathbf{e},$$

where $\hat{\boldsymbol{\alpha}} = (\boldsymbol{\alpha}, \mathbf{0})$, $S_i = -A_i^{-1} C_i, i = 0, 1, 2$; $X_{i,j} = -A_i^{-1} B_j$.
Mean number of interruptions to a customer at the main server is given by

$$\epsilon_i = \sum_{j=0}^{M} j y_j.$$

3.4 Performance Measures

Here a number of performance measures are listed to bring out the qualitative aspects of the present model. These measures and their formulae are given below. For this, we partition the vectors $\boldsymbol{z}_n$, $n \geq 0$ as

$\boldsymbol{z}_0 = (\zeta, \boldsymbol{z}_{00}), 1 \leq k \leq L$, $\boldsymbol{z}_1 = (\boldsymbol{z}_{10}, \boldsymbol{z}_{11}, \boldsymbol{z}_{12}, \boldsymbol{z}_{13})$ and $\boldsymbol{z}_n = (\boldsymbol{z}_{n0}, \boldsymbol{z}_{n1}, \boldsymbol{z}_{n21}, \boldsymbol{z}_{n22}, \boldsymbol{z}_{n3}, \boldsymbol{z}_{n4})$, for $n \geq 2$. Note that ζ is only a scalar, $\boldsymbol{z}_{00}$, $\boldsymbol{z}_{10}, \boldsymbol{z}_{11}, \boldsymbol{z}_{12}, \boldsymbol{z}_{13}$, $\boldsymbol{z}_{n0}, \boldsymbol{z}_{n1}, \boldsymbol{z}_{n21}, \boldsymbol{z}_{n22}, \boldsymbol{z}_{n3}, \boldsymbol{z}_{n4}$, for $n \geq 2$ are vectors of dimensions $L_1 - 1$, $2L_1 q$, $2Lq$, $L_2 p$, $2Lpq$, $L_1 q^2$, $4Lq^2$, $2LMpq^2$, $2L(M-1)pq^2$, $4Lpq^2$ and $2Lpq^2$, respectively.

We use some abbreviations which are given below:

$$\gamma_0(j) = \sum_{j=1}^{L} j \boldsymbol{z}_{00j},\ \gamma_1(j) = \sum_{n=1}^{1} \sum_{l=1}^{2} \sum_{j=0}^{L-1} j \boldsymbol{z}_{n1lt_1j}, n = 1$$

$$\gamma_2(j) = \sum_{m=0, m\neq 1}^{3} \sum_{l=1}^{2} \sum_{t_1=1}^{q} \sum_{j=1}^{L} j \boldsymbol{z}_{nmlt_1j} \mathbf{e}, n = 1,$$

$$\gamma_3(j) = \sum_{n=2}^{\infty} \sum_{t_1, t_2=1}^{q} \sum_{j=1}^{L} j \boldsymbol{z}_{n0t_1t_2j} \mathbf{e},$$

$$\gamma_4(j) = \nu_{14(L-1)}(j),\ \gamma_5(j) = \nu_{24L}(j) + \nu_{34L}(j),\ \gamma_6(j) = \nu_{42L}(j),$$

$$\gamma_7(j) = \sum_{n=1}^{1} \sum_{l=1}^{2} \sum_{t_1=1}^{q} \boldsymbol{z}_{n0lt_10},\ \gamma_8(j) = \sum_{n=2}^{\infty} \sum_{t_1, t_2=1}^{q} n \boldsymbol{z}_{n10t_1t_20},$$

where $\nu_{abc}(j) = \sum_{n=2}^{\infty}\sum_{l=1}^{b}\sum_{t_1,t_2=1}^{q}\sum_{j=c-L+1}^{c} j\boldsymbol{z}_{nalt_1t_2j}\mathbf{e}.$

ω_k is obtained from $\gamma_{k(1)}$ by replacing the last summation $\sum_{j=c-L+1}^{c}$ by $\sum_{j=c}^{c}$.

(1) Mean number of customers in the system $\epsilon = \gamma_0(j) + \sum_{k=1}^{8}\gamma_k(n+j)$.

(2) Mean number of type 1 customers in the system

$$\epsilon_1 = \gamma_0(j) + \sum_{k=1}^{6}\gamma_k(j).$$

(3) Mean number of type 2 customers in the system $\epsilon_2 = \sum_{n=1}^{\infty} n\boldsymbol{z}_n\mathbf{e}$.

(4) Mean number of type 2 customers in the queue $\tilde{\epsilon}_2 = \sum_{n=3}^{\infty}(n-2)\boldsymbol{z}_n\mathbf{e}$.

(5) Probability that an arriving Type 1 customer is lost on seeing the buffer is full $\eta = \sum_{k=0}^{6}\omega_k(1)$.

(6) Probability that the system is idle $\delta_1 = \zeta$.

(7) Probability that the main server is idle $\delta_2 = \zeta + \gamma_1(1) + \gamma_8(1)$

(8) Probability that both the regular servers are idle $\delta_3 = \boldsymbol{z}_0\mathbf{e}$.

4 Numerical Analysis

In this section we analyse the effect of the λ_1 and λ_2 on the performance measures. Let U, V, $\boldsymbol{\alpha}$, $\boldsymbol{\beta}$ and ξ are as in example **4.1.** Choose $\theta_1 = 1$, $\theta_2 = 1$, $M = 2$ and $L = 3$.

The above mentioned values, vectors and matrices are chosen so as to satisfy the stability condition $\rho < 1$.

Table 1 shows that if λ_1 increases, then the buffer is filled in an increased rate. Thus there is an increase in η. Since there are more Type 1 customers, the main server's busy period increases and this in turn reduces the main server's idle time δ_2. So there will be a slight delay for the regular server to get consultations. Then the mean number of type 2 customers ϵ_2 increases slightly, even if ϵ_2 is not depending upon λ_1 directly. Thus there is a slight decrease in δ_3. All these results in a decrease in system's idle time.

From Table 2, we see that as λ_2 increases, more and more type 2 customers accumulate in both the system the queue. Thus ϵ_2 and $\tilde{\epsilon}_2$ increase. So the main server is forced to spend more time in consultation. During this time the number of type 1 customers waiting at main server increases and thus there will be frequent loss of type 1 customers due to lack of room in buffer, so η increases.

Table 1. Effect of λ_1 on the performance measures

λ_1	1	1.5	2	2.5	3	3.5
ρ	0.4627	0.5406	0.6184	0.6962	0.7740	0.8518
ϵ_2	0.7094	0.7100	0.7105	0.7109	0.7113	0.7116
$\tilde{\epsilon}_2$	0.1268	0.1272	0.1275	0.1278	0.1280	0.1282
η	0.0074	0.0208	0.0395	0.0626	0.0895	0.1197
δ_1	0.4522	0.3994	0.3484	0.3000	0.2551	0.2144
δ_2	0.6050	0.5465	0.4873	0.4290	0.3728	0.3201
δ_3	0.5570	0.5569	0.5568	0.5567	0.5567	0.5566

Table 2. Effect of λ_2 on the performance measures

λ_2	1	1.5	2	2.5	3	3.5
ρ	0.4638	0.5411	0.6184	0.6957	0.7730	0.8503
ϵ_2	0.2984	0.4820	0.7105	1.0087	1.4125	1.972
$\tilde{\epsilon}_2$	0.0137	0.0497	0.1275	0.2725	0.5213	0.9251
η	0.0169	0.0271	0.0395	0.0542	0.0708	0.089
δ_1	0.4876	0.4136	0.3484	0.2909	0.2399	0.1947
δ_2	0.6004	0.5408	0.4873	0.4390	0.395	0.3546
δ_3	0.7538	0.6501	0.5568	0.4723	0.3953	0.3249

Thus the busy period of all the three servers increases whereas the idle times δ_1, δ_2 and δ_3 decrease.

Conclusion

In this paper, we discussed a three server queueing system equipped with consultation by main server with different arrival process and a finite buffer. We established an explicit formula for the number of interruptions to a customer at the main server. Some other performance measures are studied numerically. In this paper, we do not consider any limit for the number of consultations possible to the regular servers during the service of a particular customer. This may lead to the impatience of that customer and he may leave the system at all. So we can impose some maximum possible value for the number of consultations. In addition to the maximum number of interruptions, a super clock can be set to control the duration of interruptions spent by a customer at the main server.

Acknowledgement. Authors thank A. Krishnamoorthy for the inspiring ideas, guidance and the valuable time spent for us.

References

1. Avi-Itzhak, B., Naor, P.: Some queueing problems with the service station subject to breakdowns. Oper. Res. **11**(3), 303–320 (1963)

2. Boxma, O., Mandjes, M., Kella, O.: On a queueing model with service interruptions. Prob. Eng. Inf. Sci. **22**, 537–555 (2008)
3. Chakravarthy, S.R.: A multi-server queueing model with server consultations. Eur. J. Oper. Res. **233**(3), 625–639 (2014)
4. Fiems, D., Maertens, T., Brunee, H.: Queueing systems with different types of interruptions. Eur. J. Oper. Res. **188**(3), 838–845 (2008)
5. Gaver, D.P.: A waiting line with interrupted service including priority. J. R. Stat. Soc. **B24**(1), 73–90 (1962)
6. Ibe, O.C., Trivedi, K.S.: Two queues with alternating service and server breakdown. Queueing Syst. **7**(3), 253–268 (1990). https://doi.org/10.1007/BF01154545
7. Keilson, J.: Queues subject to service interruptions. Ann. Math. Stat. **33**(4), 1314–1322 (1962)
8. Krishnamoorthy, A., Pramod, P.K., Chakravarthy, S.: A note on characterizing service interruptions with phase type distribution. Stoch. Anal. Appl. **31**(4), 671–683 (2013)
9. Krishnamoorthy, A., Pramod, P.K., Deepak, T.G.: On a queue with interruptions and repeat/resumption of service. Non Linear Anal. Theory Methods Appl. **71**, e1673–e1683 (2009)
10. Neuts, M.F.: Matrix-Geometric Solutions in Stochastic Models: An Algorithmic Approach. The Johns Hopkins University Press, Baltimore (1981)
11. Takine, T., Sengupta, B.: A single server queue with service interruptions. Queueing Syst. **26**, 285–300 (1997). https://doi.org/10.1023/A:1019189326131
12. White, H., Christie, L.S.: Queueing with preemptive priorities or with breakdown. Oper. Res. **6**(1), 79–95 (1958)

Simulation of Railway Marshalling Yards Based on Four-Phase Queuing Systems

Maksim Zharkov[1(✉)], Anna Lempert[1,2], and Michael Pavidis[3]

[1] Matrosov Institute for System Dynamics and Control Theory of Siberian Branch of Russian Academy of Sciences, Irkutsk, Russia
[2] Irkutsk National Research Technical University, Irkutsk, Russia
lempert@icc.ru
[3] Irkutsk State Transport University, Irkutsk, Russia
http://idstu.irk.ru, http://www.istu.edu/eng/, https://www.irgups.ru

Abstract. Design and study of mathematical models of marshalling yards in order to increase productivity and ensure their smooth operation is relevant, since these objects are key elements for the organization of freight transport on the railway network. In this work, we develop a mathematical model for the operation of a marshalling yard in the form of a four-phase queuing system with *BMAP* flow and group service of requests. Each phase is a non-Markov multichannel queuing system with a finite queue and group service of requests in the channel. For its numerical study, we create and implement a simulation model. The proposed mathematical apparatus and software are tested on for the operating marshalling yard, which is typical and located on the East Siberian Railway. We demonstrate that it allows us to assess the current level of operation, determine the maximum permissible load and find bottlenecks in the structure of the selected station and then eliminate them.

Keywords: Mathematical modeling · Simulation model · Multiphase queuing system · *BMAP* · Railway marshalling yard

1 Introduction

The paper is devoted to construction and study of mathematical models of marshalling yards to increase their productivity and ensure their uninterrupted operation. This problem is relevant, since the volume of transported cargo (goods), the stability of the transportation process and the efficiency of the railway network as a whole depend on the quality of these facilities [1,2]. Various approaches and methods are used to solve this problem, among which deterministic models, both optimization and predictive play a key role [2–4]. However, their use is often not effective enough when it is necessary to take into account the possibility of

The reported study was funded by RFBR, project number 20-010-00724, and by RFBR and the Government of the Irkutsk Region, project number 20-47-383002.

A. Dudin et al. (Eds.): ITMM 2020, CCIS 1391, pp. 143–154, 2021.
https://doi.org/10.1007/978-3-030-72247-0_11

failure of signaling and communication devices, breakdowns of operating locomotives, technological windows, and many other random factors. In this case, useful research tools are various kinds of probabilistic (stochastic) models [4–6]. In particular, these can be queuing systems (QS) that are well suited for describing objects where the same type of actions are regularly repeated. Queuing systems are most often used to simulate the operation of information [7,8] and telecommunication systems [9–11]. The features of such systems are the random nature of the data receipt and non-deterministic processing, regardless of their semantic meaning. Some technical objects from other fields, such as transport systems, have similar properties. Hence, the apparatus of queuing theory is suitable for describing the railway marshalling yards operating.

Previously, using the queuing theory, we performed a mathematical description of the operation of specific passenger transport systems with a multi-level hierarchical structure and complex incoming passenger traffic [12,13]. To describe the system structure, we use a three-phase QS, in which each phase corresponds to a separate functional subsystem (service level). To model the transport flow, we use the Batch Markovian Arrival Process (*BMAP*) [14–16], which allows us to take into account the presence of several sub-flows with different characteristics and combine them. Note that multiphase QS with *BMAP* are often used to describe fragments of communication systems and networks [17,18]. However, as far as we know, they have not been previously used in the transport sector.

In this paper, we improve this approach by using four-phase QS and apply it to simulate the operation of a new class of objects marshalling yards. The resulting QS has a complex structure, and in the general case, it seems to be not possible to determine its characteristics analytically. Therefore, we study it numerically using simulation methods. Based on the results of computational experiments, we give recommendations for improving the parameters of the marshalling yard.

2 Subject of Research

Marshalling yards are intended for mass disbanding of trains into separate groups of cars, sorting these groups in accordance with the further direction of movement as well as forming and dispatching new trains. The main technological subsystems of marshalling yards, on which its performance depends, are as follows. A receiving yard is a place where trains are received, and locomotives are uncoupled. A hump is intended for disbanding the train and sending separate groups of cars to the sorting bowl. A sorting bowl is responsible for accumulating cars in accordance with the direction and moving them to the departure yard. In a departure yard, technical and commercial inspection of trains is performed, and they are prepared for further departure from the system.

You can see that the yards are separate functional elements. First, they perform various operations with cars. Secondly, each yard has its own service devices that have different processing capacity. Third, parks can be located rather far from each other, depending on the structure of the station. Thus, it is logical to describe these functional elements as separate phases.

We consider marshalling yard IrS that located on the East Siberian Railway. It is a two-system system with a sequential arrangement of yards. Both systems belong to the general type of railway marshalling yards, i.e. include an receiving yard, a sorting bowl yard with a hump, and a departure yard. They are almost the same, so we study the operation of only the odd system, the scheme of which is shown in Fig. 1.

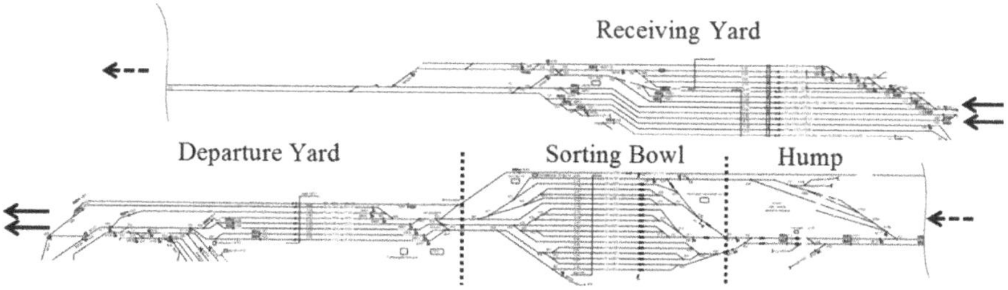

Fig. 1. Diagram of the odd IrS station system

3 Field Research Results

In the period from 1.09.2020 to 30.10.2020, a full-scale survey of the odd IrS station system was performed. We obtained normative and actual (statistical) indicators of the being of cars in each subsystem and technical characteristics of the station, such as the number of shunting locomotives and working crews, the number and capacity of tracks, and the station layout. The sample size for each subsystem was 951 elements. Each element is the service time of a group of cars in the subsystem. We also collected statistical data on incoming trains for this period, including category, arrival time and a number of cars. The sample size was 2,886 elements. Based on this information, the technical and time parameters of car servicing, as well as the characteristics of the incoming train flow, are found.

The parameters of the considered system are as follows. The receiving yard has 7 tracks with an average capacity of 74 conventional cars (conv. cars) each and two locomotives that move trains to the hump. The hump has two railway tracks with a capacity of 85 conv. cars and a device for disbanding trains with a processing capacity of up to 3500 conv. cars per day. The sorting bowl has 18 tracks with an average capacity of 60 conv. cars, and three locomotives are used to complete the formation of trains. The departure yard has 8 tracks of 75 service cars, and two technical inspection teams send trains in two directions.

Table 1 shows the planned and actual average indicators of cars' being in each subsystem. Where ***Tp*** is scheduled service time (h), ***Tf*** is actual average service time (h).

Table 1. Characteristics of the odd system of IrS station

	Receiving yard	Hump	Sorting bowl	Departure yard
Tp	1.02	0.40	1.30	2.04
Tf	0.97	0.45	1.23	2.03

Operation characteristics of service devices in each yard are obtained using statistical processing of data on time spent by cars in each subsystem. These results are presented in Table 2. Note that the truncated normal distribution turned up to be the most suitable distribution law for describing the service time in subsystems [19]. This is natural since the station staff tends to bring the duration of technical operations closer to the standard values. At the same time, the duration of technical operations cannot be zero or take negative values. But anyway various deviations arise due to the influence of random factors. For example, if a train consists of cars with only two or three destination stations, then it can be disbanded by the hump in 10 min instead of 24 min with respect to the standard.

Table 2. Operation characteristics of service devices of the odd system of IrS station

	Receiving yard	Hump	Sorting bowl	Departure yard
F	*N*(0.97, 0.15)	*N*(0.45, 0.10)	*N*(1.23, 0.30)	*N*(2.03, 0.30)
X	0.4–1.5	0.2–1.0	0.1–3.1	0.8–3.3
W	*B*(75, 0.91)	*B*(75, 0.91)	*B*(71, 0.90)	*B*(71, 0.90)
X	58–75	58–75	53–71	53–71

Here ***F*** is a distribution of service time, ***W*** is a distribution of a serviced group size; $N(\mu, \sigma)$ is the truncated normal distribution, its probability density function

$$f(x) = \begin{cases} 0, & \text{if } x \leq x_1, \\ \dfrac{1}{\sigma} \dfrac{\phi\left(\frac{x-\mu}{\sigma}\right)}{\Phi\left(\frac{x_2-\mu}{\sigma}\right) - \Phi\left(\frac{x_1-\mu}{\sigma}\right)}, & \text{if } x_1 < x < x_2, \\ 0, & \text{if } x \geq x_2, \end{cases}$$

where μ is a mathematical expectation, σ is a standard deviation of a normal distribution, $\phi(x) = \frac{1}{\sqrt{2\pi}} \exp\left(-\frac{1}{2}x^2\right)$, $\Phi(x) = \frac{1}{2}\left(1 + \text{erf}(x/\sqrt{2})\right)$ [19]; $B(n, p)$ is a binomial distribution, where n is a number of trials, p is a probability of success; $X = \{x_1 < x < x_2\}$ an interval that contains all values of the random variable.

Three categories of trains arrive at the station: transit trains with processing, transit trains without processing, and local trains. Their arrival is regulated by the dispatchers' office in shifts. However, due to intense intra-station work, it is not always possible to distribute evenly incoming trains between day and night shifts. Based on statistical data, we found that the average number of trains under disbanding (local and transit with processing) in the day and night shifts are 11 and 6, respectively; and non-disbanding (transit without processing) ones are 1.5 trains per shift.

A significant part of the incoming train flow is made up of transit trains that run throughout the territory of the Russian Federation. This group of trains is sensitive to random factors due to the huge distance of the route, so the dispatching department can not effectively plan the schedule for all categories of trains on a separate section of the railway network. As a result, significant deviations of train traffic from the schedule can appear. Thus, the average deviation from the schedule is two hours for the considered station. For this reason, we assumed that the arrival time of trains is a random variable. Based on the results of statistical processing of the available data, we found that the time between train arrivals is distributed exponentially with different parameters for each train flow.

The number of cars in a train obeys the binomial distribution $B(75, 0.91)$, their average number is 68.25, and the maximum one is 75 for all categories of trains. The maximum recorded number of trains received for disbandment per day was 23. With such a train traffic, the station operates in a "forced" mode.

4 Mathematical Model

4.1 Generalized Marshalling Yard Model

We propose to use four-phase queuing systems to model the operation of a marshalling yard. This allows us to describe the four-stage servicing of the flow of cars at the station, as well as determine the capacity of each of the subsystems separately. The construction of a mathematical model of a marshalling yard operation in the form of a QS occurs in two stages. At first, we describe the incoming car flow; at second, we deal with its servicing.

The station receives trains of three categories with different parameters for day and night shifts. Trains of different categories can arrive with a significant deviation from the schedule, i.e. randomly. For the same reason, trains that are scheduled to arrive in the defined shift may arrive at the station in the next one. We consider the arriving train as a group of requests since the cars are serviced independently and occupy a certain place in the system. Based on the results of statistical processing of data on incoming car traffic, it can be assumed that the size of groups of cars and the time of their arrival in the system are random variables [5,6]. Thus, the total incoming car flow is correlated, non-stationary (piecewise stationary) and allows group arrival. To formalize it we use the *BMAP* model [14–16].

In addition, the *BMAP* flow provides significant potential for the development of the model, since it can be used to describe a more complex transport flow. Due to the peculiarities of marshalling yard operation, it is possible not to divide cars into types, since all operations are performed the same for all cars. However, if we model freight stations, the type of car affects on the choice of a loading front that will serve the car, as well as the duration of loading. In this case, the *BMAP* flow turns out to be one of the most suitable mathematical model. Since we are planning to consider the operation of such a station, we assume using *BMAP* for marshalling yard modeling as well.

BMAP (Batch Markovian Arrival Process) is a generalization of the group Poisson process, allowing the change in the intensity of the arrival of request groups, but keeping the basic Markov structure.

We have a Markov chain v_t with continuous time and state space $\{0, 1, ..., W\}$. The intensity of arrival request groups λ_v depends on the state number of the Markov chain v_t. The residence time in each state is exponentially distributed with parameter λ_v. With probability $p_k(v, v')$ the chain can go to state v'. This generates a group of random size $k \geq 0$. The normalization condition is satisfied,

$$\sum_{k=0}^{\infty} \sum_{v=0}^{W} p_k(v, v') = 1.$$

The transitions intensities are written in matrix form

$$\begin{aligned} (D_0)_{v,v} &= -\lambda_v, \quad v = \overline{0, W}, \\ (D_0)_{v,v'} &= \lambda_v p_0(v, v'), \quad v, v' = \overline{0, W}, \\ (D_k)_{v,v'} &= \lambda_v p_k(v, v'), \quad v, v' = \overline{0, W}, k \geq 1. \end{aligned} \tag{1}$$

To model the operation of marshalling yards, we propose using a special open four-phase QS. Each phase corresponds to the subsystem of the marshalling yard and has the following characteristics [5]:

- each phase has n service devices (channels) and m places in a queue, $n, m < \infty$;
- requests from the queue are serviced in accordance with FIFO (first in, first out);
- group service of requests in each channel is allowed;
- the length of the queue, the distribution of service time for a group of requests in the channel, and its size may differ for each phase;
- channels of the first three phases are temporarily blocked if there are no available places for accepting requests in the queue of the next phases;
- groups of requests are accepted according to the discipline of complete rejection; if there is not enough space for at least one request, the arrived group is lost [5,15].

In terms of the queuing theory [5], the marshalling yard operation model can be written as

$$BMAP/G^{X1}/n_1/m_1 \rightarrow */G^{X2}/n_2/m_2 \rightarrow */G^{X3}/n_3/m_3 \rightarrow */G^{X4}/n_4/m_4$$

where G means an arbitrary distribution law of the service time of requests in the channel; Xi is distribution of the group size, accepted for servicing in the channel of phase i; m_i is a queue length, n_i is a number of channels for phase i.

For parametric identification of a specific station model in the form of a multiphase QS, it is necessary, first, to determine the parameters of the *BMAP* flow, such as the number of sub-flows, the intensity of the arrival of request groups, and the distribution of the group size. Second, for each phase, we need to set the number of places in the queue, the number of channels, the distribution of service time, and the size of the served request groups.

The following parameters and components are not taken into account in the presented generalized model of the marshalling yard: 1) train formation plan; 2) specialization of railway tracks; 3) administrative staff; 4) duty station staff; 5) equipment breakdowns and failures; 6) expenses related to the downtime of cars in the station's yards.

Train formation plan regulates the order of formation of all categories of car and train traffic. The model considers only the car flow arriving at the station. Specialization of railway tracks taken into account indirectly by the capacity of the yards. The capacity is calculated based on the tracks' specialization for train flows with and without disbanding, even and odd directions. Administrative staff does not directly affect the operation of the service devices. Duty station staff is an integral part of the service devices (locomotives, service crews, etc.), so no need to consider it separately. The purpose of the simulation is to determine the capacity of the station in a regular operating regime, so events (5) are not taken into account. The financial and economic indicators of the station (6) are beyond the scope of the study.

4.2 Model of the Odd IrS Station System

The station normally processes only freight car traffic. Passenger trains pass by the station and do not affect its operation, therefore this flow is not considered. Also, we do not take into account freight trains that are not under the disbanding for the following reasons. First, these trains run past the first three subsystems of the station and only stop at the departure yard. Second, they are served on separate tracks and do not significantly affect the processing of other categories of trains. The model takes into account only the oncoming car traffic with processing. We consider transit trains with processing and local trains as the same type of train, since the number of cars in these types of trains is described by the same distribution. For its mathematical description, we use the *BMAP*.

Trains arrive in the system in day and night shifts with different parameters. Hence, there are two sub-flows, and the control Markov chain has two states $\{0, 1\}$. Since the duration of the shifts is the same, the probabilities of the chain transition from one state to another are $p_0 = p_1 = p = 0,5$. The maximum number of cars in a train is 75, therefore the *BMAP* flow includes 76 matrices of size $2 \times 2, D_k, k = \overline{0,75}$. Their elements are calculated by formulas (1), where $\lambda_0 = 11/12 = 0,92$, $\lambda_1 = 6/12 = 0,5$ and the transition probabilities $p_k(v, v') = pf(k)$, $v, v' = \overline{0,1}$, $k = \overline{0,75}$, $f(k)$ is an arrival probability of a group of k cars, the group size obey binomial distribution $B(75, 0.91)$.

The model of the odd system of IrS station is as follows. The system has four servicing phases. Phase 1 corresponds to the arrival yard. We assume that the two thrust locomotives are channels, the tracks of the yard are the queue with a capacity of 518 cars. Phase 2 describes the operation of the hump. Let one thrust track and the sorting device be one channel, the second track is an queue with 85 places. Phase 3 simulates the operating of the sorting bowl. The channel are three thrust locomotives, the queue is the tracks of the park with 1080 places. Phase 4 describes the operation of the departure yard, where we consider two departure tracks (main course) as channels, the queue is also the tracks of the yard with 600 places. The distributions of the service time and the sizes of the served groups of requests in the channels at each phase correspond to the characteristics of the service devices, which are presented in Table 2.

In terms of the queuing theory, the model of the odd IrS station system takes the form

$$BMAP/G^B/2/518 \rightarrow */G^B/1/85 \rightarrow */G^B/3/1080 \rightarrow */G^B/2/600,$$

where B is binomial distribution. Figure 2 shows the scheme of the described system.

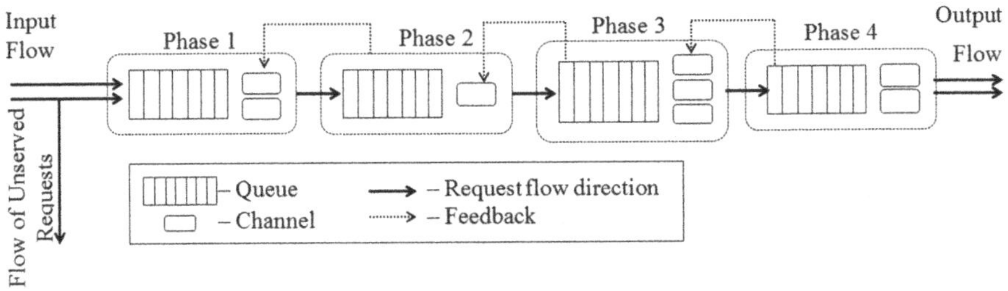

Fig. 2. Scheme of four-phase QS

We study the resulting four-phase QS numerically using a simulation model [20]. The desired characteristics, i.e. performance indicators are the probabilities of servicing both the request and the group of requests, the average time the request stays in the system, the average queue length, and the channel blocking time for each phase.

5 Computational Experiment

The simulation model of a multiphase QS is based on a discrete-event modeling approach and is implemented as a software module [12,13]. It is intended for the numerical finding of the probabilities of states and, on their basis, for determining the performance indicators of the QS, as well as for carrying out multivariate scenario calculations. The user can set the number of phases, the number of channels and the maximum queue length, as well as the parameters of the channels and *BMAP* flow. The channels work independently, it is permissible to have a different distribution of service time for each channel. As soon as the channel is released, a request or group of requests from the queue can be received. If the queue is empty, the channel goes into standby mode. The data of the module operation process are displayed in tabular form. The results of the simulation model can be saved in MS Excel format.

We now turn to the description of the computational experiment. We carry out scenario modeling of the operation of the proposed four-phase QS (see Fig. 2) for various parameters of the incoming car flow and servicing process. Model time is five weeks for all experiments. We chose this value because it is the minimum time for which the simulation model allows calculating the stationary characteristics of the studied QS. The main indicator that the station can cope with the load is equal to zero loss probability, i.e. the ability of the QS to service all requests.

Experiment 1. Table 3 shows the results of simulation of the four-phase QS operation at the intensity of sub-flows $\lambda_0 = 0,92$ and $\lambda_1 = 0,5$ groups of requests per hour for day and night shifts.

Table 3. Results of experiment

	Arrived	Losses	T_{sist} (m)	P_G	P_R
Groups	566.90	0	585.48	0	0
Requests	36755.80	0			
	$\overline{k}$	l	T_{ph} (m)	T_{lock} (m)	P_{lock}
Phase 1	0.95	10.13	79.57	1071.20	0.0213
Phase 2	0.60	12.88	57.67	0	0
Phase 3	1.98	13.43	118.20	3471.90	0.0689
Phase 4	1.87	138.22	330.04	–	–

Here and further T_{sist} is an average residence time of a request in the system, P_G is a loss probability for a group of requests, P_R is a loss probability for a separate request, $\overline{k}$ is an average number of busy channels, l is an average queue length, T_{ph} is an average residence time of a request in the phase, T_{lock} is the total blocking time of the phase, P_{lock} is a channel blocking probability.

You can see that in this case the average time spent by one request (car) in the system is just under nine hours. The average queue length in Phases 1–3 is small, and the loss probability is zero. According to the observed indicators, we can conclude that the odd system of IrS station is operating normally and coping with the current load.

Next, we study the operation of the station when the incoming car traffic increases. In the first step (experiment 2), we increase the average number of incoming trains by two for each shift relative to the observed car traffic, i.e. up to 21 trains per day. On the second (experiment 3) it is increased up to 23 trains per day, which corresponds to the maximum observed value.

Experiment 2. Table 4 shows the results of simulation of the four-phase QS operation at the intensity of sub-flows $\lambda_0 = 13/12 = 1,08$ and $\lambda_1 = 8/12 = 0,67$ groups of requests per hour for day and night shifts, respectively.

Table 4. Results of experiment 2

	Arrived	Losses	T_{sist} (m)	P_G	P_R
Groups	669.45	8.00	1398.78	0.0120	0.0118
Requests	45713.64	541.55			
	$\overline{k}$	l	T_{ph} (m)	T_{lock} (m)	P_{lock}
Phase 1	1.27	47.54	130.55	9729.27	0.0965
Phase 2	0.75	31.27	75.03	4202.80	0.0834
Phase 3	2.65	355.48	557.73	42188.27	0.2790
Phase 4	1.97	433.39	635.48	–	–

According to the results of calculations, there is a non-zero probability of blocking channels in all phases, which is especially high in Phase 3 (28%). At the same time, in Phase 4 (departure yard), the average number of busy channels and the average queue length approach the maximum possible values (2 and 600), and the average time spent by the request in the system is the highest among all phases. Therefore, the departure yard is a bottleneck. Its capacity is clearly insufficient. A further increase in the intensity of train arrivals will lead to a sharp increase in the loss probability.

Increasing the number of channels in the departure yard is difficult, as this requires creating new tracks on the main course. Adding new technical inspection teams is economically unprofitable, since some of them will be idle if the traffic is low. Therefore, to reduce the downtime of trains in the departure yard, it is advisable, first, to increase the number of people in the teams, and secondly, to organize the timely issuance of travel locomotives for trains. In the latter case, additional locomotive crews are required to ensure that these locomotives are fully equipped. Based on a consultation with the station's chief engineer, we found that applying these recommendations will reduce the average service time in the departure yard by 25%.

We now study the operation of the station while reducing the average service time in Phase 4 channels by 25%. With the new operating parameters, we assume that the station is able to serve up to 23 trains per day in normal mode, i.e. 3 more trains for each shift compared to the average observed values.

Experiment 3. Now let the service time in the channels of Phase 4 have the normal distribution $N(1.52, 0.50)$; in the day shift, groups of requests arrive with an intensity of $\lambda_0 = 14/12 = 1,17$ per hour, in the night shift with $\lambda_1 = 9/12 = 0,75$ per hour. The results of calculations are presented in Table 5.

In this case, the loss probability is non-zero, which indicates that the processing capacity of the station is insufficient. However, at first, it is quite small (less

Table 5. Results of experiment 3

	Arrived	Losses	T_{sist} (m)	P_G	P_R
Groups	795.18	0.27	757.95	0.0003	0.0003
Requests	54275.27	18.64			
	$\overline{k}$	l	T_{ph} (m)	T_{lock} (m)	P_{lock}
Phase 1	1.35	30.01	96.65	3942.09	0.0652
Phase 2	0.79	26.95	65.20	112.09	0.0037
Phase 3	2.61	91.08	196.54	15752.64	0.1736
Phase 4	1.96	309.86	399.57	–	–

than 0.05%). Second, the simulation model allows the arrival of four or more trains in one hour. If the queue for Phase 1 (arrival yard) is more than half full during this time period, some trains may be rejected. But for a real object, the dispatcher unit distributes the load more evenly during the shift. Therefore, the observed loss probability can be ignored.

Thus, reducing the average service time by 25% in the departure yard will increase the capacity of the station by 33.6%. To further improve the system's performance, a change in the station's operational technology will be required, which is acceptable only if it is completely modernized.

6 Conclusions

The paper presents a generalized model of railway marshalling yards based on four-phase QS, which allows us to describe four functional subsystems that are the stages of car flow processing. The model is highly complex and unsuitable for analytical research, but modern simulation tools allow us to analyze it numerically.

The created software and mathematical apparatus have been applied for studying an operating marshalling yard located on the East Siberian railway. We have shown that it allows us to assess the current level of operation, determine the maximum permissible load and find bottlenecks in the structure of the selected station and then eliminate them.

We suppose further improvement of the proposed approach due to the application of queuing networks apparatus for modeling railway stations. They make it possible to investigate not only objects with a linear structure but also to describe the circular movement, which is typical for marshalling yards.

The authors are grateful to Doctor of Technical Sciences, Professor of Tomsk State University A.A. Nazarov for fruitful discussion and helpful comments.

References

1. Pyrgidis, C.: Railway Transportation Systems. Design, Construction and Operation, CRC Press, Boca Raton (2016)

2. Borndorfer, R., Klug, T., Lamorgese, L. (eds.): Handbook of Optimization in the Railway Industry. Springer, Heidelberg (2018). https://doi.org/10.1007/978-3-319-72153-8
3. Milenkovic, M., Bojovic, N.: Optimization Models for Rail Car Fleet Management. Elsevier Inc., Amsterdam (2019)
4. Potthoff, G.: Verkehrs stromungs lehre. Transpress, Berlin (1980)
5. Kerner, B.: Introduction to Modern Traffic Flow Theory and Control. Springer, New York (2009). https://doi.org/10.1007/978-3-642-02605-8
6. Bolch, G., Greiner, S., De Meer, H., Trivedi, K.: Queueing Networks and Markov Chains: Modeling and Performance Evaluation with Computer Science Applications. Wiley, New York (2006)
7. Lisovskaya, E.Y., Moiseev, A.N., Moiseeva, S.P., Pagano, M.: Modeling of mathematical processing of physics experimental data in the form of a non-Markovian multi-resource queuing system. Russ. Phys. J. **61**(12), 2188–2196 (2019). https://doi.org/10.1007/s11182-019-01655-6
8. Dudin, A., Klimenok, V., Vishnevsky, V.: The Theory of Queuing Systems with Correlated Flows. Springer, Cham (2019). https://doi.org/10.1007/978-3-030-32072-0
9. Mauro, M., Liotta, A.: Statistical assessment of ip multimedia subsystem in a softwarized environment: a queueing networks approach. IEEE Trans. Netw. Serv. Manage. **16**(4), 1493–1506 (2019)
10. Bushkova, T., Danilyuk, E., Moiseeva, S., Pavlova, E.: Resource queueing system with dual requests and their parallel service. Commun. Comput. Inf. Sci. **6**(2), 364–374 (2019)
11. Nazarov, A., Nikitina, M.: Application of Markov chain ergodicity conditions to the study of the existence of stationary regimes in communication networks. Autom. Control Comput. Sci. **1**(1), 50–55 (2003)
12. Lempert, A., Kazakov, A., Zharkov, M.: A stochastic model of a transport hub and multi-phase queueing systems. Adv. Intell. Syst. Res. **158**, 117–123 (2018). https://doi.org/10.2991/iwci-18.2018.21
13. Zhuravskaya, M., Lempert, A., Anashkina, N., Zharkov, M.: Issues of sustainable urban mobility simulation. Bus. Logist. Mod. Manage. 439–452 (2018)
14. Banik, A.D.: Single server queues with a batch Markovian arrival process and bulk renewal or non-renewal service. J. Syst. Sci. Syst. Eng. **24**(3), 337–363 (2015). https://doi.org/10.1007/s11518-015-5268-y
15. Artalejo, J., Gomez-Corral, A., He, Q.-M.: Markovian arrivals in stochastic modelling: a survey and some new results. Stat. Oper. Res. Trans. **34**(2), 101–156 (2010)
16. Cordeiro, J., Kharoufeh, J.: Batch Markovian Arrival Processes (BMAP). Wiley Encyclopedia of Operations Research and Management Science, Wiley, New York (2011). https://doi.org/10.1002/9780470400531.eorms0096
17. Kim, Ch., Park, S. Dudin, A. (eds.): Investigation of the $BMAP/G/1 \rightarrow \bullet/PH/1/M$ tandem queue with retrials and losses. Appl. Math. Model. **34**(10), 2926–2940 (2010). https://doi.org/10.1007/978-3-642-31217-5_39
18. Kim, Ch., Dudin, A., Klimenok, V., Taramin, A.: A tandem $BMAP/G/1 \rightarrow \bullet/M/N/0$ queue with group occupation of servers at the second station. Math. Probl. Eng. **2012**, 1–26 (2012). https://doi.org/10.1155/2012/324604
19. Burkardt, J.: The truncated normal distribution. Florida State University, Tallahassee (2018)
20. Law, A., Kelton, W.: Simulation Modelling and Analysis. McGraw-Hill, New York (2000)

Numerical Methods to Analyses of Queuing Systems with Instantaneous Feedback, Positive Server Setup Time and Impatient Calls

Agassi Melikov[1(✉)], V. Divya[2], and Sevinc Aliyeva[3]

[1] Institute of Control Systems, National Academy of Science of Azerbaijan, Baku, Azerbaijan
[2] Mathematics, N.S.S. College, Cherthala, Palakkad, India
[3] Baku State University, Baku, Azerbaijan
s@aliyeva.info

Abstract. The Markovian model of single server queuing system with instantaneous feedback and positive server setup time for servicing of feedback calls is proposed. Arrival rate of calls from outside depends on server status which might be in two regimes: working regime and setup regime. Calls are impatient when server is in setup regime. Two approaches are developed to study the system: matrix-geometric method and space merging method. Results of numerical experiments are demonstrated.

Keywords: Queuing system · Instantaneous feedback · Server setup time · Impatient calls · Calculation methods

1 Introduction and Related Work

In this paper generalization of model of queuing systems with feedback (QSwFB) in Melikov *et al.* (2020) [1] is proposed by introducing the effect of impatience of calls when server is in setup time.

Let's briefly consider the state of art the problem. Takac's work have been a pioneer [2,3] where models of queuing systems with instantaneous [2] and delayed feedback [3] are investigated. Such kind of models are intensively investigated in last two decades. Kumari [4] gives a detailed review of work done in feedback queuing models until 2011. Further developments could be accessed from Koroliuk *et al.* (2016) [5] and Melikov *et al.* (2015, 2019) [6,7].

In Krishna Kumar *et al.* (2009) [8] a multi-server feedback retrial queueing system with finite waiting room $M/M/c/N+c$ and constant retrial rate is analyzed. It is shown that the mathematical model of the investigated system is a quasi-birth-and-death (QBD) process and the necessary and sufficient condition for stability of the system is obtained. Performance measures of the system are

A. Dudin et al. (Eds.): ITMM 2020, CCIS 1391, pp. 155–170, 2021.
https://doi.org/10.1007/978-3-030-72247-0_12

calculated by using matrix geometric method and the impact of various parameters on the system performance measures are illustrated numerically. Do (2010) [9] propose a more efficient computation method to calculate the steady state probabilities when $N+c$ is large.

In Krieger *et al.* (2005) [10] the feedback queue model of the type BMAP/PH/1 operating in a Markov random environment that has a finite state space is investigated. It is assumed that changing state of random environment causes instantaneous changes of the parameters of the BMAP input, the PH service processes and the feedback probabilities. The stationary distribution of multidimensional continuous time Markov chain (CTMC) describing the behavior of the system is calculated by means by reduction to discrete time Markov chain (DTMC) at transition epochs. The Laplace-Stieltjes transform of the calls sojourn time distribution is calculated and practicability of the developed algorithms is illustrated by numerical experiments.

Rajadurai *et al.* (2018) [11] considered a single server feedback retrial queueing system with multiple working vacations and vacation interruption. An arriving call may balk the system at some particular times and as soon as orbit becomes empty at regular service completion instant, the server goes for a working vacation (WV). The server works at a lower service rate during WV period. After completion of regular service, the unsatisfied customer may rejoin into the orbit to get another service as feedback call. The normal busy server may get to breakdown and the service channel will fail for a short interval of time. The probability generating function (PGF) for the system size is obtained by using the supplementary variable method and system performance measures are determined. Similar model of retrial queueing model with feedback and working breakdown services has been investigated in Rajadurai *et al.* (2020) [12]. The regular busy server may become defective by negative customers that arrive only at the service time of a positive customer and remove the positive customer from the service. At the instant of failure, the main server is sent for repair and the repair period begins immediately. During the repair period, the server gives service at a lower rate (as in previous paper with WV [11]). The PGF for system size and orbit size are obtained using the method of supplementary variable as in [11]. By using PGF analytical expressions for performance measures are calculated.

The M/G/1 retrial queue with Bernoulli feedback and single vacation where the server is subjected to starting failure is analyzed by Mokaddis *et al.* (2007) [13]. The server leaves for a vacation as soon as the system becomes empty. The sojourn time in orbit is assumed to follow an arbitrary distribution and when the server returns from the vacation and finds no calls, he/she waits free for the call to arrive from outside the system. The system size distribution at random points and various performance measures are derived.

A discrete-time feedback queueing system of the type Geo/G/1/∞ under an (m, N)-policy is analyzed in Hernández-Díaz *et al.* (2009) [14]. The system operates under an N-policy with an early setup where the startup period begins when m ($\leq N$) calls accumulate in the system. Moreover, it is assumed that the i-th service of each call is either unsuccessful (and then the call feedback for another service) with probability (w.p.) α_i or successful (and then the call leaves the system forever) with complementary probability $1-\alpha_i$. The joint PGF of the server state and the system length as well as the main performance measures are obtained. The distributions of the lengths of the idle, setup, standby and busy periods, as well as the distribution of the number of calls served during a busy period, are also derived.

Zhao *et al.* (2020) [15] consider the model of cognitive radio networks in form of feedback queuing system. In this paper, in order to reduce possible packet loss of the primary users (PUs) it is assumed that there is a buffer with a finite capacity for the PU packets. At the same time, focusing on the packet interruptions of the secondary users (SUs), feedback scheme for the interrupted SU packets is proposed. For evaluate the influence of the finite buffer setting and the feedback probability to the system performance, DTMC model is developed. The expressions of some performance measures of the PU packets and the SU packets are obtained and numerical results to evaluate how the buffer setting of the PU packets and the feedback probability influence the system performance are demonstrated. For solving the problems, the matrix-geometric method is used.

The models of feedback queuing systems with two types of calls are considered in Lee (2005) [16] and Krishnamoorthy *et al.* (2018) [17]. Lee (2005) [16] consider the model of feedback retrial queue with two types of calls where after being served each call either joins the retrial group or departs the system forever. If an arriving priority call finds the server idle, he/she immediately starts to receive service; if call finds the server busy, he/she is queued in the priority group and then served in accordance with some conservative discipline. On the other hand, when an arriving non-priority call finds the server idle, he/she obtains service immediately; if call finds the server busy, he/she joins the retrial group in order to seek service again after a random amount of time (delayed feedback). The retrial time is exponentially distributed and is independent of all previous retrial times and all other stochastic process in the system. Both kinds of calls who has received service either departs the system or feedback to appropriate groups in accordance to Bernoulli scheme. In M/G/1 retrial queueing system with two types of calls and feedback, the joint PGF of the number of calls in two groups are derived by using the supplementary variable method. Krishnamoorthy *et al.* (2018) [17] consider the model with two-priority queueing system according to a marked Poisson process. Both waiting rooms have infinite capacity. Calls are served one at a time according to FIFO discipline on priority basis: those in waiting line 1 ($P1$) are given priority over the ones in line 2 ($P2$). The service time is class-dependent phase type. After completion of service, high priority calls may feedback for service according to a Bernoulli process. Feedback calls are sent to the low priority queue. When at a service completion epoch of a $P1$ call, if there is none left behind in $P1$ line, then the server goes to serve $P2$ class. For the two-priority queueing system, we assume that $P2$ calls are not allowed an additional feedback. Both preemptive and non-preemptive service disciplines are analyzed. Waiting time distribution of both type of calls is derived.

The models of feedback tandem queuing systems are investigated in some papers as well. So, in Kim (2007) [18] the problem of calculation of the stationary state distribution for the BMAP/G/1→·PH/1/M tandem queue with losses and feedback of calls is solved. Expressions for some performance measures of the system are derived and the numerical examples are presented. Similar model was used to study an open tandem communication network where each node consists of a buffer and a transmitter in Raghavendran, *et al.* (2014) [19]. The two buffers are $Q1$, $Q2$ and transmitters are $S1$, $S2$ connected in tandem. The arrival of packets at the first node follows non-homogeneous Poisson processes with a mean arrival rate. It is also assumed that the packets are transmitted through the transmitters and the mean service rate in the transmitter is linearly dependent on the content of the buffer connected to it. It is assumed that the packet after getting transmitted through first transmitter may join the second buffer which is in series connected to $S2$ or may be returned back buffer connected to $S1$ for retransmission (feedback) with certain probabilities. The packets delivered from the first node and arrived at the second node may be transmitted out of the network or returned back (feedback) to $Q2$ for retransmission. After getting transmitted from the first transmitter the packets are forwarded to $Q2$ for forward transmission w.p. 1-θ or returned back to the $Q1$ w.p. θ. The packets arrived from the first transmitter are forwarded to $Q2$ for transmission and exit from the network w.p. 1-π or returned back to the $Q2$ w.p. π. The service completion in both the transmitters follows Poisson processes with the different parameters. The transmission rate of each packet is adjusted just before transmission depending on the content of the buffer connected to the transmitter. The transient analysis of the model is capable of capturing the changes in the performance measures of the network explicitly. It is observed that the feedback probabilities have significant influence on the overall performance of the network.

Varalakshmi *et al.* (2016) [20] consider a single server retrial tandem queueing system with immediate Bernoulli feedbacks, single vacation and starting failures. The PGF for number of calls in the system when it is idle, busy, on vacation or under repair is found by the use of supplementary variable technique. The performance measures such that mean number of calls in the system/orbit and mean waiting time of a call in the system/orbit were deduced. The analytical results are validated with the help of numerical illustrations. Melikov *et al.* (2016) [21] consider the Markov model of a two-stage queueing network with feedback. Poisson flows arrive to both stages from outside. A part of already serviced calls at the first node instantaneously enter the second node (if there is free space here) while the other calls leave the network. After the service is completed at the second node, there are three possibilities: (i) the call leaves the network; (ii) it instantaneously feeds back to the first node (if there is free space here); (iii) it feeds back to the first node after some delay in orbit. All feedbacks are determined by known probabilities. Both nodes have finite capacities. The mathematical model of the investigated network is a three-dimensional

Markov chain, and a hierarchical space merging algorithm is developed to calculate its steady-state probabilities. The results of numerical experiments are demonstrated.

In indicated works it is assumed that server can start to service of feedback call immediately. However, this assumption sometimes is unrealistic, i.e. to start of servicing of feedback call some positive server setup time is required. Such situations are ubiquitous in production systems where to process the defective part, some positive time for server setup is required.

To our best knowledge there are not works devoted to feedback queuing model with setup time for servicing of calls that required repeated processing. Here the model of infinite queuing system with instantaneous feedback and positive setup time of server is proposed. It is assumed that calls in system are impatient if server is in setup period. Two approach are developed to calculation of steady-state probabilities and performance measures of investigated system: matrix-geometric method of Neuts [22] and space merging method [21].

The rest of the paper is arranged as follows. The system description and construction of the generating matrix of the appropriate two-dimensional Markov chain (2D MC) are given in Sect. 2. Section 3 provides steady state analysis of the model by means of matrix-geometric method. Application of the space merging method is considered in Sect. 4. In Sect. 5 total cost function is constructed and results of numerical experiments are demonstrated. Conclusion remarks are given in Sect. 6.

2 Description of the System and Construction of the Generating Matrix

Consider a QS with one server and an infinite size of waiting room, which receives a Poisson flow of primary calls (p-calls) from the outside. The service times of these calls are independent and identically distributed (iid) r.v. and have a common exponential cumulative distribution function (cdf) with an average value μ^{-1}. After the completion of the servicing process, p-calls, independently of each other and according to the Bernoulli scheme, either leave the system w.p. σ or immediately require repeated servicing with complementary probability $1-\sigma$. Calls that require re-servicing will be called feedback calls (f-calls). The service times of f-calls are also iid r.v. with the same average μ^{-1}.

To start the process of servicing f-calls, the server needs some positive random time, which is called the server setup time. It has an exponential cdf with an average value θ^{-1}, i.e. at any time the server can be in one of two states: in working mode or in setup mode. It is not allowed to interrupt the server setup time, i.e. the f-call, which initiates the server setup process, remains on it for the entire period of its setup.

The p-calls are considered to have information about the server status. This means that if the server is in switch mode, then their intensity is equal to λ_0, otherwise it is equal to λ_1. Calls in the queue are expected to become impatient during the time the server is in setup mode and only the call at the head of the queue being impatient. In other words, during the server setup period, the call that is at the head of the queue leaves the system after a random time, which has an exponential distribution with an average value τ^{-1}.

It is considered that random processes of call arrival, their servicing, server setup time and allowable waiting time in the queue for calls during the server setup period are independent of each other.

It is required to find the joint distribution of the number of calls in the system and the state of the server, as well as the following performance measures of the system: the average number of calls in the system (L_s), fraction of the time the system is in working (T_{wr}) and in setup mode (T_{sr}), as well as the intensity of the server switching from setup mode to working mode (R_{sw}).

The state of the system at an arbitrary moment in time is determined by the two-dimensional vector $(n,\ k)$, where n is the number of calls in the system, k is the state of server, i.e.

$$k = \begin{cases} 0, \text{ if the server is in setup mode,} \\ 1, \text{ if the server is in working mode.} \end{cases}$$

The system operation is described by a two-dimensional MC (2D MC) with the following state space:

$$E = E_0 \bigcup_{n=1}^{\infty} E_n, \tag{1}$$

where $E_0 = \{(0,\ 1)\}$, $E_n = \{(n,\ 0),\ (n,\ 1)\}$, $n = 1, 2, ...$

Elements of the generating matrix Q of the given 2D MC are denoted by $q((n,k),(n',k'))$, i.e. quantities $q((n,k),(n',k'))$ indicates the intensity of the transition from the state $(n,\ k)$ to state $(n',\ k')$. These quantities are defined as:

$$q\left((n,\ 1),\ (n+1,\ 1)\right) = \lambda_1; \tag{2}$$

$$q\left((n,\ 1),\ (n-1,\ 1)\right) = \mu\sigma\ ,\ n > 0; \tag{3}$$

$$q\left((n,\ 1),\ (n,\ 0)\right) = \mu\left(1-\sigma\right),\ n > 0; \tag{4}$$

$$q\left((n,\ 0),\ (n,\ 1)\right) = \theta; \tag{5}$$

$$q\left((n,\ 0),\ (n+1,\ 0)\right) = \lambda_0; \tag{6}$$

$$q\left((n,\ 0),\ (n-1,\ 0)\right) = \tau,\ n > 1. \tag{7}$$

Let us denote by $p(n,k)$ the steady state probability of $(n,k) \in E$ (the ergodicity condition is established below). These quantities satisfy the system of equilibrium equations (SEE), which is compiled on the basis of relations (2)–(7) (due to the obviousness of the compilation, the explicit form of this SEE is not given here). Using the method of generating functions to find stationary probabilities of states for a similar model with patient calls is considered in [1]. Numerical methods for solving this problem are considered below.

3 Matrix-Geometric Method

To apply the matrix-geometric method, the states of the system are numbered lexically, i.e. in order $(0,\,1)\,,\,(1,\,0)\,,\,(1,\,1)\,,\,(2,\,0)\,,\,(2,\,1)\,,\ldots$. Then the generating matrix Q of given 2D MC can be represented as follows:

$$Q = \begin{bmatrix} B_0 \; C_0 \; \ldots \; \ldots \\ B_1 \; A_1 \; A_0 \; \ldots \\ \ldots \; A_2 \; A_1 \; A_0 \\ \ldots \; \ldots \; \ldots \; \ldots \end{bmatrix},$$

where block matrices are defined as follows: $B_0 = (-\lambda_1)$ is matrix of dimension 1×1; $C_0 = \begin{pmatrix} 0 \; \lambda_1 \end{pmatrix}$ is matrix of dimension 1×2; A_0, A_1 and A_2 are matrices of dimension 2×2, i.e.

$$B_1 = \begin{pmatrix} \tau \\ \mu(1-\sigma) \end{pmatrix};$$

$$A_2 = \begin{pmatrix} \tau & 0 \\ 0 & \mu\,(1-\sigma) \end{pmatrix}; A_1 = \begin{pmatrix} -\left(\lambda_0 + \theta + \tau\right) & \theta \\ \mu\sigma & -\left(\lambda_1 + \mu\right) \end{pmatrix}; A_0 = \begin{pmatrix} \lambda_0 & 0 \\ 0 & \lambda_1 \end{pmatrix}.$$

Let be $\nu = (\nu_0,\ \nu_1\ )$ is a vector of state probabilities of a MC with two states and an infinitesimal generator A, where

$$A = A_0 + A_1 + A_2 = \begin{pmatrix} -\theta & \theta \\ \mu\sigma & -\mu\sigma \end{pmatrix},$$

i.e. specified vector $\boldsymbol{\nu} = (\nu_0,\ \nu_1\ )$ is found from the following system of equations:

$$\boldsymbol{\nu}\, A = \mathbf{0}, \quad \boldsymbol{\nu}\, \mathbf{e} = 1, \tag{8}$$

where $\boldsymbol{e} = (1,1)^{\mathrm{T}}$.

From the system of Eqs. (8) we find that, $\nu_0 = \mu\sigma/\left(\theta + \mu\sigma\right)$, $\nu_1 = \theta/\left(\theta + \mu\sigma\right)$.

The quasi birth-death (QBD) process is stable if and if the following condition is satisfied [22]:

$$\boldsymbol{\nu} A_0 \mathbf{e} < \boldsymbol{\nu} A_2 \mathbf{e}. \tag{9}$$

From (9) we find the following condition for the ergodicity of the system:

$$\lambda_1\theta + \lambda_0\mu\sigma < \mu\left(\theta\sigma + \tau\left(1-\sigma\right)\right). \tag{10}$$

When the ergodicity condition (10) is fulfilled, the steady state probabilities of the system are calculated according to the following algorithm. Let us express the steady state probabilities $p(n,k)$, $(n,\,k)\, \in\, E$ in vector form $\mathbf{p} = (\mathbf{p}_0,\ \mathbf{p}_1,\ \mathbf{p}_2, \ldots)$, where $\mathbf{p}_0$ has the dimension 1, i.e. $\mathbf{p}_0 = (p\,(0,\,1))$, while p_n, $n \geq 1$, are two-dimensional vectors, i.e. $\mathbf{p}_n = (p\,(n,0)\,,\,p\,(n,\,1))$.

According to the algorithm of the matrix-geometric method [22], the indicated quantities satisfy the equation $\boldsymbol{p}_n = \boldsymbol{p}_1 R^{n-1}$, $n \geq 2$, where R is the minimal non-negative solution of the following quadratic matrix equation:

$$R^2 A_2 + R A_1 + A_0 = 0. \tag{11}$$

Probabilities of boundary states p_0 and p_1 are found from the following system of equations:

$$p_0 B_0 + p_1 B_1 = 0;$$
$$p_0 C_0 + p_1 (A_1 + R A_2) = 0. \tag{12}$$

System of Eqs. (12) is solved taking into account the normalization condition:

$$p(0, 1) + p_1 (I - R)^{-1} \, e = 1, \tag{13}$$

where I is the 2×2 identity matrix.

Remark 1. In [22–24] proposed efficient algorithms for solving equations of the type (11)–(13), and therefore we do not dwell on this issue.

After finding the steady state probabilities, the desired performance measures of the system are calculated as follows:

$$L_s = \sum_{n=1}^{\infty} \sum_{k=0}^{1} np(n, k). \tag{14}$$

$$T_{sr} = \sum_{n=1}^{\infty} p(n, 0). \tag{15}$$

$$T_{wr} = \sum_{n=0}^{\infty} p(n, 1). \tag{16}$$

$$R_{sw} = \theta \sum_{n=1}^{\infty} p(n, 0). \tag{17}$$

4 Space Merging Method

Note that for certain ratios of the values of the system parameters, the space merging method (SMM) can be correctly used to solve the problem under study. Application of this method allows one to propose explicit formulas for finding the stationary distribution of the studied 2D MC as well as for calculating the performance measures of the original system.

Thus, this method, in particular, can be correctly applied to a model in which the server setup time is significantly less than the intervals between call arrivals, i.e. it is assumed that the relation $\theta >> \max(\lambda_0, \lambda_1)$ is true. In addition, assume that the proportion of calls requiring re-service is much larger than the proportion of calls that do not require re-service, i.e. the following relation is fulfilled: $\sigma << 0.5$. If these conditions are satisfied, then we obtain that the intensities of transitions inside the classes E_n (see (1)) turn out to be much larger than the intensities of transitions between states from different classes.

Then, according to the method proposed in [1], each class E_n, $n = 0, 1, 2, ...$, is represented as a separate merged state $<n>$, $n = 0, 1, 2, ...$, and probabilities of states inside classes E_n, $n = 0, 1, 2, ...$ are denoted by $\rho_n(k)$, $k = 0, 1$. The class E_0 contains only one state (0, 1), therefore it is assumed that $\rho_0(1) = 1$.

Based on relations (4) and (5), we conclude that the probabilities of states within each class E_n, $n > 0$ do not depend on the index n and are calculated as follows:

$$\rho(0) = \frac{\mu(1-\sigma)}{\theta + \mu(1-\sigma)}, \ \rho(1) = \frac{\theta}{\theta + \mu(1-\sigma)}. \tag{18}$$

Consequently, from relations (2), (3), (6), (7), taking into account (18), we find that the intensities of transitions between merged states are calculated as follows:

$$q_1(<0>, <1>) = \lambda_1; \ q_1(<1>, <0>) = \mu_1; \tag{19}$$

$$q_1(<n>, <n+1>) = \tilde{\lambda}, \ n \geq 1; \ q_1(<n>, <n-1>) = \tilde{\mu}, \ n \geq 2. \tag{20}$$

Hereinafter,

$$\mu_1 = \frac{\mu\theta\sigma}{\theta + \mu(1-\sigma)}, \tilde{\lambda} = \frac{1}{\theta + \mu(1-\sigma)}(\lambda_1\theta + \lambda_0\mu(1-\sigma)),$$

$$\tilde{\mu} = \frac{\mu}{\theta + \mu(1-\sigma)}(\theta\sigma + \tau(1-\sigma)).$$

From relations (19) and (20), we conclude that the merged model is described by an infinite birth-death process with variable parameters. The ergodicity condition for this process is $\tilde{\lambda} < \tilde{\mu}$, i.e. we obtain condition (10).

Remark 2. The ergodicity condition (10) has the following probabilistic interpretation. So, when the server is in working mode, the rate of incoming calls is $\lambda_1\rho(1)$, and when the server is in setup mode, this value is $\lambda_0\rho(0)$, i.e. $\tilde{\lambda}$ is the total rate of incoming calls for different operating modes of the server. On the other hand, calls are served only when the server is in working mode, i.e. the service intensity is $\mu\sigma\rho(1)$. In addition, calls leave the system not serviced if the server is in setup mode, while the rate of call leaving the system is $\tau\rho(0)$. In other words, the total rate of calls leaving the system is $\tilde{\mu}$. Hence, we conclude that the ergodicity condition (10) has a simple probabilistic interpretation: the total intensity of incoming calls for different operating modes of the server should be less than the intensity of calls leaving the system.

From relations (19) and (20) we find that the probabilities of the merged states $\pi(<n>)$, $n \geq 0$ are calculated as follows:

$$\pi(<n>) = \frac{\lambda_1}{\mu_1}\left(\frac{\tilde{\lambda}}{\tilde{\mu}}\right)^{n-1} \pi(<0>), \ n \geq 1, \tag{21}$$

where $\pi(<0>) = \left(1 + \frac{\lambda_1}{\mu_1}\frac{1}{1-\alpha}\right)^{-1}$, $\alpha = \frac{\tilde{\lambda}}{\tilde{\mu}}$.

Finally, from relations (18) and (21), the approximate values of the steady state probabilities of the initial 2D MC are calculated as follows:

$$\tilde{p}(0,\,1) = \pi\,(<0>)\,; \tag{22}$$

$$\tilde{p}(n,\,0) = \rho\,(0)\,\pi\,(<n>)\,; \tag{23}$$

$$\tilde{p}(n,\,1) = \rho\,(1)\,\pi\,(<n>)\,. \tag{24}$$

Approximate values of performance measures (14)–(17) are defined as follows:

$$L_s \approx \sum_{n=1}^{\infty}\sum_{k=0}^{1} n\tilde{p}(n,\,k) = \sum_{n=1}^{\infty} n\pi\,(<n>) \;= \frac{\lambda_1}{\mu_1}\frac{1}{(1-\alpha)^2}\pi\,(<0>)\,; \tag{25}$$

$$T_{sr} \approx \sum_{n=1}^{\infty} \tilde{p}(n,\,0) = \rho\,(0)\sum_{n=1}^{\infty}\pi\,(<n>) = \rho\,(0)\,(1-\pi\,(<0>))\,; \tag{26}$$

$$T_{wr} \approx \sum_{n=0}^{\infty} \tilde{p}(n,\,1) = \pi\,(<0>) + \rho\,(1)\,(1-\pi\,(<0>))\,; \tag{27}$$

$$R_{sw} \approx \theta\sum_{n=1}^{\infty} \tilde{p}(n,\,0) \;= \theta T_{sr}. \tag{28}$$

Now let's consider the case when the server setup time is significantly longer than the intervals between call arrivals, i.e. it is assumed that the relation $\theta << \max(\lambda_0,\,\lambda_1)$ is hold. Under this condition, the following splitting of the initial state space E is considered:

$$E = X_0 \bigcup X_1,\; X_0 \bigcap X_1 = \emptyset, \tag{29}$$

where $X_0 = \{(n,\,0) : n \geq 1\}\,,\, X_1 = \{(n,\,1)\,:\, n \geq 0\}\,.$

In this case, the intensities of transitions within the classes X_0 and X_1 are significantly bigger than the intensity of transitions between states from different classes.

According to the algorithm of the SMM, the classes X_0 and X_1 are represented as merged states <0> and <1> respectively.

State probabilities $(n,\,k)$ inside classes X_k are denoted by $\chi_k\,(n)\,,\; k = 0, 1$. From the relations (2)–(7) we conclude that these quantities are determined from the classical formulas for one-dimensional birth-death process. Moreover, the birth rate for the model with state space X_0 is equal to λ_0, and the death rate is τ; corresponding parameters for the model with state space X_1 is equal to λ_1, and the death rate is $\mu\sigma$. Therefore, under the ergodicity conditions $\lambda_0 < \tau$ and $\lambda_1 < \mu\sigma$ the indicated probabilities are calculated as

$$\begin{aligned} \chi_0\,(n) &= \alpha^{n-1}\,(1-\alpha)\,,\, n \geq 1; \\ \chi_1\,(n) &= \beta^{n}\;\,(1-\beta)\,,\, n \geq 0, \end{aligned} \tag{30}$$

where $\alpha = \lambda_0/\tau,\; \beta = \lambda_1/\mu\sigma\,.$

Remark 3. If the above conditions are met for $\alpha < 1$ and $\beta < 1$, then the ergodicity condition (10) is also satisfied.

Taking into account relations (30), we find that the intensities of transitions between the merged states $<0>$ and $<1>$ are defined as follows:

$$q_2\left(<0>,\ <1>\right) = \theta\ ;$$

$$q_2\left(<1>,\ <0>\right) = \ \mu\left(1-\sigma\right)\left(1-\beta\right). \tag{31}$$

From relations (31), the probabilities of the merged states $\psi\left(<n>\right),\ n = 0, 1$ are calculated as follows:

$$\psi\left(<0>\right) = \frac{\mu\left(1-\sigma\right)\left(1-\beta\right)}{\theta + \mu\left(1-\sigma\right)\left(1-\beta\right)},\ \psi\left(<1>\right) = \frac{\theta}{\theta + \mu\left(1-\sigma\right)\left(1-\beta\right)}. \tag{32}$$

Taking into account relations (32) for this case, we find the approximate values of the steady state probabilities of the initial 2D MC:

$$\tilde{p}\left(n,\ k\right) = \chi_k\left(n\right)\psi\left(<k>\right),\ k = 0, 1. \tag{33}$$

Then, similarly to (25)–(28), when using partition (29), the approximate values of the desired performance measures (14)–(17) are defined as follows:

$$\begin{aligned} L_s &\approx \sum_{n=1}^{\infty}\sum_{k=0}^{1} n\tilde{p}\left(n, k\right) = \sum_{k=0}^{1}\psi\left(<k>\right)\sum_{n=1}^{\infty} n\chi_k\left(n\right) = \\ &= \frac{\beta}{1-\beta}\psi\left(<1>\right) + \frac{1}{1-\alpha}\psi\left(<0>\right); \end{aligned} \tag{34}$$

$$T_{sr} \approx \sum_{n=1}^{\infty}\tilde{p}\left(n,\ 0\right) = \psi\left(<0>\right); \tag{35}$$

$$T_{wr} \approx \sum_{n=0}^{\infty}\tilde{p}\left(n,\ 1\right) = \psi\left(<1>\right); \tag{36}$$

$$R_{sw} \approx \theta\sum_{n=1}^{\infty}\tilde{p}\left(n,\ 0\right) \ = \theta T_{sr}. \tag{37}$$

5 Numerical Results

Below we study the effect of different parameters on various performance measures and the cost function. The cost function includes the following components: penalties associated with calls being in the system, penalties for lost calls due to impatience, the cost of servicing a call during server working mode, and penalties for server switch to working mode.

For this purpose, the following coefficients are introduced: c_h is penalty per unit of time spent by one call in the system; c_{sr} is penalty for losing one call due to his impatience; c_{wr} is the cost of servicing a call in the working mode of the server; c_{sw} is penalty for one server switch to working mode.

Then the objective function of the problem under study is written as follows:

$$C = c_h L_s + c_{sr} T_{sr} + c_{wr} T_{wr} + c_{sw} R_{sw}. \tag{38}$$

Consequently, the task is to minimize functional (38). Since this functional depends on many variables and has a complex form, an analytical solution to this problem is not possible. Based on these facts, here we study the problems of the influence of system parameters on the performance measures of systems and functional (38).

First, we study the effect of parameter λ_0.

Table 1. Effect of parameter λ_0: $\lambda_1 = 1.2, \mu = 10, \sigma = 0.6, \theta = 2, \tau = 1, C_h = 3, C_{sr} = 2, C_{wr} = 35$ and $C_{sw} = 10$.

λ_0	L_s	T_{sr}	T_{wr}	R_{sw}	C
0.1	0.4756	0.3049	0.6951	0.6098	32.4632
0.2	0.5034	0.3156	0.6844	0.6311	32.4080
0.3	0.5380	0.3269	0.6731	0.6539	32.3638
0.4	0.5811	0.3391	0.6609	0.6782	32.3348
0.5	0.6349	0.3522	0.6478	0.7043	**32.3266**
0.6	0.7023	0.3662	0.6338	0.7324	32.3467
0.7	0.7874	0.3813	0.6187	0.7626	32.4057
0.8	0.8957	0.3976	0.6024	0.7952	32.5184
0.9	1.0349	0.4153	0.5847	0.8305	32.7062
1	1.2165	0.4345	0.5655	0.8690	33.0011

From Table 1, we see that L_s and T_{sr} increases when λ_0 increases, as expected. As a result T_{wr} decreases. As λ_0 increases, R_{sw} increases, since the expected number of calls in the system when the server is in set up regime increases. The cost function first decreases to a minimum value and after that it increases. In all tables, the number in bold letter denotes the minimum value of cost function.

The effect of parameter λ_1 is shown in Table 2. From this table, we see that $L_s$increases when λ_1 increases as expected. As λ_1 increases T_{wr} decreases, since customers get service when the server is in working regime and as a result,T_{sr} and hence R_{sw} increases. The cost function first decreases to a minimum value and after that it increases.

Table 2. Effect of parameter λ_1: $\lambda_0 = 1, \mu = 10, \sigma = 0.6, \theta = 2, \tau = 1, C_h = 3, C_{sr} = 2, C_{wr} = 35$ and $C_{sw} = 10$.

λ_1	L_s	T_{sr}	T_{wr}	R_{sw}	C
0.2	0.2296	0.1082	0.8918	0.2163	34.2827
0.4	0.4368	0.1969	0.8031	0.3938	33.7504
0.6	0.6315	0.2710	0.7290	0.5421	33.3712
0.8	0.8218	0.3338	0.6662	0.6677	33.1254
1	1.0145	0.3877	0.6123	0.7755	33.0030
1.2	1.2165	0.4345	0.5655	0.8690	**33.0011**
1.4	1.4347	0.4754	0.5246	0.9508	33.1237
1.6	1.6774	0.5115	0.4885	1.0231	33.3821
1.8	1.9545	0.5437	0.4563	1.0873	33.7962
2	2.2796	0.5724	0.4276	1.1448	34.3978

Table 3. Effect of parameter μ: $\lambda_0 = 1, \lambda_1 = 1.2, \sigma = 0.6, \theta = 2, \tau = 1, C_h = 3, C_{sr} = 2, C_{wr} = 35$ and $C_{sw} = 10$.

μ	L_s	T_{sr}	T_{wr}	R_{sw}	C
3.1	52.3818	0.4718	0.5282	0.9436	186.0118
3.2	26.1638	0.4701	0.5299	0.9402	107.3803
3.3	17.4434	0.4684	0.5316	0.9369	81.2404
3.4	13.0959	0.4669	0.5331	0.9338	68.2178
3.5	10.4965	0.4655	0.5345	0.9309	60.4384
3.6	8.7704	0.4641	0.5359	0.9282	55.2777
3.7	7.5427	0.4628	0.5372	0.9257	51.6114
3.8	6.6261	0.4616	0.5384	0.9232	48.8773
3.9	5.9165	0.4605	0.5395	0.9209	46.7635
4	5.3516	0.4594	0.5406	0.9188	45.0829

The effect of parameter μ is shown in Table 3. From Table 3, we can see that L_s decreases when μ increases, as expected. Also we see that T_{sr} and R_{sw} decreases and T_{wr} increases in slow rate, when μ increases. This is due the effect of feedback of customers to the system after service completion. Also, the cost function decreases as μ increases.

Effect of parameter θ is shown in Table 4. From this table, we see that L_s decreases when θ increases. This happens because when θ increases, server shifts from setup regime to working regime at a faster rate. Also T_{sr} decreases and T_{wr}increases when θ increases, as expected. R_{sw} increases as θ increases since the shifting rate of server from setup regime to working regime increases. The cost function first decreases to a minimum value and after that it increases.

Table 4. Effect of parameter θ: $\lambda_0 = 1, \lambda_1 = 1.2, \sigma = 0.6, \mu = 10, \tau = 1, C_h = 3, C_{sr} = 2, C_{wr} = 35$ and $C_{sw} = 10$.

θ	L_s	T_{sr}	T_{wr}	R_{sw}	C
0.2	10.5267	0.8691	0.1309	0.1738	39.6373
0.4	5.2509	0.7741	0.2259	0.3096	28.3052
0.6	3.5251	0.7006	0.2994	0.4204	**26.6586**
0.8	2.6784	0.6416	0.3584	0.5133	26.9952
1	2.1793	0.5928	0.4072	0.5928	27.9034
1.2	1.8519	0.5516	0.4484	0.6619	28.9721
1.4	1.6216	0.5163	0.4837	0.7228	30.0559
1.6	1.4511	0.4855	0.5145	0.7768	31.0997
1.8	1.3201	0.4585	0.5415	0.8253	32.0833
2	1.2165	0.4345	0.5655	0.8690	33.0011

6 Conclusion

The paper studies a model of a service system with one server, an infinite queue, instantaneous feedback and impatient calls. After the completion of the service, some of the calls according to the Bernoulli scheme either leave the system or immediately require repeated service. To start the process of reserving calls, the server needs some random setup time, which has an exponential distribution function. It is considered that when the server is in the setup state, it cannot handle calls and interruption of the setup period are not allowed. Calls in queue are impatient when server is in setup regime. The rate of incoming calls depends on the server status.

It is shown that the mathematical model of the system under study is a 2D MC with an infinite state space. The ergodicity condition for the model is found and two approach for studying the corresponding 2D MC are proposed: the approaches based on matrix-geometric method and on space merging method. The first of them is exact method and the second is approximate one. Results of minimization of total cost are demonstrated.

References

1. Melikov, A.Z., Aliyeva, S.H., Shahmaliyev, M.O.: Methods to calculate the system with instantaneous feedback and varying arrival rate. Autom. Remote Contr. **81**(9), 1637–1648 (2020)
2. Takacs, L.: A single-server queue with feedback. Bell Syst. Tech. J. **42**, 505–519 (1963)
3. Takacs, L.: A queuing model with feedback. Oper. Res. **11**, 345–354 (1977)
4. Kumari, N.: An analysis of some continuous time queuing systems with feedback. Ph.D. thesis. Punjabi University, 235 p. (2011)

5. Koroliuk, V.S., Melikov, A.Z., Ponomarenko, L.A., Rustamov, A.M.: Methods for analysis of multi-channel queuing models with instantaneous and delayed feedbacks. Cyber. Syst. Anal. **52**(1), 58–70 (2016)
6. Melikov, A., Ponomarenko, L., Rustamov, A.: Methods for analysis of queueing models with instantaneous and delayed feedbacks. In: Dudin, A., Nazarov, A., Yakupov, R. (eds.) ITMM 2015. CCIS, vol. 564, pp. 185–199. Springer, Cham (2015). https://doi.org/10.1007/978-3-319-25861-4_16
7. Melikov, A.Z., Aliyeva, S.H., Sztrik, J.: Analysis of queuing system MMPP/M/K/K with delayed feedback. Mathematics. **7**, 1128 (2019). https://doi.org/10.3390/math7111128
8. Kumar, B.K., Rukmani, R., Thangaraj, V.: On multiserver feedback retrial queue with finite buffer. Appl. Math. Model. **33**(4), 2062–2083 (2009)
9. Do, T.V.: An efficient computation algorithm for a multiserver feedback retrial queue with a large queueing capacity. Appl. Math. Model. **34**, 2272–2278 (2010)
10. Krieger, U., Klimenok, V.I., Kazimirsky, A.V., Breuer, L., Dudin, A.N.: A BMAP/PH/1 queue with feedback operating in a random environment. Math. Comp. Model. **41**, 867–882 (2005)
11. Rajadurai, P., Saravanarajan, M.C., Chandrasekaran, V.M.: A study on M/G/1 feedback retrial queue with subject to server breakdown and repair under multiple working vacation policy. Alex. Eng. J. **57**, 947–962 (2018)
12. Rajadurai, P., Sundararaman, M., Narasimhan, D.: Performance analysis of an M/G/1 retrial G-queue with feedback under working breakdown services. Songklanakarin J. Sci. Technol. **42**, 236–247 (2020)
13. Mokaddis, G.S., Metwally, S.A., Zaki, B.M.: A feedback retrial queuing system with starting failures and single vacation. Tamkang J. of Sci. Eng. **10**, 183–192 (2007)
14. Hernández-Díaz, A.G., Moreno, P.: A discrete-time single-server queueing system with an N-policy, an early setup and a generalization of the Bernoulli feedback. Math. Comp. Model. **49**, 977–990 (2009)
15. Zhao, Y., Yue, W.: Performance analysis and optimization for cognitive radio networks with a finite primary user buffer and a probability returning scheme. J. Ind. Manag. Opt. **16**, 1119–1134 (2020)
16. Lee, Y.W.: The M/G/1 feedback retrial queue with two types of customers. Bull. Korean Math. Soc. **42**, 875–887 (2005)
17. Krishnamoorthy, A., Manjunath, A.S.: On queues with priority determined by feedback. Calcutta Stat. Assos. Bul. **70**(1), 33–56 (2018)
18. Kim, C.S., Klimenok, V.I., Tsarenkov, G., Breuer, L., Dudin, A.N.: The BMAP/G/1→·PH/1/M tandem queue with feedback and losses. Perform. Eval. **64**, 802–818 (2007)
19. Raghavendran, C.V., Satish, G.N., Rama Sundari, M.V., Suresh, Varma P.: A two node tandem communication network with feedback having DBA and NHP arrivals. Int. J. Comp. Electr. Eng. **6**, 422–435 (2014)
20. Varalakshmi, M., Rajadurai, P., Saravanarajan, M.C., Chandrasekaran, V.M.: An M/G/1 retrial queuing system with two phases of service, immediate Bernoulli feedback, single vacation and starting failures. Int. J. Math. Oper. Res. **9**, 302–328 (2016)
21. Melikov, A.Z., Ponomarenko, L.A., Rustamov, A.M.: Hierarchical space merging algorithm for the analysis of open tandem queuing networks. Cyber. Syst. Anal. **52**, 867–877 (2016)
22. Neuts, M.F.: Matrix-Geometric Solutions in Stochastic Models: An Algorithmic Approach, p. 332. John Hopkins University Press, Baltimore (1981)

23. Latouche, G., Ramaswami, V.: Introduction to Matrix Analytic Methods in Stochastic Modelling, 334 p. SIAM (1999)
24. Naoumov, V., Krieger, U.R., Wagner, D.: Analysis of a multi-server delay-loss system with a general Markovian arrival process. In: Chakravarthy, S.R., Alfa, A.S. (eds.) Matrix-Analytic Methods in Stochastic Models, pp. 43–66. Marcel Dekker, New York (1997)

Waiting Time Asymptotic Analysis of a M/M/1 Retrial Queueing System Under Two Types of Limiting Condition

Anatoly Nazarov and Maria Samorodova(✉)

Institute of Applied Mathematics and Computer Science,
National Research Tomsk State University,
36 Lenina Avenue, 634050 Tomsk, Russian Federation

Abstract. In our paper, the waiting time analysis of a M/M/1 retrial queueing system is presented and the asymptotic distribution of the number of returns of the tagged request to the orbit is driven since they are connected to each other. The research was conducted by the use of asymptotic analysis method. Two different cases are considered. First we conduct analysis under a heavy load condition and then under a low rate of retrials condition. Two different characteristic functions of the waiting time were obtained. The analysis was carried out using asymptotic distributions of the number of requests in the orbit under a heavy load condition and a low rate of retrials condition, which were also obtained. To show the effectiveness of asymptotic results for the considered retrial queuing system, the approximation of the distribution of the number of returns of the tagged request to the orbit in prelimit situation, numerical illustrations and results are given.

Keywords: Retrial queue · Asymptotic analysis · Waiting time · Number of returns · Number of retrials

1 Introduction

Retrial queue (RQ) systems are adequate models of processes arising from real world applications. Main fields where such models are used are telecommunication networks, computer networks, call centers, wireless communication systems, cognitive networks, cloud computing. An extensive review of recent developments and methods related to RQ systems one can find in, for example, Artalejo, Falin [2], Artalejo, Gomez-Corral [3], Gomez-Corral, Phung-Duc [16], Falin, Templeton [13], Kim, Kim [17], Nobel [24], Phung-Duc [26].

The waiting time distribution, the time a request spends in the orbit, is very complicated problem in the retrial queue system theory. Different approaches to the investigation of the waiting time can be found in Artalejo, Chakravarthy, Lopez-Herrero [1], Artalejo, Gomez-Corral [4], Choi, Chang [6], Falin, Artalejo [8], Gharbi, Dutheillet [14], Sudyko, Nazarov, Sztrik [27], Tóth, Bérczes, Sztrik

A. Dudin et al. (Eds.): ITMM 2020, CCIS 1391, pp. 171–185, 2021.
https://doi.org/10.1007/978-3-030-72247-0_13

[28], Zhang, Feng, Wang [29] for finite retrial group of sources and Chakravarthy, Dudin [7], Falin, Fricker [9], Falin [1,10,12], Gomez-Corral, Ramalhoto [15], Neuts [20], Nobel, Tijms [25], Lee, Kim, Kim [18] for infinite number of sources.

Since a request waiting time distribution and the number of returns distribution are connected to each other, in this paper we investigate both of them. We use the method of asymptotic analysis under a heavy load condition and a low rate of retrials condition following approach applied in, for example, Moiseeva, Nazarov [19], Nazarov, Moiseeva [21], Nazarov, Semenova [22], Nazarov, Sztrik, Kvach [23].

The rest of the paper is organized as follows. In Sect. 2 the RQ-system mathematical model and the waiting time characteristic function are presented. In Sect. 3 we derive Kolmogorov's equations for the system states. Section 4 is connected with the asymptotic analysis of the distribution of the number of requests in the orbit, which is needed for further research. In Sect. 5 asymptotic analysis of characteristic functions for the number of request returns to the orbit is provided. In Sect. 6 the asymptotic distribution of the waiting time in the orbit is derived for two different conditions. In Sect. 7 we found the approximation of the waiting time distribution in prelimit situation. Also, several sample results obtained by numerical methods showing that proposed approximation is effective. The paper ends with a Conclusion.

2 Mathematical Model

Let us consider a M/M/1 retrial queuing (RQ) system. The system input receives a Poisson flow of requests which is given by a scalar intensity λ. When incoming request arrive at the system the server can be idle or busy. In the first case this request occupies the server and the service starts immediately. Served request leaves the system. In second case, the request joins to the orbit. Each request from the orbit after a random delay retries to get accesses to the server. At the retrial moment server again can be idle or busy. In the first case this request occupies the server for a random service time; otherwise, it instantly returns to the orbit for a next random delay. Service time and random delay time are independent and exponentially distributed with parameters μ and σ, respectively.

We assume the system being in stationary mode. Let's define W – waiting time of the tagged request in the orbit as the length of the interval from the moment the request arrives in the system till the start of the service. Let's denote by $\tilde{\nu}$ the number of transitions of the tagged request to the orbit. Also we denote by r the probability that the server is busy at the moment the request arrives at the system. Obviously $\tilde{\nu} = 0$, with the probability $(1 - r)$ that the request finds the server idle at the moment of the arrival to the system. In addition, we denote by $\nu(t)$ the number of returns of the tagged request to the orbit from the moment t until the start of the service. Using above notations:

$$\tilde{\nu} = \begin{cases} 0, & \text{with probability } (1 - r), \\ 1 + \nu(t), & \text{with probability } r. \end{cases}$$

The characteristic function for W can be written as follows:

$$G(u) = E\left\{e^{juW}\right\} = (1-r) + r\sum_{n=0}^{\infty} E\left\{e^{juW}/\nu = 1+n\right\} P\left\{\nu(t) = n\right\}$$
$$= (1-r) + r\sum_{n=0}^{\infty}\left(\frac{\sigma}{\sigma - ju}\right)^{1+n} P\left\{\nu(t) = n\right\}. \qquad (1)$$

The aim of our study is to find the asymptotic distribution of W the waiting time of the tagged request in the orbit. As can be seen from (1), for that purpose it is enough to find the probability r and the probability distribution $P\{\nu(t) = n\}$ under limiting conditions. First, we conduct our research under a heavy load condition and then under a low rate of retrials condition. As a result, two different asymptotic distributions of W were obtained.

3 Kolmogorov's Equations

Let's denote by $i(t)$ the number of requests in the orbit at time t and by $k(t)$ - the state of the server at time t:

$$k(t) = \begin{cases} 0, & \text{if the server is idle,} \\ 1, & \text{if the server is busy.} \end{cases}$$

The system state at time t can be described by means of a markov chain $\{k(t), i(t)\}$ with stationary probability distribution:

$$P_k(i) = P\left\{k(t) = k, i(t) = i\right\}$$

Probability distribution $P_k(i)$ satisfy the following Kolmogorov equations:

$$\begin{aligned} (\lambda + i\sigma)P_0(i) &= \mu P_1(i), \\ (\lambda + \mu)P_1(i) &= \lambda P_0(i) + (i+1)\sigma P_0(i+1) + \lambda P_1(i-1). \end{aligned} \qquad (2)$$

Steady-state partial characteristic functions $H_k(u)$ for $i(t)$ can be written in the following form:

$$H_k(u) = \sum_{i=0}^{\infty} e^{jui} P_k(i), \qquad (3)$$

where $j = \sqrt{-1}$ is the imaginary unit.

According to (2) and (3) we obtain the system of equations for $H_k(u)$:

$$\begin{aligned} -\lambda H_0(u) + j\sigma\frac{\partial H_0(u)}{\partial u} + \mu H_1(u) &= 0, \\ -(\lambda + \mu)H_1(u) + \lambda H_0(u) - j\sigma e^{-ju}\frac{\partial H_0(u)}{\partial u} + \lambda e^{ju} H_1(u) &= 0. \end{aligned} \qquad (4)$$

Steady-state characteristic functions for $\nu(t)$ can be written in the following form:

$$G(u) = E\left\{e^{ju\nu(t)}\right\} = \sum_{i=0}^{\infty}\left[G_0(i,u)P_0(i) + G_1(i,u)P_1(i)\right].$$

Let's consider:

$$G_k(i,u,t) = E\left\{e^{juv(t)}/k(t) = k, i(t) = i\right\},$$

where $G_k(i,u,t)$ - conditional partial characteristic functions for $\nu(t)$. Taking into account that the system is in a stationary mode, for $G_k(i,u)$ we obtain the system of inverse Kolmogorov equations:

$$-(\lambda + i\sigma)G_0(i,u) + \lambda G_1(i,u) + (i-1)\sigma G_1(i-1,u) + \sigma = 0, \quad (5)$$

$$-(\lambda + \mu + \sigma)G_1(i,u) + \mu G_0(i,u) + \lambda G_1(i+1,u) + e^{ju}\sigma G_1(i,u) = 0. \quad (6)$$

4 Asymptotic Analysis of the Number of Requests in the Orbit

Heavy Load Condition. Denote $\rho = \frac{\lambda}{\mu}$, $\varepsilon = 1-\rho$, making substitutions $u = \varepsilon w$, $H_0(u) = \varepsilon F_0(\omega,\varepsilon)$, $H_1(u) = F_1(\omega,\varepsilon)$ and deviding (4) by μ we get:

$$\begin{aligned} F_1(w,\varepsilon) - (1-\varepsilon)\varepsilon F_0(w,\varepsilon) + j\frac{\sigma}{\mu}\frac{\partial F_0(w,\varepsilon)}{\partial w} &= 0,\\ (1-\varepsilon)\varepsilon F_0(w,\varepsilon) + \left[(1-\varepsilon)(e^{jw\varepsilon}-1) - 1\right]F_1(w,\varepsilon)&\\ -j\frac{\sigma}{\mu}e^{-jw\varepsilon}\frac{\partial F_0(w,\varepsilon)}{\partial w} &= 0. \end{aligned} \quad (7)$$

The beforelimited characteristic function under a heavy load condition can be determined approximately by the equation: $H(u) = H_0(u) + H_1(u) \approx h(w) = F_1(w)$.

Step 1. Let $\varepsilon \to 0$ in (7). Denote

$$\lim_{\varepsilon\to 0} F_k(\omega,\varepsilon) = F_k(\omega).$$

For functions $F_k(w)$ we obtain the system of equations:

$$\begin{aligned} F_1(w) + j\frac{\sigma}{\mu}\frac{\partial F_0(w)}{\partial w} &= 0,\\ -F_1(w) - j\frac{\sigma}{\mu}\frac{\partial F_0(w,\varepsilon)}{\partial w} &= 0. \end{aligned}$$

This system consists of two equivalent equations.

Step 2. Let's rewrite $F_k(w,\varepsilon)$ from (7) as follows:

$$F_k(w,\varepsilon) = F_k(w) + \varepsilon f_k(w) + o(\varepsilon^2),$$

and devide equations of the system (7) by ε. With respect to step 1 results and after taking $\varepsilon \to 0$, we obtain:

$$\begin{aligned} f_1(w) - F_0(w) + j\frac{\sigma}{\mu}\frac{\partial f_0(w)}{\partial w} &= 0, \\ -F_1(w) + f_1(w) + j\frac{\sigma}{\mu}\frac{\partial f_0(w)}{\partial w} + \frac{\sigma}{\mu}w\frac{\partial F_0(w)}{\partial w} &= 0. \end{aligned}$$

Combining resulting equations of steps 1 and 2, we get the following system:

$$\begin{aligned} F_1(w) + j\frac{\sigma}{\mu}\frac{\partial F_0(w)}{\partial w} &= 0, \\ f_1(w) - F_0(w) + j\frac{\sigma}{\mu}\frac{\partial f_0(w)}{\partial w} &= 0, \\ -F_1(w) + f_1(w) + j\frac{\sigma}{\mu}\frac{\partial f_0(w)}{\partial w} + \frac{\sigma}{\mu}w\frac{\partial F_0(w)}{\partial w} &= 0. \end{aligned}$$

Solving the obtained system as it is shown in [19] the following expression can be got:

$$h(w) = (1 - jw)^{-\left(\frac{\mu+\sigma}{\sigma}\right)}, \tag{8}$$

where $h(w)$ is a characteristic function of gamma distribution $\gamma(x)$ with parameters $\beta = \frac{\mu+\sigma}{\sigma}$ and $\alpha = 1$.

Low Rate of Retrials Condition. Making substitutions $\sigma = \varepsilon$, $u = \varepsilon w$, $H_k(u) = F_k(\omega, \varepsilon)$ in (4), we obtain:

$$\begin{aligned} -\lambda F_0(\omega,\varepsilon) + j\frac{\partial F_0(\omega,\varepsilon)}{\partial \omega} + \mu F_1(\omega,\varepsilon) &= 0, \\ -(\lambda+\mu)F_1(\omega,\varepsilon) + \lambda F_0(\omega,\varepsilon) - je^{-j\varepsilon\omega}\frac{\partial F_0(\omega,\varepsilon)}{\partial \omega} + \lambda e^{j\varepsilon\omega}F_1(\omega,\varepsilon) &= 0. \end{aligned} \tag{9}$$

Step 1. Let $\varepsilon \to 0$ in (9). Denote

$$\lim_{\varepsilon\to 0} F_k(\omega,\varepsilon) = F_k(\omega).$$

For functions $F_k(w)$ we obtain the system of equations:

$$\begin{aligned} -\lambda F_0(\omega) + j\frac{\partial F_0(\omega)}{\partial \omega} + \mu F_1(\omega) &= 0, \\ \lambda F_0(\omega) - j\frac{\partial F_0(\omega)}{\partial \omega} - \mu F_1(\omega) &= 0. \end{aligned}$$

Note that equations of this system are equivalent.

Step 2. Adding equations in system (9), we get:

$$j(1-e^{-j\varepsilon\omega})\frac{\partial F_0(\omega,\varepsilon)}{\partial\omega}+\lambda(e^{j\varepsilon\omega}-1)F_1(\omega,\varepsilon)=0.$$

Let's decompose exponential function and rewrite above expression:

$$j(j\varepsilon\omega+o(\varepsilon))\frac{\partial F_0(\omega,\varepsilon)}{\partial\omega}+\lambda(j\varepsilon\omega+o(\varepsilon))F_1(\omega,\varepsilon)=0.$$

Taking $\varepsilon\to 0$, we obtain:

$$j\frac{\partial F_0(\omega)}{\partial\omega}+\lambda F_1(\omega)=0.$$

Combining with the result of Step 1, we get the system similar to system obtained in [22]:

$$\begin{aligned}-\lambda F_0(\omega)+j\frac{\partial F_0(\omega)}{\partial\omega}+\mu F_1(\omega)=0,\\ j\frac{\partial F_0(\omega)}{\partial\omega}+\lambda F_1(\omega)=0.\end{aligned}\qquad(10)$$

Lets find the solution in the following form:

$$F_k(\omega)=R(\kappa)e^{jwk},\qquad(11)$$

where $R(\kappa)=P(k=\kappa)$ - stationary probabilities of Markov chain $k(t)$, $\kappa=0,1$.

Substituting (11) in (10), we obtain a homogeneous system of two equations for the probability distribution $R(k)$:

$$\begin{aligned}-(\lambda+\kappa)R(0)+\mu R(1)=0\\ -\kappa R(0)+\lambda R(1)=0.\end{aligned}\qquad(12)$$

Equation

$$\lambda(\lambda+\kappa)-\mu\kappa=0\qquad(13)$$

determines the value of κ for (11). Solving (13) we obtain:

$$\kappa=\frac{\lambda^2}{\mu-\lambda}\qquad(14)$$

Taking into account the normalization condition $R(0)+R(1)=1$ and (12) the probability distribution $R(k)$ can be calculated as:

$$R(0)=\frac{\mu-\lambda}{\mu},R(1)=\frac{\lambda}{\mu}.$$

5 Asymptotic Analysis of the Number of Returns of the Tagged Request to the Orbit

Heavy Load Condition. Denote $\rho=\frac{\lambda}{\mu}$, $\varepsilon=1-\rho$, making substitutions $u=\varepsilon w$, $i\varepsilon=x$, $G_k(i,u)=g_k(x,w,\varepsilon)$ and multiplying (5) by ε we obtain:

$$\begin{aligned}-(\varepsilon\lambda+\sigma)g_0(,w,\varepsilon)+\varepsilon\lambda g_1(,w,\varepsilon)+(x-\varepsilon)\sigma g_1(-\varepsilon,w,\varepsilon)+\varepsilon\sigma=0\\ -(\lambda+\mu+\sigma(1+e^{j\varepsilon w}))g_1(,w,\varepsilon)+\mu g_0(,w,\varepsilon)+\lambda g_1(+\varepsilon,w,\varepsilon)=0\end{aligned}\qquad(15)$$

Step 1. Let $\varepsilon \to 0$ in (15). Denote

$$\lim_{\varepsilon \to 0} g_k(x, w, \varepsilon) = g_k(x, w).$$

We obtain the following system for $g_k(x, w)$:

$$\begin{aligned} -x\sigma g_0(x, w) + x\sigma g_1(x, w) = 0, \\ -(\mu + \sigma)g_1(x, w) + (\mu + \sigma)g_0(x, w) = 0. \end{aligned} \tag{16}$$

Note that equations of this system are equivalent and $g_0(x, w) = g_1(x, w) = g(x, w)$.

Step 2. Let's rewrite $g_k(x, w, \varepsilon)$ from (15) as follows:

$$g_k(x, w, \varepsilon) = g(x, w) + \varepsilon f_k(x, w) + o(\varepsilon), \tag{17}$$

and then rewrite (15):

$$\begin{aligned} -(\varepsilon\lambda + x\sigma)g_0(x, w, \varepsilon) + (\varepsilon\lambda + x\sigma)\, g_1(x, w, \varepsilon) \\ - \varepsilon\frac{\partial\left[x\sigma g_1(x, w, \varepsilon)\right]}{\partial x} + \varepsilon\sigma = O(\varepsilon^2), \\ -(\mu + \sigma)g_1(x, w, \varepsilon) + \mu g_0(x, w, \varepsilon) \\ + \varepsilon\frac{\partial\left[\lambda g_1(x, w, \varepsilon)\right]}{\partial x} + e^{j\varepsilon w}\sigma g_1(x, w, \varepsilon) = O(\varepsilon^2). \end{aligned} \tag{18}$$

Substituting decomposition (17) in (18), taking $\varepsilon \to 0$ and after performing some actions on the equations, we get the following system:

$$x\left[f_1(x, w) - f_0(x, w)\right] = g(x, w) + x\frac{\partial g(x, w)}{\partial x} - 1,$$

$$\left[f_1(x, w) - f_0(x, w)\right] = jw\frac{\sigma}{\mu}g(x, w) + \frac{\partial g(x, w)}{\partial x} - \frac{\sigma}{\mu}f_1(x, w).$$

Adding equations in the system (18) and taking $\varepsilon \to 0$, we get the following expression:

$$(x - \frac{\mu}{\sigma})\left[f_1(x, w) - f_0(x, w)\right] = (1 - jw)\, g(x, w) + \left(x - \frac{\mu}{\sigma}\right)\frac{\partial g(x, w)}{\partial x} - 1.$$

Combining obtained expressions we get the system of three equations:

$$x\left[f_1(x, w) - f_0(x, w)\right] = g(x, w) + x\frac{\partial g(x, w)}{\partial x} - 1,$$

$$\left[f_1(x, w) - f_0(x, w)\right] = jw\frac{\sigma}{\mu}g(x, w) + \frac{\partial g(x, w)}{\partial x} - \frac{\sigma}{\mu}f_1(x, w),$$

$$(x - \frac{\mu}{\sigma})\left[f_1(x, w) - f_0(x, w)\right] = (1 - jw)\, g(x, w) + \left(x - \frac{\mu}{\sigma}\right)\frac{\partial g(x, w)}{\partial x} - 1.$$

Eliminating $f_1(x,w), f_0(x,w)$ from this system, we obtain the equation for the function $g(x,w)$, solving which we find:

$$g(x,w) = \frac{\frac{\mu}{\sigma}}{\frac{\mu}{\sigma} - jw}, \tag{19}$$

where $g(x,w)$ is a conditional characteristic function of exponential distribution with parameter $\alpha = \frac{\mu}{\sigma}$.

Let's pass from the conditional characteristic function $g(x,w)$ to the characteristic function $g(w)$:

$$g(w) = \int\limits_0^{+\infty} g(x,w)\gamma(x)\,dx.$$

It is easy to show that the inverse Fourier transform has the form of the probability distribution density of the limiting value of $\nu(t)$:

$$P(z) = \int\limits_0^{\infty} \gamma(x)\frac{\mu}{x\sigma}\exp\left\{-\frac{\mu}{x\sigma}z\right\}dx. \tag{20}$$

Low Rate of Retrials Condition. Making substitutions $\sigma = \varepsilon$, $i\varepsilon = x$, $G_k(i,u) = g_k(x,u,\varepsilon)$ in (5), (6) we obtain:

$$\begin{aligned} -(\lambda + x)g_0(x,u,\varepsilon) + \lambda g_1(x,u,\varepsilon) \\ + (x-\varepsilon)g_1(x-\varepsilon,u,\varepsilon) + \varepsilon = 0, \\ -(\lambda+\mu+\varepsilon)g_1(x,u,\varepsilon) + \mu g_0(x,u,\varepsilon) \\ + \lambda g_1(x+\varepsilon,u,\varepsilon) + e^{ju}\varepsilon g_1(x,u,\varepsilon) = 0. \end{aligned} \tag{21}$$

Step 1. Let $\varepsilon \to 0$ in (21). Denote

$$\lim_{\varepsilon\to 0} g_k(x,w,\varepsilon) = g_k(x,w).$$

We obtain the following system for $g_k(x,w)$:

$$\begin{aligned} -(\lambda + x)g_0(x,u) + (\lambda +)g_1(x,u) = 0, \\ -\mu g_1(x,u) + \mu g_0(x,u) = 0. \end{aligned}$$

This system consist of two equivalent equations and $g_0(x,w) = g_1(x,w) = g(x,w)$.

Step 2. Let's rewrite $g_k(x,w,\varepsilon)$ from (21) as follows:

$$g_k(x,w,\varepsilon) = g(x,w) + \varepsilon f_k(x,w) + o(\varepsilon), \tag{22}$$

and then rewrite (21):

$$\begin{aligned} &-(\lambda + x)g_0(x,u,\varepsilon) + \lambda g_1(x,u,\varepsilon) + x g_1(x,u,\varepsilon) \\ &\qquad - \varepsilon \frac{\partial [x g_1(x,u,\varepsilon)]}{\partial x} + \varepsilon = o(\varepsilon), \\ &-(\lambda + \mu + \varepsilon)g_1(x,u,\varepsilon) + \mu g_0(x,u,\varepsilon) + \lambda g_1(x,u,\varepsilon) \\ &\qquad + \varepsilon \frac{\partial [\lambda g_1(x,u,\varepsilon)]}{\partial x} + e^{ju} \varepsilon g_1(x,u,\varepsilon) = o(\varepsilon). \end{aligned} \tag{23}$$

Substituting decomposition (22) in (23), taking $\varepsilon \to 0$ and after performing some actions on the equations, we get the following system:

$$\begin{aligned} (\lambda + x)(f_1(x,u) - f_0(x,u)) &= \frac{\partial [x g(x,u)]}{\partial x} - 1 \\ \mu(f_1(x,u) - f_0(x,u)) &= \frac{\partial [\lambda g(x,u)]}{\partial x} + (e^{ju} - 1)g(x,u). \end{aligned}$$

Thus, for the function $g(x,u)$ we obtain the following equation:

$$[\lambda(\lambda + x) - \mu] \frac{\partial g(x,u)}{\partial x} + \left[(\lambda + x)(e^{ju} - 1) - \mu\right] g(x,u) + \mu = 0 \tag{24}$$

In limiting case $k = i\varepsilon = x$ and k is the solution of the Eq. (13). This equation is equal to the coefficient of the derivative $\frac{\partial g(x,u)}{\partial x}$. Then the coefficient is zero in (24) and $k = x$:

$$\left[(\lambda + k)(e^{ju} - 1) - \mu\right] g(x,u) + \mu = 0.$$

Solving this system for $g(x,u)$ we obtain:

$$g(u) = \frac{1 - \frac{\lambda}{\mu}}{1 - e^{ju}\frac{\lambda}{\mu}}.$$

Thus, probability distribution of $\nu(t)$ the number of returns of the tagged request to the orbit is geometric

$$P(\nu(t) = n) \approx (1 - p)p^n, \tag{25}$$

where $\rho = \frac{\lambda}{\mu}$, $n = 0, 1, 2, \ldots$

6 Distribution of the Waiting Time in the Orbit

Using results obtained in previous sections, we derive expressions for waiting time asymptotic distributions under considered conditions.

Heavy Load Condition. Using the found distribution density, let's get an approximation of the asymptotic discrete probability distribution of the number of returns of the tagged request to the orbit:

$$P_1(n) = P((1 - \rho)n) \cdot \left(\sum_{m=0}^{\infty} P((1 - \rho)m) \right)^{-1}. \tag{26}$$

Substituting above distribution into (1), we obtain:

$$G(u) = E\left\{e^{juW}\right\} = (1-r) + r\sum_{n=0}^{\infty} E\left\{e^{juW}/\nu = 1+n\right\}P\left\{\nu(t) = n\right\}$$
$$= (1-r) + r\sum_{n=0}^{\infty}\left(\frac{\sigma}{\sigma - ju}\right)^{1+n} P_1(n).$$

As a result, we have found the limiting characteristic function of the waiting time of the request in a M/M/1 RQ system under a heavy load condition. Applying the inverse Fourier transform of the obtained $G(u)$, we find the asymptotic distribution of the waiting time of the request in the orbit.

Low Rate of Retrials Condition. In previous section we found that probability distribution of $\nu(t)$ under low rate of retrials condition is geometric with parameter $\rho = \frac{\lambda}{\mu}$. Substituting this distribution in (1), we obtain:

$$G(u) = E\left\{e^{juW}\right\} = (1-r) + r\sum_{n=0}^{\infty} E\left\{e^{juW}/\nu = 1+n\right\}P\left\{\nu(t) = n\right\}$$
$$= (1-r) + r\sum_{n=0}^{\infty}\left(\frac{\sigma}{\sigma - ju}\right)^{1+n}(1-p)p^n = (1-r)$$
$$+ r\frac{\sigma}{\sigma - ju}(1-p)\sum_{n=0}^{\infty}\left(\frac{\sigma}{\sigma - ju}p\right)^n = (1-r) + r\frac{\sigma(1-p)}{\sigma(1-p) - ju}$$

Obtained $G(u)$ is the limiting characteristic function of the waiting time of the request in a M/M/1 RQ system under a low rate of retrials condition. The distribution of W in this case is two-phase hyper exponential distribution with an infinite parameter in the first phase and a parameter $\sigma(1-p)$ in the second phase.

7 Numerical Results

Probability distributions $P_1(n)$ and $P(\nu(t) = n)$ given in (26) and (25) have been obtained by the method of asymptotic analysis under a heavy load and a low rate of retrials conditions respectively.

We compare the resulting asymptotic distribution and the numerical solution of the system of Eq. (5), (6) in order to investigate the applicability of these asymptotic results for prelimit situations. Also, for that purpose Kolmogorov distances between distributions were found:

$$\Delta_1 = \max_{0 \le n < \infty}\left|\sum_{j=0}^{n} P_1(j) - \sum_{j=0}^{n}\pi(j)\right|,$$

$$\Delta = \max_{0 \le n < \infty}\left|\sum_{j=0}^{n} P(j) - \sum_{j=0}^{n}\pi(j)\right|.$$

Let us denote the prelimit probability distribution of the number of returns of the tagged request to the orbit by $\pi(n)$. We obtain it by numerical methods similar as it is shown in [27]. Using the law of total probability, $\pi(n)$ can be written in the following form:

$$\pi(n) = \sum_{i=0}^{\infty} [\Pi_0(n,i)\, P_0(i) + \Pi_1(n,i)\, P_1(i)], \tag{27}$$

where unconditional probabilities $P_k(i)$ are solutions of (2) and normalization condition. In order to find conditional probabilities $\Pi_k(n,i) = P(\nu(t) = n/k(t) = k, i(t) = i)$, let's write conditional characteristic function $G_k(i,u)$ as follows:

$$G_k(i,u) = \sum_{n=0}^{\infty} e^{jun}\Pi_k(n,i),$$

and substitute this representation into inverse Kolmogorov Eq. (5), (6) for $G_k(i,u)$:

$$-(\lambda + i\sigma)\sum_{n=0}^{\infty} e^{jun}\Pi_0(n,i) + \lambda\sum_{n=0}^{\infty} e^{jun}\Pi_1(n,i)$$
$$+(i-1)\sigma\sum_{n=0}^{\infty} e^{jun}\Pi_1(n,i-1) + \sigma = 0$$
$$-(\lambda+\mu+\sigma)\sum_{n=0}^{\infty} e^{jun}\Pi_1(n,i) + \mu\sum_{n=0}^{\infty} e^{jun}\Pi_0(n,i)$$
$$+\lambda\sum_{n=0}^{\infty} e^{jun}\Pi_1(n,i+1) + \sigma\sum_{n=1}^{\infty} e^{jun}\Pi_1(n-1,i) = 0.$$

Equating coefficients of corresponding powers of the exponent the following systems of equations for probabilities $\Pi_k(n,i)$ were obtained:

For $n = 0$:

$$\begin{aligned} -(\lambda+i\sigma)\Pi_0(0,i) + \lambda\Pi_1(0,i) + (i-1)\sigma\Pi_1(0,i-1) + \sigma = 0 \\ -(\lambda+\mu+\sigma)\Pi_1(0,i) + \mu\Pi_0(0,i) + \lambda\Pi_1(0,i+1) = 0, \end{aligned} \tag{28}$$

For $n \geq 1$:

$$\begin{aligned} -(\lambda+i\sigma)\Pi_0(n,i) + \lambda\Pi_1(n,i) + (i-1)\sigma\Pi_1(n,i-1) = 0, \\ -(\lambda+\mu+\sigma)\Pi_1(n,i) + \mu\Pi_0(n,i) + \lambda\Pi_1(n,i+1) + \sigma\Pi_1(n-1,i) = 0. \end{aligned} \tag{29}$$

We find exact values of $P_k(i)$ and $\Pi_k(0,i)$ by solving (2) and (28) for some given parameters λ, μ, σ using numerical methods. Then we substitute found $P_k(i)$ and $\Pi_k(0,i)$ values into (27) and get the probability $\pi(0)$. Solving (29) by numerical methods for some given parameters λ, μ, σ we similarly find $\Pi_k(n, i)$ for $n = 1, 2, 3, \ldots$ (Table 2).

Table 1. The difference between the numerical distribution $\pi(n)$ and asymptotic distribution $P(\nu(t) = n)$ under a low rate of retrials condition for various parameters

Parameters	Δ
$\lambda = 0.95,\ \mu = 1,\ \sigma = 0.1$	0.024
$\lambda = 0.95,\ \mu = 1,\ \sigma = 0.01$	0.009
$\lambda = 0.8,\ \mu = 1,\ \sigma = 0.1$	0.019
$\lambda = 0.8,\ \mu = 1,\ \sigma = 0.01$	0.002
$\lambda = 0.5,\ \mu = 1,\ \sigma = 0.1$	0.026
$\lambda = 0.5,\ \mu = 1,\ \sigma = 0.01$	0.002

Table 2. The difference between the numerical distribution $\pi(n)$ and asymptotic distribution $P_1(n)$ under a heavy load condition for various parameters

Parameters	Δ
$\lambda = 0.95,\ \mu = 1,\ \sigma = 0.1$	0.0095
$\lambda = 0.95,\ \mu = 1,\ \sigma = 1$	0.0189
$\lambda = 0.9,\ \mu = 1,\ \sigma = 0.1$	0.0196
$\lambda = 0.9,\ \mu = 1,\ \sigma = 1$	0.0695
$\lambda = 0.8,\ \mu = 1,\ \sigma = 0.1$	0.0416
$\lambda = 0.8,\ \mu = 1,\ \sigma = 1$	0.2674

By substituting these values in (27) we can find the numerical probability distribution $\pi(n)$.

We consider different parameters setup of λ, μ, σ for each asymptotic distribution. In Fig. 1 prelimit probabilities $\pi(n)$ and asymptotic probabilities $P_1(n)$ are compared to each other. In Fig. 2 prelimit probabilities $\pi(n)$ and asymptotic probabilities $P(\nu(t) = n)$ are shown. Table 1 present Kolmogorov distances between the numerical and asymptotic distributions for various parameters. The analysis of obtained numerical results shows that under these parameters obtained asymptotic distributions and prelimit distributions are very close to each other and asymptotic method is very effective.

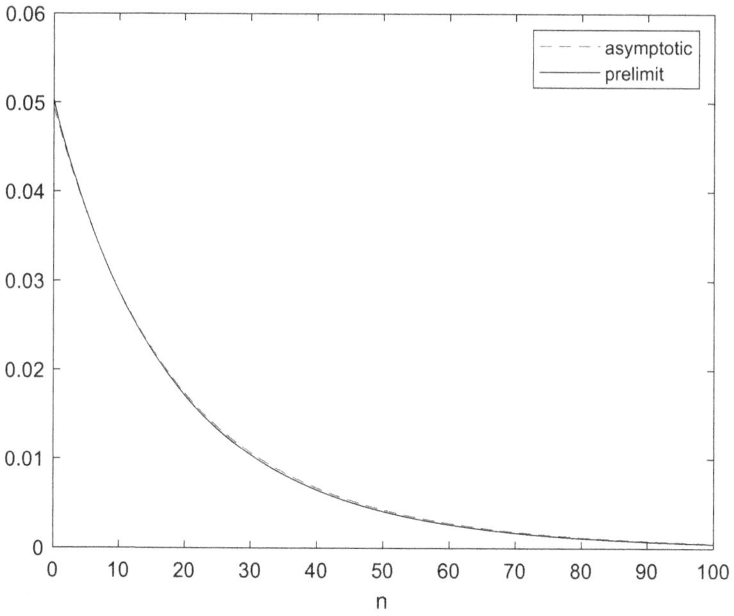

Fig. 1. The difference between the numerical $\pi(n)$ and asymptotic $P_1(n)$ distributions under a heavy load condition, $\lambda = 0.95$, $\mu = 1$, $\sigma = 0.1$

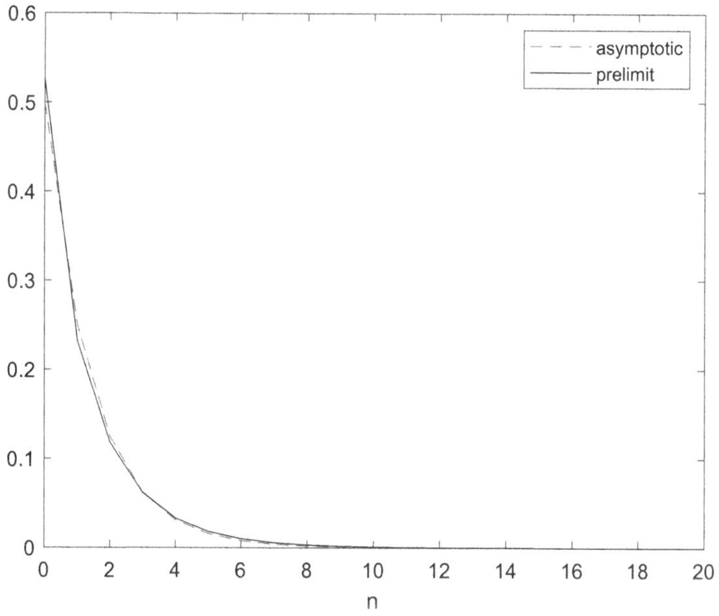

Fig. 2. The difference between the numerical $\pi(n)$ and asymptotic $P(\nu(t) = n)$ distributions under a low rate of retrials condition, $\lambda = 0.5$, $\mu = 1$, $\sigma = 0.1$

8 Conclusion

In this paper was presented an asymptotic analysis of the waiting time and the number of returns of a M/M/1 retrial queueing system. Two different cases were considered. First we conducted analysis under a heavy load condition and then under a low rate of retrials condition. Numerical illustrations and results show the effectiveness of asymptotic method for the considered retrial queuing system.

References

1. Artalejo, J.R., Chakravarthy, S.R., Lopez-Herrero, M.J.: The busy period and the waiting time analysis of a MAP/M/c queue with finite retrial group. Stoch. Anal. Appl. **25**(2), 445–469 (2007)
2. Artalejo, J.R., Falin, J.I.: Standard and retrial queueing systems: a comparative analysis. Rev. Mat. Complutense **15**, 101–129 (2002)
3. Artalejo, J.R., Gomez-Corral, A.: Retrial Queueing Systems: A Computational Approach. Springer, Berlin (2008). https://doi.org/10.1007/978-3-540-78725-9
4. Artalejo, J.R., Gómez-Corral, A.: Waiting time analysis of the M/G/1 queue with finite retrial group. Naval Res. Logist. (NRL) **54**(5), 524–529 (2007)
5. Artalejo, J.R., Lopez-Herrero, M.J.: On the distribution of the number of retrials. Appl. Math. Model. **31**(3), 478–489 (2007)
6. Choi, B.D., Chang, Y.: MAP1, MAP2/M/c retrial queue with the retrial group of finite capacity and geometric loss. Math. Comput. Model. **30**(3–4), 99–113 (1999)
7. Chakravarthy, S., Dudin, A.: Analysis of a retrial queuing model with MAP arrivals and two types of customers. Math. Comput. Model. **37**, 343–363 (2003)
8. Falin, G., Artalejo, J.: A finite source retrial queue. Eur. J. Oper. Res. **108**, 409–424 (1998)
9. Falin, G., Fricker, C.: On the virtual waiting time in an M/G/1 retrial queue. J. Appl. Probab. **28**(2), 446–460 (1991)
10. Falin, G.I.: Waiting time in a single-channel queuing system with repeated calls. Moscow Univ. Comput. Math. Cybern. **4**, 66–69 (1977)
11. Falin, G.I.: On the waiting-time process in a single-line queue with repeated calls. J. Appl. Probab. **23**, 185–192 (1986)
12. Falin, G.I.: Virtual waiting time in systems with repeated calls. Moscow Univ. Math. Bull. **43**(6), 6–10 (1988)
13. Falin, G.I., Templeton, J.G.C.: Retrial Queues. Chapman and Hall, London (1997)
14. Gharbi, N., Dutheillet, C.: An algorithmic approach for analysis of finite-source retrial systems with unreliable servers. Comput. Math. Appl. **62**(6), 2535–2546 (2011)
15. Gomez-Corral, A., Ramalhoto, M.: On the waiting time distribution and the busy period of a retrial queue with constant retrial rate. Stochast. Model. Appl. **3**, 37–47 (2000)
16. Gómez-Corral, A., Phung-Duc, T.: Retrial queues and related models. Ann. Oper. Res. **247**(1), 1–2 (2016). https://doi.org/10.1007/s10479-016-2305-2
17. Kim, J., Kim, B.: A survey of retrial queueing systems. Ann. Oper. Res. **247**(1), 3–36 (2015). https://doi.org/10.1007/s10479-015-2038-7
18. Lee, S.W., Kim, B., Kim, J.: Analysis of the waiting time distribution in M/G/1 retrial queues with two way communication. Ann. Oper. Res. (2020). (Accepted/In press). https://doi.org/10.1007/s10479-020-03717-2

19. Moiseeva, E., Nazarov, A.: Asymptotic analysis of RQ-systems M/M/1 on heavy load condition. In: Proceedings of the IV International Conference Problems of Cybernetics and Informatics (PCI 2012), pp. 164–166. IEEE (2012)
20. Neuts, M.: The joint distribution of the virtual waiting time and the residual busy period for the M/G/1 queue. J. Appl. Probab. **5**, 224–229 (1968)
21. Nazarov, A., Moiseeva, S.P.: Methods of asymptotic analysis in queueing theory (in Russian). NTL Publishing House of Tomsk University, Tomsk (2006)
22. Nazarov, A.A., Semenova, I.A.: Asymptotic analysis of retrial queueing systems. Optoelectron. Instr. Proc. **47**, 406 (2011)
23. Nazarov, A., Sztrik, J., Kvach, A., Tóth, Á.: Asymptotic sojourn time analysis of finite-source M/M/1 retrial queueing system with collisions and server subject to breakdowns and repairs. Ann. Oper. Res. **288**(1), 417–434 (2019). https://doi.org/10.1007/s10479-019-03463-0
24. Nobel, R.: Retrial queueing models in discrete time: a short survey of some late arrival models. Ann. Oper. Res. **247**(1), 37–63 (2015). https://doi.org/10.1007/s10479-015-1904-7
25. Nobel, R., Tijms, H.: Waiting-time probabilities in the M/G/1 retrial queue. Stat. Neerl. **60**(3), 73–78 (2006)
26. Phung-Duc, T.: Retrial queueing models: a survey on theory and applications. arXiv preprint arXiv:1906.09560 (2019)
27. Sudyko, E., Nazarov, A., Sztrik, J.: Asymptotic waiting time analysis of a finite-source M/M/1 retrial queueing system. Probab. Eng. Inf. Sci. **33**(3), 387–403 (2018)
28. Tóth, A., Bérczes, T., Sztrik, J.: The simulation of finite-source retrial queueing systems with collisions and blocking. J. Math. Sci. **246**, 548–559 (2020)
29. Zhang, F., Wang, J.: Performance analysis of the retrial queues with finite number of sources and service interruptions. J. Korean Stat. Soc. **42**, 117–131 (2012). https://doi.org/10.1016/j.jkss.2012.06.002

On a Single Server Queueing Inventory System with Common Life Time for Inventoried Items

Khamis Abdullah K. AL Maqbali(✉), Varghese C. Joshua, and Achyutha Krishnamoorthy

Department of Mathematics, CMS College Kottayam, Kottayam, Kerala, India
{khamis,vcjoshua,krishnamoorthy}@cmscollege.ac.in

Abstract. We consider a single server queueing inventory model. The customers arrive according to Markovian arrival Process (MAP). The service is assumed to follow Phase type(PH) distribution. An inventory of commodities is attached to the service station. The common life time (CLT) for inventoried items is assumed to follow Phase type(PH) distribution. The inventoried items perish all together. In this case, the supply of items is immediately in local purchase to bring the inventory level to maximum inventory level S. The inventory is not allowed to go down to zero because of local purchase. Each service requires a unit of commodity for service. This unit is instantaneously taken at the beginning of the service. The replenishment of inventory follows (s, S) policy with lead time positive. The lead time follows exponential distribution. In the case of local purchase, the outstanding order of the normal purchase (wait until replenishment) is cancelled. Service of a customer begins only when the server is free. Otherwise, the arriving customer joins the buffer. Steady state analysis of the queueing inventory model is performed. Some performance measures are computed under steady state. A numerical example is presented.

Keywords: Queueing inventory · Lead time · Common life time · Local purchase · Phase type distribution · Markovian arrival process · Matrix analytic method

1 Introduction

In many real life situations, customer, who needs inventoried items to complete his service, may arrive to service station according to Markovian arrival process. After that, he may go through different phases to complete his service in order to get the inventory. Moreover, common life time (CLT) for inventoried items may go through different phases until perishing, before they are taken by

Supported by the Indian Council for Cultural Relations (ICCR) and Ministry of Higher Education in Sultanate of Oman.

A. Dudin et al. (Eds.): ITMM 2020, CCIS 1391, pp. 186–197, 2021.
https://doi.org/10.1007/978-3-030-72247-0_14

customers. The Markovian Arrival Process (MAP) is more general than Poisson process. MAP keeps the memoryless property of the Poisson process(partial memoryless) [3].

Many papers studied queueing inventory models. For example, Krishnamoorthy et al. [4] studied a PH/PH/1 queueing inventory system under (s, S) policy when the lead time is zero. AL Maqbali, Joshua and Krishnamoorthy [1] studied M/PH/1 queueing inventory system under (s, S) policy with lead time positive. Also, Krishnamoorthy and Shajin [5] studied the MAP/PH/1 queueing inventory system under (s, S) policy with lead time positive. In addition, Divya et al. [2] studied MAP/PH/1 queueing inventory system with processing of service items under vacation and N-policy with impatient customers. In their study, customers arrive according to MAP and service time follows two different phase type distribution. The inventory processing time follows phase type distribution. Moreover. Nair and Jose [8] studied the MAP/PH/1 production inventory model with varying service rates under (s, S) policy with lead time positive.

Some papers studied queueing inventory systems with common life time. For instance, Shajin et al. [10] studied a MAP/PH/1 queueing inventory system with Markovian lead time to bring the inventory level to its maximum. In their study, the common life time of inventoried items follows independent exponential distribution. Besides this, their study provided an interesting application of queueing inventory model with common life time as medicines with the same expiry date. Moreover, Shajin et al. [9] studied a MAP/M/1 and M/M/1 queueing inventory system with advanced reservation and cancellation for the next K time frames a head in the case of overbooking. In their study, the common life time (CLT) of inventoried items follows Phase type distribution.

Some papers studied queueing inventory systems with local purchase. Local purchase was introduced by Krishnamoorthy and Raju [6]. Krishnamoorthy, Varghese and Lakshmy [7] studied an (s, S) production inventory model with positive service time under local purchase.

As mentioned above, Krishnamoorthy and Shajin [5] and Divya et al. [2] studied MAP/PH/1 queueing inventory system. Then Shajin et al. [10] studied a MAP/PH/1 queueing inventory system with common life time. In this paper, we consider an MAP/PH/1 queueing inventory model under (s, S) policy with lead time positive. Besides this, we consider PH distributed common life time (CLT) for inventoried items. According to the common life time (CLT), the inventoried items perish all together. The supply of items is immediately in local purchase to bring the inventory level to the maximum inventory level S.

This model can be described as follows: customers arrive according to Markovian Arrival Process (MAP) with representation (D_0, D_1) of order y. The service is assumed to follow PH-distribution with representation (β, T) of order m. An inventory of commodities is attached to the service station. The inventoried items have common life time (CLT) which follows PH-distribution with representation (α, W) of order l. We assume that the inventoried items perish all together. In this model, the inventory is not allowed to go down to zero because of local purchase. In order to keep customer goodwill during stock out, the supply of items is immediately in local purchase to bring inventory level to S. Local purchases are purchased at a higher cost than the regular order (wait until replenishment) procedure.

Each service requires a unit of commodity for service. This unit is instantaneously taken at the beginning of the service. The replenishment of inventory follows (s, S) policy with lead time positive. The lead time follows exponential distribution with rate θ. When $1 \leq i \leq s$, the replenishment occurs to bring the inventory level i to S according to the rate of lead time. In the case of local purchase, the outstanding order of the normal purchase (wait until replenishment) is cancelled. According to (MAP), the first arriving customer instantaneously takes one item of the inventory at the beginning of his service. Then, the service of this customer immediately follows Phase type (PH) distribution. When service station is available, the next arriving customer takes one time at the beginning of his service and the service of this customer instantaneously follows Phase type (PH) distribution. Otherwise, this customer must wait in the buffer until the availability of service station. This process goes on.

According to types of blood group, blood bank has store for each blood group. The motivation for the model comes from the inventory management of one store in bank blood. For example, patients deal with one type of blood group in this store. They arrive according to Markovian Arrival Process (MAP). When the service station is available, the service of this patient follows Phase type (PH) distribution in the hospital and one blood bag is immediately taken from store to the patient at the beginning of his service. The common life time (CLT) for blood bags may go through different phases until perishing, before patients take the blood bags. In this case, the supply of blood bag is immediately in local purchase.

2 Mathematical Description of the Model

The model discussed above can be studied as a level Independent Qusi-Birth-Death (LIQBD) process. We introduce the following notations.

At time t:

$N(t)$: the number of customers in the system.
$I(t)$: the number of items in the inventory and these items are the same type.
$L(t)$: the phase of common life time.
$M(t)$: the phase of service.
$Y(t)$: the phase of the arrival process.

$X(t) = \{(N(t), I(t), L(t), M(t), Y(t)); t \geq 0\}$ is a continuous time Markov Chain (CTMC) with state space

$\Omega = \{(0, i, l_1, y_1); 1 \leq i \leq S; 1 \leq l_1 \leq l; 1 \leq y_1 \leq y\} \cup \{(n, i, l_1, m_1, y_1); n \geq 1; 1 \leq i \leq S; 1 \leq l_1 \leq l; 1 \leq m_1 \leq m; 1 \leq y_1 \leq y\}$.

The terms of transitions of the states are shown in the Table 1.

The infinitesimal generator Q of the continuous time Markov Chain (CTMC) is given by

Table 1. Intensities of transitions

From	To		Transition rate
$(0,1,l_1,y_1)$	$(1,S,l_1,m_1,y_1^{'})$	$1\le l_1\le l; 1\le m_1\le m$	$d_{y_1y_1^{'}}(1)\ \beta_{m_1}$
$(0,i,l_1,y_1)$	$(1,i-1,l_1,m_1,y_1^{'})$	$2\le i\le S$	$d_{y_1y_1^{'}}(1)\ \beta_{m_1}$
(n,i,l_1,m_1,y_1)	$(n+1,i,l_1,m_1,y_1^{'})$	$1\le n; 1\le i\le S$	$d_{y_1y_1^{'}}(1)$
$(0,i,l_1,y_1)$	$(0,i,l_1,y_1^{'})$	$1\le i\le S; y_1\ne y_1^{'}$	$d_{y_1y_1^{'}}(0)$
(n,i,l_1,m_1,y_1)	$(n,i,l_1,m_1,y_1^{'})$	$1\le n; 1\le i\le S; y_1\ne y_1^{'}$	$d_{y_1y_1^{'}}(0)$
$(0,i,l_1,y_1)$	$(0,S,l_1,y_1)$	$1\le i\le s$	θ
(n,i,l_1,m_1,y_1)	(n,S,l_1,m_1,y_1)	$1\le n; 1\le i\le s$	θ
$(1,i,l_1,m_1,y_1)$	$(0,i,l_1,y_1)$	$1\le i\le S$	$\tau^0_{m_1}$
(n,i,l_1,m_1,y_1)	$(n-1,i-1,l_1,m_1^{'},y_1)$	$2\le i\le S\ ;\ 2\le n$	$\tau^0_{m_1}\beta_{m_1^{'}}$
$(n,1,l_1,m_1,y_1)$	$(n-1,S,l_1,m_1^{'},y_1)$	$2\le n$	$\tau^0_{m_1}\beta_{m_1^{'}}$
(n,i,l_1,m_1,y_1)	$(n,i,l_1,m_1^{'},y_1)$	$1\le n; m_1\ne m_1^{'}; 1\le i\le S$	$\tau_{m_1m_1^{'}}$
$(0,i,l_1,y_1)$	$(0,S,l_1^{'},y_1)$	$1\le i\le S$	$w^0_{l_1}\alpha_{l_1^{'}}$
(n,i,l_1,m_1,y_1)	$(n,S,l_1^{'},m_1,y_1)$	$1\le n; 1\le i\le S$	$w^0_{l_1}\alpha_{l_1^{'}}$
$(0,i,l_1,y_1)$	$(0,i,l_1^{'},y_1)$	$1\le i\le S; l_1\ne l_1^{'}$	$w_{l_1l_1^{'}}$
(n,i,l_1,m_1,y_1)	$(n,i,l_1^{'},m_1,y_1)$	$1\le n; 1\le i\le S; l_1\ne l_1^{'}$	$w_{l_1l_1^{'}}$

$$Q=\begin{pmatrix} B_{00} & B_{01} & & & \\ B_{10} & A_1 & A_0 & & \\ & A_2 & A_1 & A_0 & \\ & & A_2 & A_1 & A_0 \\ & & & \ddots & \ddots & \ddots \end{pmatrix};$$

where

$$B_{00}=\begin{pmatrix} \Upsilon_1 & O_{(yls)\times[(S-s-1)yl]} & \Upsilon_2 \\ O_{((S-1-s)yl)\times(syl)} & \Upsilon_3 & \Upsilon_4 \\ O_{(yl)\times(syl)} & O_{(yl)\times((S-1-s)yl)} & \Upsilon_5 \end{pmatrix};$$

B_{00} is a square matrix of order (Syl);

where

$$\Upsilon_1=I_{(s)}\otimes((I_l\otimes D_0)+(W_l\otimes I_l)-[\theta I_{(yl)}]);$$
$$\Upsilon_2=e_s\otimes[[(\alpha\otimes W_l^0)\otimes I_y]+\theta I_{ly}];$$
$$\Upsilon_3=I_{(S-1-s)}\otimes((I_l\otimes D_0)+(W_l\otimes I_l));$$
$$\Upsilon_4=e_{(S-1-s)}\otimes[(\alpha\otimes W_l^0)\otimes I_y]\text{ and}$$
$$\Upsilon_5=(I_l\otimes D_0)+(W_l\otimes I_l)+[(\alpha\otimes W_l^0)\otimes I_y].$$

$$B_{01}=\begin{pmatrix} O_{(yl)\times((Syml)-(yml))} & I_l\otimes(\beta\otimes D_1) \\ I_{(S-1)}\otimes(I_l\otimes(\beta\otimes D_1)) & O_{([S-1]yl)\times(ylm)} \end{pmatrix};$$

B_{01} is a matrix of order $(Syl) \times (Syml)$.

$$B_{10} = \left(I_S \otimes [I_l \otimes (T_m^0 \otimes I_y)] \right);$$

B_{10} is a matrix of order $(Syml) \times (Syl)$.

$$A_2 = \begin{pmatrix} O_{(yml)\times[(S-1)yml]} & [I_l \otimes [T_m^0 \otimes (\beta \otimes I_y)]] \\ \Gamma & O_{[(S-1)yml]\times(yml)} \end{pmatrix};$$

A_2 is a square matrix of order $(Syml)$;
where $\Gamma = (I_{(S-1)} \otimes [I_l \otimes [T_m^0 \otimes (\beta \otimes I_y)]])_{[(S-1)yml]\times[(S-1)yml]}$.

$$A_0 = \left(I_{(mlS)} \otimes D_1 \right);$$

A_0 is a square matrix of order $(Syml)$.

$$A_1 = \begin{pmatrix} \varphi_1 & \varphi_2 \\ (O_{[(S-1-s)yml]\times(ylms)}, \varphi_3) & \varphi_4 \\ O_{[yml]\times([S-1]ylm)} & \varphi_5 \end{pmatrix};$$

A_1 is a square matrix of order $(Syml)$. where

$\varphi_1 = \left(I_s \otimes ([\{I_l \otimes [(I_m \otimes D_0) + (T_m \otimes I_y)]\} + (W_l \otimes I_{(ym)})] - \theta I_{(yml)}) \right);$
φ_1 is a matrix of order $(syml) \times (ylms)$;
$\varphi_2 = \left(O_{(syml)\times[(S-s-1)ylm]} \; (e_s \otimes [[(\alpha \otimes W_l^0) \otimes I_{(ym)}] + \theta I_{(yml)}])_{(syml)\times(yml)} \right);$
φ_2 is a matrix of order $(syml) \times (yml)$;
$\varphi_3 = \left([I_{(S-1-s)} \otimes [\{I_l \otimes [(I_m \otimes D_0) + (T_m \otimes I_y)]\} + (W_l \otimes I_{(ym)})]] \right).$
φ_3 is a matrix of order $((S-1-s)yml) \times ((S-1-s)yml)$;
$\varphi_4 = \left([e_{(S-1-s)} \otimes [(\alpha \otimes W_l^0) \otimes I_{(ym)}]] \right);$
φ_4 is a matrix of order $((S-1-s)yml) \times (yml)$;
$\varphi_5 = \left([\{I_l \otimes [(I_m \otimes D_0) + (T_m \otimes I_y)]\} + (W_l \otimes I_{(ym)})] + [(\alpha \otimes W_l^0) \otimes I_{(ym)}] \right);$
φ_5 is a matrix of order $(yml) \times (yml)$.

3 Steady-State Analysis

3.1 Stability Condition

Theorem 1. *The stability condition of the queueing inventory model with common life time for inventoried items under study is given by*

$$\lambda < \mu$$

Where $\lambda = (\sum_{i=0}^{(mlS)} \boldsymbol{\pi}_i) D_1 e_y$; $\boldsymbol{\pi}_i$ *is a row vector of order* (y)
and $\mu = (\sum_{i=0}^{S+1} \boldsymbol{\pi}_i) \wedge e_{(yml)}$;
where $\wedge = [I_l \otimes [T_m^0 \otimes (\beta \otimes I_y)]]$ *and* $\boldsymbol{\pi}_i$ *are row vectors of order* (yml).

Proof. Let $A = A_2 + A_1 + A_0$. We can realize that A is an irreducible matrix. Thus, there exists the stationary vector $\boldsymbol{\pi}$ of A such that

$$\boldsymbol{\pi} A = 0$$

$$\boldsymbol{\pi} e = 1.$$

The Markov chain with generator Q is stable if and only if

$$\boldsymbol{\pi} A_0 e < \boldsymbol{\pi} A_2 e.$$

Recall, $A_0 = \left(I_{(mlS)} \otimes D_1\right)$; A_0 is a square matrix of order $(Syml)$.

$$\begin{aligned}
\boldsymbol{\pi} A_0 e &= (\boldsymbol{\pi_0}, \boldsymbol{\pi_1}, \boldsymbol{\pi_2}, \dots, \boldsymbol{\pi_{(mlS)}}) \begin{pmatrix} D_1 & & & \\ & D_1 & & \\ & & \ddots & \\ & & & D_1 \end{pmatrix} \begin{pmatrix} e_y \\ e_y \\ \vdots \\ e_y \end{pmatrix}_{(mlS)} ; \\
&= (\boldsymbol{\pi_0} D_1, \boldsymbol{\pi_1} D_1, \boldsymbol{\pi_2} D_1, \dots, \boldsymbol{\pi_{(mlS)}} D_1) \begin{pmatrix} e_y \\ e_y \\ \vdots \\ e_y \end{pmatrix}_{(mlS)} ; \\
&= (\boldsymbol{\pi_0} D_1, \boldsymbol{\pi_1} D_1, \boldsymbol{\pi_2} D_1, \dots, \boldsymbol{\pi_{(mlS)}} D_1) \begin{pmatrix} 1 \\ 1 \\ \vdots \\ 1 \end{pmatrix}_{(mlS)} e_y; \\
&= (\boldsymbol{\pi_0} D_1 + \boldsymbol{\pi_1} D_1 + \boldsymbol{\pi_2} D_1 + \dots + \boldsymbol{\pi_{(mlS)}} D_1) e_y; \\
&= (\boldsymbol{\pi_0} + \boldsymbol{\pi_1} + \boldsymbol{\pi_2} + \dots + \boldsymbol{(mlS)} D_1 e_y; \\
&= (\sum_{i=0}^{(mlS)} \boldsymbol{\pi_1}) D_1 e_y \\
&= \lambda.
\end{aligned}$$

Recall,

$$A_2 = \begin{pmatrix} O_{(yml)\times[(S-1)yml]} & [I_l \otimes [T_m^0 \otimes (\beta \otimes I_y)]] \\ \Gamma & O_{[(S-1)yml]\times(yml)} \end{pmatrix};$$

A_2 is a square matrix of order $(Syml)$.
where $\Gamma = (I_{(S-1)} \otimes [I_l \otimes [T_m^0 \otimes (\beta \otimes I_y)]])_{[(S-1)yml]\times[(S-1)yml]}$.

To be more clear, we rewrite matrix A_2 as following:

$$A_2 = \begin{pmatrix} O & \cdots & & & \wedge \\ \wedge & O & \cdots & & O \\ O & \wedge & O & \ldots & \vdots \\ \vdots & O & \wedge & & \\ \vdots & \vdots & & \ddots & \vdots \\ & & & \wedge & O \end{pmatrix};$$

where $\wedge = [I_l \otimes [T_m^0 \otimes (\beta \otimes I_y)]]$.

$$\begin{aligned}
\boldsymbol{\pi} A_2 e &= (\boldsymbol{\pi_0}, \boldsymbol{\pi_1}, \boldsymbol{\pi_2}, \ldots, \boldsymbol{\pi_{(S+1)}}) \begin{pmatrix} O & \cdots & & & \wedge \\ \wedge & O & \cdots & & O \\ O & \wedge & O & \ldots & \vdots \\ \vdots & O & \wedge & & \\ \vdots & \vdots & & \ddots & \vdots \\ & & & \wedge & O \end{pmatrix} \begin{pmatrix} e_{yml} \\ e_{yml} \\ e_{yml} \\ \vdots \\ e_{yml} \\ e_{yml} \\ e_{yml} \end{pmatrix}_{(S)} ; \\
&= (\boldsymbol{\pi_0}\wedge, \boldsymbol{\pi_1}\wedge, \boldsymbol{\pi_2}\wedge, \ldots, \boldsymbol{\pi_{(S+1)}}\wedge) \begin{pmatrix} e_{yml} \\ e_{yml} \\ e_{yml} \\ \vdots \\ e_{yml} \\ e_{yml} \\ e_{yml} \end{pmatrix}_{(S)} ; \\
&= (\boldsymbol{\pi_0}\wedge, \boldsymbol{\pi_1}\wedge, \boldsymbol{\pi_2}\wedge, \ldots, \boldsymbol{\pi_{(S+1)}}\wedge) \begin{pmatrix} 1 \\ 1 \\ \vdots \\ 1 \\ 1 \end{pmatrix}_{(S)} e_{(yml)}; \\
&= (\boldsymbol{\pi_0} \wedge + \boldsymbol{\pi_1} \wedge + \boldsymbol{\pi_2} \wedge + \cdots + \boldsymbol{\pi_{(S+1)}}\wedge) e_{(yml)}; \\
&= (\boldsymbol{\pi_0} + \boldsymbol{\pi_1} + \boldsymbol{\pi_2} + \cdots + \boldsymbol{\pi_{(S+1)}}) \wedge e_{(yml)}; \\
&= (\sum_{i=0}^{S+1} \boldsymbol{\pi_1}) \wedge e_{(yml)}; \\
&= \mu.
\end{aligned}$$

Then, $\boldsymbol{\pi} A_0 e = (\sum_{i=0}^{(mlS)} \boldsymbol{\pi_1}) D_1 e_y = \lambda$ *and*

$$\boldsymbol{\pi} A_2 e = (\sum_{i=0}^{S+1} \boldsymbol{\pi_1}) \wedge e_{(yml)} = \mu.$$

Since $\pi A_0 e = \lambda$ and $\pi A_2 e = \mu$, then the queueing inventory model under study is stable if and only if

$$\lambda < \mu$$

3.2 Stationary Distribution

According to Stewart [11], we can obtain the stationary distribution of the Markov chain under study by solving the set of Eqs. 1 and 2.

$$\mathbf{X}Q = 0 \tag{1}$$

$$\mathbf{X}e = 1. \tag{2}$$

Let $\mathbf{X}$ be decomposed with Q as following :

$$\begin{aligned}
\mathbf{X} &= (\mathbf{X}_0, \mathbf{X}_1, \dots) \text{ where } \mathbf{X}_0 = (\mathbf{X}_{01}, \mathbf{X}_{02}, \dots, \mathbf{X}_{0S});\\
\mathbf{X}_{0k} &= (\mathbf{X}_{0k1}, \mathbf{X}_{0k2}, \mathbf{X}_{0k3}, \dots, \mathbf{X}_{0kl}) \text{ for } k = 1, 2, 3, \cdots, S;\\
\mathbf{X}_{0kr} &= (x_{0kr1}, x_{0kr2}, x_{0kr3}, \dots, x_{0kry}) \text{ for } r = 1, 2, 3, \cdots, l;\\
\mathbf{X}_i &= (\mathbf{X}_{i1}, \mathbf{X}_{i2}, \dots, \mathbf{X}_{iS}) \text{ for } i = 1, 2, 3, \cdots;\\
\mathbf{X}_{ik} &= (\mathbf{X}_{ik1}, \mathbf{X}_{ik2}, \mathbf{X}_{ik3}, \dots, \mathbf{X}_{ikl});\\
\mathbf{X}_{ikr} &= (\mathbf{X}_{ikr1}, \mathbf{X}_{ikr2}, \mathbf{X}_{ikr3}, \dots, \mathbf{X}_{ikrm});\\
\mathbf{X}_{ikrj} &= (x_{ikrj1}, x_{ikrj2}, x_{ikrj3}, \dots, x_{ikrjy}) \text{ for } j = 1, 2, 3, \cdots, m.
\end{aligned}$$

From Eq. 1, we get set of equations as following.

$$\mathbf{X}_0 B_{00} + \mathbf{X}_1 B_{10} = 0; \tag{3}$$

$$\mathbf{X}_0 B_{01} + \mathbf{X}_1 A_1 + \mathbf{X}_2 A_2 = 0; \tag{4}$$

$$\vdots$$

$$\mathbf{X}_{i-1} A_0 + \mathbf{X}_i A_1 + \mathbf{X}_{i+1} A_2 = 0 \text{ for } i \geq 2.$$

where i is a positive integer number.

There exists a constant matrix R such that

$$\mathbf{X}_i = \mathbf{X}_{i-1} R \text{ for } i \geq 2\ . \tag{5}$$

We can rewrite the Eq. 5 as following

$$\mathbf{X}_i = \mathbf{X}_1 R^{i-1} \text{ for } \ i \geq 2.$$

We can use the matrix quadratic Eq. 6 to obtain the matrix R.

$$R^2 A_2 + R A_1 + A_0 = 0. \tag{6}$$

The matrix R can be obtained from $R_{k+1} = -V - R_k^2 W$ and $R_0 = 0$; where $V = A_0 A_0^{-1}$ and $W = A_2 A_1^{-1}$. Then, we can find X_0 and X_1 by solving Eqs. 3 and 4. After that, we must normalize X_0 and X_1 by using the normalizing condition $\mathbf{X_0} + \mathbf{X}_1 (I - R)^{-1} e = 1$. Then, we use $\mathbf{X}_i = \mathbf{X}_1 R^{i-1}$ for $\ i = 2, 3, \dots$.

4 Performance Measures

Under steady state, some performance measures of this queueing inventory model can be obtained as following:

1. Expected number of customers in the system

$$E[N] = \sum_{i=0}^{\infty} i\mathbf{X}_i e.$$

2. Expected number of items in inventory.

$$E[I] = \sum_{i=0}^{\infty}\sum_{k=1}^{S} k\mathbf{X}_{ik} e.$$

3. Probability that the server is idle

$$b_0 = \sum_{k=1}^{S} \mathbf{X}_{0k} e.$$

5 Numerical Example

For the arrival process, we consider Markovian arrival process (MAP) with representation (D_0, D_1) of order $y = 3$, where

$$D_0 = \begin{pmatrix} -8 & 1.5 & 1 \\ 1.5 & -6 & 1.5 \\ 1 & 1 & -7 \end{pmatrix} \text{ and } D_1 = \begin{pmatrix} 1.5 & 1.5 & 2.5 \\ 0.5 & 1 & 1.5 \\ 2.5 & 1.5 & 1 \end{pmatrix}.$$

For the service process, we consider PH-representation (β, T) of order $m = 3$, where

$$\beta = (0.2, 0.5, 0.3),$$
$$T = \begin{pmatrix} -12 & 3 & 4 \\ 6 & -13 & 3 \\ 5 & 3 & -14 \end{pmatrix} \text{ and } T^0 = -Te = \begin{pmatrix} 5 \\ 4 \\ 6 \end{pmatrix}.$$

For the common life time (CLT) for inventoried items, we consider PH-representation (α, W) of order $l = 3$, where

$$\alpha = (0.3, 0.4, 0.3),$$
$$W = \begin{pmatrix} -0.35 & 0.1 & 0.2 \\ 0.3 & -0.41 & 0.1 \\ 0.3 & 0.2 & -0.52 \end{pmatrix} \text{ and } W^0 = -We = \begin{pmatrix} 0.05 \\ 0.01 \\ 0.02 \end{pmatrix}.$$

We fix the rate of lead time $\theta = 0.6$ and $s = 3$.

Now, we analyze the effect of S on the performance measures of the system in the Table 2.

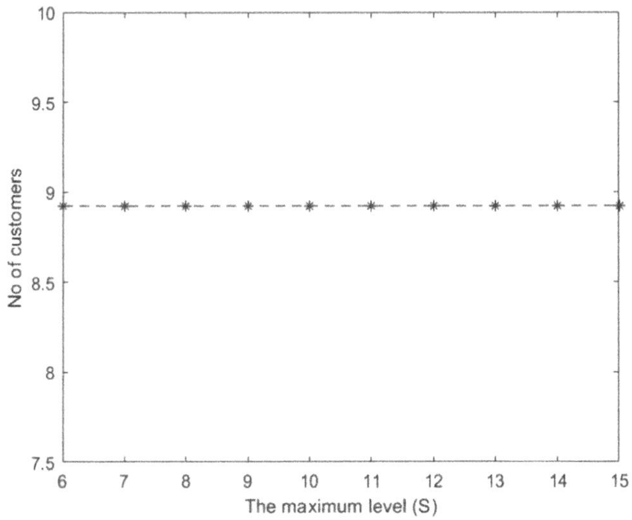

Fig. 1. Effect of S on expected number of customers

Table 2. Effect of S on various performance measures

S	E[N]	E[I]	b_0
6	8.9228	3.7312	0.1014
7	8.9228	4.2546	0.1014
8	8.9228	4.7748	0.1014
9	8.9228	5.2932	0.1014
10	8.9228	5.8109	0.1014
11	8.9228	6.3283	0.1014
12	8.9228	6.8460	0.1014
13	8.9228	7.3640	0.1014
14	8.9228	7.8826	0.1014
15	8.9228	8.4019	0.1014

From Figs. 1, 2 and 3, we can realize the effect of S on performance measures as following:

1. The expected number of customers in the system $E[N]$ has no change when the maximum inventory level S increases.
2. The expected number of items in inventory $E[I]$ increases when the maximum inventory level S increases.
3. The probability that the server is idle b_0 has no change when the maximum inventory level S increases.

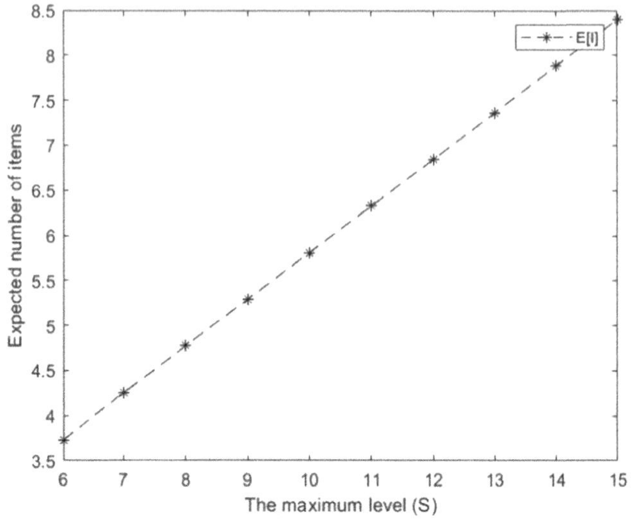

Fig. 2. Effect of S on expected number of items in inventory

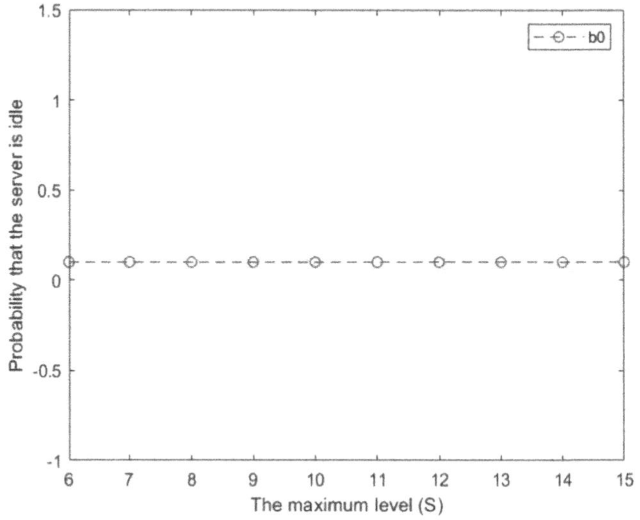

Fig. 3. Effect of S on probability that the server is idle

6 Conclusion

In this paper, we analyse an $MAP/PH/1$ queueing inventory model under (s, S) policy with lead time positive and with common life time for inventoried items. In the case of expiry of the common life time for the inventoried items, the supply of items is immediately in local purchase to bring the inventory level to the maximum inventory level S. Different performance measures are estimated under Steady state condition. In this paper, we study the effect of maximum inventory level S on the performance measures of the system numerically. We realize that firstly, the expected number of customers in the system $E[N]$ has no change when the maximum inventory level S increases. Secondly, the expected number of items in inventory $E[I]$ increases when the maximum inventory level S increases. Finally, the probability that the server is idle b_0 has no change when the maximum inventory level S increases.

Acknowledgement. The first author acknowledges the Indian Council for Cultural Relations (ICCR) (Order No: 2019-20/838) and Ministry of Higher Education (MOHE), Order No: 2019/35 in Sultanate of Oman for their supports.

References

1. Maqbali, K.A.K.A.L., Joshua, V.C., Krishnamoorthy, A.: On a single server queueing inventory system. In: Vishnevskiy, V.M., Samouylov, K.E., Kozyrev, D.V. (eds.) DCCN 2020. LNCS, vol. 12563, pp. 579–588. Springer, Cham (2020). https://doi.org/10.1007/978-3-030-66471-8_44
2. Divya, V., Krishnamoorthy, A., Vishnevsky, V.M., Kozyrev, D.V.: On a queueing system with processing of service items under vacation and N-policy with impatient customers. Queueing Models Serv. Manage. **3**(2), 167–201 (2020)
3. He, Q.M.: Fundamentals of Matrix-Analytic Methods. Springer, New York (2014). https://doi.org/10.1007/978-1-4614-7330-5
4. Krishnamoorthy, A., Jose, K.P., Narayanan, V.C.: Numerical investigation of a PH/PH/1 inventory system with positive service time and shortage. Neural Parallel Sci. Comput. **16**, 579–592 (2008)
5. Krishnamoorthy, A., Shajin, D.: MAP/PH/1 retrial queueing-inventory system with orbital search and reneging of customers. In: Rykov, V.V., Singpurwalla, N.D., Zubkov, A.M. (eds.) ACMPT 2017. LNCS, vol. 10684, pp. 158–171. Springer, Cham (2017). https://doi.org/10.1007/978-3-319-71504-9_15
6. Krishnamoorthy, A., Raju, N.: N-policy for (s, S) perishable inventory system with positive lead time. Korean J. Comput. Appl. Math. **5**(1), 253–261 (1998)
7. Krishnamoorthy, A., Varghese, R., Lakshmy, B.: Production inventory system with positive service time under local purchase. In: Dudin, A., Nazarov, A., Moiseev, A. (eds.) ITMM 2019. CCIS, vol. 1109, pp. 243–256. Springer, Cham (2019). https://doi.org/10.1007/978-3-030-33388-1_20
8. Nair, S.S., Jose, K.P.: A MAP/PH/1 production inventory model with varying service rates. Int. J. Pure Appl. Math. **117**(12), 373–381 (2017)
9. Shajin, D., Krishnamoorthy, A., Dudin, A.N., Joshua, V.C., Jacob, V.: On a queueing-inventory system with advanced reservation and cancellation for the next K time frames ahead: the case of overbooking. Queueing Syst. **94**(1), 3–37 (2020)
10. Shajin, D., Krishnamoorthy, A., Manikandan, R.: On a queueing-inventory system with common life time and Markovian lead time process. Oper. Res. 1–34 (2020)
11. Stewart, W.J.: Probability, Markov Chains, Queues, and Simulation: The Mathematical Basis of Performance Modeling. Princeton University Press, Princeton (2009)

Resource Queueing System for Analysis of Network Slicing Performance with QoS-Based Isolation

Faina Moskaleva[1(✉)], Ekaterina Lisovskaya[1], and Yuliya Gaidamaka[1,2]

[1] Peoples' Friendship University of Russia (RUDN University), 6 Miklukho-Maklaya Street, Moscow 117198, Russian Federation
{moskaleva-fa,lisovskaya-eyu,gaydamaka-yuv}@rudn.ru

[2] Federal Research Center "Computer Science and Control" of the Russian Academy of Sciences (FRC CSC RAS), 44-2 Vavilov Street, Moscow 119333, Russian Federation

Abstract. Network slicing is defined as one of the main components of fifth-generation mobile communications that can solve the problem of colossal growth in data volume traffic in cellular networks. A key feature of slicing is to limit the effect of one slice on another to provide a high quality of service. Therefore, in this paper, a model for resource sharing in slicing using the queueing theory methods is constructed. The main aim is to determine how radio resources should be fairly shared between different slices in the system. The proposed algorithm ensures the isolation of slices according to the quality of service. The resource sharing problem is formulated as an optimization problem. Analysis of the system's performance characteristics will allow us to conclude that the isolation parameter has a significant effect on metrics of interest.

Keywords: Network slicing · Isolation · Slice · Resource allocation · Optimization problem

1 Introduction

The new 5G networks and their operators have to manage a wide range of services with very varied connection requirements, targeting new market segments and vertical industries. Network slicing is a key technology that allows network operators to provide their physical infrastructure to support various services with different requirements [10]. Different sets of services may be associated with logically independent end-to-end networks, i.e. slices. The slice is a logical network that provides functional capabilities and network characteristics [3]. Slices are

This paper has been supported by the RUDN University Strategic Academic Leadership Program (recipient Yu.Gaidamaka, mathematical model development). The reported study was partially funded by RFBR, projects 19-07-00933 and 20-07-01064 (recipient F.Moskaleva, numerical analysis).

A. Dudin et al. (Eds.): ITMM 2020, CCIS 1391, pp. 198–211, 2021.
https://doi.org/10.1007/978-3-030-72247-0_15

configured and managed by tenants, to whom operators delegate control over resource utilization and service performance. Network slicing properties [4] are the automation of creating and setting slice, the slice isolation (slice independence from traffic in other slices, but also security, etc.), the elasticity of slicing (fair [6] and efficient use of resources, adaptation to conditions), management (self-management in a slice) and ability to assign priority slices.

The fundamental base for implementing the future, as well as keeping current, 5G application scenarios is network slicing [12]. This approach allows us to consider the network as a service, not as an infrastructure, as shown in the paper [5], which lets us maximize the long-term utility of the network. This technology facilitates the economical deployment and operation of complex logical networks in a joint physical network infrastructure.

The key feature of network slicing for ensuring performance and high quality of service is isolation, which limits the influence of slices on each other. Using isolation and resource sharing strategies on the radio interface is a rather entangled process [11]. Actually, it is necessary to take into account the stochastic nature of the wireless environment and the high variability of traffic in time and space, for example, using the Markov chain methods [15] and [13].

However, as shown in [8], it is possible to achieve an optimal state between isolation and efficiency by setting some network parameters. It allows us to prioritize and configure slices in accordance with the specific tasks for which it is used. Nevertheless, resource management inside the slice and between different slices should guarantee not only slice isolation but also the fair sharing of resources between users, as noted in [2].

Important in future fifth-generation networks is the economic aspect. The issue of pricing for services was considered in [18], and such an optimization structure allows us to find a compromise between the interests of communication service providers and the social welfare of the network, without violating the interests of users. Maximizing revenue using the access control mechanism for the network slice is presented in [16]. Infrastructure providers have the ability to rent network slices, both one-time and on a periodic basis. In conditions of heavy traffic for infrastructures with a large volume of resources, such a solution can provide high performance. In this regard, tenants need to coordinate common resources in the market in real-time, based on the instant needs of the slices. Based on the theory of games, the authors [7] propose a model that allows tenants to optimize their service strategies, receiving resources when and where it is necessary, in accordance with the level of quality and reliability requested by specific types of traffic.

This paper proposes the use of queueing theory and optimization theory methods. So it is a combination and extension of the research conducted in [1,9,14,17]. In [9] the authors consider one of the potential radio resource allocation schemes for multiservice wireless networks with Network Slicing technology. They describe it by using the retrial queue with the orbit to wait for sessions. In [1] the simulation architecture of VRRM (Virtual radio resource management) in terms of queueing systems is presented. In addition, the authors developed a simulator that allows you to analyze scenario with three VNOs (Virtual Network Operators) and various types of SLA (Service Level Agreement).

The paper organizes as follows. In the next section, we construct a system model of a network with two active slices. The third section is a detailed description of the mathematical model for network slicing with isolation. In the next two sections, we present an algorithm for the sharing of resources, ensuring their fair and efficient use, and demonstrate it on a small dimension example. Then we present the numerical analysis for metrics of interest. Finally, we conclude with some remarks on open issues.

2 System Model

Suppose that two slices are activated in a base station of the fifth-generation wireless system with New Radio access technology (Fig. 1). The slicing module divides between them radio resource, the total amount of which is equal to C resource units. In each slice, a certain communication service is provided to users, which involves the continuous file transfer to it on a certain number of resource units, at least b_{min}, $d_{min} \geq 0$ and not more than b_{max}, $d_{max} \geq 0$ for the first and second slices, respectively. Moreover, the transmission rate is variable and depends on the number of user connections (sessions) at every instant in time for each slice. We assume that the resources of each slice are divided equally among its users.

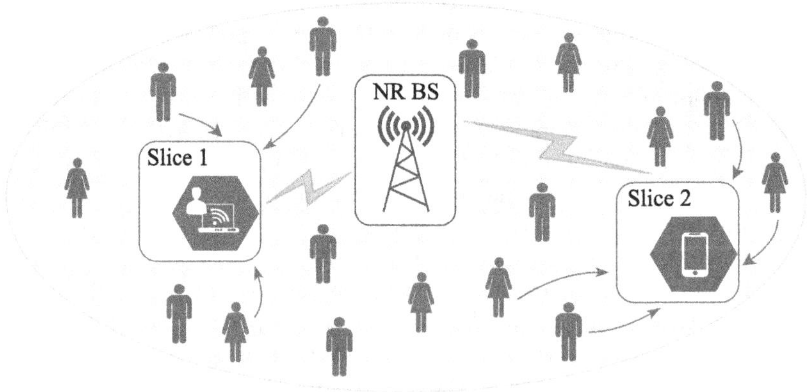

Fig. 1. System model

The radio resource slicing algorithm should ensure fair and efficient use of resources, and should also be aimed at ensuring the slices isolation according to the quality of service (QoS-based isolation).

3 Queueing System

Let two Poisson processes, corresponding to requests for data transmission from users of two different slices, arrive at the multi-server resource queueing system (QS). The intensities of arrivals are constant and equal to λ_1 and λ_2, respectively. Service durations per resource unit are independent random variables distributed exponentially with the parameters μ_1 and μ_2 for the first and second slices, respectively.

Let the total amount of QS resources for customers servicing be equal to C. The amount of resources allocated to the request depends on the system load and varies in the ranges $[b_{min}, b_{max}]$ and $[d_{min}, d_{max}]$ for first and second slices, respectively.

We define a stochastic process $\mathbf{X}(t) = \{M(t), N(t), t > 0\}$, where $M(t) = m$ is the number of customers in the first slice at the time t, $N(t) = n$ is the number of customers in the second slice at the time t. Moreover

$$m \in \{0, 1, \ldots, \lfloor C/b_{min} \rfloor\}, \quad n \in \{0, 1, \ldots, \lfloor C/d_{min} \rfloor\}.$$

Then the states space of the two-dimensional process has the form:

$$\mathbb{X} = \{(m, n) : b_{min}m + d_{min}n \leq C\}.$$

Denote the amount of the allocated resource to one request in the first slice in the state (m, n) by $b(m, n)$, and in the second slice by $d(m, n)$. Then the service intensities for the first and second slices are defined as $mb(m, n)\mu_1$ and $nd(m, n)\mu_2$.

To ensure the isolation of slices, we introduce the isolation parameters $\overline{M}$ and $\overline{N}$ which control the acceptance of arrival customers into the system and the interruption of servicing previously received customers in the following way. When shortage of resources, if one of the slices exceeds a predetermined threshold value, then the arrival request of the second slice can preempt one or more requests of the first to get service, the number of which is calculated as

$$k(m, n) = \left\lceil \frac{(m+1)b_{min} + nd_{min} - C}{d_{min}} \right\rceil,$$

$$s(m, n) = \left\lceil \frac{mb_{min} + (n+1)d_{min} - C}{b_{min}} \right\rceil.$$

If none of the slices exceeds the threshold value or exceeds both in case of resource shortage, the arrival customer will be dropped.

Thus, we write down the conditions for dropping (offloading) arriving requests and interrupting servicing requests due to resource preemption for the first and the second slices. The states space of customers dropping:

$$\mathbb{B}_1^{arr} = \{(m, n) : ((m+1, n) \notin \mathbb{X}) \cap ((m \geq \overline{M}) \cup (n \leq \overline{N}))\},$$

$$\mathbb{B}_2^{arr} = \{(m, n) : ((m, n+1) \notin \mathbb{X}) \cap ((m \leq \overline{M}) \cup (n \geq \overline{N}))\}.$$

The states space of interruption customers servicing:

$$\mathbb{B}_1^{pr} = \{(m,n) : ((m,n+1) \notin \mathbb{X}) \cap (m > \overline{M}) \cap (n < \overline{N}))\},$$
$$\mathbb{B}_2^{pr} = \{(m,n) : ((m+1,n) \notin \mathbb{X}) \cap (m < \overline{M}) \cap (n > \overline{N})\}.$$

To find the state vector of probabilities $\mathbf{p} = [p(m,n)]_{(m,n)\in\mathbb{X}}$, it is necessary to solve the system of linear equations:

$$\begin{cases} \mathbf{pQ} = \mathbf{0}, \\ \mathbf{pe} = 1, \end{cases} \tag{1}$$

where $\mathbf{Q}$ is the generator matrix of the two-dimensional stochastic process $\mathbf{X}(t)$, and $\mathbf{e}$ is the column vector of ones.

We write the generator matrix, its elements have the form:

$$q((m,n)(m',n')) = \begin{cases} \lambda_1, & m' = m+1, [(n' = n, (m+1,n) \in \mathbb{X}) \cup \\ & \cup (n' = n - k(m,n), (m+1,n) \notin \mathbb{X}, \\ & m < \overline{M}, n > \overline{N})], \\ \lambda_2, & n' = n+1, [(m' = m, (m,n+1) \in \mathbb{X}) \cup \\ & \cup (m' = m - s(m,n), (m,n+1) \notin \mathbb{X}, \\ & m > \overline{M}, n < \overline{N})], \\ mb(m,n)\mu_1, & m' = m-1, n' = n, m > 0, \\ nd(m,n)\mu_2, & m' = m, n' = n-1, n > 0, \\ Q, & m' = m, n' = n, \end{cases}$$

where

$$Q = -[\lambda_1 \cdot I((m+1,n) \in \mathbb{X}) + \lambda_1 \cdot I((m+1,n) \notin \mathbb{X}, m < \overline{M}, n > \overline{N})$$
$$+ \lambda_2 \cdot I((m,n+1) \in \mathbb{X}) + \lambda_2 \cdot I((m,n+1) \notin \mathbb{X}, m > \overline{M}, n < \overline{N})$$
$$+ mb(m,n)\mu_1 \cdot I(m > 0) + nd(m,n)\mu_2 \cdot I(n > 0)].$$

Having obtained the probability distribution, we can find some metrics characterizing the system performance:

- probabilities of customers dropping and service interruption:

$$B_s^{arr} = \sum_{(m,n)\in\mathbb{B}_s^{arr}} p(m,n), \quad B_s^{pr} = \sum_{(m,n)\in\mathbb{B}_s^{pr}} p(m,n), \quad s = 1,2;$$

- loss probabilities:

$$B_s = B_s^{arr} + B_s^{pr}, \quad s = 1,2;$$

- average service time:

$$S_1 = \frac{N_1}{\lambda_1 (1 - B_1^{arr}) - \lambda_2 \sum\limits_{(m,n)\in\mathbb{B}_1^{arr}} s(m,n)\, p(m,n)},$$

$$S_2 = \frac{N_2}{\lambda_2 (1 - B_2^{arr}) - \lambda_1 \sum\limits_{(m,n)\in\mathbb{B}_2^{arr}} k(m,n)\, p(m,n)};$$

- probability of violation (the states when the number of customers is greater than the isolation parameter):

$$V_1 = \sum_{(m,n):m>\overline{M}} p(m,n), \quad V_2 = \sum_{(m,n):n>\overline{N}} p(m,n).$$

We now consider a subspace of states where each request in both slices can be allocated the maximum required resource amount, i.e. $\Omega_1 = \{(m,n) : mb_{max} + nd_{max} \leq C\}$, which we call the states subspace of excess resources. Note that the states subspace $\Omega_0 = \mathbb{X} \backslash \Omega_1$ are the states of limited resources, for which the optimization problem can be formulated and solved.

4 Resource Sharing

In order for the resource sharing algorithm to satisfy the requirement of fairness and efficient resource use, we will solve the optimization problem. Let the data transfer rate to the user of the first and second slices corresponding to the allocated resource $b(m,n)$ and $d(m,n)$ in the state (m,n) respectively have the utility functions

$$U_1(a) = \ln a, \tag{2}$$

or

$$U_2(a, a_{min}, a_{max}) = \frac{\ln a - \ln a_{min}}{\ln a_{max} - \ln a_{min}}. \tag{3}$$

The weights $w_1(m,n)$ and $w_2(m,n)$ are calculated using the following formulas:

$$w_1(m,n) = \begin{cases} 1, & m \leq \overline{M}, \\ \frac{1}{m-\overline{M}+1}, & m > \overline{M}, \end{cases} \quad w_2(m,n) = \begin{cases} 1, & n \leq \overline{N}, \\ \frac{1}{n-\overline{N}+1}, & n > \overline{N}. \end{cases}$$

For the utility function (2) the optimal resource amounts correspond to solving the following optimization problem:

$$\begin{gathered} \max \left[w_1(m,n) m U_1(b(m,n)) + w_2(m,n) n U_1(d(m,n)) \right], \\ \text{s.t. } mb(m,n) - nd(m,n) = C, \\ \text{over} \begin{cases} (m,n) \in \Omega_0, \\ b_{min} \leq b(m,n) \leq b_{max}, \\ d_{min} \leq d(m,n) \leq d_{max}. \end{cases} \end{gathered} \tag{4}$$

Then the objective function of Lagrange problem for the state (m,n) has the form:

$$f(b(m,n), d(m,n)) = w_1(m,n) m \ln (b(m,n)) + w_2(m,n) n \ln (d(m,n)). \tag{5}$$

The functional constraint of the proposed problem is:

$$g(b(m,n), d(m,n)) = C - mb(m,n) - nd(m,n) = 0 \tag{6}$$

and corresponds to the fact that all system resources in the states (m, n) of limited resources are used. Direct constraint has the form

$$b_{min} \leq b(m,n) \leq b_{max} \text{ and } d_{min} \leq d(m,n) \leq d_{max}. \tag{7}$$

Thus, the stationary point of the objective function (5) with utility function (2) has the coordinates:

$$b(m,n) = \frac{w_1(m,n)C}{w_1(m,n)m + w_2(m,n)n}, \quad d(m,n) = \frac{w_2(m,n)C}{w_1(m,n)m + w_2(m,n)n} \tag{8}$$

and is located in the intersection point of lines $mb(m,n) + nd(m,n) = C$ and $w_2(m,n)b(m,n) = w_1(m,n)d(m,n)$.

For the utility function (3) the optimal resource amounts correspond to solving the following optimization problem:

$$\begin{gathered} \max\left[w_1(m,n)mU_2(b(m,n), b_{min}, b_{max}) + w_2(m,n)nU_2(d(m,n), d_{min}, d_{max})\right], \\ \text{s.t. } mb(m,n) - nd(m,n) = C, \\ \text{over } \begin{cases} (m,n) \in \Omega_0, \\ b_{min} \leq b(m,n) \leq b_{max}, \\ d_{min} \leq d(m,n) \leq d_{max}. \end{cases} \end{gathered} \tag{9}$$

Then the objective function of Lagrange problem for the state (m, n) has the form:

$$\begin{gathered} f(b(m,n), d(m,n)) \\ = w_1(m,n)m\frac{\ln b(m,n) - \ln b_{min}}{\ln b_{max} - \ln b_{min}} + w_2(m,n)n\frac{\ln d(m,n) - \ln d_{min}}{\ln d_{max} - \ln d_{min}}. \end{gathered} \tag{10}$$

with functional constraint (6) and direct constraint (7).

The stationary point of the objective function (10) for utility function (3) has the coordinates:

$$\begin{aligned} b(m,n) &= \frac{Cw_1(m,n)(\ln d_{max} - \ln d_{min})}{mw_1(m,n)(\ln d_{max} - \ln d_{min} + nw_2(m,n)(\ln b_{max} - \ln b_{min})}, \\ d(m,n) &= \frac{Cw_2(m,n)(\ln b_{max} - \ln b_{min})}{mw_1(m,n)(\ln b_{max} - \ln b_{min}) + nw_2(m,n)(\ln d_{max} - \ln d_{min})} \end{aligned} \tag{11}$$

and is located in the intersection point of lines $mb(m,n) + nd(m,n) = C$ and $w_2(m,n)b(m,n) = w_1(m,n)d(m,n)$.

However, the obtained point may not satisfy the direct constraints of $b_{min} \leq b(m,n) \leq b_{max}$ and $d_{min} \leq d(m,n) \leq d_{max}$. Then the solution should be sought in the intersection point of the straight line corresponding to the functional constraint and the borders of the coordinate rectangle corresponding to direct constraints.

5 Demonstration of Resource Sharing Algorithm

To demonstrate the obtained results, consider a numerical example of a small dimension. Let the system have $C = 10$ resource units for use by two slices, and the resource requirements are characterized by the parameters: $b_{min} = 1, b_{max} = 3, d_{min} = 2, d_{max} = 5$, slice isolation parameters: $\overline{M} = 3, \overline{N} = 2$.

To obtain the performance parameters of the system, it is necessary to calculate the amount of the allocated resource for one request in each state according to formulas (8) taking into account the direct constraints (7). Let us consider in detail the solution of the optimization problem for several states, demonstrating all possible cases of the location of the stationary point of the optimization problem (5), (6).

5.1 State $(m, n) = (2, 2)$

For this state, the weight functions are $w_1(2,2) = 1, w_2(2,2) = 1$. In Fig. 2 the arguments $(2, 2)$ are omitted for the sake of compactness, in Fig. 2a shows: red dotted line corresponds to straight line $b(2,2) = d(2,2)$, solution (8); green rectangle is direct constraints $1 \leq b(2,2) \leq 3$ and $2 \leq d(2,2) \leq 5$; blue line corresponds to $2b(2,2) + 2d(2,2) = 10$, functional limitation; blue curves – level lines of the objective function (5). Here, the range of valid values is the red line inside the green rectangle. While the stationary point of the optimization problem

$$b(2,2) = \frac{1 \cdot 10}{1 \cdot 2 + 1 \cdot 2} = 2.5, \quad d(2,2) = \frac{1 \cdot 10}{1 \cdot 2 + 1 \cdot 2} = 2.5,$$

belongs to this area, therefore it is a solution to the optimization problem. The objective function value at this point is $f(b(2,2), d(2,2)) \approx 3.66$. Figure 2b shows the graph of the objective function of one variable, in which the variable $d(2,2)$ is expressed from the functional constraint through the variable $b(2,2)$. Let us display the direct constraints: it is known that $2 \leq \frac{10-2b(2,2)}{2} \leq 5$, whence we get:

$$\begin{cases} 1 \leq b(2,2) \leq 3, \\ 0 \leq b(2,2) \leq 3, \end{cases}$$

the solution to the system of inequalities is $1 \leq b(2,2) \leq 3$. The task is to find the maximum of the objective function on the segment $[1; 3]$. The solution will be the extremum point of the function $(b(2,2); f(b(2,2))) = (2.5; \approx 3.66)$, then $d(2,2) = 2.5$, which is consistent with the solution in Fig. 2a.

5.2 State $(m, n) = (1, 2)$

For this state, the weight functions are $w_1(1,2) = 1, w_2(1,2) = 1$, and the stationary point of the optimization problem without taking into account direct constraints has coordinates

$$b(1,2) = \frac{1 \cdot 10}{1 \cdot 1 + 1 \cdot 2} = \frac{10}{3}, \quad d(1,2) = \frac{1 \cdot 10}{1 \cdot 1 + 1 \cdot 2} = \frac{10}{3},$$

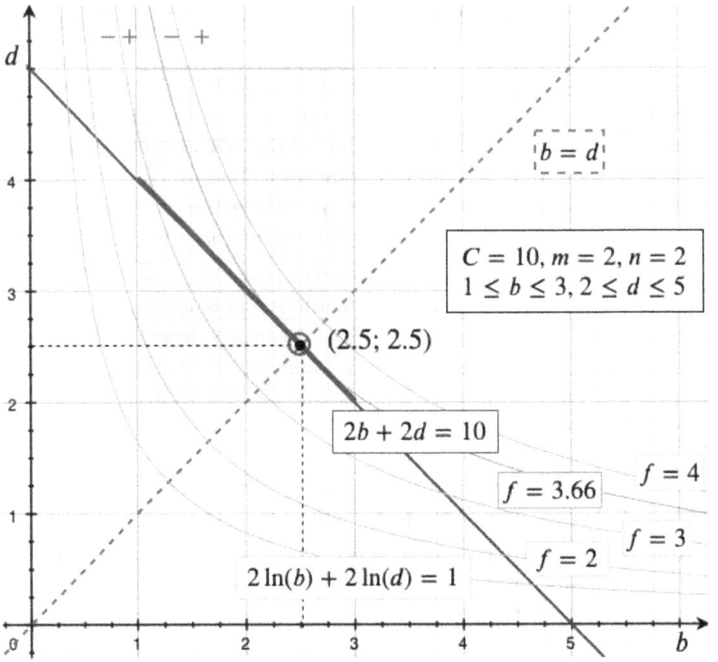

a) Maximum of the objective function of two arguments

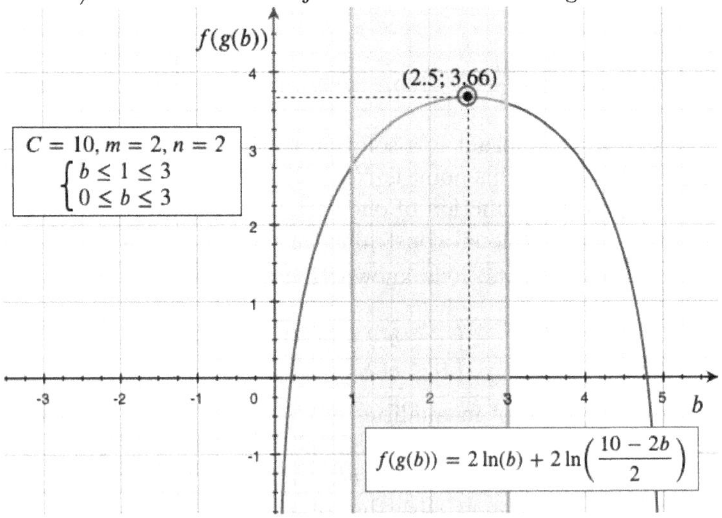

b) Maximum of the objective function of one argument

Fig. 2. Solution of the optimization problem for the state $(2, 2)$

(on Fig. 3a) and does not belong to the range of valid values, because $\frac{10}{3} > 3$, that is, $b(1,2) > b_{max} = 3$. Therefore, the solution to the problem should be sought in the intersection points of the straight line $b(1,2) + 2d(1,2) = 10$ and

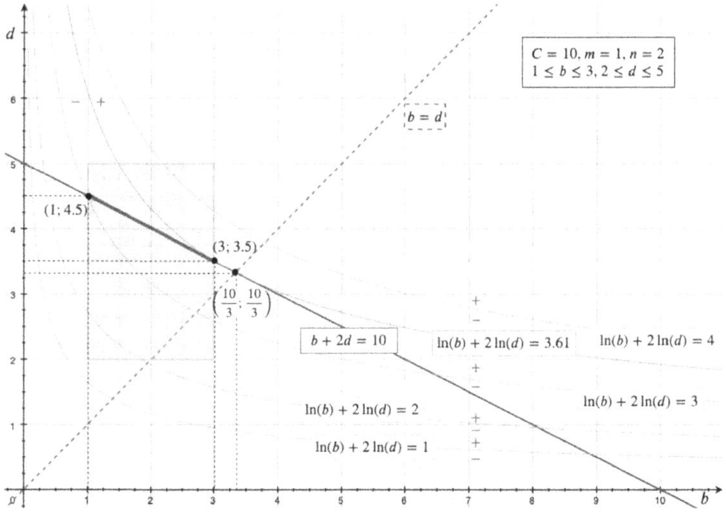

a) Maximum of the objective function of two arguments

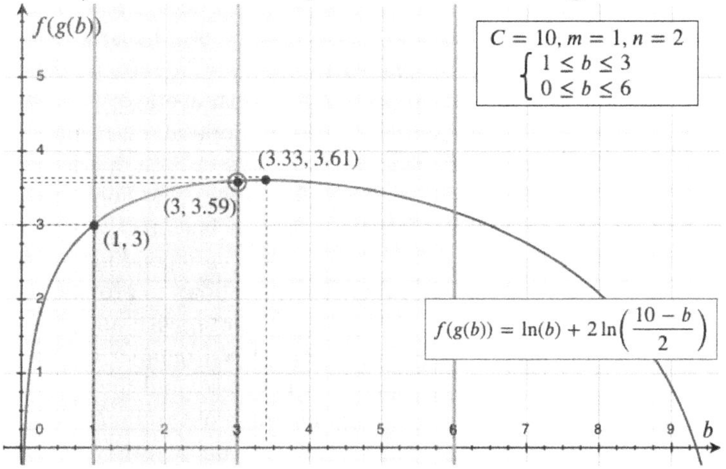

b) Maximum of the objective function of one argument

Fig. 3. Solution of the optimization problem for the state (1, 2) (Color figure online)

direct constraints (green rectangle). We observe the behavior of the lines of the objective function level, we see that at the point $(3; \frac{10-1\cdot 3}{2}) = (3; 3.5)$ it distributes the maximum value $f(b(1,2), d(1,2)) \approx 3.61$. Figure 3b shows a graph of the objective function of one variable. Let us display the direct constraints: it is known that $2 \leq \frac{10-b(1,2)}{2} \leq 5$, whence we get:

$$\begin{cases} 1 \leq b(1,2) \leq 3, \\ 0 \leq b(1,2) \leq 6, \end{cases}$$

the solution to the system of inequalities is $1 \leq b(1,2) \leq 3$. The task is to find the maximum of the objective function on the segment $[1;3]$. The solution will be the maximum point of the function on the boundary of the segment $(b(1,2); f(b(1,2))) = (3; \approx 3.61)$, then $d(1,2) = 3.5$, which is consistent with the solution in Fig. 3a.

The solution of such an optimization problem will make it possible to divide the resource between customers in such a way as to provide isolation of slices, efficient use of the system resource and fairness in serving users.

6 Numerical Analysis

Let us illustrate the dependence of system characteristics on the isolation of the network slices for data close to real [19]. So we consider the following services: buffered HD video streaming for the first slice and file download (software updates) for the second slice. We assume that two operators share 50 Mbps ($C = 50$) according to the solution of the optimization problem (5). The minimum data transfer rates are 5 Mbps ($b_{min} = 5$) and 1 Mbps ($d_{min} = 1$), maximum – 8 Mbps ($b_{max} = 8$) and 50 Mbps ($d_{max} = 50$). Isolation parameters: $\overline{M} = 5$ and $\overline{N} = 25$ customers, arrival rates are $\lambda_1 = 1/150$ and $\lambda_2 = 1/120$, service rates are $\mu_1 = 1/1920$ and $\mu_2 = 1/4000$, with average file sizes 1.2 GB and 500 MB for the services of the first and second operators, respectively.

Figure 4 shows the graphs of the loss probabilities of both operators for both utility functions (red lines correspond to the first slice, blue lines to the second slice, solid lines to the first utility function, dashed lines to the second utility function). It is possible to change the utility function depending on the operator's needs, for example, the second utility function allows to provide the greater resource to customers, number of which is bigger. Note, with an increase in the isolation parameter of the first slice, i.e. a guaranteed number of received customers $\overline{M}$, the loss probability for the first slice decreases and the loss probability for the second slice increases. The tipping point turns out to be $\overline{M} = 5$, at which the system resource becomes insufficient to simultaneously satisfy the guarantees of the first and second slice, i.e. $\overline{M} b_{min} + \overline{N} d_{min} > C$.

Figure 5 shows that the isolation parameter has a significant effect on the probability V_s of a slice being in the violation state ($m > \overline{M}$ for the first slice, $n > \overline{N}$ for the second). With an increase in the isolation parameter $\overline{M}$, the probability V_1 for the first slice tends to zero, and the probability V_2 for the second slice changes insignificantly, which indicates that isolation is provided.

Thus, we can conclude that changing the isolation parameter has a significant effect on the performance characteristics of the system, while the constructed model is able to provide isolation of slices.

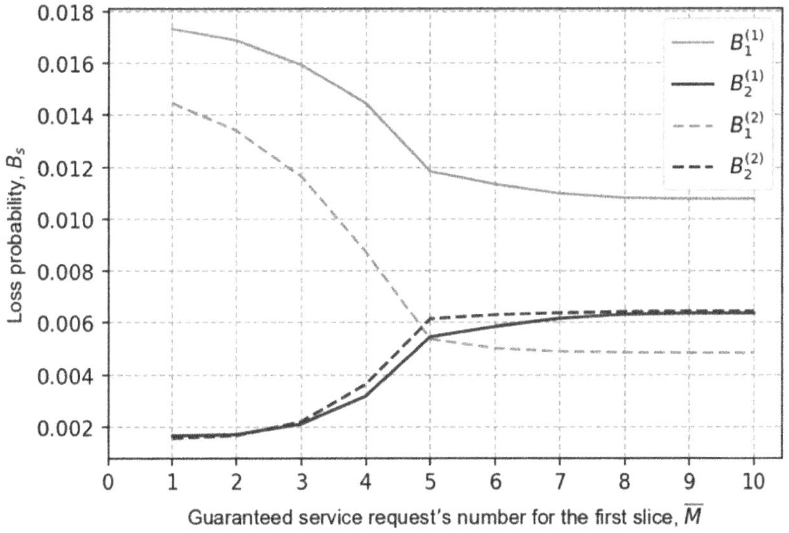

Fig. 4. Loss probability

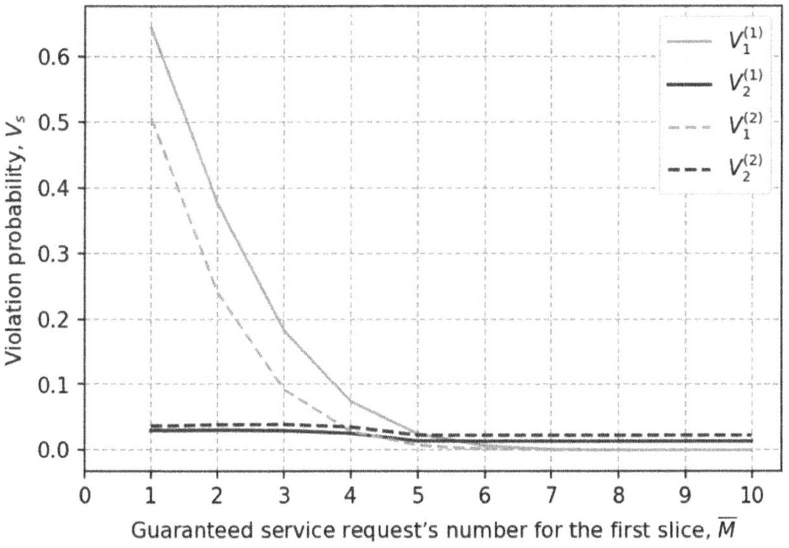

Fig. 5. Violation probability

7 Conclusion

This paper presents a model of resource sharing for network slicing to ensure the isolation of slices. The algorithm for sharing resources is developed that takes into account the features of the presented mathematical model. The resource sharing problem is formulated as an optimization problem and its solution is obtained. The resource sharing algorithm allows us to share network resources fairly and effectively. The objective of further tasks is to compare the proposed algorithm with complete sharing and complete isolation algorithms. In addition, an assessment of the influence of isolation parameters in the proposed algorithm on the main characteristics of system performance will be made. In addition, it is necessary to extend the model to the case of an arbitrary number of slices, as well as consider the model with zero minimum and unlimited maximum resource requirements.

References

1. Ageev, K., et al.: Modelling of virtual radio resources slicing in 5G networks. Commun. Comput. Inf. Sci. **1109**, 150–161 (2019). https://doi.org/10.1007/978-3-030-33388-1_13
2. Alfoudi, A.S.D., Newaz, S.H.S., Otebolaku, A., Lee, G.M., Pereira, R.: An efficient resource management mechanism for network slicing in a LTE network. IEEE Access **7**, 89441–89457 (2019). https://doi.org/10.1109/ACCESS.2019.2926446
3. ETSI: 5G: System architecture for the 5G System (Release 15). 3GPP TS 23.501 V 15.2.0 (2018)
4. GSMA: Network Slicing - Use Case Requirements. Technical report, April 2018
5. Han, B., Lianghai, J., Schotten, H.D.: Slice as an evolutionary service: genetic optimization for inter-slice resource management in 5G networks. IEEE Access **6**, 33137–33147 (2018). https://doi.org/10.1109/ACCESS.2018.2846543
6. Jain, R., Chiu, D.M., Hawe, W.R.: A quantitative measure of fairness and discrimination for resource allocation in shared computer systems. CoRR cs.NI/9809099, January 1998
7. Lieto, A., Moro, E., Malanchini, I., Mandelli, S., Capone, A.: Strategies for network slicing negotiation in a dynamic resource market. In: 2019 IEEE 20th International Symposium on "A World of Wireless, Mobile and Multimedia Networks" (WoWMoM), pp. 1–9 (2019). https://doi.org/10.1109/WoWMoM.2019.8792999
8. Marabissi, D., Fantacci, R.: Highly flexible RAN slicing approach to manage isolation, priority, efficiency. IEEE Access **7**, 97130–97142 (2019). https://doi.org/10.1109/ACCESS.2019.2929732
9. Markova, E., Adou, Y., Ivanova, D., Golskaia, A., Samouylov, K.: Queue with retrial group for modeling best effort traffic with minimum bit rate guarantee transmission under network slicing. In: Vishnevskiy, V.M., Samouylov, K.E., Kozyrev, D.V. (eds.) DCCN 2019. LNCS, vol. 11965, pp. 432–442. Springer, Cham (2019). https://doi.org/10.1007/978-3-030-36614-8_33
10. NGMN: 5G White Paper. Technical report (2015)
11. Richart, M., Baliosian, J., Serrat, J., Gorricho, J.: Resource slicing in virtual wireless networks: a survey. IEEE Trans. Netw. Serv. Manage. **13**(3), 462–476 (2016). https://doi.org/10.1109/TNSM.2016.2597295
12. Sallent, O., Perez-Romero, J., Ferrus, R., Agusti, R.: On radio access network slicing from a radio resource management perspective. IEEE Wirel. Commun. **24**(5), 166–174 (2017). https://doi.org/10.1109/MWC.2017.1600220WC
13. Tang, L., Tdan, Q., Shi, Y., Wang, C., Chen, Q.: Adaptive virtual resource allocation in 5G network slicing using constrained Markov decision process. IEEE Access **6**, 61184–61195 (2018). https://doi.org/10.1109/ACCESS.2018.2876544

14. Vikhrova, O., Suraci, C., Tropeano, A., Pizzi, S., Samouylov, K., Araniti, G.: Enhanced radio access procedure in sliced 5G networks. In: 2019 11th International Congress on Ultra Modern Telecommunications and Control Systems and Workshops (ICUMT), pp. 1–6 (2019). https://doi.org/10.1109/ICUMT48472.2019.8970776
15. Vilà, I., Pérez-Romero, J., Sallent, O., Umbert, A.: Characterization of radio access network slicing scenarios with 5G QoS provisioning. IEEE Access **8**, 51414–51430 (2020). https://doi.org/10.1109/ACCESS.2020.2980685
16. Vincenzi, M., Lopez-Aguilera, E., Garcia-Villegas, E.: Maximizing infrastructure providers' revenue through network slicing in 5G. IEEE Access **7**, 128283–128297 (2019). https://doi.org/10.1109/ACCESS.2019.2939935
17. Vlaskina, A., Polyakov, N., Gudkova, I.: Modeling and performance analysis of elastic traffic with minimum rate guarantee transmission under network slicing. In: Galinina, O., Andreev, S., Balandin, S., Koucheryavy, Y. (eds.) NEW2AN/ruSMART -2019. LNCS, vol. 11660, pp. 621–634. Springer, Cham (2019). https://doi.org/10.1007/978-3-030-30859-9_54
18. Wang, G., Feng, G., Qin, Sh., Wen, A.R., Sun, S.: Optimizing network slice dimensioning via resource pricing. IEEE Access **7**, 30331–30343 (2019). https://doi.org/10.1109/ACCESS.2019.2902432
19. Yarkina, N., Gaidamaka, Y., Correia, L.M., Samouylov, K.: An analytical model for 5G network resource sharing with flexible SLA-oriented slice isolation. Mathematics **8**(7), 1177 (2020). https://doi.org/10.3390/math8071177

Infinite Markings Method in Queueing Systems with the Infinite Variance of Service Time

Vladimir N. Zadorozhnyi[1], Michele Pagano[2], and Tatiana R. Zakharenkova[1](✉)

[1] Omsk State Technical University, 11, Mira Pr., Omsk 644050, Russian Federation
[2] University of Pisa, Via Caruso, 16, 56126 Pisa, Italy

Abstract. Possible applications of the developed method with assignment of preemptive priorities determined by infinite markings of the service time semiaxis are analyzed. In queueing systems without priority the stationary average queue length is infinite when the service time has infinite variance even if the utilization is less than one. In case of Poisson arrival flow and Pareto distributed service times with finite mean and infinite variance, the stationary distribution of the queue length is shown to be asymptotically power-law with the exponent less than one; consequently, the stationary average queue length is infinite. But if we use the preemptive priorities in accordance with infinite markings method, the average waiting time in the queue and, hence, its average length become finite. For the case when the request length is proportional to its service time, we introduce an indicator defined as the sum of request lengths in the queue. Generalized Little's formula for calculating and estimating of average value of the indicator in systems with finite and infinite stationary average queue length is derived. We consider the case when actual service time of arriving requests is unknown. For this situation it is proposed to employ a version of infinite markings method based on dynamically configurable preemptive priorities. In this case, it is found that the infinite markings method provides finite average waiting time.

Keywords: Queueing systems · Fractal traffic · Heavy-tailed distributions · Service discipline · Infinite markings method

1 Introduction

We consider possible applications of the method proposed in [1,2] to networks with fractal traffic [3] for reducing the average queue length in queueing systems with infinite variance of the service time x. The method is based on introducing preemptive priorities assigned in accordance with infinite markings of service time semiaxis. From Pollaczek–Khinchine formula it follows that M/GI/1 systems with infinite variance of the service time have an infinite average queue length as shown in detail for Pareto distributed service time in [4]. In [1] it is

A. Dudin et al. (Eds.): ITMM 2020, CCIS 1391, pp. 212–224, 2021.
https://doi.org/10.1007/978-3-030-72247-0_16

proven that assignment of preemptive priorities in accordance with the infinite markings method (IMM) makes the average waiting time finite.

We consider the following cases:

- when the service time x is known and the queue length is defined not only by the number of requests in the queue, but also by requests length (as when transferring files over the Internet),
- when the queue length is defined only by the number of requests in the queue and the service time of arrival requests is unknown.

The first case, already investigated in [1,3], was the original application of IMM. In this paper we focus on both cases and suggest to apply an IMM version based on dynamically configurable preemptive priorities. It is established that our approach allows the average queue length to be finite.

2 Mathematical Model

In the rest of the paper we assume that requests service time x is Pareto distributed:

$$F(t) = P(x \leq t) = 1 - \left(\frac{K}{t}\right)^{\alpha}, \qquad \alpha > 0, \qquad t \geq K,$$

where $K > 0$ is the least value of the random variable (r.v.) and simultaneously the scale parameter of Pareto distribution; α is shape parameter.

The first two moments of the r.v. x and its variance are determined as follows:

$$\mathrm{E}(x) = \frac{\alpha K}{\alpha - 1}, \qquad (\alpha > 1),$$

$$\mathrm{E}(x^2) = \frac{\alpha K^2}{(\alpha - 1)^2(\alpha - 2)}, \qquad (\alpha > 2),$$

$$\mathrm{D}(x) = \frac{\alpha K^2}{(\alpha - 1)^2(\alpha - 2)}, \qquad (\alpha > 2).$$

When $\alpha < 1$ all moments of x are infinite, while for $1 < \alpha \leq 2$ we have $\mathrm{E}(x) < \infty$, $\mathrm{D}(x) = \infty$. This variant is highly demanded in fractal traffic modeling. In M/Pa/1/∞ systems (with Pareto distributed service time x) with $1 < \alpha \leq 2$ according to Pollaczek–Khinchine formula [5] the average queue length is infinite:

$$L = \frac{\lambda^2 \mathrm{E}(x^2)}{2(1 - \rho)} = \infty, \tag{1}$$

where $\rho = \lambda \mathrm{E}(x) < 1$ is the utilization and λ is the arrival rate.

According to Little's formula, the average waiting time is $W = L/\lambda$. Consequently, $W = \infty$ for $L = \infty$. Since $\rho < 1$ in (1), a stationary regime in the system exists, and the infinite average queue length can be explained by the fact that the stationary queue length distribution [6–15] is heavy tailed. Let us verify this hypothesis by system simulation.

3 Simulation of M/Pa/1 System at $D(x) = \infty$

Let us consider a test M/Pa/1 with the following parameters: $\lambda = 1/6$, $K = 1$, $\alpha = 1.5$. Then, the service time has finite mean, $E(x) = 3$ and infinite variance $D(x) = \infty$, utilization $\rho = \lambda\ E(x) = 0.5 < 1$, and, therefore, a stationary distribution of the queue length l exists. Let us determine its properties by means of simulation experiments with different lengths (in all cases the test system is empty at $t = 0$). The simulation length T_M represents the time during which on average N requests enter into the system, i.e. $T_M = N/\lambda$. The actual number of arriving requests (random number) deviates from N by only percent fractions in system with Poisson flows when N is several millions.

Figures 1, 2 and 3 illustrate probabilities $r_j = P(l = j)$ for the queue length $l(j = 0, 1, 2, \ldots)$ calculated at increasing (finite) values of N. When $N \to \infty$ the calculated values (empirical probabilities) converge to the probabilities of interest $r_j = P(l = j)$ in accordance with the strong law of large numbers. This may tell us about the actual queue length l distribution.

When $N = 10^4$ (Fig. 1) the maximum length $l_{\max}$ is equal to 35 and the estimate for L is 1.099. The obtained empirical distribution (continuous curve in the figure) is already well approximated by the power function $y = 0.147x^{-1.667}$ for integer variable $x \in \{0, 1, 2, ...\}$ (the graphs are represented by continuous lines for sake of clarity); indeed, the coefficient of determination R^2 is close to one. The graph of the approximating power function $y(x)$ (red solid line – a trend line) has a straight-line form, since we use logarithmic axes. The value 0 on logarithmic scales is unrepresentable, therefore, the probability r_0 for zero queue length, estimate of which is 0.764, on the graph in Fig. 1 is missing. The estimate 0.4564 of the utilization is not far from the exact value $\rho = 0.5$.

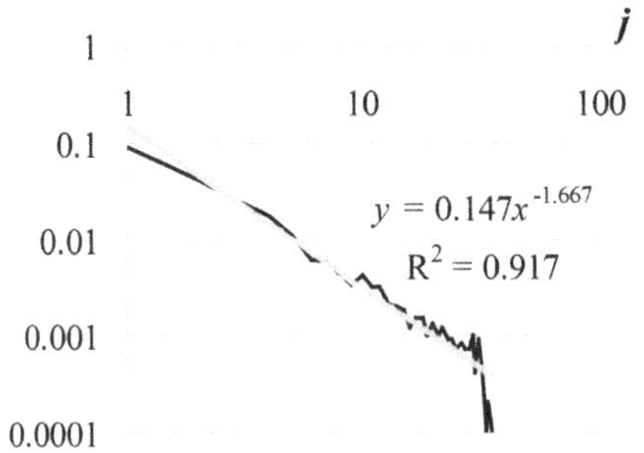

Fig. 1. Calculation of distribution r_j in the test system at simulation run $N = 10^4$ (Color figure online)

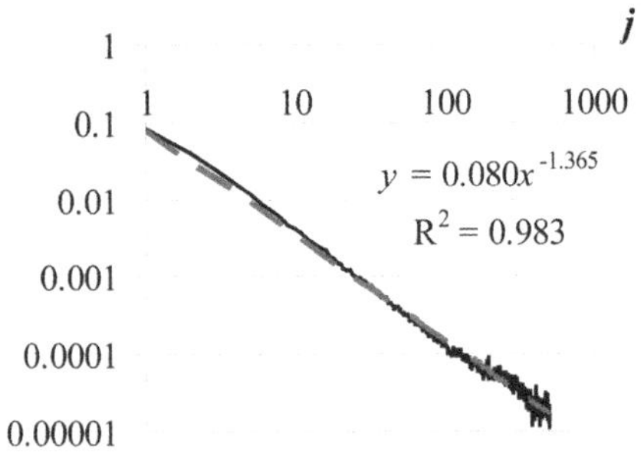

Fig. 2. Calculation of distribution r_j in the test system at simulation run $N = 10^6$

Figure 2 and Fig. 3 show similar results obtained for simulation run lengths $N = 10^6$ and $N = 10^8$, respectively. Since the trend line at long simulation runs begins to merge with the empirical probability distribution, it is depicted as a thickened dashed line.

The position of the trend line and the coefficient of determination R^2, which is much closer to one, in Fig. 2 demonstrate that the empirical distribution of the queue length converges to a power distribution and the fitting is adequate over a bigger range (up to $j = 500$). The corresponding estimate of the probability r_0 is 0.707, while the estimates for l_{max} and L increased significantly and reached the values of 7194 and 64.49, respectively. The estimate of the utilization 0.5052 noticeably approached the exact value $\rho = 0.5$.

Figure 3 shows simulation results of the testing system at simulation run $N = 100$ mln.

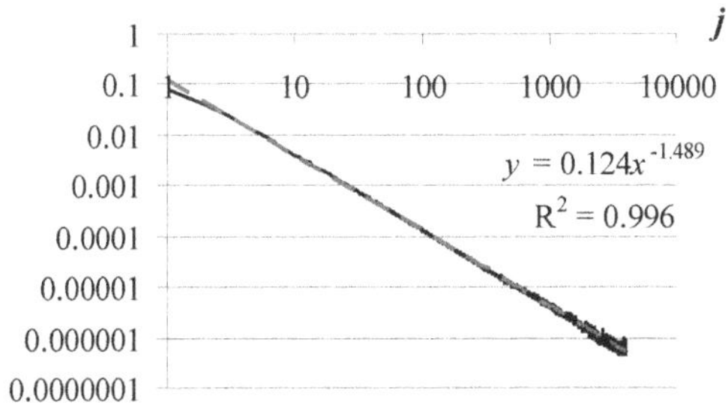

Fig. 3. Distribution calculation r_j in the test system at $N = 10^8$

Simulation results in Fig. 1, 2 and 3 confirm the above assumption about the asymptotically power-law form of the queue length distribution. The "straight" range of the distribution in Fig.3 increased and reached the value $j = 4000$, the maximum queue length reached 62314, and the estimate for L increased up to 85.42. The obtained estimate of the probability r_0 is equal to 0.715 and the utilization estimate is 0.4991. The coefficient of determination has become even closer to one. The exponent of the power-law distribution approaches the value (-1.5). So, by dropping the estimates of probabilities r_j in the empirical distribution at $j > 1000$ (they are obtained from a small number of realizations of the corresponding events and have large confidence intervals), and also estimates for r_j at $j < 3$ (where the distribution r_j is noticeably different from its power asymptotic), we obtain the approximating function $r_j = P(l = j) = 0.138j^{-1.498}$ characterized by $R^2 = 0.999$. With such power-law distribution for discrete r.v. j the sum defining its mean diverges to infinity:

$$L = \sum_{i=1}^{\infty} j r_j \approx \sum_{i=1}^{\infty} j \cdot 0.138 j^{-1.498} = \sum_{i=1}^{\infty} 0.138 j^{-0.498} = \infty.$$

Therefore, the results of the performed experiments agree with the above assumption explaining the infinite average queue length in stationary regime.

4 A Brief Description of the Infinite Markings Method (IMM)

The discipline of preemptive priorities in accordance with IMM is defined by splitting the range $K \le t < \infty$ of possible values of the service time x, using an infinite sequence of points (markup) into intervals

$$[t_0, t_1),\ [t_1, t_2),\ ...,\ [t_{k-1}, t_k),\ ...$$

where $t_0 = K$ (Fig. 4).

Fig. 4. Infinite markup of a range of possible vales x

If the entering request has service time belonging to the $[t_0, t_1)$ interval, to this request the P_1 priority is assigned, if $x \in [t_1, t_2)$, then the assigned priority is $P_1 - 1$ priority, if $x \in [t_2, t_3)$, then its priority is $P_1 - 2$, and so on. With such preemptive priority assignment [1, 2] the average queue length L and the average waiting time W become finite.

Further consideration will be given of the exponential marking with

$$t_0 = K, \qquad t_k = K + ce^{ak}, \qquad (k = 1, 2, ...). \tag{2}$$

The values of the parameters a, c are optimized according to the criterion $W \to \min$. To this aim we use the known formulas to calculate W in systems with preemptive priorities [5].

To determine the request class (marking interval number) by its service time x, the following formula from (2) is used

$$k = \frac{1}{a}\left\lceil \log\left(\frac{x - K}{c}\right)\right\rceil.$$

5 Influence of IMM on the Queue Determined by Request Length

If the request service time is proportional to its length – as when file transmission occurs, then the queue length can be calculated in two ways. The first way is the traditional one when the queue length l is determined as the number of requests in it. In this case, applying IMM to a system with an infinite variance of the service time leads to finite queue length L. In the second way, the queue length is determined as the sum of the request lengths in the queue. In order to distinguish such queue length from the traditional one, we will call it the queue duration, denote it by ω, and denote the average queue length by Ω. Moreover, assuming that the request service time is proportional to request length with a proportionality coefficient of one, we obtain

$$\omega = \sum_{i=1}^{l} x_i, \tag{3}$$

where x_i is the service time of a request in the queue.

Let us consider again the M/Pa/1/∞ introduced in Sect. 3 in which the average queue length L (1) is infinite (and hence $\Omega = \infty$). The optimized values of parameters a, c of exponential marking (2) are $a \approx 1$, $c \approx 0.5$. Table 1 shows the simulation results without the application of IMM for different lengths of the simulation runs. Recall, that a run length is determined by the simulated time period, during which the system receives on average N requests. Estimates $\hat{L}$, $\hat{\Omega}$ and $\hat{W}$ of parameters L, Ω and W converge to their stationary values (to infinity) as the simulation length increases. Table 2 shows the simulation results of the same test system M/Pa/1/∞ when IMM is used. Estimates for L and W stabilize near the stationary values of these indicators, close to 0.22 and 1.3, respectively. At the same time, the estimate of the average duration Ω is growing along with simulation run length, which indicates that $\Omega = \infty$. Taking into account the small average number $L \approx 0.22$ of summands in formula (3), the stationary average $\Omega = \infty$ is caused by requests with high service time. This is quite consistent with the physical meaning of IMM, which makes it possible for "short" requests to be serviced without delays due to their high priority.

Table 1. Estimated values $\hat{L}$, $\hat{\Omega}$, $\hat{W}$ with growing N in the model without priorities

Simulation run N	$\hat{L}$	$\hat{\Omega}$	$\hat{W}$
10^4	1.093	2.773	3.737
10^5	3.025	7.884	15.222
10^6	64.486	161.834	383.375
10^7	129.524	332.182	773.538

Table 2. Estimated values $\hat{L}$, $\hat{\Omega}$, $\hat{W}$ with growing N in the model with IMM

Simulation run N	$\hat{L}$	$\hat{\Omega}$	$\hat{W}$
10^4	0.133	3747	0.903
10^5	0.176	45688	1.070
10^6	0.233	514686	1.391
10^7	0.222	5055579	1.329

The increase of simulation estimates of the average queue duration with the simulation length growth is explained by fact that the average queue duration is infinite when using IMM in contrast to the average queue length. To show this, it is enough to use again Little's formula.

In fact, in continuous time t the average queue duration Ω is determined by the formula

$$\Omega = \lim_{T\to\infty} \frac{1}{T} \int_0^T \omega(t)dt, \quad (4)$$

where $\omega(t)$ is the stochastic process describing the evolution of the queue duration in time. At the same time, we cannot apply Little's formula directly, but we can use a technique of deriving Little's formula with corresponding adjustment. Formula (4) corresponds to the process shown in Fig. 5 and depicting the time-averaged value $l(t)$.

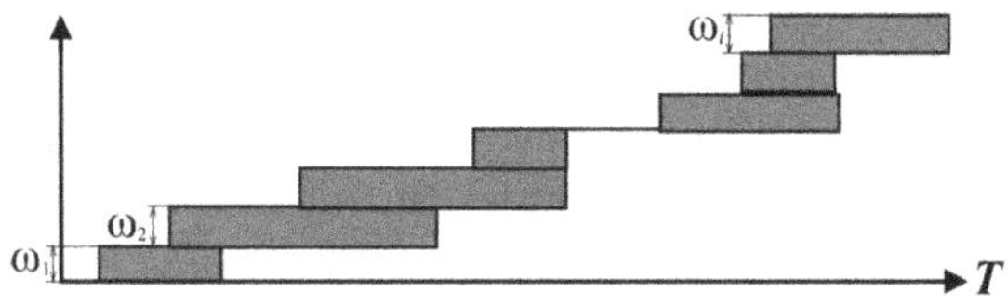

Fig. 5. Deriving Little's formula

When deriving the classic Little's formula, to establish a connection between the average number of requests L in the queue and the average waiting time W, all rectangles in Fig. 5 have unit width and length equal to the queue occupancy time t_i of the corresponding request. The figure clearly shows that just $N \sim \lambda T$ requests pass through the queue during large time T, and average queue occupancy time is $W = \frac{1}{N}\sum\limits_i t_i = \frac{1}{\lambda T}\sum\limits_i t_i$ with the sum over all requests arrived during time T. The average number L of requests in the queue (i.e., the filled area divided by T) is $L = \frac{1}{T}\sum\limits_i t_i$. Comparing the expressions obtained for W and L, we get $W = L/\lambda$, i.e. Little's formula.

A similar formula for the IMM case with preemptive priorities can be obtained as follows.

Since the average queue duration Ω is determined taking into account the request lengths ω_i, Fig. 5 must be modified introducing rectangles with different width, equal to the length ω_i of the corresponding request. Therefore, the duration Ω, defined as the time-averaged sum of the areas of the filled rectangles, now takes the following form:

$$\Omega = \frac{1}{T}\sum_{i=1}^{N} t_i\omega_i \sim \frac{1}{T}N\mathrm{E}(t\omega) = \frac{1}{T}\lambda T\mathrm{E}(t_i\omega_i) = \lambda\mathrm{E}(t_i\omega_i), \quad (5)$$

where t_i is the queue occupancy time of the i-th requests and ω_i its length. Multipliers under the expectation sign E in (5) are dependent and, therefore, the mean is calculated in the context of correlation coefficient for multipliers:

$$\Omega = \lambda r_{t\omega}\sigma_t\sigma_\omega + \lambda \mathrm{E}(t)\mathrm{E}(\omega) = \lambda r_{t\omega}\sigma_t\sigma_\omega + \lambda W \mathrm{E}(\omega), \quad (6)$$

where $r_{t\omega}$ is the correlation coefficient of r.v. t and ω, while σ_t, σ_ω are their standard deviations.

The correlation coefficient on the right hand side of (6) is positive, since the lower the priority of a request (i.e., the bigger its length), the longer it remains in the queue and, hence, the more times it is counted in the queue duration. Moreover, at least σ_ω is infinite (since $\alpha < 2$). Thus, the average queue duration Ω remains infinite when using IMM. Obviously, this also leads to infinite average waiting time in such queues.

It is worth noticing that form (6) of Little's formula coincides with the usual Little's formula when $r_{t\omega} = 0$ and $\mathrm{E}(\omega) = 1$.

6 Application of IMM to Computer Networks with Packet Switching

A possible application of IMM to packet switching networks is based on the fact that the relative priorities of transmitted files, split into packets, function almost in the same manner as the absolute priorities. Figure 6 shows how packets transmission of a certain file (unpainted rectangles in the channel queue) is temporarily interrupted when packets of higher priority file arrive (black rectangles).

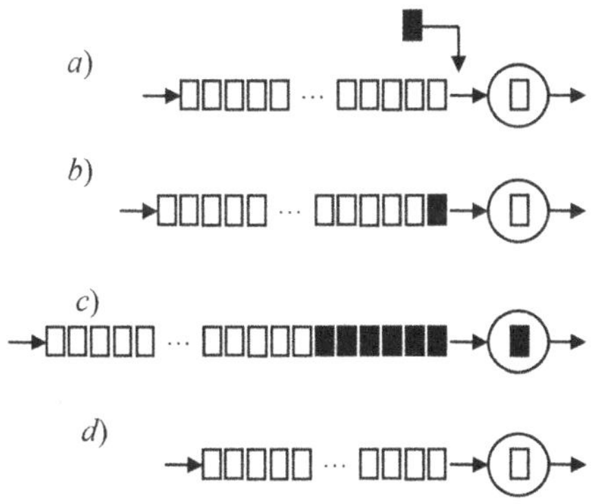

Fig. 6. Transmission of higher priority file split into packets (black rectangles) at relative priorities discipline

Figures 6 a), b) illustrate the arrival of the priority "black" packet and how it waits for the end of transmission of the "white" packet in the channel. At the end of the "white" packet transmission, the "black" packet begins to be transmitted, and at the same time the remaining "black" packets of the priority file arrive at the queue (Fig. 6 *c*). After the priority file transmission, the transmission of "white" packets of the lower priority file resumes (Fig. 6 *d*). Since the transmitted files consist of several packets, the delay in the start of servicing the priority file, which is a fraction of transmission time of one packet, is relatively small. This leads to a slight difference in the file transmission between relative and preemptive priorities.

The possible application of IMM to file transmission is further investigated by means of simulation.

7 Simulation of a Typical Network Fragment

The simulated network fragment is shown in Fig. 7. Arriving at inputs 1, 2, and 3 file flows are Poissonian and transmission time x (determined by the file size) is Pareto distributed with the parameters $K = 1$, $\alpha = 1.5$; consequently, $\mathrm{E}(x) = 3$, $\mathrm{D}(x) = \infty$.

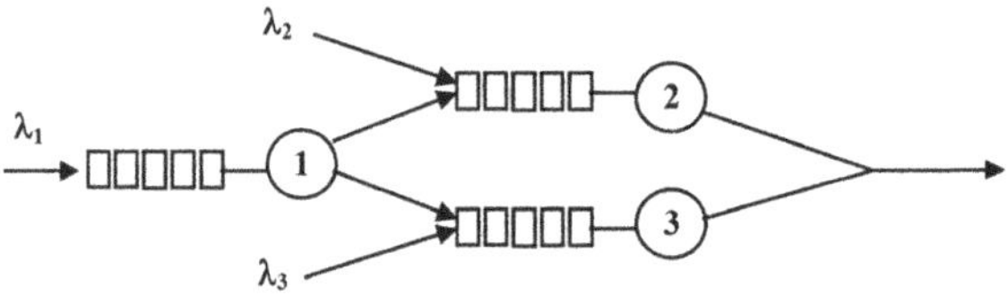

Fig. 7. Simulated network fragment

The average times between file arrivals at inputs 1, 2, and 3 are 6, 12, and 12 units, respectively. Priorities are assigned to files according to IMM with the following difference from the basic version of IMM – these priorities are not preemptive, but relative. We use an exponential markup (2) during priority assignment. As a result of the optimization procedure, it was found that at all inputs the optimal parameters are approximately the same: $a = 1$, $c = 0.5$.

Before transmission, files are split into packets of unit duration, which inherit the priorities of the files. Packets are free to move on different routes. A packet from node 1, for instance, arrives at node 2 if the queue there is shorter than that in node 3, otherwise, the packet arrives at node 3. At the exit from the network the packets are reassembled into files. A file is considered to be transmitted when assembly is complete.

Table 3 shows the simulation results obtained for approximately 10 million files at the first input (and, consequently, approximately 5 million at the second and third inputs). The only difference between the two network scenarios is the use of IMM in one of them. Consequently, the estimates of the utilizations ρ_1, ρ_2, and ρ_3 of channels 1, 2, and 3 for the compared networks turned out to be approximately the same. However, the estimates of the average transmission time T_1, T_2, and T_3, taking into account the waiting time of arriving files at inputs 1, 2, and 3, differ significantly. For the network with IMM the estimates converge to the limit values shown in the last row of Table 3, while for the non-priority network the corresponding estimates do not converge and grow with increasing simulation time. The simulation thereby confirms the assumption that the relative priorities of files in the packet-switched network behave in approximately the same way as the preemptive priorities. This allows the use of IMM, and thus, we can provide finite average waiting time even when the variances of the file transmission time x are infinite.

Table 3. Simulation results (60 million units of modelling time)

Network	ρ_1	ρ_2	ρ_3	T_1	T_2	T_3
Non-priority network	0.502	0.364	0.634	798.4	23.5	50.5
Network with IMM	0.503	0.362	0.636	5.82	3.70	3.92

Application of IMM does not lead to a significant decrease in the average queue lengths L_1, L_2, and L_3, although it does not worsen these indicators (Table 4).

Table 4. Average queue lengths (in packets) after 60 million units of modelling time

Network	ρ_1	ρ_2	ρ_3	T_1	T_2	T_3	L_1	L_2	L_3
Non-priority network	0.752	0.635	0.862	3422.2	118.7	156.8	2591.5	84.45	111.2
Non-priority network	0.600	0.466	0.732	1368.4	50.6	93.8	1265.3	41.9	80.1
Non-priority network	0.502	0.364	0.634	798.4	23.5	50.5	876.14	21.5	49.59
Non-priority network	0.251	0.153	0.346	136.2	6.83	10.8	282.63	10.85	19.10
Non-priority network	0.100	0.0544	0.145	18.2	1.27	2.36	89.3	3.41	8.94
Network with IMM	0.754	0.632	0.865	13.402	7.52	7.69	2262.8	61.25	105.8
Network with IMM	0.601	0.463	0.735	7.91	4.86	5.11	1127.0	48.1	64.4
Network with IMM	0.503	0.362	0.636	5.817	3.702	3.92	806.53	19.49	44.26
Network with IMM	0.251	0.153	0.347	2.32	1.63	1.68	266.78	13.06	16.68
Network with IMM	0.100	0.054	0.145	0.843	0.615	0.628	85.72	4.97	8.74

Indicators L_1, L_2, and L_3 are growing with time, since their stationary values are infinite – see formula (6). In the considered model with infinite buffers, this fact does not matter, but in practice it will lead to the problem of high loss probabilities. The results presented in Table 4 show that the rate of queue growth is sensitive to channel utilization. Figure 8 summarizes these results.

For both scenarios (i.e., non-priority network and network with IMM), Table 4 reports the main performance parameters for different utilization levels (the

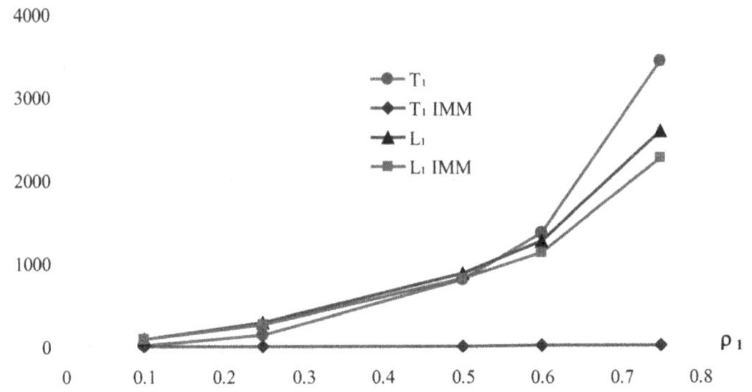

Fig. 8. Dependence of the total average time of files transmission on network utilization (T_1 is the average time for transmission of files arriving at the first input in network and L_1 is the average number of packets in the queue at the first node)

original test case corresponds to the utilization $\rho_1 = 0.5$). As we can see, indicators L_i and, therefore, the loss probabilities (with finite buffers) can be significantly reduced by increasing the link capacity. For instance, decreasing the channel utilizations by five times w.r.t. the original version of network, after the transmission of 10 million files, the average queue lengths are still far from values threatening packet losses for reasonable buffer sizes.

8 Application IMM with Unknow Service Time

Another important case of IMM application is associated with queue length determined only by the number of requests in it (for example, as when we send mathematical tasks to the wolframalpha.com site). In this case, the request service time depends only on its features (on the complexity of the transferred mathematical task). At the same time, as a rule, the service time of arriving requests (the time for task solving) is unknown.

We can propose such version of IMM, in which priorities assignments occur during their servicing. At the beginning the highest priority P_1 is assigned to every request arriving to the system. If a request is completely serviced during a time in the $[t_0, t_1)$ interval, the request leaves the system and, as a result, the highest priority was assigned correctly. If a request is serviced longer, i.e. time x reaches the threshold t_1, the request priority is reduced by one. Further, if its service is completed by t_2, the request leaves the system and, as a result, the priority was assigned correctly. If request service time x reaches the value t_3, the request priority is reduced by one more, and so on.

Table 5 shows the simulation results for a test system having the IMM version with dynamically configurable preemptive priorities. The average queue length in this IMM version is approximately 0.53, and the average waiting time is approximately 3.14 time units. When the service time is known, the IMM provides 0.22 and 1.3 for these indicators, respectively (see Table 1), which, of course,

Table 5. Characteristics of IMM with unknown request service time

Simulation run N	$\hat{L}$	$\hat{W}$
10^4	0.366	2.337
10^5	0.419	2.534
10^6	0.556	3.325
10^7	0.525	3.144

are somewhat better. At the same time, with unknown service time the IMM guarantees an improvement of these indicators (due to the dynamically configurable preemptive priorities) in comparison with the non-priority system (see Table 1), where these indicators are infinite. The relatively high efficiency of the configurable priorities with unknown service time is caused by the exponential markup (2). Since the length of the marking intervals grows rapidly with the growth of the interval number, the request service time in the initial intervals (with inadequately high priority) takes a relatively small fraction of the total time of their service.

Note that in the particular case when there is an inverse order of request service, applying such an order of servicing (with preemption [16]) reduces the average queue length and average waiting time greatly then IMM. Besides, the implementation of such a service order is usually more complex than the implementation of IMM. The inverse order of service also does not solve the problem of infinite average queue duration Ω for $1 < \alpha \leq 2$.

9 Conclusion

The paper highlighted the effectivness of the infinite markings method in presence of highly variable service time. Indeed, the analytical and simulation studies carried out in this article point out the following results:

1. In M/Pa/1 systems with infinite variance of the service time the stationary queue length distribution is power-law and the average queue length is infinite.
2. The introduction in such M/Pa/1 systems of preemptive priorities assigned in accordance with the infinite markings method makes the average queue length finite. However, the Ω indicator (the average queue duration) remains infinite. The Ω indicator was determined as the average sum of request lengths in the queue.
3. The average queue duration and the average waiting time are related through generalized Little's formula (6).
4. The assignment of relative priorities according to infinite markings method in packet switching networks with infinite variance of file sizes provides drastic reduction the average file transmission time.
5. When the queue duration equals to its length, infinite markings method can be effectively used to reduce drastically the queue length and waiting time even with unknown request service time.

References

1. Zadorozhnyi, V.N., Zakharenkova, T.R.: Methods to reduce loss probability in systems with infinite service time dispersion. In: Dudin, A., Nazarov, A., Moiseev, A. (eds.) ITMM 2019. CCIS, vol. 1109, pp. 296–311. Springer, Cham (2019). https://doi.org/10.1007/978-3-030-33388-1_24
2. Zadorozhnyi, V.N., Zakharenkova, T.R., Pagano, M.: Queue normalization methods in systems GI/GI/1/m with infinite variance of service time. J. Phys.: Conf. Ser. **1441**, 1–11 (2020). 012180. https://doi.org/10.1088/1742-6596/1441/1/012180
3. Leland, W.E., Willinger, W., Taqqu, M.S., Wilson, D.V.: On the self-similar nature of ethernet traffic. In: ACM SIGCOMM 1993, San Francisco, pp. 183–193 (1993)
4. Likhanov, N., Tsybakov, B., Georganas, N.: Analysis of an ATM buffer with self-similar ("Fractal") input traffic. In: IEEE INFOCOM 1995, vol. 3, pp. 985–992. IEEE Publisher, Boston (1995)
5. Kleinrock, L.: Queueing Systems, vol. II, Computer Applications. Wiley Inter-Science, New York (1976). 576 p
6. Zwart, A.P.: Queueing Systems with Heavy Tails. Eindhoven University of Technology, Eindhoven (2001). 227 p
7. Mandelbrot, B.: The Fractal Geometry of Nature. W. H. Freeman and Co., San Francisco (1982). 460 p
8. Paxon, V., Floyd, S.: Wide area traffic: the failure of Poisson modeling. In: IEEE/ACM Transactions on Networking, vol. 3, pp. 226–244. IEEE Publisher (1995)
9. Park, K., Willinger, W.: Self-Similar Network Traffic and Performance Evaluation. Wiley-Interscience, New York (2000). 558 p
10. Erramilli, A., Narayan, O., Willinger, W.: Experimental queueing analysis with long range dependent packet traffic. IEEE/ACM Trans. Netw. **4**, 209–223 (1996). https://doi.org/10.1109/90.491008
11. Chistyakov, V.P.: Theorem on sums of independent random positive variables and its applications to branching processes. Theory Probab. Appl. **9**(4), 640–648 (1964)
12. Asmussen, S., Binswanger, K., Hojgaard, B.: Rare events simulation for heavy-tailed distributions. Bernoulli **6**, 303–322 (2000)
13. Boots, N.K., Shahabuddin, P.: Simulating GI/GI/1 queues and insurance risk processes with subexponential distributions. Unpublished manuscript, Free University, Amsterdam. Shortened version. In: Proceedings of the 2000 Winter Simulation Conference, pp. 656–665 (2000)
14. Blanchet, J., Li, C.: Efficient Rare Event Simulation for Heavy-tailed Compound Sums (2008). http://www.columbia.edu/jb2814/papers/RGSFinalJan08B.pdf. Accessed 30 May 2015
15. Zadorozhnyi, V.N., Zakharenkova, T.R.: Methods of simulation queueing systems with heavy tails. In: Dudin, A., Gortsev, A., Nazarov, A., Yakupov, R. (eds.) ITMM 2016. CCIS, vol. 638, pp. 382–396. Springer, Cham (2016). https://doi.org/10.1007/978-3-319-44615-8_33
16. Nazarov, A.A., Terpugov, A.F.: Teoriya massovogo obsluzhivaniya [Queueing theory]. NTL, Tomsk (2004). 228 p. (in Russ.)

Analysis of the Queuing Systems with Processor Sharing Service Discipline and Random Serving Rate Coefficients

Eduard Sopin[1,2](✉) and Maksym Korshikov[1]

[1] Peoples' Friendship University of Russia (RUDN University), Miklukho-Maklaya str. 6, Moscow 117198, Russia
{sopin-es,korshikov-mv}@rudn.ru
[2] Institute of Informatics Problems, FRC CSC RAS, Vavilova 44-2, Moscow 119333, Russia

Abstract. In this paper, we develop the queuing system model with processor sharing discipline with random serving rate coefficients. Each arriving customer is characterized by its length (job volume) and a serving rate coefficient determined by some probability distributions. The coefficients remain constant during the service process of customers. The proposed model is aimed to model the service process of elastic sessions in wireless networks, in which each session is assigned to one of modulation and coding schemes (MCSs) according to the state of the radio channel. Each MCS is characterized by its value of the spectral efficiency, which is modeled by serving rate coefficients.

First, we analyze the proposed model. Then we apply a simplification that significantly reduces the complexity of the analysis and allows us to deduce formulas for the blocking probability and the average sojourn time. Finally, we conduct a numerical analysis of the considered model.

Keywords: Queuing system · Limited processor sharing · Serving rate coefficients · Blocking probability · Sojourn time

1 Introduction

In the modern wireless networks, the bitrate achieved by elastic sessions depends not only on the number of current sessions but also on the state of the radio channel. Based on the signal-to-noise ratio, a base station chooses one of the Modulation and Coding Schemes (MCS) [1,2]. Each MCS is characterized by a certain value of spectral efficiency, which determines the bitrate achieved on the unit spectrum bandwidth. Consequently, even if the spectrum bandwidth

The reported study was funded by RSF, project no. 20-71-00124 (recipient Sopin E., mathematical model and analysis). This paper has been supported by the RUDN University Strategic Academic Leadership Program (recipient Korshikov M., numerical analysis).

A. Dudin et al. (Eds.): ITMM 2020, CCIS 1391, pp. 225–235, 2021.
https://doi.org/10.1007/978-3-030-72247-0_17

is divided equally between all current sessions, the bitrates of the sessions vary significantly [3]. So, to analyze the performance measures of a base station that serves elastic sessions these peculiarities should be taken into account.

Queuing system with processor sharing service discipline [4], as well as their special case - limited processor sharing queues [5], are widely implemented in the analysis of telecommunication systems with elastic traffic. In this paper, we develop the queuing system with a single server and processor sharing (PS) discipline that can serve no more than N customers simultaneously. The main difference from other PS queues is that each customer is characterized not only by its length (or job volume) but also by the serving rate coefficient. Both length of a customer and its serving rate coefficient are independent random variables with certain distribution functions. The coefficients remain constant during the service process of customers.

In the classic PS queueing systems, if there are n customers in the system, then each customer receives $\frac{1}{n}$-th fraction of the server's performance. In the proposed model, the serving rate received by a customer is obtained by the multiplication of the $\frac{1}{n}$-th fraction of the server's rate and the serving rate coefficient associated with the customer. The considered system may be interpreted as the $M/G/1/PS$ queuing system (by incorporating the serving rate coefficients into the distribution function of the customers' length), however, the proposed approach has some advantages. The main advantage is as follows: the considered system may be described in terms of stochastic lists [6], hence the simplification method applied to them, and resource queuing systems [7] may be applied here also. As a result, the simplification method allows us to derive tractable expressions for the cumulative distribution function (CDF) of the customer's sojourn time.

The rest of the paper is as follows. Section 2 briefly describes the considered queueing system together with the system of equations for stationary distribution. Section 3 presents the simplification approach and provides an analysis of the simplified system. Section 4 continues the analysis on the part of the sojourn time's CDF. Section 5 presents the results of the numerical analysis, while Sect. 6 concludes the paper.

2 Model Description

Consider a single server queuing system that serves customers according to the PS discipline, but no more than N customers simultaneously. Customers arrive according to a Poisson process with intensity λ, the length of customers is exponentially distributed with parameter μ. Besides, a customer is characterized by a serving rate coefficient ν, which may take values from the set $V = \{\nu_1, \nu_2, ..., \nu_L\}$, $\nu_1 < \nu_2 < ... < \nu_L$. The serving rate coefficients of different customers are independent of each other, independent of the arrival process and the customer's length. The distribution of the serving rate coefficients is denoted by $\{p_i\}, i \geq 0$, where $p_i = P\{\nu = \nu_i\}$.

So, if there are n customers in the system and their serving rate coefficients are represented by the vector $(u_1, u_2, ..., u_n)$, then each customer is allocated $\frac{1}{n}$

of the processing time. The service rate of the i-th customer is a product of the allocated processing time ratio and its serving rate coefficient, that is $\frac{u_i}{n}$. Once the job associated with a customer is done, the customer leaves the system. If an arriving customer finds that there are N customers in the system already, then it is lost.

The behavior of the system may be described by the Markov process $X_1(t) = \{\xi(t), (\nu_1(t), \nu_2(t), ..., \nu_{\xi(t)}(t))\}$, where $\xi(t)$ is the number of customers in the system at time t, while $(\nu_1(t), \nu_2(t), ..., \nu_{\xi(t)}(t))$ is the vector of the serving rate coefficients. The set of states is expressed by the following formula:

$$S_1 = \{(n, u_1, u_2, ..., u_n) : 0 \le n \le N, u_i \in V, i = 1, 2, .., n\} \tag{1}$$

Denote $q_n(u_1, u_2, ..., u_n)$ the stationary probability that the system is in state $(n, u_1, u_2, ..., u_n)$:

$$q_n(u_1, u_2, ..., u_n) = \lim_{t\to\infty} P\{\xi(t) = n, \nu_1(t) = u_1, ..., \nu_n(t) = u_n\}. \tag{2}$$

Then the system of equilibrium equations can be written as follows.

$$\lambda q_0 = \mu \sum_{l=1}^{L} \nu_1 q_1(\nu_1), \tag{3}$$

$$\left(\lambda + \frac{\mu}{n}\sum_{i=1}^{n} u_i\right) q_n(u_1, u_2, ..., u_n) = \frac{\lambda}{n}\sum_{i=1}^{n} p_{u_i} q_{n-1}(u_1, .., u_{i-1}, u_{i+1}, ..., u_n)$$

$$+ \frac{\mu}{n+1}\sum_{l=1}^{L}\sum_{i=1}^{n+1}\left(\nu_l + \sum_{j=1}^{n} u_j\right) q_{n+1}(u_1, ..., u_{i-1}, \nu_l, u_i, , ..., u_n), \tag{4}$$

$n = 1, 2, .., N-1,$

$$\left(\frac{\mu}{N}\sum_{i=1}^{N} u_i\right) q_N(u_1, u_2, ..., u_N) = \frac{\lambda}{N}\sum_{i=1}^{N} p_{u_i} q_{N-1}(u_1, .., u_{i-1}, u_{i+1}, ..., u_N). \tag{5}$$

The left-hand side of (4) represents the exit intensity from the state $(n, u_1, ..., u_n)$. The first sum on the right-hand side of (4) stands for transition intensity to the state $(n, u_1, ..., u_n)$ induced by the arrival of a customer. Note that arriving customer is assigned any possible number in the list equiprobably (with probability $\frac{1}{n}$). The second sum on the right-hand side represents transitions induced by the departure of a customer. Equations (3) and (5) are the special cases of (4) for the number of customers n in the system equal to 0 and N respectively.

Computational complexity of the numerical solution of the system (3)–(5) increases very fast, since the number of states in set S_1 is $\frac{L^{N+1}-1}{L-1}$. So, in the next section we propose a simplified model based on the state aggregation technique [7].

3 Analysis of the Simplified Model

To simplify the model, we propose to trace only the sum of the serving rate coefficients of all customers in the system. Consider the simplified process $X_2(t) = \{\xi(t), \nu(t)\}$, where $\xi(t)$ is the number of customers in the system at time t, as before, and $\nu(t)$ is the sum of their serving rate coefficients. The set of states is given by

$$S_2 = \{(n, u) : 0 \le n \le N, p_u^{(n)} > 0\}, \tag{6}$$

where distribution $\{p_u^{(n)}\}$ is the probability that the sum of n serving rate coefficients is equal to u, which can be derived from the n-fold convolution of the distribution $\{p_i\}$:

$$p_u^{(1)} = \begin{cases} p_i, \text{if } \nu_i = u; \\ 0, \text{otherwise}; \end{cases} \tag{7}$$

$$p_u^{(n)} = \sum_{v \in V} p_v^{(1)} p_{u-v}^{(n-1)}. \tag{8}$$

As a result of the simplification, we cannot exactly determine the decrease of the $\nu(t)$ after a customer's departure from the system. To handle the problem, we introduce the probabilities $\varphi_i(n, u)$ that a customer's serving rate coefficient is equal to i given that the sum of the n customers' coefficients is u. The latter can be estimated by the Bayes formula:

$$\varphi_i(n, u) = \frac{p_i^{(1)} p_{u-i}^{(n-1)}}{p_u^{(n)}}, i \le u, (n, u) \in S_2. \tag{9}$$

The stationary probabilities $q_n(u)$

$$q_n(u) = \lim_{t \to \infty} P\{\xi(t) = n, \nu(t) = u\}$$

of the process $X_2(t)$ satisfy the following equilibrium equations:

$$\lambda q_0 = \mu \sum_{l=1}^{L} \nu_l q_1(\nu_l); \tag{10}$$

$$\left(\lambda + \frac{\mu u}{n}\right) q_n(u) = \lambda \sum_{l=1}^{L} p_l q_{n-1}(u - \nu_l)$$

$$+ \frac{\mu}{n+1} \sum_{l=1}^{L} (u + \nu_l) q_{n+1}(u + \nu_l) \varphi_{\nu_l}(n + 1, u + \nu_l), \tag{11}$$

$$n = 1, 2, .., N - 1, (n, u) \in S_2,$$

$$\frac{\mu u}{n} q_N(u) = \lambda \sum_{l=1}^{L} p_l q_{N-1}(u - \nu_l), (N, u) \in S_2. \tag{12}$$

The system (10)–(12) is obtained similarly to the equilibrium Eqs. (3)–(5) for the initial queuing system. The system (10)–(12) along with the normalizing condition can be solved using any appropriate numerical method to obtain the stationary probabilities. Note that the transitions between states are possible if and only if the difference in the number of customers is equal to 1. So, with the proper order of states, the generator matrix of $X_2(t)$ can be represented in block-tridiagonal form. Thus, one may choose numerical methods that can benefit from the special structure of the generator matrix.

Performance metrics of the considered queuing system may be evaluated using the stationary distribution obtained from the system (10)–(12). Since a customer is blocked upon arrival if there are N customers in the system already, then the blocking probability π is

$$\pi = \sum_{(N,u)\in S_2} q_N(u). \tag{13}$$

Another important performance metric is the customer's sojourn time. The next section is devoted to its analysis.

4 Analysis of the Mean Sojourn Time

According to [8], customers' sojourn time in the considered system has a phase-type CDF $F(x)$. Earlier we employed the proposed phase method to the analysis of the sojourn time in the limited processor sharing queuing systems with server vacations [9]. The same approach is used here, for the considered system with random service coefficients. So, we introduce a Markov process $Y_v(t)$, $v \in V$ that describes the system's behavior from the arrival to the departure of a customer with the serving rate coefficient v. The set of states of $Y_v(t)$ is given by

$$S_v = \{(1, v)\} \cup \{(n, u) : 2 \leq n \leq N, p_{u-v}^{(n-1)} > 0\} \cup \{\omega\}, \tag{14}$$

where ω is the absorbing state, which is reached at the departure of the considered customer.

We assign the following order of states in S_v. If $n_1 < n_2$, then $(n_1, u_1) \prec (n_2, u_2)$. In the case $n_1 = n_2$, the order of states is defined by the second component: if $u_1 < u_2$, then $(n, u_1) \prec (n, u_2)$. The last state in S_v is the absorbing state ω. Then, the generator matrix of $Y_v(t)$ has the following form:

$$\mathbf{A} = \begin{pmatrix} \mathbf{G} & \mathbf{g} \\ \mathbf{0} & 0 \end{pmatrix}, \tag{15}$$

where $\mathbf{g}$ is an exit vector to the absorbing state. The matrix $\mathbf{G}$ has block-tridiagonal structure with diagonal $\mathbf{D}_n$, $n = 1, 2, .., N$, upper-diagonal $\mathbf{\Lambda}_n$, $n =$

$1, 2, ..., N-1$ and lower-diagonal blocks $\mathbf{M}_n$, $n = 2, 3, ..., N$.

$$\mathbf{G} = \begin{pmatrix} \mathbf{D}_1 & \mathbf{\Lambda}_1 & 0 & \cdots & \cdots & \cdots & 0 \\ \mathbf{M}_2 & \mathbf{D}_2 & \mathbf{\Lambda}_2 & 0 & \cdots & \cdots & 0 \\ 0 & \mathbf{M}_3 & \mathbf{D}_3 & \mathbf{\Lambda}_3 & 0 & \cdots & 0 \\ 0 & 0 & \ddots & \ddots & \ddots & 0 & \cdots \\ 0 & \cdots & \cdots & 0 & \mathbf{M}_{N-1} & \mathbf{D}_{N-1} & \mathbf{\Lambda}_{N-1} \\ 0 & \cdots & \cdots & \cdots & 0 & \mathbf{M}_N & \mathbf{D}_N \end{pmatrix}. \tag{16}$$

Diagonal elements represent the exit intensities from the corresponding states. So, block $\mathbf{D}_1$ consist of only one element, $\mathbf{D}_1 = (-\lambda - \mu v)$, while other diagonal blocks are diagonal matrices:

$$\mathbf{D}_n = diag\left[-\left(\lambda + \frac{\mu(v + (n-1)\nu_1)}{n}\right), ..., -\left(\lambda + \frac{\mu(v + (n-1)\nu_L)}{n}\right)\right], \tag{17}$$
$$n = 2, 3, ..., N.$$

The upper-diagonal blocks stand for the transitions induced by the arrival of a customer. So, block $\mathbf{\Lambda}_1$ is a row vector:

$$\mathbf{\Lambda}_1 = (\lambda p_{\nu_1}, ..., \lambda p_{\nu_L}), \tag{18}$$

while other upper-diagonal blocks are given by

$$\mathbf{\Lambda}_n = \begin{pmatrix} \lambda p_{\nu_1} & \cdots & \lambda p_{\nu_L} & 0 & \cdots & \cdots & 0 \\ 0 & \lambda p_{\nu_1} & \cdots & \lambda p_{\nu_L} & 0 & \cdots & 0 \\ 0 & 0 & \ddots & \ddots & \ddots & 0 & \cdots \\ 0 & \cdots & \cdots & 0 & \lambda p_{\nu_1} & \cdots & \lambda p_{\nu_L} \end{pmatrix}, n = 2, 3, ..., N-1. \tag{19}$$

The lower-diagonal blocks represent transitions induced by the departure of any customer, except the considered one. Block $\mathbf{M}_2$ is a column vector:

$$\mathbf{M}_2 = \begin{pmatrix} \frac{\mu\nu_1}{n} \\ \vdots \\ \frac{\mu\nu_L}{n} \end{pmatrix}, \tag{20}$$

and other lower-diagonal blocks have the following form:

$$\mathbf{M}_n = \begin{pmatrix} \frac{\mu(n-1)\nu_1}{n} & 0 & \cdots\cdots & 0 \\ \frac{\mu((n-2)\nu_1+\nu_2)}{n} \times \varphi_{\nu_2}(n-1, (n-2)\nu_1+\nu_2) & \frac{\mu((n-2)\nu_1+\nu_2)}{n} \times \varphi_{\nu_1}(n-1, (n-2)\nu_1+\nu_2) & 0 \ \cdots & 0 \\ \vdots & \ddots & \ddots \ 0 \ \cdots & 0 \\ 0 & \cdots & 0 \ \cdots \ 0 & \frac{\mu(N-1)\nu_L}{n} \end{pmatrix},$$
$$n = 3, 4, ..., N. \tag{21}$$

The exit vector $\mathbf{g}$ to the absorbing state ω consist of departure intensities of the considered customer with the service rate coefficient v. If there are n customers in the system, then the received processing time of the considered customer is $\frac{1}{n}$. And the serving rate of the considered customer is expressed by $\frac{\mu v}{n}$. Thus, the exit vector $\mathbf{g}$ can be expressed as follows.

$$\mathbf{g} = \begin{pmatrix} \mu v \\ \frac{\mu v}{2} \\ \vdots \\ \frac{\mu v}{2} \\ \vdots \\ \frac{\mu v}{N} \\ \vdots \\ \frac{\mu v}{N} \end{pmatrix}. \tag{22}$$

According to the PASTA property [10], the stationary distribution of $X_2(t)$ is equal to the stationary distribution of the Markov chain $X_2(t_n - 0)$ embedded at the moments $t_n, n = 1, 2, ...$ just before arrivals. Since the initial distribution θ of the Markov process $Y_v(t)$ is equal to the distribution of the Markov chain $X_2(t_n + 0)$, we obtain

$$\theta_{n,u} = \begin{cases} \dfrac{q_0}{1-\pi}, & n = 1, u = v; \\ \dfrac{q_{n-1}(u-v)}{1-\pi}, & n = 2, 3, .., N, p_{u-v}^{(n-1)} > 0. \end{cases} \tag{23}$$

Then, according to [8], the CDF $F(x)$ of the customer's sojourn time is

$$F(x) = 1 - \theta e^{\mathbf{G}x}\mathbf{1}, \tag{24}$$

where $\mathbf{1}$ is a column-vector of ones with the appropriate size. The k-th moment $w_v^{(k)}$ of the sojourn time is given by the following expression.

$$w_v^{(k)} = (-1)^k k! \theta \mathbf{G}^{-k}\mathbf{1}. \tag{25}$$

Particularly, the mean sojourn time $w_v^{(1)}$ for a customer the with serving rate coefficient v is

$$w_v^{(1)} = -\theta \mathbf{G}^{-1}\mathbf{1}. \tag{26}$$

Finally, averaging over all possible serving rate coefficients, we obtain the formula for the mean sojourn time $\bar{w}$ of an arbitrary customer:

$$\bar{w} = \sum_{l=1}^{L} p_{\nu_l} w_{\nu_l}^{(1)}. \tag{27}$$

5 Numerical Analysis

In this section, we summarize the results of the numerical analysis. We evaluated the performance of the considered system under the following initial values. The server serves customers with the unit rate ($\mu = 1$), the maximum number of simultaneously served customers $N = 10$. Assume there are $L = 5$ different values of the serving rate coefficients, $V = \{1, 2, 3, 4, 5\}$.

We consider two different distributions of the serving rate coefficients:

$$p_1 = (0.1, 0.2, 0.3, 0.3, 0.1);$$
$$p_2 = (0.3, 0.2, 0, 0.1, 0.5).$$

Note that the average serving rate for both distributions is equal to 3.1, but their variances are different.

First, we analyze the dependence of the blocking probability π on the arrival intensity λ. Figure 1 depicts the dependence with λ varying from 2.5 to 3.5. One can note that the blocking probability for the distribution p_2 with greater variance is higher.

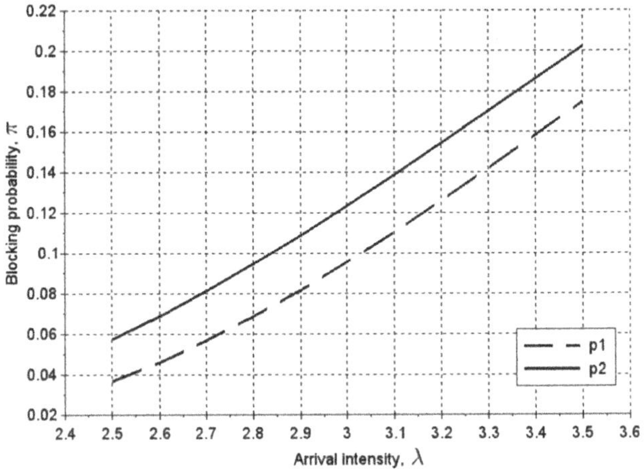

Fig. 1. The blocking probability as a function of the arrival intensity λ

Figures 2 and 3 show the dependence of the average sojourn time $\bar{w}$ on the arrival intensity λ for the distributions p_1 and p_2, respectively. Besides the average sojourn time itself, the figures also depict the average sojourn times $w_v^{(1)}$ for customers with different serving rate coefficients v.

As one can see from the Figs. 2 and 3, the average sojourn time for the distribution p_2 is greater than for the p_1, although the accepted load for p_2 is smaller. So, the type of the serving rate distribution and its higher-order moments have a significant impact on the system performance.

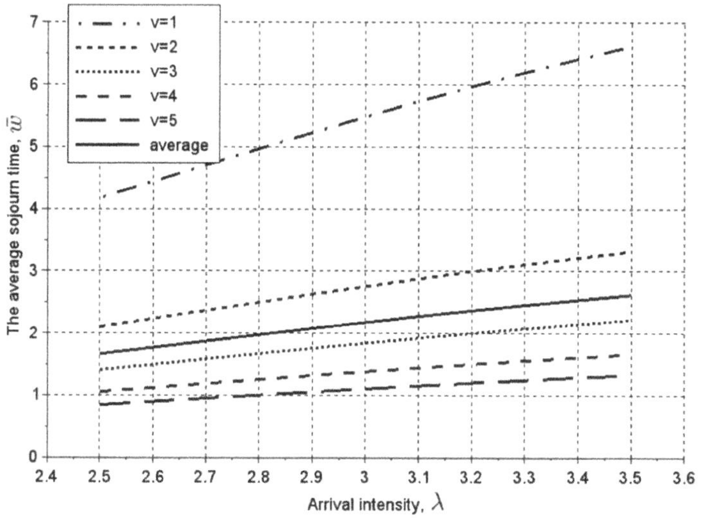

Fig. 2. The average sojourn time $\bar{w}$ as a function of the arrival intensity λ for distribution p_1

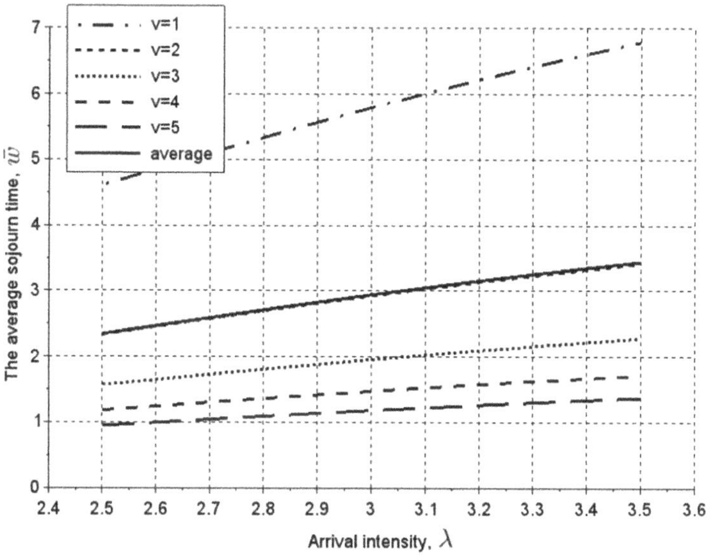

Fig. 3. The average sojourn time $\bar{w}$ as a function of the arrival intensity λ for distribution p_2

Figures 4 and 5 depict the dependences of the performance metrics on the maximum number of customers in the system N for the arrival intensity $\lambda = 3$. The general behavior of the lines is expected. With the increase of the maximum

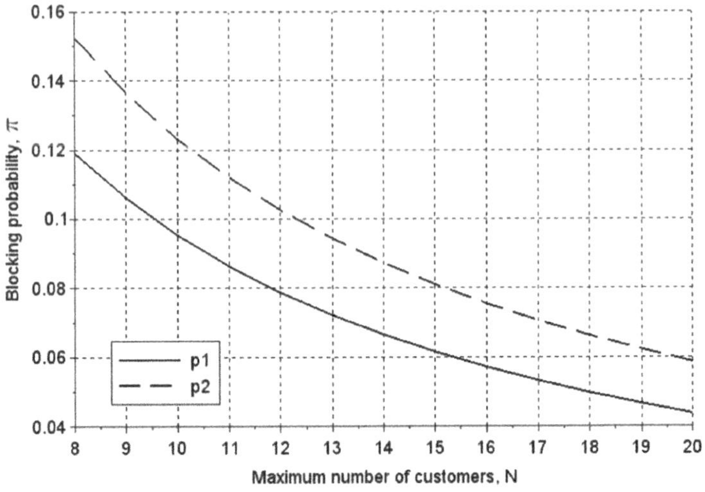

Fig. 4. The blocking probability as a function of the maximum number of customers N

Fig. 5. The average sojourn time $\bar{w}$ as a function of the maximum number of customers N

number of customers, the blocking probability decreases. As a consequence, the increase in the number of served customers without a boost of the server's performance, leads to the increase of the sojourn time.

6 Conclusion

In the paper, we developed a queuing system model with PS discipline and random serving rate coefficients. We also proposed a simplification of the model that allowed to decrease the dimension of the set of states. For the simplified model, we obtained the equilibrium equations for the stationary distribution, as well as formulas for the blocking probability and the mean sojourn time. In the numerical analysis section, we studied the dependences of the performance metrics on the offered load and the maximum number of customers in the system.

The developed model will be used in the analysis of the optimal coverage radius of a base station in a wireless network. The decrease of the coverage radius will reduce the probability that a customer has a low serving rate coefficient, hence it will increase the performance of the base station. Another direction of further research is the development of efficient approximate methods for the performance measures evaluation.

References

1. 3GPP TS 36.213 V13.15.0. Evolved Universal Terrestrial Radio Access (E-UTRA); Physical layer procedures (Release 13). https://www.3gpp.org/ftp/Specs/archive/36_series/36.213/36213-df0.zip. Accessed 25 Jan 2021
2. 3GPP TS 38.211 V16.4.0. NR; Physical channels and modulation (Release 16). https://www.3gpp.org/ftp//Specs/archive/38_series/38.211/38211-g40.zip. Accessed 25 Jan 2021
3. Sopin, E., Samouylov, K., Vikhrova, O., Kovalchukov, R., Moltchanov, D., Samuylov, A.: Evaluating a case of downlink uplink decoupling using queuing system with random requirements. In: Galinina, O., Balandin, S., Koucheryavy, Y. (eds.) NEW2AN/ruSMART -2016. LNCS, vol. 9870, pp. 440–450. Springer, Cham (2016). https://doi.org/10.1007/978-3-319-46301-8_37
4. Yashkov, S.F., Yashkova, A.S.: Processor sharing: a survey of the mathematical theory. Autom. Remote Control **68**(9), 1662–1731 (2007). https://doi.org/10.1134/S0005117907090202
5. Zhang, J., Zwart, B.: Steady state approximations of limited processor sharing queues in heavy traffic. Queueing Syst. **60**, 227–246 (2008). https://doi.org/10.1007/s11134-008-9095-4
6. Samouylov, K., Naumov, K.: Stochastic lists of multiple resources. In: The 5th International Conference on Stochastic Methods (ICSM-2020). Springer Proceedings in Mathematics and Statistics (in print)
7. Naumov, V.A., Samuilov, K.E., Samuilov, A.K.: On the total amount of resources occupied by serviced customers. Autom. Remote Control **77**(8), 1419–1427 (2016). https://doi.org/10.1134/S0005117916080087
8. Asmussen, S.: Applied Probability and Queues, 2nd edn. Springer, New York (2003). https://doi.org/10.1007/b97236
9. Samouylov, K., Naumov, V., Sopin, E., Gudkova, I., Shorgin, S.: Sojourn time analysis for processor sharing loss system with unreliable server. In: Wittevrongel, S., Phung-Duc, T. (eds.) ASMTA 2016. LNCS, vol. 9845, pp. 284–297. Springer, Cham (2016). https://doi.org/10.1007/978-3-319-43904-4_20
10. Wolff, R.W.: Poisson arrivals see time averages. Oper. Res. **30**(2), 223–231 (1982)

Asymptotic-Diffusion Analysis of Multiserver Retrial Queueing System with Priority Customers

Anatoly Nazarov[1], Tuan Phung-Duc[2,3], and Yana Izmailova[1(✉)]

[1] Institute of Applied Mathematics and Computer Science, National Research Tomsk State University, 36 Lenina ave., 634050 Tomsk, Russian Federation
[2] Faculty of Engineering Information and Systems, University of Tsukuba, 1-1-1 Tennodai, Tsukuba, Ibaraki 305-8573, Japan
[3] VNU Vietnam Japan University, Hanoi, Vietnam

Abstract. This paper considers a priority multi-server retrial queue with two classes of customers. Primary customers have preemptive priority over secondary users. The dynamics of primary customers is the same as that of an Erlang loss system with Poisson input and exponential service time distribution. Secondary users can cognitively use the channels when they are not used by primary users. Secondary users that see all the channels occupied upon arrival join the orbit and retry later. Upon arrival, if a primary user is lost if it sees all the channels occupied by other primary users. Upon the arrival of a primary customer, if all the channels are occupied but some channels are occupied by secondary users, one of these ongoing secondary users is interrupted by the primary user and the interrupted secondary user enters the orbit. Secondary users from the orbit retry to occupy an idle server until they are successfully occupying one. For this model, we consider an asymptotic regime in which the retrial rate is extremely low. While the number of secondary users in the orbit explodes in this regime, we prove that a scaling version of the number of users in the orbit weakly converges to a diffusion process whose drift and diffusion coefficients are constructed.

Keywords: Retrial queueing system · Asymptotic-diffusion analysis · Priority customers

1 Introduction

Priority queue is one of the main streams in queueing theory because the model naturally arises in various applications from services to telecommunication and computer systems. In practice, service differentiation is required in various situations. In service systems such as airline, first class customer has priority over economy class. In telecommunication system, voice packets have higher priority than other data packets such as email etc.

A. Dudin et al. (Eds.): ITMM 2020, CCIS 1391, pp. 236–250, 2021.
https://doi.org/10.1007/978-3-030-72247-0_18

Recently, mobile traffic has explosively increased leading to the shortage of wireless spectrums. Cognitive radio networks are promising technologies for this spectrum shortage problem. Two types of users in cognitive radio networks are primary users and secondary users. Primary users are granted some spectrum bands. Secondary users can cognitively utilize these bands when they are not used by primary users. To this end, secondary users must use these band in such a way that does not interfere primary users. Once primary users arrive, ongoing secondary users must evacuate and then sense to occupy a channel in a later time. Motivated by this application, we consider a multiserver priority retrial queue. Primary users and secondary users arrive at the system of multiple servers according to Poisson processes. Service time distributions of primary and secondary users are exponential distributions with distinct parameters. From the viewpoint of primary users, the dynamics of the number of primary users is the same as that of an Erlang loss system.

Upon arrival, if there are some idle servers, the secondary user can use one of these idle ones, otherwise they join the orbit and sense to find an idle one in an exponentially distributed time. Upon arrival, if there are some idle servers, the primary user chooses one of them. Otherwise, the primary user is either lost if all the channels busy with other primary users, or interrupts one of ongoing secondary users. The interrupted ongoing secondary user returns to the orbit and sense to occupy an idle server. The sensing process of blocked secondary users and the interrupted ones is the same as the orbit in retrial queues.

As a related work, Akutsu and Phung-Duc [1] study a similar model where secondary users must sense before accessing the channel using simulation. In the study [1], analytical solution is derived for the single server case and simulation is carried out for the multi-server case. Furthermore, Phung-Duc et al. [6] analytically study the same model and derive the stability condition. Morozov et al. [4] study the model where secondary can wait at a buffer in case an idle server is not available or in case of being interrupted. In paper [7] consider the effect of retrial phenomenon on in performance modeling of radio by using a finite-source queueing model. Nemouchi and Sztrik [8] used stochastic simulation for performance evaluation of cognitive radio network. In this paper, in contrast to the models in [1,4,6], we consider a new model where fresh secondary users can occupy an empty channel upon arrival without sensing. Numerical solutions for priority retrial queues can be found in [9,10,12]. Especially the model in [12] is more general than the one presented in the current paper.

In this paper, rather than finding a numerical solution, we consider asymptotic analysis under the condition that the delay in the orbit is extremely long using the asymptotic diffusion method [2,3,5]. Under that condition, the number of customers in the orbit explosively increases. However, using an appropriate scaling, we prove that the scaled number of customers in the orbit weakly converges to a diffusion process. The results allow us to derive an approximation for the distribution of the number of customers in the orbit.

The remainder of the paper is organized as follows. In Sect. 2, we present the model in detail. Section 3 presents the first order asymptotics while second order asymptotic analysis is presented in Sect. 4. An approximation of the distribution of the stationary number of customers in the orbit is presented in Sect. 5. Concluding remarks are presented in Sect. 6.

2 Mathematical Model

We consider multi-server retrial queueing system with two incoming streams of primary users and secondary users. We assume that arrival flows to the system follow the stationary Poisson processes with intensity λ_1 and λ_2, respectively. The first arrival flow of primary users has preemptive priority over that of secondary users. The system has N servers. Service times of customers of the first and the second flows are exponentially distributed with rate μ_1 and μ_2, respectively. If upon arrival, the customer (of either the first or the second flow) finds free servers, then he gets on one of these free servers and starts the service. If all servers are busy with priority customers, the incoming primary customer is lost otherwise, the primary customer will preempt a non-priority customer and the preempted customer joins the orbit. Non-priority customers that see all the servers busy upon arrival join the orbit. From the orbit non-priority customers retry in to occupy a free server after an exponentially distributed time with rate σ.

Let $n_1(t)$ and $n_2(t)$ denote the number of busy servers with priority customers and that with non-priority ones at time t, respectively and let $i(t)$ denote the number of customers in the orbit at the time t.

The random process $\{n_1(t), n_2(t), i(t)\}$ forms a continuous time Markov chain. Let

$$P(n_1, n_2, i, t) = P\{n_1(t) = n_1, n_2(t) = n_2, i(t) = i\}$$

denote the probability distribution of the process $\{n_1(t), n_2(t), i(t)\}$. The system of Kolmogorov differential equations for $P(n_1, n_2, i, t), n_1 + n_2 = n, n \leq N$ is given as follows.

$$\begin{aligned}
\frac{\partial P(0,0,i,t)}{\partial t} &= -(\lambda_1 + \lambda_2 + i\sigma)P(0,0,i,t) + \mu_1 P(1,0,i,t) \\
&\quad + \mu_2 P(0,1,i,t), n = 0, \\
\frac{\partial P(n_1,n_2,i,t)}{\partial t} &= -(\lambda_1 + \lambda_2 + \mu_1 n_1 + \mu_2 n_2 + i\sigma)P(n_1,n_2,i,t) \\
&+\lambda_1 P(n_1 - 1, n_2, i, t) + (i+1)\sigma P(n_1, n_2 - 1, i+1, t) + \lambda_2 P(n_1, n_2 - 1, i, t) \\
&+\mu_1 (n_1 + 1) P(n_1 + 1, n_2, i, t) + \mu_2 (n_2 + 1) P(n_1, n_2 + 1, i, t), 1 \leq n < N, \\
\frac{\partial P(n_1,n_2,i,t)}{\partial t} &= -(\lambda_1 + \lambda_2 + \mu_1 n_1 + \mu_2 n_2)P(n_1,n_2,i,t) \\
&+\lambda_1 P(n_1 - 1, n_2 + 1, i - 1, t) + \lambda_2 P(n_1, n_2 - 1, i, t) + \lambda_1 P(n_1 - 1, n_2, i, t) \\
&+\lambda_2 P(n_1, n_2, i - 1, t) + (i+1)\sigma P(n_1, n_2 - 1, i + 1, t), n = N, n_1 < N, \\
\frac{\partial P(N,0,i,t)}{\partial t} &= -(\lambda_2 + N\mu_1)P(N,0,i,t) + \lambda_1 P(N-1,0,i,t) \\
&+\lambda_1 P(N-1,1,i-1,t) + \lambda_2 P(N,0,i-1,t), n_1 = N.
\end{aligned} \quad (1)$$

We consider the partial characteristic function with $j = \sqrt{-1}$

$$H(n_1, n_2, u, , t) = \sum_{i=0}^{\infty} e^{jui} P(n_1, n_2, i, t).$$

The system (1) is transformed to

$$\begin{aligned}
\frac{\partial H(0,0,u,t)}{\partial t} &= -(\lambda_1+\lambda_2)H(0,0,u,t) + \mu_1 H(1,0,u,t) \\
&+ \mu_2 H(0,1,u,t) + j\sigma \frac{\partial H(0,0,u,t)}{\partial u}, n = 0, \\
\frac{\partial H(n_1,n_2,u,t)}{\partial t} &= -(\lambda_1+\lambda_2+\mu_1 n_1+\mu_2 n_2)H(n_1,n_2,u,t) \\
&+ j\sigma \frac{\partial H(n_1,n_2,u,t)}{\partial u} + \lambda_1 H(n_1-1,n_2,u,t) \\
&- j\sigma e^{-ju} \frac{\partial H(n_1,n_2-1,u,t)}{\partial u} + \lambda_2 H(n_1,n_2-1,u,t) \\
&+ \mu_1(n_1+1)H(n_1+1,n_2,u,t) + \mu_2(n_2+1)H(n_1,n_2+1,u,t), 1 \le n < N, \\
\frac{\partial H(n_1,n_2,u,t)}{\partial t} &= -(\lambda_1+\lambda_2+\mu_1 n_1+\mu_2 n_2)H(n_1,n_2,u,t) \\
&+ \lambda_1 e^{ju} H(n_1-1,n_2+1,u,t) + \lambda_2 H(n_1,n_2-1,u,t) + \lambda_1 H(n_1-1,n_2,u,t) \\
&+ \lambda_2 e^{ju} H(n_1,n_2,u,t) - j\sigma \frac{\partial H(n_1,n_2-1,u,t)}{\partial u}, n = N, n_1 < N, \\
\frac{\partial H(N,0,u,t)}{\partial t} &= -(\lambda_2+N\mu_1)H(N,0,u,t) + \lambda_1 H(N-1,0,u,t) \\
&+ \lambda_1 e^{ju} H(N-1,1,u,t) + \lambda_2 e^{ju} H(N,0,u,t), n_1 = N.
\end{aligned} \tag{2}$$

Summing up the equations of system (2) yields an additional equation that we need for further research:

$$\begin{aligned}
\sum_{n_1+n_2 \le N} \frac{\partial H(n_1,n_2,u,t)}{\partial t} &= (e^{ju}-1)\left(j\sigma e^{-ju} \sum_{n_1+n_2<N} \frac{\partial H(n_1,n_2,u,t)}{\partial u} \right. \\
&\left. + (\lambda_1+\lambda_2) \sum_{n_1+n_2=N} H(n_1,n_2,u,t) - \lambda_1 H(N,0,u,t) \right).
\end{aligned} \tag{3}$$

The main contribution of the paper is threefold: i) the probability distribution of the number of customers in the orbit, ii) the joint stationary probability distribution that n_1 servers are busy with serving priority customers and n_2 servers are busy with non-priority customers, iii) the blocking probability of priority customers. The system of Eqs. (2) seems to be impossible to solve by a direct way, so we will consider it under the condition of long delay customers in the orbit, i.e., when $\sigma \to 0$.

3 First Order Asymptotic Analysis

In system of Eqs. (2) and Eq. (3), we consider the following substitutions:

$$\sigma = \epsilon, \tau = \epsilon t, u = \epsilon w, H(n_1, n_2, u, t) = F(n_1, n_2, w, \tau, \epsilon).$$

We can transform system (2) and (3) to the form:

$$\begin{aligned}
\epsilon\frac{\partial F(0,0,w,\tau,\epsilon)}{\partial \tau} &= -(\lambda_1+\lambda_2)F(0,0,w,\tau,\epsilon) + \mu_1 F(1,0,w,\tau,\epsilon)\\
&+ \mu_2 F(0,1,w,\tau,\epsilon) + j\frac{\partial F(0,0,w,\tau,\epsilon)}{\partial w}, n=0,\\
\epsilon\frac{\partial F(n_1,n_2,w,\tau,\epsilon)}{\partial \tau} &= -(\lambda_1+\lambda_2+\mu_1 n_1+\mu_2 n_2)F(n_1,n_2,w,\tau,\epsilon)\\
&+ j\frac{\partial F(n_1,n_2,w,\tau,\epsilon)}{\partial w} + \mu_1(n_1+1)F(n_1+1,n_2,w,\tau,\epsilon)\\
&- je^{-j\epsilon w}\frac{\partial F(n_1,n_2-1,w,\tau,\epsilon)}{\partial w} + \lambda_2 F(n_1,n_2-1,w,\tau,\epsilon)\\
&+ \lambda_1 F(n_1-1,n_2,w,\tau,\epsilon) + \mu_2(n_2+1)F(n_1,n_2+1,w,\tau,\epsilon), 1 \le n < N,\\
\epsilon\frac{\partial F(n_1,n_2,w,\tau,\epsilon)}{\partial \tau} &= -(\lambda_1+\lambda_2+\mu_1 n_1+\mu_2 n_2)F(n_1,n_2,w,\tau,\epsilon)\\
&+ \lambda_1 e^{j\epsilon w}F(n_1-1,n_2+1,w,\tau,\epsilon) + \lambda_2 F(n_1,n_2-1,w,\tau,\epsilon)\\
&+ \lambda_1 F(n_1-1,n_2,w,\tau,\epsilon) + \lambda_2 e^{j\epsilon w}F(n_1,n_2,w,\tau,\epsilon)\\
&- j\frac{\partial F(n_1,n_2-1,w,\tau,\epsilon)}{\partial w}, n=N, n_1<N,\\
\epsilon\frac{\partial F(N,0,w,\tau,\epsilon)}{\partial \tau} &= -(\lambda_2+N\mu_1)F(N,0,w,\tau,\epsilon) + \lambda_1 F(N-1,0,w,\tau,\epsilon)\\
&+ \lambda_1 e^{j\epsilon w}F(N-1,1,w,\tau,\epsilon) + \lambda_2 e^{j\epsilon w}F(N,0,w,\tau,\epsilon), n_1=N,
\end{aligned} \tag{4}$$

$$\begin{aligned}
\epsilon\sum_{n_1+n_2\le N}\frac{\partial F(n_1,n_2,w,\tau,\epsilon)}{\partial \tau} &= \left(e^{j\epsilon w}-1\right)\Bigg(-\lambda_1 F(N,0,w,\tau,\epsilon)\\
+(\lambda_1+\lambda_2)\sum_{n_1+n_2=N}F(n_1,n_2,w,\tau,\epsilon) &+ je^{-j\epsilon w}\sum_{n_1+n_2<N}\frac{\partial F(n_1,n_2,w,\tau,\epsilon)}{\partial w}\Bigg).
\end{aligned} \tag{5}$$

We denote $F(n_1, n_2, w, \tau) = \lim_{\epsilon\to 0} F(n_1, n_2, w, \tau, \epsilon)$.

Theorem 1. *Let $F(n_1, n_2, w, \tau)$ denote the solution of (4) by taking the limit as $\epsilon \to 0$. Then function $F(n_1, n_2, w, \tau)$ has the following form*

$$F(n_1, n_2, w, \tau) = exp\{jwx(\tau)\}R(n_1, n_2, x).$$

Here the probabilities $R(n_1, n_2, x)$ are solutions to the following system of equations:

$$
\begin{gathered}
-(\lambda_1 + \lambda_2 + x)R(0,0,x) + \mu_1 R(1,0,x) + \mu_2 R(0,1,x) = 0, n = 0, \\
-(\lambda_1 + \lambda_2 + \mu_1 n_1 + \mu_2 n_2 + x)R(n_1,n_2,x) + \lambda_2 R(n_1,n_2-1,x) \\
+\lambda_1 R(n_1-1,n_2,x) + xR(n_1,n_2-1,x) + \mu_1(n_1+1)R(n_1+1,n_2,x) \\
+\mu_2(n_2+1)R(n_1,n_2+1,x) = 0, 1 \le n < N, \\
-(\lambda_1 + \mu_1 n_1 + \mu_2 n_2)R(n_1,n_2,x) + (\lambda_2 + x)R(n_1,n_2-1,x) \\
+\lambda_1 R(n_1-1,n_2+1,x) + \lambda_1 R(n_1-1,n_2,x) = 0, n = N, n_1 < N, \\
-N\mu_1 R(N,0,x) + \lambda_1(R(N-1,0,x) + R(N-1,1,x)) = 0, n_1 = N,
\end{gathered}
\tag{6}
$$

and $x(\tau)$ is the solution of the following ordinary differential equation

$$
x'(\tau) = -x(\tau)\sum_{n \le N-1} R(n_1,n_2,x) + (\lambda_1+\lambda_2)\sum_{n=N} R(n_1,n_2,x) - \lambda_1 R(N,0,x).
\tag{7}
$$

Proof. Taking $\epsilon \to 0$ in the system (4), we find the solution of (4) in the form:

$$
F(n_1,n_2,w,\tau) = R(n_1,n_2,x)\Phi(w).
\tag{8}
$$

Substituting the product (8) into the system of Eq. (4), we obtain

$$
\begin{gathered}
-(\lambda_1+\lambda_2)R(0,0,x)\Phi(w) + jR(0,0,x)\frac{\partial \Phi(w)}{\partial w} \\
+\mu_1 R(1,0,x)\Phi(w) + \mu_2 R(0,1,x)\Phi(w) = 0, n = 0,
\end{gathered}
$$

$$
\begin{gathered}
-(\lambda_1+\lambda_2+\mu_1 n_1+\mu_2 n_2)R(n_1,n_2,x)\Phi(w) + \mu_1(n_1+1)R(n_1+1,n_2,x)\Phi(w) \\
+jR(n_1,n_2,x)\frac{\partial \Phi(w)}{\partial w} - jR(n_1,n_2-1,x)\frac{\partial \Phi(w)}{\partial w} + \lambda_2 R(n_1,n_2-1,x)\Phi(w) \\
+\lambda_1 R(n_1-1,n_2,x)\Phi(w) + \mu_2(n_2+1)R(n_1,n_2+1,x)\Phi(w) = 0, 1 \le n < N,
\end{gathered}
$$

$$
\begin{gathered}
-(\lambda_1+\lambda_2+\mu_1 n_1+\mu_2 n_2)R(n_1,n_2,x)\Phi(w) + \lambda_1 R(n_1-1,n_2+1,x)\Phi(w) \\
+\lambda_2 R(n_1,n_2-1,x)\Phi(w) + \lambda_1 R(n_1-1,n_2,x)\Phi(w) + \lambda_2 R(n_1,n_2,x)\Phi(w) \\
-jR(n_1,n_2-1,x)\frac{\partial \Phi(w)}{\partial w} = 0, n = N, n_1 < N,
\end{gathered}
$$

$$
\begin{gathered}
-(\lambda_2+N\mu_1)R(N,0,x)\Phi(w) + \lambda_1 R(N-1,0,x)\Phi(w) \\
+\lambda_1 R(N-1,1,x)\Phi(w) + \lambda_2 R(N,0,x)\Phi(w) = 0, n_1 = N,
\end{gathered}
\tag{9}
$$

Here $\Phi(w)$ is given in the form

$$
\Phi(w) = exp\{jwx(\tau)\}.
\tag{10}
$$

Substituting expression (10) into (9), we obtain the system of equations, which coincides with (6).

Let's use the following expansion

$$e^{jw\epsilon} = 1 + jw\epsilon + O(\epsilon^2).$$

Substituting this into (5) and sending $\epsilon \to 0$, we obtain

$$\sum_{n_1+n_2\leq N} \frac{\partial F(n_1,n_2,w,\tau)}{\partial \tau} = jw\,(-\lambda_1 F(N,0,w,\tau) + (\lambda_1+\lambda_2)\sum_{n_1+n_2=N} F(n_1,n_2,w,\tau) + j\sum_{n_1+n_2<N} \frac{\partial F(n_1,n_2,w,\tau)}{\partial w}\Bigg). \quad (11)$$

Substituting (8) and (10) into Eq. (11), we get (7).

The system (6) is a system of linear equations for $R(n_1, n_2, x)$. Equation (7) is an ordinary differential equation for function $x(\tau)$. We can find a solution of the system of Eqs. (6), which depends on $x(\tau)$, and substitute it into Eq. (7). Solving Eq. (7), we get an expression for finding $x(\tau)$.

Taking into account that $\sum\limits_{n_1+n_2=n\leq N} R(n_1, n_2, x) = 1$, we denote $r(x) = \sum\limits_{n<N} R(n_1, n_2, x)$, $1 - r(x) = \sum\limits_{n=N} R(n_1, n_2, x)$. Denoting the right-hand side of Eq. (7) as $a(x)$, we can write following equality

$$a(x) = -xr(x) + (\lambda_1 + \lambda_2)(1 - r(x)) - \lambda_1 R(N, 0, x). \quad (12)$$

In stationary regime, Eq. (7) transforms to the following equation

$$-xr(x) + (\lambda_1 + \lambda_2)(1 - r(x)) - \lambda_1 R(N, 0, x) = 0.$$

Here x is the fixed point. Let us denote the solution of this equation as κ. Substituting $x = \kappa$ into the system Eq. (6), we obtain stationary probabilities $R(n_1, n_2, \kappa) = R(n_1, n_2)$ of the states of the servers. Probability $R(N, 0)$ is the blocking probability of priority customers.

4 Second Order Asymptotic Analysis

In the system Eqs. (2) and Eq. (3), we consider the characteristic functions of $i(t) - \frac{x(\sigma t)}{\sigma}$ as follows.

$$H(n_1, n_2, u, t) = e^{j\frac{u}{\sigma}x(\sigma t)} H^{(2)}(n_1, n_2, u, t).$$

By putting $\sigma = \epsilon^2$, $\tau = \epsilon t$, $u = \epsilon w$, $H^{(2)}(n_1, n_2, u, t) = F^{(2)}(n_1, n_2, w, \tau, \epsilon)$, we get

$$\epsilon^2 \frac{\partial F^{(2)}(0,0,w,\tau,\epsilon)}{\partial \tau} + j\epsilon w a(x) F^{(2)}(0,0,w,\tau,\epsilon) = \mu_1 F^{(2)}(1,0,w,\tau,\epsilon)$$
$$- (\lambda_1 + \lambda_2 + x) F^{(2)}(0,0,w,\tau,\epsilon) + \mu_2 F^{(2)}(0,1,w,\tau,\epsilon)$$
$$+ j\epsilon \frac{\partial F^{(2)}(0,0,w,\tau,\epsilon)}{\partial w}, n = 0,$$

$$\epsilon^2 \frac{\partial F^{(2)}(n_1,n_2,w,\tau,\epsilon)}{\partial \tau} + j\epsilon w a(x) F^{(2)}(n_1,n_2,w,\tau,\epsilon)$$
$$= - (\lambda_1 + \lambda_2 + \mu_1 n_1 + \mu_2 n_2 + x) F^{(2)}(n_1,n_2,w,\tau,\epsilon) + j\epsilon \frac{\partial F^{(2)}(n_1,n_2,w,\tau,\epsilon)}{\partial w}$$
$$+ \mu_1 (n_1+1) F^{(2)}(n_1+1,n_2,w,\tau,\epsilon) - j\epsilon e^{-j\epsilon w} \frac{\partial F^{(2)}(n_1,n_2-1,w,\tau,\epsilon)}{\partial w}$$
$$+ \lambda_2 F^{(2)}(n_1,n_2-1,w,\tau,\epsilon) + \lambda_1 F^{(2)}(n_1-1,n_2,w,\tau,\epsilon)$$
$$+ \mu_2 (n_2+1) F^{(2)}(n_1,n_2+1,w,\tau,\epsilon), 1 \le n < N,$$

$$\epsilon^2 \frac{\partial F^{(2)}(n_1,n_2,w,\tau,\epsilon)}{\partial \tau} + j\epsilon w a(x) F^{(2)}(n_1,n_2,w,\tau,\epsilon)$$
$$= - (\lambda_1 + \lambda_2 + \mu_1 n_1 + \mu_2 n_2 + \lambda_2 e^{j\epsilon w}) F^{(2)}(n_1,n_2,w,\tau,\epsilon)$$
$$+ \lambda_1 e^{j\epsilon w} F^{(2)}(n_1-1,n_2+1,w,\tau,\epsilon) + (\lambda_2 + x e^{-j\epsilon w}) F^{(2)}(n_1,n_2-1,w,\tau,\epsilon)$$
$$+ \lambda_1 F^{(2)}(n_1-1,n_2,w,\tau,\epsilon) - j\epsilon e^{-j\epsilon w} \frac{\partial F(n_1,n_2-1,w,\tau,\epsilon)}{\partial w}, n = N, n_1 < N,$$

$$\epsilon^2 \frac{\partial F^{(2)}(N,0,w,\tau,\epsilon)}{\partial \tau} + j\epsilon w a(x) F^{(2)}(N,0,w,\tau,\epsilon)$$
$$= - (\lambda_2(1 - e^{j\epsilon w}) + N\mu_1) F^{(2)}(N,0,w,\tau,\epsilon) + \lambda_1 F^{(2)}(N-1,0,w,\tau,\epsilon) \tag{13}$$
$$+ \lambda_1 e^{j\epsilon w} F^{(2)}(N-1,1,w,\tau,\epsilon), n_1 = N,$$

$$\epsilon^2 \sum_{n_1+n_2 \le N} \frac{\partial F^{(2)}(n_1,n_2,w,\tau,\epsilon)}{\partial \tau} + j\epsilon w a(x) \sum_{n_1+n_2 \le N} F^{(2)}(n_1,n_2,w,\tau,\epsilon)$$
$$= (e^{j\epsilon w} - 1) \Bigg(-\lambda_1 F^{(2)}(N,0,w,\tau,\epsilon) + j\epsilon e^{-j\epsilon w} \sum_{n_1+n_2 < N} \frac{\partial F^{(2)}(n_1,n_2,w,\tau,\epsilon)}{\partial w}$$
$$+ (\lambda_1 + \lambda_2) \sum_{n_1+n_2 = N} F^{(2)}(n_1,n_2,w,\tau,\epsilon) - x e^{-j\epsilon w} \sum_{n_1+n_2 < N} F^{(2)}(n_1,n_2,w,\tau,\epsilon) \Bigg). \tag{14}$$

We denote $F^{(2)}(n_1,n_2,w,\tau) = \lim_{\epsilon \to 0} F^{(2)}(n_1,n_2,w,\tau,\epsilon)$. Function $F^{(2)}$ (n_1,n_2,w,τ) has the following form:

$$F^{(2)}(n_1,n_2,w,\tau) = \Phi(w,\tau) R(n_1,n_2,x).$$

$\Phi(w,\tau)$ is the characteristic function of $y(\tau)$, where $y(\tau) = \lim_{\sigma \to 0} \sqrt{\sigma}$ $\left(i(\frac{\tau}{\sigma}) - \frac{x(\tau)}{\sigma} \right)$.

Theorem 2. *Function $\Phi(w,\tau)$ is a solution of the differential equation:*

$$\frac{\partial\Phi(w,\tau)}{\partial\tau} = a'(x)w\frac{\partial\Phi(w,\tau)}{\partial w} - \frac{w^2}{2}b(x)\Phi(w,\tau), \tag{15}$$

where $a(x)$ is determined by (12), $b(x)$ has the form

$$b(x) = a(x)+2(xr(x)-x\sum_{n<N} g(n_1,n_2,x)+(\lambda_1+\lambda_2)\sum_{n=N} g(n_1,n_2,x)-g(N,0,x)). \tag{16}$$

Here functions $g(n_1,n_2,x)$ are defined by the following system of equations

$$-(\lambda_1+\lambda_2+x)g(0,0,x)+\mu_1 g(1,0,x)+\mu_2 g(0,1,x) = a(x)R(0,0,x), n=0,$$

$$\begin{gathered}-(\lambda_1+\lambda_2+\mu_1 n_1+\mu_2 n_2+x)g(n_1,n_2,x)+(\lambda_2+x)g(n_1,n_2-1,x)\\ +\lambda_1 g(n_1-1,n_2,x)+\mu_1(n_1+1)g(n_1+1,n_2,x)+\mu_2(n_2+1)g(n_1,n_2+1,x)\\ = a(x)R(n_1,n_2,x)+xR(n_1,n_2-1,x), 1\le n\le N,\end{gathered}$$

$$\begin{gathered}-(\lambda_1+\mu_1 n_1+\mu_2 n_2)g(n_1,n_2,x)+(\lambda_2+x)g(n_1,n_2-1,x)\\ +\lambda_1 g(n_1-1,n_2+1,x)+\lambda_1 g(n_1-1,n_2,x) = xR(n_1,n_2-1,x)\\ +(a(x)-\lambda_2)R(n_1,n_2,x)-\lambda_1 R(n_1-1,n_2+1,x), n=N, n_1<N,\end{gathered}$$

$$\begin{gathered}-N\mu_1 g(N,0,x)+\lambda_1(g(N-1,0,x)+g(N-1,1,x))) =\\ = a(x)R(N,0,x)-\lambda_2 R(N,0,x)-\lambda_1 R(N-1,1,x), n_1=N.\end{gathered} \tag{17}$$

Proof. For the solution of the system of Eqs. (13), we write in the form

$$F^{(2)}(n_1,n_2,w,\tau,\epsilon) = \Phi(w,\tau)\left(R(n_1,n_2,x)+j\epsilon wf(n_1,n_2,x)\right)+O(\epsilon^2). \tag{18}$$

We will substitute this expression into the system of Eqs. (13). Taking (6) into account and sending $\epsilon\to 0$, we obtain

$$\begin{gathered}wa(x)R(0,0,x) = -(\lambda_1+\lambda_2+x)wf(0,0,x)+\mu_2 wf(0,1,x)\\ +\mu_1 wf(1,0,x)+R(0,0,x)\frac{\partial\Phi(w,\tau)}{\Phi(w,\tau)\partial w}, n=0,\end{gathered}$$

$$\begin{gathered}wa(x)R(n_1,n_2,x) = -(\lambda_1+\lambda_2+\mu_1 n_1+\mu_2 n_2+x)wf(n_1,n_2,x)\\ +R(n_1,n_2,x)\frac{\partial\Phi(w,\tau)}{\Phi(w,\tau)\partial w}+\mu_1(n_1+1)wf(n_1+1,n_2,x)\\ -R(n_1,n_2-1,x)\frac{\partial\Phi(w,\tau)}{\Phi(w,\tau)\partial w}-xwR(n_1,n_2-1,x)+\lambda_2 wf(n_1,n_2-1,x)\\ +\lambda_1 wf(n_1-1,n_2,x)+\mu_2(n_2+1)wf(n_1,n_2+1,x), 1\le n<N,\end{gathered}$$

$$\begin{gathered}wa(x)R(n_1,n_2,x) = -(\lambda_1+\mu_1 n_1+\mu_2 n_2)wf(n_1,n_2,x)\\ +\lambda_1 w\left(R(n_1-1,n_2+1,x)+f(n_1-1,n_2+1,x)\right)\\ +\lambda_2 wR(n_1,n_2,x)+(\lambda_2+x)wf(n_1,n_2-1,x)-xwR(n_1,n_2-1,x)\\ +\lambda_1 wf(n_1-1,n_2,x)-R(n_1,n_2-1,x)\frac{\partial\Phi(w,\tau)}{\Phi(w,\tau)\partial w}, n=N, n_1<N,\end{gathered}$$

$$\begin{aligned} wa(x)R(N,0,x) = &-N\mu_1 wf(N,0,x) + \lambda_1 wf(N-1,0,x) \\ &+\lambda_1 w(R(N-1,1,x) + f(N-1,1,x)) + \lambda_2 wR(N,0,x), n_1 = N. \end{aligned} \tag{19}$$

For the solution $f(n_1, n_2, x)$ of the system (19), we present in the following form

$$f(n_1,n_2,x) = CR(n_1,n_2,x) + g(n_1,n_2,x) - \phi(n_1,n_2,x)\frac{\partial\Phi(w,\tau)}{w\Phi(w,\tau)\partial w}. \tag{20}$$

Substituting this expression into the system Eq. (19), we obtain the following system of equations:

$$\begin{aligned}
&-(a(x) + C(\lambda_1+\lambda_2+x))R(0,0,x) + C\mu_1 R(1,0,x) + C\mu_2 R(0,1,x) \\
&-(\lambda_1+\lambda_2+x)g(0,0,x) + \mu_1 g(1,0,x) + \mu_2 g(0,1,x) + ((\lambda_1+\lambda_2+x))\phi(0,0,x) \\
&-\mu_1\phi(1,0,x) - \mu_2\phi(0,1,x) + R(0,0,x))\frac{\partial\Phi(w,\tau)}{w\Phi(w,\tau)\partial w} = 0, n = 0, \\
&\\
&-(a(x) + C(\lambda_1+\lambda_2+\mu_1 n_1+\mu_2 n_2+x))R(n_1,n_2,x) \\
&+(\lambda_2+x)CR(n_1,n_2-1,x) + \lambda_1 CR(n_1-1,n_2,x) - xR(n_1,n_2-1,x) \\
&+\mu_1(n_1+1)CR(n_1+1,n_2,x) + \mu_2(n_2+1)CR(n_1,n_2+1) \\
&+(\lambda_2+x)g(n_1,n_2-1,x) + \lambda_1 g(n_1-1,n_2,x) + \mu_1(n_1+1)g(n_1+1,n_2,x) \\
&+\mu_2(n_2+1)g(n_1,n_2+1) - (\lambda_1+\lambda_2+\mu_1 n_1+\mu_2 n_2+x)g(n_1,n_2,x) \\
&+((\lambda_1+\lambda_2+\mu_1 n_1+\mu_2 n_2+x)\phi(n_1,n_2,x) + R(n_1,n_2,x) \\
&-R(n_1,n_2-1,x) - (\lambda_2+x)\phi(n_1,n_2-1,x) \\
&-\lambda_1\phi(n_1-1,n_2,x) - \mu_1(n_1+1)\phi(n_1+1,n_2,x) \\
&-\mu_2(n_2+1)\phi(n_1,n_2+1))\frac{\partial\Phi(w,\tau)}{w\Phi(w,\tau)\partial w} = 0, 1 \le n < N, \\
&\\
&-(a(x) + C(\lambda_1+\mu_1 n_1+\mu_2 n_2))R(n_1,n_2,x) + C\lambda_1 R(n_1-1,n_2+1,x) \\
&+\lambda_1 R(n_1-1,n_2+1,x) + \lambda_2 R(n_1,n_2,x) + (\lambda_2+x)CR(n_1,n_2-1,x) \\
&-xR(n_1,n_2-1,x) + C\lambda_1 R(n_1-1,n_2,x) \\
&-(\lambda_1+\mu_1 n_1+\mu_2 n_2)g(n_1,n_2,x) + \lambda_1 g(n_1-1,n_2+1,x) \\
&+(\lambda_2+x)g(n_1,n_2-1,x) + \lambda_1 g(n_1-1,n_2,x) + (-\lambda_1\phi(n_1-1,n_2+1,x) \\
&+(\lambda_1+\mu_1 n_1+\mu_2 n_2)\phi(n_1,n_2,x) - (\lambda_2+x)\phi(n_1,n_2-1,x) \\
&-\lambda_1\phi(n_1-1,n_2,x) - R(n_1,n_2-1,x))\frac{\partial\Phi(w,\tau)}{w\Phi(w,\tau)\partial w} = 0, n = N, n_1 < N, \\
&\\
&-(a(x) + CN\mu_1)R(N,0,x) + \lambda_1 CR(N-1,0,x) \\
&+\lambda_1(R(N-1,1,x) + CR(N-1,1,x)) + \lambda_2 R(N,0,x) + (\lambda_1 g(N-1,0,x) \\
&-N\mu_1 g(N,0,x) + \lambda_1 g(N-1,1,x)) + (N\mu_1\phi(N,0,x) - \lambda_1\phi(N-1,0,x) \\
&-\lambda_1\phi(N-1,1,x))\frac{\partial\Phi(w,\tau)}{w\Phi(w,\tau)\partial w} = 0, n_1 = N.
\end{aligned} \tag{21}$$

Considering the system of Eqs. (6), the coefficient C before the constant is equal to zero. As a result, we obtain the following two systems of equations:

For finding the functions $\phi(n_1, n_2, x)$:

$$
\begin{aligned}
&(\lambda_1+\lambda_2+x)\phi(0,0,x) - \mu_1\phi(1,0,x) - \mu_2\phi(0,1,x) = -R(0,0,x), n=0,\\
&(\lambda_1+\lambda_2+\mu_1 n_1+\mu_2 n_2+x)\phi(n_1,n_2,x) - (\lambda_2+x)\phi(n_1,n_2-1,x)\\
&\quad -\lambda_1\phi(n_1-1,n_2,x) - \mu_1(n_1+1)\phi(n_1+1,n_2,x)\\
&\quad -\mu_2(n_2+1)\phi(n_1,n_2+1) = R(n_1,n_2-1,x) - R(n_1,n_2,x), 1\le n<N,\\
&(\lambda_1+\mu_1 n_1+\mu_2 n_2)\phi(n_1,n_2,x) - \lambda_1\phi(n_1-1,n_2+1,x) - \lambda_1\phi(n_1-1,n_2,x)\\
&\quad -(\lambda_2+x)\phi(n_1,n_2-1,x) = R(n_1,n_2-1,x), n=N, n_1<N,\\
&N\mu_1\phi(N,0,x) - \lambda_1\phi(N-1,0,x) - \lambda_1\phi(N-1,1,x) = 0, n_1=N.
\end{aligned}
\tag{22}
$$

For finding the functions $g(n_1, n_2, x)$:

$$
\begin{aligned}
&-(\lambda_1+\lambda_2+x)g(0,0,x) + \mu_1 g(1,0,x) + \mu_2 g(0,1,x) = a(x)R(0,0,x), n=0,\\
&-(\lambda_1+\lambda_2+\mu_1 n_1+\mu_2 n_2+x)g(n_1,n_2,x) + (\lambda_2+x)g(n_1,n_2-1,x)+\\
&+\lambda_1 g(n_1-1,n_2,x) + \mu_1(n_1+1)g(n_1+1,n_2,x) + \mu_2(n_2+1)g(n_1,n_2+1,x)\\
&\quad = a(x)R(n_1,n_2,x) + xR(n_1,n_2-1,x), 1\le n\le N,\\
&-(\lambda_1+\mu_1 n_1+\mu_2 n_2)g(n_1,n_2,x) + (\lambda_2+x)g(n_1,n_2-1,x)\\
&\quad +\lambda_1 g(n_1-1,n_2+1,x) + \lambda_1 g(n_1-1,n_2,x) = xR(n_1,n_2-1,x)\\
&\quad +(a(x)-\lambda_2)R(n_1,n_2,x) - \lambda_1 R(n_1-1,n_2+1,x), n=N, n_1<N,\\
&-N\mu_1 g(N,0,x) + \lambda_1(g(N-1,0,x) + g(N-1,1,x)))\\
&\quad = a(x)R(N,0,x) - \lambda_2 R(N,0,x) - \lambda_1 R(N-1,1,x), n_1=N.
\end{aligned}
\tag{23}
$$

Now, consider system of Eqs. (6). Differentiating it with respect to x, we derive

$$
\begin{aligned}
&(\lambda_1+\lambda_2+x)\frac{\partial R(0,0,x)}{\partial x} + R(0,0,x) - \mu_1\frac{\partial R(1,0,x)}{\partial x} - \mu_2\frac{\partial R(0,1,x)}{\partial x} = 0, n=0,\\
&(\lambda_1+\lambda_2+\mu_1 n_1+\mu_2 n_2+x)\frac{\partial R(n_1,n_2,x)}{\partial x} + R(n_1,n_2,x)\\
&\quad -(\lambda_2+x)\frac{\partial R(n_1,n_2-1,x)}{\partial x} - R(n_1,n_2-1,x)\\
&\quad -\lambda_1\frac{\partial R(n_1-1,n_2,x)}{\partial x} - \mu_1(n_1+1)\frac{\partial R(n_1+1,n_2,x)}{\partial x}\\
&\quad -\mu_2(n_2+1)\frac{\partial R(n_1,n_2+1,x)}{\partial x} = 0, 1\le n<N,\\
&(\lambda_1+\mu_1 n_1+\mu_2 n_2)\frac{\partial R(n_1,n_2,x)}{\partial x} - \lambda_1\frac{\partial R(n_1-1,n_2+1,x)}{\partial x}\\
&\quad -\lambda_1\frac{\partial R(n_1-1,n_2,x)}{\partial x} - (\lambda_2+x)\frac{\partial R(n_1,n_2-1,x)}{\partial x}\\
&\quad -R(n_1,n_2-1,x) = 0, n=N, n_1<N,
\end{aligned}
$$

$$N\mu_1 \frac{\partial R(N,0,x)}{\partial x} - \lambda_1 \frac{\partial R(N-1,0,x)}{\partial x} - \lambda_1 \frac{\partial R(N-1,1,x)}{\partial x} = 0, n_1 = N. \quad (24)$$

Comparing systems of Eqs. (22) and (24), we can conclude that $\phi(n_1,n_2,x) = \dfrac{\partial R(n_1,n_2,x)}{\partial x}$, where $\sum\limits_{n\leq N} \phi(n_1,n_2,x) = 0$.

Functions $g(n_1,n_2,x)$ are particular solution to (19), we choose the one satisfying $\sum\limits_{n\leq N} g(n_1,n_2,x) = 0$. Notice that system of Eq. (23) for functions $g(n_1,n_2,x)$ coincides with the system of Eqs. (17).

Then we consider the scalar Eq. (14). Using the expansion

$$e^{j\epsilon w} = 1 + j\epsilon w + \frac{(j\epsilon w)^2}{2} + O(\epsilon^3)$$

and applying this expansion, we obtain

$$\begin{aligned}
&-j\epsilon \sum_{n\leq N} \frac{\partial F^{(2)}(n_1,n_2,w,\tau,\epsilon)}{\partial \tau} + wa(x) \sum_{n\leq N} F^{(2)}(n_1,n_2,w,\tau,\epsilon) \\
&= (w - \frac{j\epsilon w^2}{2}) \Bigg(-\lambda_1 F^{(2)}(N,0,w,\tau,\epsilon) + j\epsilon \sum_{n<N} \frac{\partial F^{(2)}(n_1,n_2,w,\tau,\epsilon)}{\partial w} \\
&- (j\epsilon)^2 w \sum_{n<N} \frac{\partial F^{(2)}(n_1,n_2,w,\tau,\epsilon)}{\partial w} + (\lambda_1+\lambda_2) \sum_{n=N} F^{(2)}(n_1,n_2,w,\tau,\epsilon) \\
&- x(1 - j\epsilon w + \frac{(j\epsilon w)^2}{2}) \sum_{n<N} F^{(2)}(n_1,n_2,w,\tau,\epsilon) \Bigg) + O(\epsilon^3).
\end{aligned}$$

We substitute the solution (18) into this equation. Considering Eq. (7), we divide by $j\epsilon$ and take the limit $\epsilon \to 0$. Then, substituting expression (20), we get

$$\begin{aligned}
-\frac{\partial \Phi(w,\tau)}{\partial \tau} &= w \Bigg(r(x) + x \sum_{n<N} \frac{\partial R(n_1,n_2,x)}{\partial x} - (\lambda_1+\lambda_2) \sum_{n=N} \frac{\partial R(n_1,n_2,x)}{\partial x} \\
&+ \lambda_1 \frac{\partial R(N,0,x)}{\partial x} \Bigg) \frac{\partial \Phi(w,\tau)}{\partial w} + \frac{w^2}{2} (xr(x) + (\lambda_1+\lambda_2)(1-r(x)) - \lambda_1 R(N,0,x) \\
&- 2x \sum_{n<N} g(n_1,n_2,x) + 2(\lambda_1+\lambda_2) \sum_{n=N} g(n_1,n_2,x) - 2\lambda_1 g(N,0,x)) \Phi(w,\tau).
\end{aligned}$$

Now, let us consider the Eq. (12). Differentiating it with respect to x, we derive the equation

$$a'(x) = -r(x) - x \sum_{n<N} \frac{\partial R(n_1,n_2,x)}{\partial x} + (\lambda_1+\lambda_2) \sum_{n=N} \frac{\partial R(n_1,n_2,x)}{\partial x} - \lambda_1 \frac{\partial R(N,0,x)}{\partial x}. \quad (25)$$

We denote

$$b(x) = xr(x) + (\lambda_1 + \lambda_2)(1 - r(x)) - \lambda_1 R(N, 0, x)$$
$$- 2x \sum_{n<N} g(n_1, n_2, x) + 2(\lambda_1 + \lambda_2) \sum_{n=N} g(n_1, n_2, x) - 2\lambda_1 g(N, 0, x) = a(x)$$
$$+ 2\left(r(x) - x \sum_{n<N} g(n_1, n_2, x) - (\lambda_1 + \lambda_2) \sum_{n=N} g(n_1, n_2, x) + \lambda_1 g(N, 0, x) \right).$$

Given this expression and Eq. (25), we will write the following equation

$$\frac{\partial \Phi(w, \tau)}{\partial \tau} = wa'(x) \frac{\partial \Phi(w, \tau)}{\partial w} + \frac{w^2}{2} b(x) \Phi(w, \tau),$$

which coincides with Eq. (15).

So, the theorem is proved.

Applying the inverse Fourier transform in the system (15), we obtain Fokker-Plank equation for probability density function $P(y, \tau) = \frac{\partial P\{y(\tau)<y\}}{\partial y}$ of process $y(\tau)$:

$$\frac{\partial P(y, \tau)}{\partial \tau} = -\frac{\partial \{ya'(x)P(y, \tau)\}}{\partial y} + \frac{1}{2} \frac{\partial^2 \{b(x)P(y, \tau)\}}{\partial y^2}.$$

Hence, $y(\tau)$ is the diffusion process whose with drift and diffusion coefficients are given by $ya'(x)$ and $b(x)$, respectively.

5 Stationary Distribution of the Diffusion Process

We consider $z(\tau) = x(\tau) + \sqrt{\sigma} y(\tau)$, which has a direct relation with $i(t)$.

We consider the stationary probability density function for $z(\tau)$:

$$\Pi(z) = \frac{\partial P\{z(\tau) < z\}}{\partial z}.$$

Theorem 3. *The stationary probability density function of $z(\tau)$ is given by*

$$\Pi(z) = \frac{C}{b(z)} exp \left(\frac{2}{\sigma} \int_0^z \frac{a(x)}{b(x)} dx \right),$$

where C is the normalizing constant.

Proof. $z(\tau)$ is a solution to

$$dz(\tau) = dx(\tau) + \epsilon dy(\tau) = a(x)d\tau + \epsilon(y(\tau)a'(x)d\tau + \sqrt{b(x)}dw(\tau))$$
$$= (a(x) + \epsilon y(\tau)a'(x))d\tau + \epsilon\sqrt{b(x)}dw(\tau),$$

where $w(\tau)$ is Wiener process.

The coefficients of the Wiener process are given in the form:

$$a(x) + \epsilon y(\tau)a'(x) = a(z) + O(\epsilon^2),$$

$$\epsilon\sqrt{b(x)} = \sqrt{\epsilon^2 b(x)} = \sqrt{\epsilon^2 b(x + \epsilon y(\tau)) + O(\epsilon^3)} = \sqrt{\epsilon^2 b(z) + O(\epsilon^3)}$$
$$= \sqrt{\epsilon^2 b(z)} + O(\epsilon^2) = \sqrt{\sigma b(z)} + O(\epsilon^2).$$

Then the stochastic differential equation for the process $z(\tau)$ can be rewritten as:

$$dz(\tau) = a(z)d\tau + \sqrt{\sigma b(z)}dw(\tau).$$

Process $z(\tau)$ is a diffusion process whose probability density

$$\Pi(z,\tau) = \frac{\partial P\{z(\tau) < z\}}{\partial z}$$

is the solution of the Fokker-Planck equation

$$\frac{\partial \Pi(z,\tau)}{\partial \tau} = -\frac{\partial\{a(z)\Pi(z,\tau)\}}{\partial z} + \frac{1}{2}\frac{\partial^2\{\sigma b(z)\Pi(z,\tau)\}}{\partial z^2}.$$

Under the stationary regime, this equation reduces to

$$-\frac{\partial\{a(z)\Pi(z)\}}{\partial z} + \frac{\sigma}{2}\frac{\partial^2\{b(z)\Pi(z)\}}{\partial z^2} = 0.$$

The solution of this differential equation is given by

$$\Pi(z) = \frac{C}{b(z)} exp\left(\frac{2}{\sigma}\int_0^z \frac{a(x)}{b(x)}dx\right). \tag{26}$$

So, the theorem is proved.

We build the approximation of the probability distribution for $i(t)$ as

$$Pd(i) = \frac{\Pi(i\sigma)}{\sum_{n=0}^{\infty} \Pi(n\sigma)}. \tag{27}$$

6 Conclusion

In this paper, we considered a multi-server retrial queueing system with priority customers. Using the method of asymptotic-diffusion analysis, we showed that the scaled version of the number of customers in the orbit converges to a diffusion process. We used this result to build an approximation for the stationary probability distribution of the number of customers in the orbit.

References

1. Akutsu, K., Phung-Duc, T.: Analysis of retrial queues for cognitive wireless networks with sensing time of secondary users. In: Phung-Duc, T., Kasahara, S., Wittevrongel, S. (eds.) QTNA 2019. LNCS, vol. 11688, pp. 77–91. Springer, Cham (2019). https://doi.org/10.1007/978-3-030-27181-7_6
2. Fedorova, E., Nazarov, A., Moiseev, A.: Asymptotic analysis methods for multi-server retrial queueing systems. In: Joshua, V.C., Varadhan, S.R.S., Vishnevsky, V.M. (eds.) Applied Probability and Stochastic Processes. ISFS, pp. 159–177. Springer, Singapore (2020). https://doi.org/10.1007/978-981-15-5951-8_11
3. Moiseev, A., Nazarov, N., Paul, S.: Asymptotic diffusion analysis of multi-server retrial queue with hyper-exponential service. Mathematics **8**(4), 531-1–531-16 (2020)
4. Morozov, E., Rogozin, S., Nguyen, Q.H., Phung-Duc, T.: Modified Erlang loss system for cognitive wireless networks (2020, submitted)
5. Nazarov, A., Phung-Duc, T., Paul, S., Lizura, O.: Asymptotic-diffusion analysis for retrial queue with batch poisson input and multiple types of outgoing calls. In: Vishnevskiy, V.M., Samouylov, K.E., Kozyrev, D.V. (eds.) DCCN 2019. LNCS, vol. 11965, pp. 207–222. Springer, Cham (2019). https://doi.org/10.1007/978-3-030-36614-8_16
6. Phung-Duc, T., Akutsu, K., Kawanishi, K., Salameh, O., Wittevrongel, S.: Queueing models for cognitive wireless networks with sensing time of secondary users (2020, submitted)
7. Sztrik, J., Bérczes, T., Nemouchi, H., Melikov, A.: Performance modeling of finite-source cognitive radio networks using simulation. In: Vishnevskiy, V.M., Samouylov, K.E., Kozyrev, D.V. (eds.) DCCN 2016. CCIS, vol. 678, pp. 64–73. Springer, Cham (2016). https://doi.org/10.1007/978-3-319-51917-3_7
8. Nemouchi, H., Sztrik, J.: Performance evaluation of finite-source cognitive networks with non-reliable services using simulation. In: Proceedings of the 10th International Conference on Applied Informatics Eger, Hungary, 30 January–1 February, pp. 225–234. (2017) https://doi.org/10.14794/ICAI.10.2017.225
9. Klimenok, V., Dudin, A., Vishnevsky, V.: Priority multi-server queueing system with heterogeneous customers. Mathematics **8**(9), 1501 (2020). https://doi.org/10.3390/math8091501
10. Dudin, A.N., Lee, M.H., Dudina, O., Lee, S.K.: Analysis of priority retrial queue with many types of customers and servers reservation as a model of cognitive radio system. IEEE Trans. Commun. **65**(1), 186–199 (2016)
11. Nazarov, A., Moiseev, A., Phung-Duc, T., Paul, S.: Diffusion limit of multi-server retrial queue with setup time. Mathematics **8**(12), 2232 (2020)
12. Sun, B., Lee, M.H., Dudin, S.A., Dudin, A.N.: Analysis of multiserver queueing system with opportunistic occupation and reservation of servers. Math. Probl. Eng. **2014**, 1–14 (2014). Article ID 178108

Stability Condition of a Multi-class Modified Erlang System

Stepan Rogozin[1,2(✉)] and Evsey Morozov[1,2,3]

[1] Institute of Applied Mathematical Research Karelian Research Centre RAS, Petrozavodsk, Russia
emorozov@karelia.ru
[2] Petrozavodsk State University, Petrozavodsk, Russia
[3] Moscow Center for Fundamental and Applied Mathematics, Moscow State University, 119991 Moscow, Russia

Abstract. We consider a modified Erlang loss system where the first priority customers (class-1) are lost if find all servers busy, while the second priority customers (class-2) form an infinite capacity queue. A new feature of this system is that sub-classes of class-1 customers are assigned on servers according to assignment probabilities. All customers follow Poisson inputs and have general class-dependent service time. We show how the product form of class-1 stationary probabilities can be used to obtain the stability condition of the whole system even when sub-classes of class-1 customers have different service rates. Also we perform discrete event simulation to confirm theoretical results.

Keywords: Modified Erlang system · Two-priority customers · Stability condition · Multi-type customers · Multi-type servers · Simulation

1 Introduction

A spectrum shortage problems appearing by the explosively increasing wireless networks can be solved in particular by using cognitive wireless networks [4,7]. In such a network, there are two main classes of users primary users and secondary users. The former users are allowed to use the transmission bandwidths only if the primary users are not present in the network at this moment. Thus, if there are no free channels, the primary users have priority interrupting transmissions of secondary users, and an interrupted secondary user resumes transmission when one of the channels becomes again available. We consider the setting when transmission rate is channel-dependent and moreover, each server in general accepts a limited set of (sub)classes of the primary users. Thus the system we study relates to a wide class of systems with *flexible servers*, see for instance, [8,9]. In such a

The research is supported by Russian Foundation for Basic Research, projects 18-07-00147, 18-07-00156 and 19-07-00303.

A. Dudin et al. (Eds.): ITMM 2020, CCIS 1391, pp. 251–263, 2021.
https://doi.org/10.1007/978-3-030-72247-0_19

system there is a possibility that a part of service capacity of a given server can be transferred to another server (or servers) to satisfy the corresponding QoS requirements. A closely related concept is the system with the so-called *cross-trained servers*, which are considered in detail, in particular, in [10–12]. In such a system a set of servers accept a limited set of customer types only, while another set of servers accept to server a wider class of customers (or all customers). Various aspects of such systems, including the problems of optimal allocation aiming minimizing a cost function are considered in the papers [8,13,14].

In this work we consider the extension of the following system with two-priority classes of customers and several identical servers. The first priority customers are lost if find all servers busy, while the second priority customers form an infinite capacity queue. All classes of customers follow Poisson input and have general independent identically distributed (iid) service times. The service discipline for class-2 customers is assumed to be FCFS (first-come-first-served). Such a modified Erlang loss system is motivated and studied in the paper [1] in which in particular the stability condition (1) of this system is obtained.

A new feature of this system considering in this work is that a *multi-type* server assignment for the priority customers is assumed. Each server can serve only a *limited set* of sub-classes of class-1 customers and different sub-classes of customers in general have different arrival rates. At the same time, class-2 customers belong to only one type and can be served by any server. In previous work [2], we have shown that this system has the same stability condition (1), provided all servers have identical service rates. In present work, we prove that this condition is also stability one for Poisson inputs and *server-dependent service rates* for the corresponding sub-classes of priority customers.

The paper is organized as follows. In Sect. 2, we describe the model in detail. In Sect. 3, the main stability result is proved using a regenerative approach and a balance equation connecting arrived and departed work. Also in this section, we describe in brief how to compute the stationary busy probabilities which are required to obtain the stability condition of the system in an explicit form. Finally, in Sect. 4, we present some numerical results illustrating and verifying theoretical results.

2 Description of the Model

We consider the extension of the following system with two-priority classes of customers and J identical servers. The first priority customers (class-1) are lost if find all servers busy, while the second priority customers (class-2) form an infinite capacity queue if find all servers busy. Customers of class-i follow Poisson input with rate $\lambda^{(i)}$. Also we assume that class-i customers have independent identically distributed (iid) service times $\{S_n^{(i)}\}$ with generic element $S^{(i)}$. The service discipline for class-2 customers is assumed to be FCFS (first-come-first-served). A new feature of this system considering in this work is that a *multi-type* server assignment for the priority customers is assumed. Each server can serve only a *limited set* of subclasses of class-1 customers and different subclasses of customers in general have different arrival rates. Suppose that we have I subclasses of class-1 customers. We assume Poisson inputs and server-dependent service rates for the sets of priority subclasses. We denote μ_j the work rate of server j and $S^{(i,j)} =_{st} S^{(i)}/\mu_j$ the service time of class-i customers arrived at server j. We also assume that $\mathsf{E}S^{(1)} = 1$, therefore $\mathsf{E}S^{(1,j)} = 1/\mu_j$. At the same time, class-2 customers belong to only one type and can be served by any server. In Fig. 1. We present the scheme of this extended system.

Also we introduce assignment probabilities of this system. We denote $S \subseteq \{1, ..., J\}$ all possible combinations of the numbers of servers. When the system is in state S an arriving class-1 customer of subclass i selects a server $j \in S$ with the probability $\mathsf{P}_{i,j}(S)$. These *assignment probabilities* are the control parameters that we can choose to obtain the stationary distribution of the system [2,3]. If subclass i customers can not be served by server j then $\mathsf{P}_{i,j}(S) = 0$. If $S = \{k\}$ and subclass i customers can be served by server j then $\mathsf{P}_{i,k}(S) = 1$.

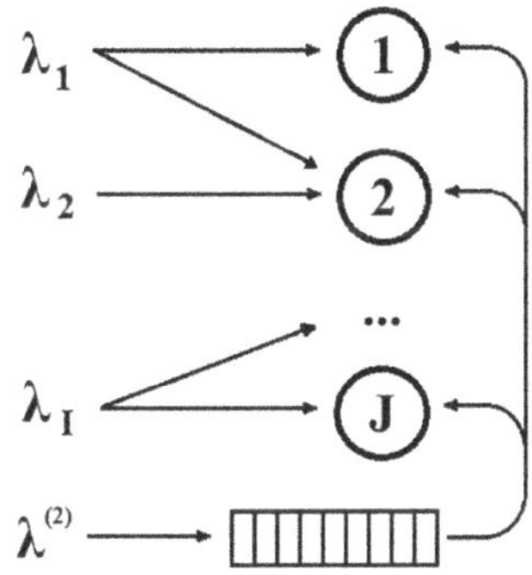

Fig. 1. Extended system

3 Stability Analysis

Denote by $\rho_2 = \lambda^{(2)}\mathsf{E}S^{(2)}$ the traffic intensity of class-2 customers and let P_i be the stationary probability that i servers are occupied by class-1 customers. Because the processes related to class-1 customers are positive recurrent and the inputs are Poisson, these stationary probabilities exist [5]. It is assumed that the first customer arrives in an empty system at instant $t_1 = 0$, and if the system is empty at this instant, we call it initially empty system. We note that in this case, the instant $T_0 = 0$ is a regeneration point and then $T_1 =_{st} T$, that is the first regeneration period is *stochastically* equivalent to generic period. In this case positive recurrence means that $\mathsf{E}T < \infty$ [6]. In this section we present in brief the proof of the following statement.

Theorem 1. *If condition*

$$\rho_2 + \sum_{i=1}^{J} i\mathsf{P}_i < J, \tag{1}$$

holds then $\mathsf{E}T < \infty$, *that is initially empty system is positive recurrent.*

Proof. In the interval of time $[0, t)$ denote: $\hat{V}_1(t)$, the total work which class-1 customers bring in the system (that is $\hat{V}_1(t)$ does not include the lost work); $V_2(t)$, the arrived work which class-2 customers bring; $B(t)$, the aggregated busy time of the servers which in turn equals the departed work in $[0, t)$. Also denote $I(t) = \sum_{i=1}^{J} I_i(t)$, where $I_i(t)$ is the idle time of server i in $[0,\ t]$. Obviously the following balance equations hold:

$$\hat{V}_1(t) + V_2(t) = W_1(t) + W_2(t) + B(t) = W(t) + Jt - I(t), \tag{2}$$

where $W_i(t)$ is the remaining workload to process class-i customers at instant t. We denote by $Q_1(t)$ the number of servers occupied by class-1 customers at instant t. First of all we note that, because class-1 customers have preemptive priority and lost when all servers are busy, then the process $\{Q_1(t)\}$ is *positive recurrent regenerative* [5]. Then, since the input is Poisson, the weak limit (limit in distribution) $Q_1(t) \Rightarrow Q_1$ exists and is the stationary number of class-1 customers in the system. Now we show that the following equality holds:

$$\lim_{t\to\infty} \frac{\mathsf{E}\hat{V}_1(t)}{t} = \mathsf{E}Q_1 = \sum_{i=1}^{J} i\mathsf{P}_i. \tag{3}$$

It is easy to see that the work $\hat{V}_1(t)$ can be presented as follows:

$$\hat{V}_1(t) = \int_0^t \sum_{i=1}^{J} i\mathbf{1}(Q_1(u) = i)du + W_1(t) = B_1(t) + W_1(t),\ \ t \ge 0, \tag{4}$$

where $\mathbf{1}(\cdot)$ denotes indicator function, $B_1(t)$ is the work spent by all servers to server class-1 customers in the time interval $[0,\ t)$. Moreover, because the process $\{Q_1(t)\}$ is positive recurrent, then with probability 1 (w.p.1) [6],

$$\lim_{t\to\infty} \frac{W_1(t)}{t} = 0.$$

Now, to find $B_1(t)$, we split indicators $\mathbf{1}(\cdot)$ into the following $2^J - 1$ disjoint subsets:

$$\begin{aligned}
\mathbf{1}(Q_1(t) = 1) &= \sum_{i=1}^{J} \mathbf{1}(\text{only the } i\text{-th server is busy at instant } t),\\
\mathbf{1}(Q_1(t) = 2) &= \sum_{i,j\in\{1,..,J\},\, i\neq j} \mathbf{1}(\text{only the } i\text{-th and the } j\text{-th servers}\\
&\qquad\qquad \text{are busy at instant } t),\\
&\ \ \dots\\
\mathbf{1}(Q_1(t) = J) &= \mathbf{1}(\text{all servers are busy at instant } t).
\end{aligned} \tag{5}$$

Therefore the arrived (and accepted) work $\hat{V}_1(t)$ can be represented as:

$$\hat{V}_1(t) = 1\int_0^t \sum_{i=1}^{J} 1(\text{only the } i\text{ -th server is busy at instant } u)du$$
$$+\, 2\int_0^t \sum_{i,j\in\{1,..,J\},\, i\neq j} 1(\text{only the } i\text{ -th and the } j\text{ -th servers are busy at instant } u)du$$
$$\dots$$
$$+\, J\int_0^t 1(\text{all servers are busy at instant } u)du + W_1(t), \quad t \geq 0. \tag{6}$$

Because, as we mentioned above, all processes related to class-1 customers are positive recurrent regenerative, then the following limit exists:

$$\lim_{t\to\infty}\frac{\mathsf{E}\hat{V}_1(t)}{t} = \lim_{t\to\infty}\frac{1}{t}\int_0^t \sum_{i=1}^{J} i\mathsf{P}(Q_1(u) = i)du$$
$$= \sum_{i=1}^{J} i \lim_{t\to\infty}\frac{1}{t}\int_0^t \mathsf{P}(Q_1(u) = i)du = \sum_{i=1}^{c} i\mathsf{P}_i. \tag{7}$$

The key observation is that the stationary probability that i servers are occupied by class-1 customers is also the limiting fraction of the corresponding busy time:

$$\mathsf{P}_i = \lim_{t\to\infty}\frac{1}{t}\int_0^t \mathsf{P}(Q_1(u) = i)du. \tag{8}$$

Note that the arrived work $\{V_2(t)\}$ is a positive recurrent *cumulative* process (see [6]), and it is easy to obtain that

$$\lim_{t\to\infty}\frac{\mathsf{E}V_2(t)}{t} = \rho_2. \tag{9}$$

Now we assume the following convergence in probability:

$$Q_2(t) \Rightarrow \infty \text{ as } t \to \infty. \tag{10}$$

Then it is easy to establish that

$$\lim_{t\to\infty}\frac{\mathsf{E}I(t)}{t} = 0. \tag{11}$$

It remains to note that

$$\liminf_{t\to\infty}\frac{\mathsf{E}W(t)}{t} = \liminf_{t\to\infty}\frac{\mathsf{E}W_2(t)}{t} \geq 0. \tag{12}$$

It now follows from balance Eq. (4), from assumption (10) and from previous limiting results that indeed a non-negative limit $\lim_{t\to\infty} \mathsf{E}W_2(t)/t \geq 0$ exists, and we obtain

$$0 \leq \lim_{t\to\infty} \frac{\mathsf{E}W_2(t)}{t} = \lim_{t\to\infty} \frac{\mathsf{E}\hat{V}_1(t)}{t} + \lim_{t\to\infty} \frac{\mathsf{E}V_2(t)}{t} - J = \sum_i i\mathsf{P}_i + \rho_2 - J, \quad (13)$$

implying a contradiction with assumption (10):

$$\rho_2 + \sum_{i=1}^{J} i\mathsf{P}_i \geq J. \quad (14)$$

Thus indeed

$$Q_2(t) \not\Rightarrow \infty, \quad (15)$$

and then one can show that it implies positive recurrence, that is $\mathsf{E}T < \infty$, and the existence of stationary regime of the system, see [6]. □

Condition (1), written as

$$\rho_2 < J - \sum_{i=1}^{J} i\mathsf{P}_i = J - \mathsf{E}Q_1, \quad (16)$$

has the following probabilistic interpretation: to have stability of the system, the traffic intensity of class-2 customers must be less than the mean number of the available servers.

Remark 1. Indeed, following [6] one can prove that condition (1) is *stability criterion* of the system under *arbitrary* initial state.

To implement our results in practice, we need to be able to calculate the probability P_i included in the stability condition (1). To do this, we present corresponding results from the paper [3]. In particular in this paper is shown that the following detailed balance equations hold:

$$\pi(S)\nu_j(S) = \pi(S\backslash\{j\})\mu_j \qquad \text{for all subset } S \subseteq \{1, ..., J\} \text{ and } j \in S, \quad (17)$$

where $\pi(S)$ is the stationary probabilities that servers S are not serving class-1 customers, $\nu_j(S)$ is the rate at which server $j \in S$ becomes busy (i.e. the arrival rate of class-1 customers arriving at server j), when the system is in state S. From these equations one can obtain the stationary distribution for $S = \{j_1, ..., j_m\}$, $j_1, ..., j_m \in \{1, ..., J\}$ and $m = 1, ..., J$:

$$\pi(S) = \pi(\emptyset) \frac{\mu_{j_1}}{\nu_{j_1}(\{j_1\})} \frac{\mu_{j_2}}{\nu_{j_2}(\{j_1, j_2\})} \frac{\mu_{j_3}}{\nu_{j_3}(\{j_1, j_2, j_3\})} \cdots \frac{\mu_{j_m}}{\nu_{j_m}(S)}, \quad (18)$$

where $\pi(\emptyset)$ normalizes the sum to 1.

Now we can calculate the stationary probabilities P_i using found probabilities $\pi(S)$:

$$\mathsf{P}_k = \sum_{j_1,\dots,j_{J-k}\in\{1,\dots,J\}} \pi(\{j_1,\dots,j_{J-k}\}) \qquad \text{for all } k \in \{1,\dots,J\}. \tag{19}$$

We note that these stationary probabilities $\pi(S)$ is correct only if the assignment probabilities satisfied some equation system obtained in the paper [3]. We present this system below for an example.

In the next section we give simulation results to demonstrate the theoretical results obtained above.

4 Simulation

In order to verify theoretical results obtained above, we present a numerical example based on discrete-event simulation of the system under consideration. Evidently, this example confirms theoretical results in a particular case only.

We consider the following example of our system. Let we have $J = 3$, $I = 2$, and we assume that servers 1 and 2 can serve all customers but server 3 cannot serve subclass-2 customers (see Fig. 2). This example is considered in details in paper [2] in which in particular the stationary probabilities are obtained. Also the system of assignment probabilities is obtained:

$$\begin{aligned}
\lambda_1\mathsf{P}_{1,1}(\{1,2\}) + \lambda_2\mathsf{P}_{2,1}(\{1,2\}) &= \frac{\lambda_1+\lambda_2}{2},\\
\mathsf{P}_{1,1}(\{1,3\}) &= \frac{\lambda_1}{2\lambda_1+\lambda_2},\\
\mathsf{P}_{1,3}(\{1,3\}) &= \frac{\lambda_1+\lambda_2}{2\lambda_1+\lambda_2},\\
\mathsf{P}_{1,2}(\{2,3\}) &= \frac{\lambda_1}{2\lambda_1+\lambda_2},\\
\mathsf{P}_{1,3}(\{2,3\}) &= \frac{\lambda_1+\lambda_2}{2\lambda_1+\lambda_2},\\
\mathsf{P}_{1,2}(\{1,2\}) &= 1 - \mathsf{P}_{1,1}(\{1,2\}),\\
\mathsf{P}_{2,2}(\{1,2\}) &= 1 - \mathsf{P}_{2,1}(\{1,2\}),\\
\lambda_1\mathsf{P}_{1,1}(\{1,2,3\}) + \lambda_2\mathsf{P}_{2,1}(\{1,2,3\}) &= \frac{(\lambda_1+\lambda_2)(2\lambda_1+\lambda_2)}{2(3\lambda_1+\lambda_2)},\\
\mathsf{P}_{1,3}(\{1,2,3\}) &= \frac{(\lambda_1+\lambda_2)}{3\lambda_1+\lambda_2},\\
\mathsf{P}_{1,2}(\{1,2,3\}) &= \frac{2\lambda_1}{3\lambda_1+\lambda_2} - \mathsf{P}_{1,1}(\{1,2,3\}),\\
\mathsf{P}_{2,2}(\{1,2,3\}) &= 1 - \mathsf{P}_{2,1}(\{1,2,3\}).
\end{aligned} \tag{20}$$

We recall that if this system is satisfied then the stationary probabilities can be found. However simulation of this example in paper [2] is performed only for identical service rates:

$$\lambda_1 = \lambda_2 = 10, \quad \mu_1 = \mu_2 = \mu_3 = 10. \tag{21}$$

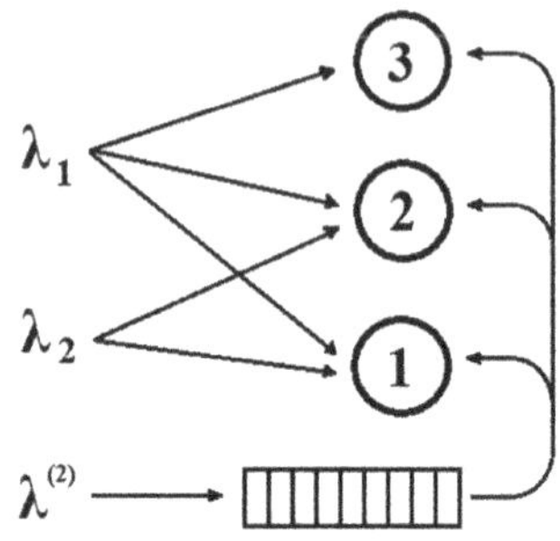

Fig. 2. Example

In the present paper we assume different work rates of servers:

$$\lambda_1 = \lambda_2 = 10, \quad \mu_1 = 15, \quad \mu_2 = 10, \quad \mu_3 = 5. \tag{22}$$

It implies different service times $S^{(i,j)}$. We present below theoretical stationary probabilities provided arrival and servers rates (22):

$$\begin{gathered} \pi(\emptyset) = 16/83, \quad \pi(\{1\}) = 12/83, \\ \pi(\{2\}) = 8/83, \quad \pi(\{3\}) = 8/83, \\ \pi(\{1,2\}) = 12/83, \quad \pi(\{1,3\}) = 9/83, \\ \pi(\{2,3\}) = 6/83, \quad \pi(\{1,2,3\}) = 12/83. \end{gathered} \tag{23}$$

Now we can calculate the stationary probabilities P_i in (19) using (23):

$$\mathsf{P}_0 = 12/83, \quad \mathsf{P}_1 = 27/83, \quad \mathsf{P}_2 = 28/83, \quad \mathsf{P}_3 = 16/83. \tag{24}$$

To compare these theoretical probabilities and calculated ones, we use the Euclidean distance between two probabilities distribution:

$$\delta(t) := \sqrt{\sum_{S \subseteq \{1,\ldots,J\}} (\pi(S) - \pi^*(S,t))^2}, \tag{25}$$

where

$$\pi^*(S,t) = \frac{T(S,t)}{t}, \quad t > 0, \tag{26}$$

and $T(S,t)$ is the computed time, in interval $[0,t)$, when the system is in the state S. To compare obtained values of $\delta(t)$ with the results from paper [2], we

Table 1. The assignment probabilities violating equation system (20): 1) uniform distribution; 2) uniform load of servers; 3) higher workload in server 2; 4) the maximum workload in server 2.

Case	$S = \{1, 2\}$				$S = \{1, 3\}$		$S = \{2, 3\}$		$S = \{1, 2, 3\}$				
	$\mathsf{P}_{1,1}$	$\mathsf{P}_{1,2}$	$\mathsf{P}_{2,1}$	$\mathsf{P}_{2,2}$	$\mathsf{P}_{1,1}$	$\mathsf{P}_{1,3}$	$\mathsf{P}_{1,2}$	$\mathsf{P}_{1,3}$	$\mathsf{P}_{1,1}$	$\mathsf{P}_{1,2}$	$\mathsf{P}_{2,1}$	$\mathsf{P}_{2,2}$	$\mathsf{P}_{1,3}$
1	0.5	0.5	0.5	0.5	0.5	0.5	0.5	0.5	1/3	1/3	0.5	0.5	1/3
2	0.5	0.5	0.5	0.5	0.0	1.0	0.0	1.0	1/6	1/6	0.5	0.5	2/3
3	1/3	2/3	1/3	2/3	2/3	1/3	2/3	1/3	1/3	2/3	1/3	2/3	1/3
4	0.0	1.0	0.0	1.0	1.0	0.0	1.0	0.0	0.0	1.0	0.0	1.0	0.0

Table 2. Comparing $\delta(t)$ for different and identical service rates, respectively: scenarios from Table 1.

Case	$\delta(t)$	$\delta_{ir}(t)$	Case	$\delta(t)$	$\delta_{ir}(t)$
1	0.0354	0.0322	3	0.1043	0.0870
2	0.0433	0.0459	4	0.1842	0.1559

denote by $\delta_{ir}(t)$ the distance between two probabilities distribution obtained in [2] for the identical service rates.

Denote by $L(t)$ the number of customer losses per time unit in interval $[0, t)$ (*loss rate*). Also we denote by $\hat{I}(t)$ the total time when servers *do not serve class-1 customers*, in interval $[0, t]$, and compute the fraction $\hat{I}(t)/t$. Note that this fraction can be larger 1 when the number of servers is bigger than 1.

We consider 13 different sets of assignment probabilities: 4 cases violate equation system (20) (see Table 1) and 9 cases satisfy it (see Table 3). To satisfy system (20), we can vary only two probabilities, $\mathsf{P}_{1,1}(\{1, 2\})$ and $\mathsf{P}_{1,1}(\{1, 2, 3\})$ (free variables, marked columns), while other probabilities depend on these free variables or are uniquely determined. The sample means of $\delta(t)$, based on 100 paths, are compared with $\delta_{ir}(t)$ given in Tables 2, 4.

Although the stationary probabilities are different for identical and different service rates, but $\delta(t)$ and $\delta_{ir}(t)$ turn out to be close. Moreover, the values of $\delta(t)$ in Table 2 is much larger than in Table 4, and it confirms that the stationary probabilities (23) are indeed found correctly for the cases given in Table 3 and satisfying system (20).

In Fig. 3 we present sample mean of $\delta(t)$ for 3 cases. It is easy to see that $\delta(t)$ converges for both in the cases when the system (20) is violated and in the cases when it is satisfied. We also show that number of class-1 customer losses $L(t)$ also converges (see Fig. 4). However for the cases when the system (20) is violated $L(t)$ is not much larger than for the cases when the system (20) is satisfied.

We also present sample mean of the ratio $\hat{I}(t)/t$ on Fig. 5. Obviously, it behaves similarly as $L(t)$. One can see on Fig. 6 that when the system (20) is satisfied, then $\hat{I}(t)/t$ converges to the upper bound of ρ_2 given by (16). It is

Table 3. The assignment probabilities satisfying equation system (20): only two (marked) probabilities can be varied.

Case	$S = \{1,2\}$				$S = \{1,3\}$		$S = \{2,3\}$		$S = \{1,2,3\}$				
	$\mathbf{P_{1,1}}$	$P_{1,2}$	$P_{2,1}$	$P_{2,2}$	$P_{1,1}$	$P_{1,3}$	$P_{1,2}$	$P_{1,3}$	$\mathbf{P_{1,1}}$	$P_{1,2}$	$P_{2,1}$	$P_{2,2}$	$P_{1,3}$
1	**0.1**	0.9	0.9	0.1	1/3	2/3	1/3	2/3	**0.1**	0.4	0.65	0.35	0.5
2	**0.1**	0.9	0.9	0.1	1/3	2/3	1/3	2/3	**0.3**	0.2	0.45	0.55	0.5
3	**0.1**	0.9	0.9	0.1	1/3	2/3	1/3	2/3	**0.5**	0.0	0.25	0.75	0.5
4	**0.5**	0.5	0.5	0.5	1/3	2/3	1/3	2/3	**0.1**	0.4	0.65	0.35	0.5
5	**0.5**	0.5	0.5	0.5	1/3	2/3	1/3	2/3	**0.3**	0.2	0.45	0.55	0.5
6	**0.5**	0.5	0.5	0.5	1/3	2/3	1/3	2/3	**0.5**	0.0	0.25	0.75	0.5
7	**0.9**	0.1	0.1	0.9	1/3	2/3	1/3	2/3	**0.1**	0.4	0.65	0.35	0.5
8	**0.9**	0.1	0.1	0.9	1/3	2/3	1/3	2/3	**0.3**	0.2	0.45	0.55	0.5
9	**0.9**	0.1	0.1	0.9	1/3	2/3	1/3	2/3	**0.5**	0.0	0.25	0.75	0.5

Table 4. Comparison $\delta(t)$ for different and identical service rates, respectively: scenarios are in Table 3. Both $\delta(t)$ and $\delta_{ir}(t)$ are much less than in Table 2.

Case	$\delta(t)$	$\delta_{ir}(t)$	Case	$\delta(t)$	$\delta_{ir}(t)$	Case	$\delta(t)$	$\delta_{ir}(t)$
1	0.0011	0.0016	4	0.0011	0.0009	7	0.0011	0.0013
2	0.0011	0.0013	5	0.0011	0.0012	8	0.0011	0.0012
3	0.0011	0.0010	6	0.0011	0.0009	9	0.0011	0.0015

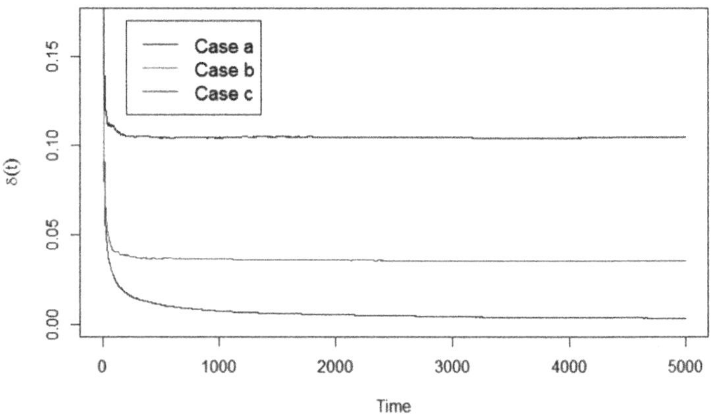

Fig. 3. Difference between theoretical and simulated stationary probabilities (Sample mean $\delta(t)$): case a) corresponds to case 3 in Table 1; case b) corresponds to case 1 in Table 1; case c) corresponds to case 1 in Table 3.

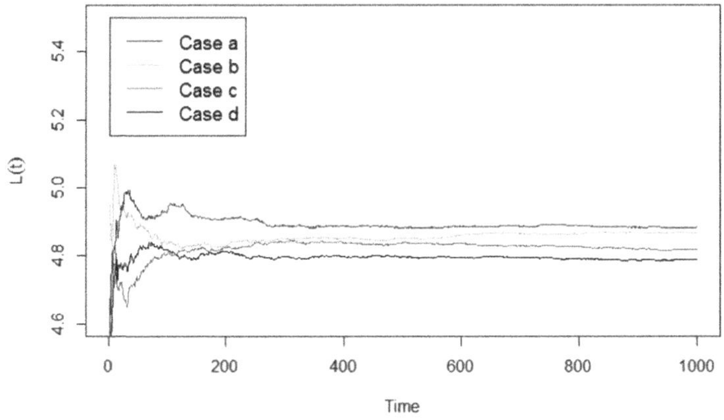

Fig. 4. Number of losses per time unit in interval $[0, t)$ (Sample mean $L(t)$): case a) corresponds to case 4 in Table 1; case b) corresponds to case 3 in Table 1; case c) corresponds to case 1 in Table 3; case d) corresponds to case 2 in Table 1.

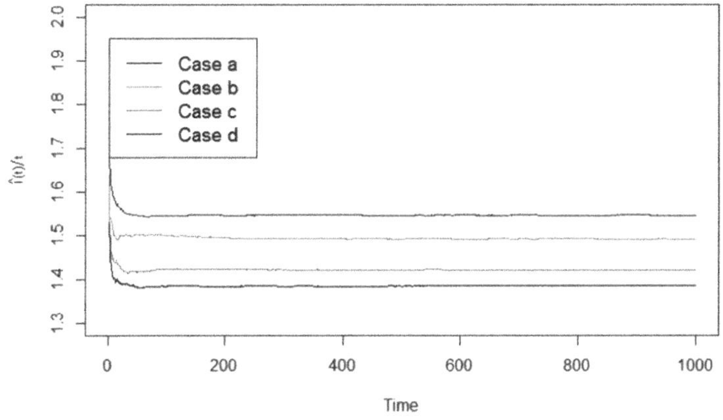

Fig. 5. Fraction time $\hat{I}(t)/t$ when servers free of class-1 customers: case a) corresponds to case 4 in Table 1; case b) corresponds to case 3 in Table 1; case c) corresponds to case 1 in Table 3; case d) corresponds to case 2 in Table 1.

intuitively clear that,

$$\lim_{t\to\infty} \frac{\hat{I}(t)}{t} = J - \mathsf{E}Q_1,$$

(also see (16)) that is, $\hat{I}(t)/t$ converges to the mean number of available servers for class-2 customers.

Finally, we compute the stability criterion (1). We obtain the upper bound for ρ_2 from (24) and (1):

$$\rho_2 < 118/83 \approx 1.4217. \tag{27}$$

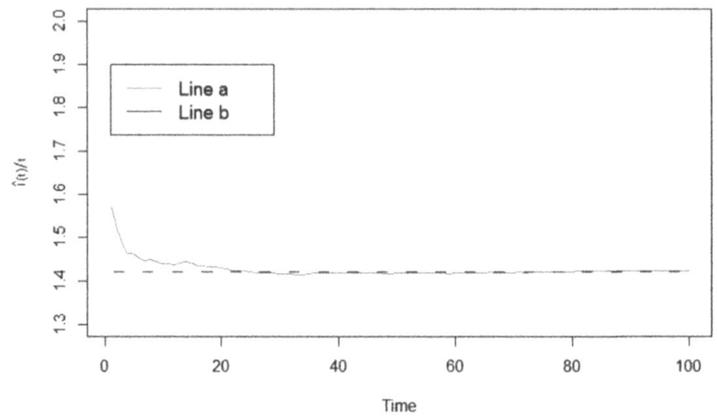

Fig. 6. Fraction time $\hat{I}(t)/t$ when servers free of class-1 customers: line a) corresponds to case 1 in Table 3; line b) corresponds to stability condition for ρ_2.

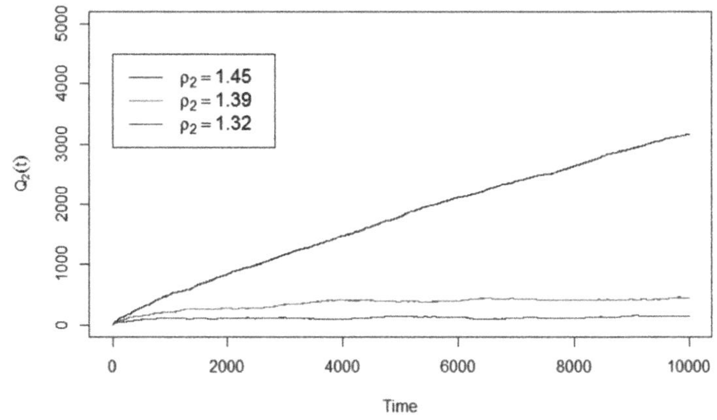

Fig. 7. Sample mean queue size of class-2 customers for Pareto service times, case 1 in Table 3. Stability condition (1) is $\rho_2 < 1.42$.

We also check stability condition by simulation queue size of class-2 customers for Pareto service times. We construct the sample mean queue size $Q_2(t)$ based on 300 paths queue size of class-2 customers for different values of ρ_2 (see Fig. 7). It is clear that, $Q_2(t)$ increases linearly to infinity, if (27) does not hold. However, when condition (27) is satisfied, we see that all paths are stable.

5 Conclusion

A modified Erlang loss system with two-priority classes of customers and multi-class customers and multi-class servers assumption of first priority customers is considered. The main aim of the work is to prove the stability condition of this system when sub-classes of class-1 customers have *different service rates.* In this system, the first priority customers (class-1) are lost if find all servers busy, while the second priority customers (class-2) form an infinite capacity queue. A new feature of this system is that sub-classes of class-1 customers are assigned on servers according to the assignment probabilities which must satisfy a preliminary condition. All customers follows Poisson inputs and have general class-dependent service times. We show how the product form of class-1 stationary probabilities can be used to obtain the stability condition of the system. Also results of discrete-event simulation to confirm theoretical results are given.

References

1. Morozov, E., Rogozin, S., Nguyen, H.Q., Phung-Duc, T.: Modified Erlang loss system for cognitive wireless networks. J. Math. Sci. (2020, submitted)
2. Rogozin, S.: Simulation a modified Erlang loss system with multi-type servers and multi-type customers. In: Proceedings of the Second International Workshop on Stochastic Modeling and Applied Research of Technology (2020, submitted)
3. Adan, I., Hurkens, C., Weiss, G.: A reversible Erlang loss system with multitype customers and multitype servers. Probab. Eng. Inf. Sci. **24**(4), 535–548 (2010)
4. Ostovar, A., Keshavarz, H., Quan, Z.: Cognitive radio networks for green wireless communications: an overview. Telecommun. Syst. **76**, 129–138 (2020). https://doi.org/10.1007/s11235-020-00703-8
5. Asmussen, S.: Applied Probability and Queues, 2nd edn. Springer, New York (2003). https://doi.org/10.1007/b97236
6. Morozov, E., Delgado, R.: Stability analysis of regenerative queues. Autom. Remote Control **70**, 1977–1991 (2009). https://doi.org/10.1134/S0005117909120066
7. Akutsu, K., Phung-Duc, T.: Analysis of retrial queues for cognitive wireless networks with sensing time of secondary users. In: Phung-Duc, T., Kasahara, S., Wittevrongel, S. (eds.) QTNA 2019. LNCS, vol. 11688, pp. 77–91. Springer, Cham (2019). https://doi.org/10.1007/978-3-030-27181-7_6
8. Down, D.G., Lewis, M.E.: Dynamic load balancing in parallel queueing systems: stability and optimal control. Eur. J. Oper. Res. **168**, 509–519 (2006)
9. Tsai, Y.C., Argon, N.T.: Dynamic server assignment policies for assembly-type queues with flexible servers. Naval Res. Logist. **55**(3), 234–251 (2008)
10. Tekin, E., Hopp, W.J., Van Oyen, M.P.: Pooling strategies for call center agent cross-training. IIE Trans. **41**(6), 546–561 (2009)
11. Ahghari, M., Balcioglu, B.: Benefits of cross-training in a skill-based routing contact center with priority queues and impatient customers. IIE Trans. **41**, 524–536 (2009)
12. Agnihothri, S.R., Mishra, A.K., Simmons, D.E.: Workforce cross-training decisions in field service systems with two job types. J. Oper. Res. Soc. **54**(4), 410–418 (2003)
13. Ahn, H.-S., Duenyas, I., Zhang, Q.R.: Optimal control of a flexible server. Adv. Appl. Probab. **36**, 139–170 (2004)
14. Stolyar, A.L., Tezcan, T.: Control of systems with flexible multi-server pools: a shadow routing approach. Queueing Syst. **66**, 1–51 (2010). https://doi.org/10.1007/s11134-010-9183-0

Methods for Analyzing Slicing Technology in 5G Wireless Network Described as Queueing System with Unlimited Buffer and Retrial Group

K. Yves Adou(✉) and Ekaterina V. Markova

Peoples' Friendship University of Russia (RUDN University),
6 Miklukho-Maklaya St., Moscow 117198, Russian Federation
{1042205051,markova-ev}@rudn.ru

Abstract. Fourth generation wireless network was very soon about to not be able to support or withstand the exponential growth of traffic in modern wireless telecommunication networks. In addition, the inflexibility of mobile networks until now is making the situation even more complicated. In response, the fifth generation (5G) wireless telecommunication network was developed, allowing flexible and quick deployment of applications and services to accommodate specific requirements of users in very diverse fields. The cornerstone of 5G is called slicing technology: a mechanism making possible the creation and configuration of multiple network services on top of the same infrastructure with help of its defined network slices. The international telecommunication union (ITU) has defined three categories of use which have been brought into conformity with the 3GPP standards. These categories are specified by three generic slices of 5G: enhanced mobile broadband (eMBB), ultra-reliable low latency communications (uRLLc) and massive machine type communications (mMTC). We consider queueing system with queue and its retrial group for modeling slicing technology in 5G wireless network.

Keywords: Queueing system · Retrial group · 5G network · Product solution · Performance measures

1 Introduction

In modern wireless telecommunication networks, the fifth generation (5G) wireless network with slicing technology [4,11,16] is being widely launched around the world. This technology was developed to address the problem of data traffic

This paper has been supported by the RUDN University Strategic Academic Leadership Program (recipient Adou Yv., mathematical model development). The reported study was funded by RFBR, project number 19-07-00933 (recipient Markova E.V., problem formulation). The reported study was funded by RFBR, project number 20-37-70079 (recipient Markova E.V., numerical analysis).

A. Dudin et al. (Eds.): ITMM 2020, CCIS 1391, pp. 264–278, 2021.
https://doi.org/10.1007/978-3-030-72247-0_20

volume exponential growth [5,7,9]. It permits a more effective utilization of the available radio resources, which are indispensable for data transfer in wireless networks. Slicing technology allows to represent the whole network infrastructure of mobile network operators (MNOs) in the form of various configurable logical networks called multi-service wireless network slices, each of which can be rented by virtual network operators (VNOs).

This paper considers the operation of a single VNO renting from one MNO a multi-service wireless network slice. We assume that VNO provides to its customers only services which generate best effort traffic with minimum guarantees (BG), characterized by the fact that a minimum bit rate is assigned for customers services. BG-services correspond to elastic or non-real time traffic such as file sharing, web browsing and social networking [1,10,18].

The methods applied in queueing theory are used to analyze the model, which is described by a queueing system (QS) with a buffer and its retrial group [2,6,8,12]. We assume that if there is free resource, the elastic session will be established upon arrival, otherwise it will await in the buffer, from where it can depart for buffer' retrial group [3,8,14,15,17] and return after a moment.

As example of physical interpretation, let us consider a web page reloading process by one user after awaiting a certain time this web page to load. In this case, user await in our model corresponds to elastic session await in buffer, and web page reloading process – to elastic session departure for buffer' retrial group and return to buffer after a moment.

Note that, model with limited storage capacities of buffer and its retrial group has already been investigated in [13], where a computed numerical solution of the equilibrium equations system was used to analyze the characteristics. However, for model to match reality, the storage capacities must be unlimited. Therefore, the purpose of this article is the study of a more complex case of that model with unlimited storage capacities of buffer and its retrial group.

2 Related Work: Single Server Queueing System with Retrial Queue

Before modeling and analyzing QS with buffer and its retrial group for describing 5G wireless network with slicing technology, we conduct analytic review of the methods used for computing solution in QSs with unlimited storage capacity. For this purpose, we consider the M/M/1/0 QS with unlimited retrial queue described in [3,8].

2.1 Mathematical Model

In this system, requests arrive according to the Poisson process with rate λ and service times are exponentially distributed with rate μ^{-1}. Storage capacity of retrial queue is unlimited.

The service of new incoming request begins immediately if the server is free, otherwise it is delayed and the request is redirected to the retrial queue, where it can await. Note that requests awaiting in retrial queue can retry indefinitely to occupy server after exponentially distributed time α^{-1}. Also, as well as the new incoming, they can leave the system after an unsuccessful attempt to occupy server with probability q. The corresponding scheme model is shown in Fig. 1.

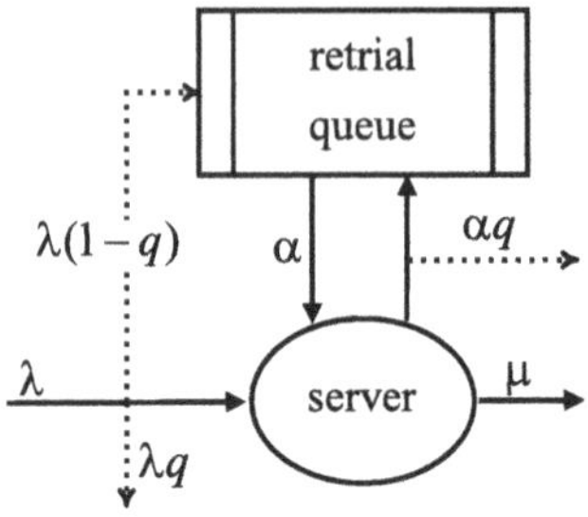

Fig. 1. Scheme model of the single server QS with retrial queue.

According to radio admission control scheme above described and since system' arrival process is Poisson distributed while service times are exponentially distributed, the system' behavior might be described using a two-dimensional vector (n, s) over state space $\mathbf{X} = \{(n, s) : n \geq 0, s = 0, 1\}$, where n represents the quantity of customers awaiting in retrial queue and s – the server' state (0 – is free, 1 – is occupied). The corresponding state transition and central state transition diagrams are shown respectively in Fig. 2 and Fig. 3.

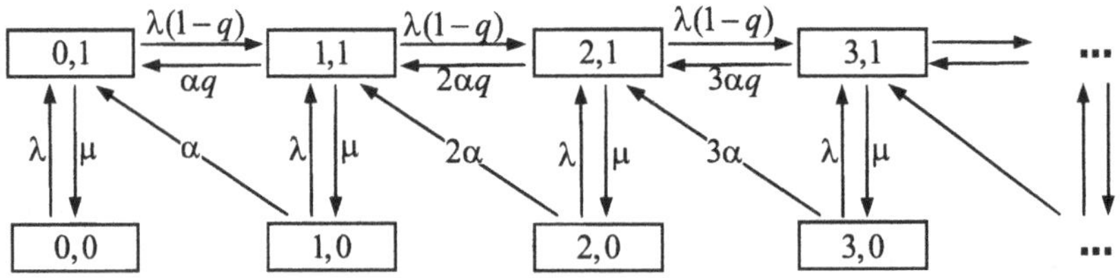

Fig. 2. State transition diagram of the single server QS with retrial queue.

The corresponding Markov process is described by the following equilibrium equations system according to central state transition diagram (Fig. 3):

$$\begin{gathered}(\lambda I\{s=0\} + \mu I\{s=1\} + (1-q)\lambda I\{s=1\} + n\alpha q I\{s=1, n>0\} \\ + n\alpha I\{s=0, n>0\})p(n,s) = \lambda I\{s=1\}p(n,s-1) + \mu I\{s=0\} \\ \times p(n,s+1) + (n+1)\alpha q I\{s=1\}p(n+1,s) + (1-q)\lambda I\{s=1, n>0\} \\ \times p(n-1,s) + (n+1)\alpha I\{s=1\}p(n+1,s-1),\end{gathered} \tag{1}$$

where $p(n, s), (n, s) \in \mathbf{X}$ represents the stationary probability distribution of requests quantity in system and I – the function indicator that equals 1, when condition is met and 0 otherwise.

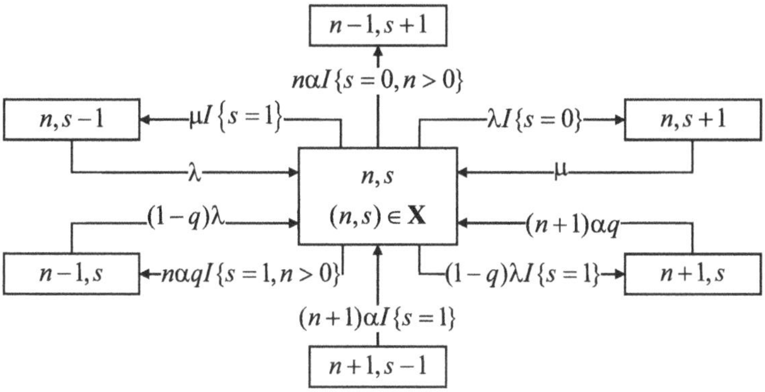

Fig. 3. Central state transition diagram of the single server QS with retrial queue.

2.2 Stationary Probability Distribution Computation

Since retrial queue has unlimited storage capacity, the probably only way for finding the stationary probability distribution $p(n, s), (n, s) \in \mathbf{X}$ would be through a generating function-based approach [3,8]. However, fixing or setting the retrial' queue storage capacity to a very huge value N_{rq}, one can also compute it either numerically or by using a product form solution obtained from equilibrium and local balance equations systems. For simplification, only the case of absolutely insistent requests is considered, i.e. requests can leave system only after getting serviced ($q = 0$).

Generating Function. The generating function-based solution is computed using formula:

$$\forall n \geq 0, p(n,s) = \begin{cases} \binom{n + 1/\xi - 1}{n} \rho^n (1-\rho)^{1+1/\xi}, & \text{if } s = 0, \\ \binom{n + 1/\xi}{n} \rho^{n+1} (1-\rho)^{1+1/\xi}, & \text{if } s = 1, \end{cases} \tag{2}$$

where $\rho = \lambda/\mu$ is the offered load of incoming requests and $\xi = \alpha/\lambda$ – the ratio coefficient of average waiting time to arrival rate of requests.

Numerical Solution. The process describing the system' behavior is not reversible Markov process. Therefore, the stationary probability distribution $p(n, s)_{(n,s)\in\mathbf{X}} = \mathbf{p}$ can be calculated using a numerical solution of the equilibrium' equations system $\mathbf{p} \cdot \mathbf{A} = 0, \mathbf{p} \cdot \mathbf{1}^T = \mathbf{1}$, where $\mathbf{A}$ is the infinitesimal generator of Markov process, elements $a((n, s)(n', s'))$ of which are defined as follows:

$$\begin{cases} \lambda, & \text{if } n' = n, s' = s{+}1, n = 0, ..., N_{rq}, s = 0, \\ & \text{or } n' = n{+}1, s' = s, n = 0, ..., N_{rq}{-}1, s = 1, \\ \mu, & \text{if } n' = n, s' = s{-}1, n = 0, ..., N_{rq}, s = 1, \\ n\alpha, & \text{if } n' = n{-}1, s' = s{+}1, n = 1, ..., N_{rq}, s = 0, \\ \psi, & \text{if } n' = n, s' = s, n = 0, ..., N_{rq}, s = 0, 1, \\ 0, & \text{otherwise,} \end{cases} \quad (3)$$

where
$\psi = -\left(\lambda I\{n \leq N_{rq}, s = 0 \| n < N_{rq}, s = 1\} + \mu I\{s = 1\} + n\alpha I\{n > 0, s = 0\}\right)$.

Product Form. The product form solution is calculated using formula:

$$\forall n \geq 0, p(n, s) = \begin{cases} p(0,0) \cdot u(n) \prod_{i=0}^{n} v(i), & \text{if } s = 0, \\ p(0,0) \cdot \prod_{i=0}^{n} v(i), & \text{if } s = 1, \end{cases} \quad (4)$$

with

$$p(0,0) = \left(\sum_{k=0}^{\infty} (u(k) + 1) \prod_{j=0}^{k} v(j) \right)^{-1},$$

$$u(i) = \frac{\mu}{\lambda + i\alpha}, i \geq 0,$$

$$v(0) = \frac{\lambda}{\mu},$$

$$v(i) = \frac{\lambda(\lambda + i\alpha)}{i\alpha\mu}, i > 0.$$

3 QS with Buffer and Retrial Group for Modeling Network Within Slicing Technology

Summarizing previous section, there are three methods for analyzing performance measures of systems with unlimited storage capacity:

- the generating function-based approach;
- the numerical solution of the equilibrium equations system;
- the product form solution.

Let us consider two of them in the modeling of the 5G wireless network with slicing technology as QS with buffer and its retrial group.

3.1 Mathematical Model

We assume that a multi-service wireless network slice has total network capacity R_{nc} MBps. Elastic sessions arrive in the system according to the Poisson process with rate λ and are parametrized by the exponentially distributed average file size θ MB and the minimum assignable bit rate b MBps. The average service time of established elastic sessions is exponentially distributed with rate $\mu^{-1} = \theta/R_{nc}$ seconds. Storage capacities of buffer and its retrial group are unlimited.

We assume that elastic session is immediately established upon arrival if there is free resource, i.e. quantity of simultaneously established elastic sessions is less than maximum $N_{res} = \lfloor R_{nc}/b \rfloor$, otherwise its establishment is delayed and it awaits free resource in buffer. Note that elastic sessions awaiting in buffer can depart for its retrial group after an exponentially distributed time β^{-1} and return after an exponentially distributed time α^{-1}. Also, each established elastic session ends successfully and retries are unlimited. The corresponding scheme model is shown in Fig. 4. Note that elastic sessions are characterized by a uniformly distributed channel rate R_{nc} between simultaneously established elastic sessions, thus service rate of single established elastic session is defined as μ/n, where $n \leq N_{res}$.

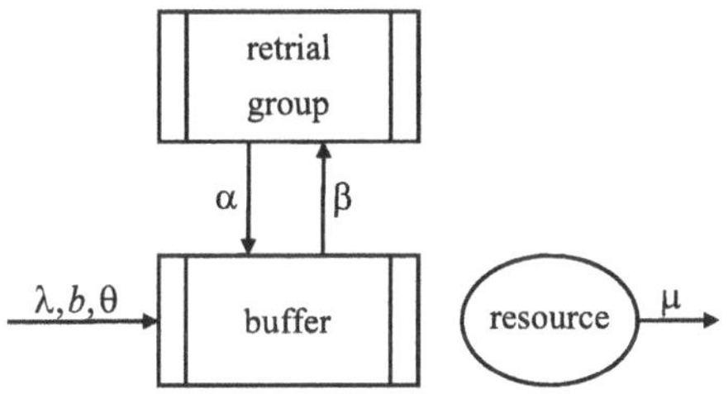

Fig. 4. Scheme model for unlimited storage capacities of buffer and its retrial group.

According to above described radio admission control scheme and taking into account that system' arrival process is Poisson distributed while average service time is exponentially distributed, the system' behavior could be described using a three-dimensional vector (n_1, n_2, s) over state space $\mathbf{Y} = \{(n_1, n_2, s) : n_1 \geq 0, n_2 \geq 0, s = 0, \cdots, N_{res}\}$, where n_1 represents the quantity of elastic sessions awaiting free resources in buffer, n_2 – the quantity of elastic sessions awaiting in buffer' retrial group and s – the quantity of simultaneously established elastic sessions. The corresponding state transition and central state transition diagrams are shown respectively in Fig. 5 and Fig. 6.

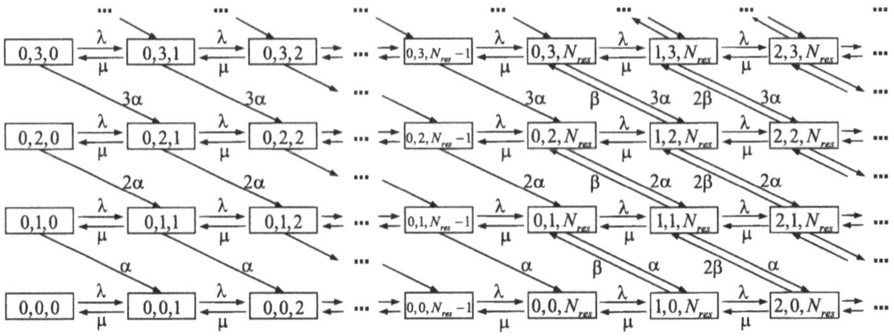

Fig. 5. State transition diagram of the model

The corresponding Markov process is described by the following equilibrium equations system according to central state transition diagram (Fig. 6):

$$\begin{aligned}
&(\lambda + \mu I\{s > 0\} + n_1\beta I\{n_1 > 0\} + n_2\alpha I\{n_2 > 0\})q(n_1, n_2, s) \\
&= \lambda I\{n_1 > 0\}q(n_1 - 1, n_2, s) + \mu I\{s = N_{res}\}q(n_1 + 1, n_2, s) \\
&+ \lambda I\{n_1 = 0, s > 0\}q(n_1, n_2, s - 1) + \mu I\{n_1 = 0, s < N_{res}\}q(n_1, n_2, s + 1) \\
&+ (n_2 + 1)\alpha I\{n_1 = 0, s > 0\}q(n_1, n_2 + 1, s - 1) + (n_2 + 1)\alpha I\{n_1 > 0\} \\
&\times q(n_1 - 1, n_2 + 1, s) + (n_1 + 1)\beta I\{n_2 > 0, s = N_{res}\}q(n_1 + 1, n_2 - 1, s),
\end{aligned} \quad (5)$$

where $q(n_1, n_2, s), (n_1, n_2, s) \in \mathbf{Y}$ represents the stationary probability distribution of elastic' sessions quantity in system and I – the function indicator that equals 1, when condition is met and 0 otherwise.

3.2 Stationary Probability Distribution Calculation

Similarly to M/M/1/0 QS model with retrial queue, since buffer and its retrial group have unlimited storage capacities, the probably only way for finding the stationary probability distribution $q(n_1, n_2, s), (n_1, n_2, s) \in \mathbf{Y}$ should be through a generating function-based approach [3,8]. However, fixing or setting the storage capacity of buffer to a very huge value N_q, one can also compute it either numerically or by using a product form solution obtained from equilibrium and local balance equations systems.

Note that, the storage capacity of buffer' retrial group, represented as N_{rg}, is always proportional to buffer' in this system (i.e. $N_{rg} = N_{res} + N_q$).

Numerical Solution. The process describing the system' behavior is not reversible Markov process. Therefore, the stationary probability distribution $q(n_1, n_2, s)_{(n_1,n_2,s)\in\mathbf{Y}} = \mathbf{q}$ can be calculated using a numerical solution [13] of the equilibrium' equations system $\mathbf{q} \cdot \mathbf{A} = 0, \mathbf{q} \cdot \mathbf{1}^T = \mathbf{1}$, where $\mathbf{A}$ is the infinitesimal generator of Markov process, elements $a((n_1, n_2, s)(n_1', n_2', s'))$ of which are defined as follows:

$$\begin{cases} \lambda, & \text{if } n_1' = n_1, n_2' = n_2, s' = s{+}1, n_1 = 0, n_2 = 0, ..., N_{rg}, s = 0, ..., N_{res}{-}1, \\ & \text{or } n_1' = n_1{+}1, n_2' = n_2, s' = s, n_1 = 0, ..., N_q{-}1, n_2 = 0, ..., N_{rg}, s = N_{res}, \\ \mu, & \text{if } n_1' = n_1, n_2' = n_2, s' = s{-}1, n_1 = 0, n_2 = 0, ..., N_{rg}, s = 1, ..., N_{res}, \\ & \text{or } n_1' = n_1{-}1, n_2' = n_2, s' = s, n_1 = 1, ..., N_q, n_2 = 0, ..., N_{rg}, s = N_{res}, \\ n_2\alpha, & \text{if } n_1' = n_1, n_2' = n_2{-}1, s' = s{+}1, n_1 = 0, n_2 = 1, ..., N_{rg}, s = 0, ..., N_{res}{-}1, \\ & \text{or } n_1' = n_1{+}1, n_2' = n_2{-}1, s' = s, n_1 = 0, ..., N_q{-}1, n_2 = 1, ..., N_{rg}, s = N_{res}, \\ n_1\beta, & \text{if } n_1' = n_1{-}1, n_2' = n_2{+}1, s' = s, n_1 = 1, ..., N_q, n_2 = 0, ..., N_{rg}{-}1, s = N_{res}, \\ \psi, & \text{if } n_1' = n_1, n_2' = n_2, s' = s, n_1 = 0, ..., N_q, n_2 = 0, ..., N_{rg}, s = 0, ..., N_{res}, \\ 0, & \text{otherwise}, \end{cases} \tag{6}$$

where
$\psi = -(\lambda I\{n_1 = 0, n_2 \leq N_{rg}, s < N_{res} \| n_1 < N_q{-}1, n_2 \leq N_{rg}, s = N_{res}\} + \mu I\{n_1 = 0, n_2 \leq N_{rg}, s > 0 \| n_1 > 0, n_2 \leq N_{rg}, s = N_{res}\} + n_2\alpha I\{n_1 = 0, n_2 > 0, s \leq N_{res}{-}1 \| n_1 \leq N_q{-}1, n_2 > 0, s = N_{res}\} + n_1\beta I\{n_1 > 0, n_2 \leq N_{rg}{-}1, s = N_{res}\})$.

Product Form. The product form solution is calculated using formula:

$$q(n_1, n_2, s) = \begin{cases} q(0,0,0) \cdot c(n_1, s), & \text{if } n_1 \geq 0, n_2 = 0, s = 0, ..., N_{res}, \\ q(0,0,0) \cdot d(n_1, n_2), & \text{if } n_1 \geq 0, n_2 \geq 0, s = N_{res}, \\ q(0,0,0) \cdot b(n_2, s), & \text{if } n_1 = 0, n_2 \geq 0, s = 0, ..., N_{res} - 1, \end{cases} \tag{7}$$

with

$$q(0,0,0) = \left(\frac{1}{1-\rho} + \sum_{k=0}^{N_{res}-1} \sum_{j=1}^{N_{rg}} b(j,k) + \sum_{i=1}^{\infty} \sum_{j=1}^{N_{res}+i} d(i,j) \right)^{-1};$$

$$c(i,j) = \rho^{i+j}, i \geq 0, j \geq 0;$$

$$d(i,j) = \delta^j \frac{(i+j)!}{i!\,j!} \rho^{N_{res}+i+j}, i \geq 0, j \geq 0;$$

$$b(i) = \delta^i \left(\frac{\mu}{\lambda + i\alpha} \right)^{N_{res}-j} \rho^{N_{res}+i},$$

where $\rho = \lambda/\mu$ represents the average offered load of incoming elastic sessions and $\delta = \beta/\alpha$ – the ratio coefficient of elastic' sessions average time spent in buffer to average time spent in buffer' retrial group.

4 Numerical Analysis

Having computed the stationary probability distribution in each system (i.e., Sects. 2 and 3), one can compute their main performance measures.

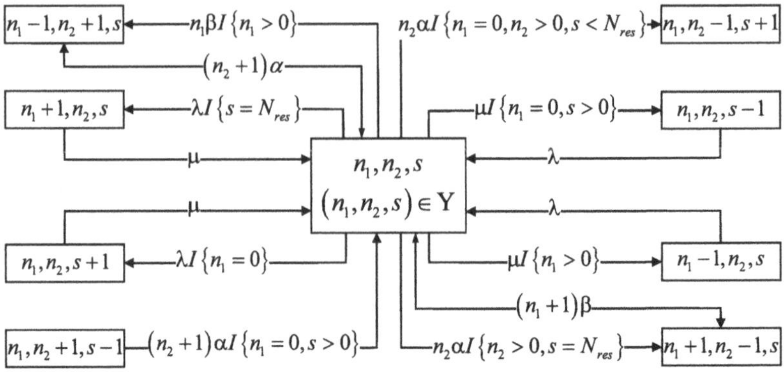

Fig. 6. Central state transition diagram of the model

To carry out numerical analysis of the QS model with buffer and its retrial group under slicing technology, we estimate the value that can represent the limit of system' storage capacity as it approaches infinity. For this purpose, we first analyze the M/M/1/0 QS with retrial queue.

4.1 Model with Server and Retrial Queue

For the single server QS model with retrial queue, the main performance metrics are the following:

– the mean number $\overline{N}$ of requests in retrial group

$$\overline{N} = \lim_{n\to\infty} \sum_{k=0}^{n} k\,(p(k,0)+p(k,1)); \tag{8}$$

– the immediate service probability Pr of incoming request in system

$$\Pr = \lim_{n\to\infty} \sum_{k=0}^{n} p(k,0). \tag{9}$$

In this system, these two measures can also be computed using formulas obtained introducing the generating function-based solution (2) in equations (8) and (9). Thus, we obtain the following expressions:

$$\overline{N} = \frac{\rho(1+\xi\rho)}{\xi(1-\rho)}; \tag{10}$$

$$\Pr = 1-\rho. \tag{11}$$

Numerical analysis was conducted to evaluate the convergence' level of computed generating function-based (2), numerical (3) and product form (4) solutions depending on retrial' queue storage capacity. We focus only on one of performance metrics – the incoming' requests immediate service probability, for which we measure the elapsed computation time through each of all three methods.

As illustrated in Fig. 7 by the plots, the results obtained using all three calculation methods tend to coincide when increasing retrial' queue storage capacity, i.e. to value 10^4. Note that, calculation through generating function-based and product form solutions generally takes less time than numerical solution as shown in Table 1 by elapsed computation time measurement.

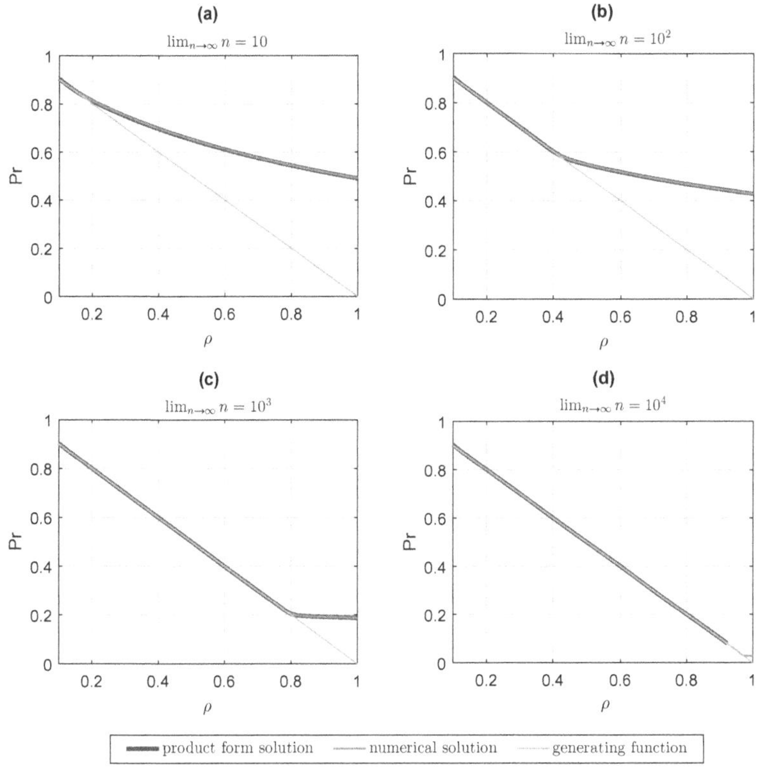

Fig. 7. Immediate service probability Pr depending on average offered load ρ of incoming requests for $\mu^{-1} = 3\text{s}$, $\alpha^{-1} = 0.01\text{s}$.

4.2 Model with Buffer and Retrial Group

For the model with buffer and its retrial group, the main performance measures are the following:

Table 1. Elapsed computation time t in seconds for incoming' requests immediate service probability calculation, $\mu^{-1} = 3$s, $\rho = 0.5$, α^{-1} in seconds.

Methods	α^{-1}	$\lim_{n\to\infty} n = 10$	$\lim_{n\to\infty} n = 10^2$	$\lim_{n\to\infty} n = 10^3$	$\lim_{n\to\infty} n = 10^4$
Generating function	0.01	0.001311	0.001256	0.001233	0.001591
	0.02	0.001208	0.001240	0.001158	0.003816
	0.03	0.001071	0.001428	0.001173	0.004740
	0.04	0.001314	0.001090	0.001251	0.003853
	0.05	0.001608	0.001187	0.001442	0.003484
Numerical solution	0.01	0.012892	0.022449	0.177820	100.837675
	0.02	0.010738	0.012267	0.177508	111.837707
	0.03	0.012182	0.015339	0.202122	120.135441
	0.04	0.010341	0.011975	0.212048	110.312899
	0.05	0.013208	0.012275	0.204793	101.917594
Product form	0.01	0.006880	0.009252	0.013336	5.045567
	0.02	0.008093	0.007329	0.015586	5.487846
	0.03	0.007528	0.010168	0.018457	5.868070
	0.04	0.007512	0.007037	0.015007	5.699569
	0.05	0.006776	0.007144	0.014051	5.941449

- the immediate establishment probability Im_1 of incoming elastic session

$$\mathrm{Im}_1 = \lim_{n_1\to\infty} \sum_{s=0}^{N_{res}-1} \sum_{n_2=0}^{N_{res}+n_1} q(0, n_2, s); \tag{12}$$

- the immediate establishment probability Im_2 of retrial elastic session

$$\mathrm{Im}_2 = \mathrm{Im}_1 - \sum_{s=0}^{N_{res}-1} q(0, 0, s); \tag{13}$$

- the emptiness probability E_1 of buffer

$$\mathrm{E}_1 = \mathrm{Im}_1 + \lim_{n_1\to\infty} \sum_{n_2=0}^{N_{res}+n_1} q(0, n_2, N_{res}); \tag{14}$$

- the emptiness probability E_2 of buffer' retrial group

$$\mathrm{E}_2 = \sum_{s=0}^{N_{res}-1} q(0, 0, s) + \lim_{n_1\to\infty} \sum_{k=0}^{n_1} q(k, 0, N_{res}). \tag{15}$$

Here, numerical analysis is performed to determine the accuracy of the obtained product form (7) compared to numerical solution (6). Let us focus only on one of the performance measures – the immediate establishment probability of incoming elastic session (12). The absolute error Δ is calculated for this metric. Note that according to previous Subsect. 4.1 with very simple restriction on buffer' storage capacity and provided that the average offered load is small, the results of characteristics' computation correspond to reality, i.e. when storage capacity of buffer is unlimited. But since buffer' storage capacity is limited, even with such average offered load there will be a small error when calculating the characteristics using the different methods. The error' estimations are presented in Tables 2 and 3.

Table 2. Absolute error Δ for incoming' elastic session immediate establishment probability calculation, $R_{nc} = 20\,\text{MBps}$, $\theta = 8\,\text{MB}$, $\alpha^{-1} = 5\,\text{s}$, $\beta^{-1} = 3\,\text{s}$.

–		$b = 2\,\text{MBps}$	$b = 5\,\text{MBps}$	$b = 8\,\text{MBps}$	$b = 11\,\text{MBps}$
$\lim\limits_{n_1 \to \infty} n_1 = 10$	$\rho = 0.1$	803.03542e−12	802.91113e−06	0.07872394731	0.61033994564
		0%	0.000008%	0.08%	0.6%
	$\rho = 0.15$	756.67250e−10	663.75605e−05	0.28360774741	1.39327955034
		0%	0.00007%	0.3%	1.4%
	$\rho = 0.2$	197.44599e−08	0.03077263160	0.72026980730	2.52439464906
		0%	0.03%	0.7%	2.5%
$\lim\limits_{n_1 \to \infty} n_1 = 10^2$	$\rho = 0.1$	803.04652e−12	802.91113e−06	0.07872394731	0.61033994566
		0%	0.000008%	0.079%	0.6%
	$\rho = 0.15$	756.67184e−10	663.75605e−05	0.28360774758	1.39327955588
		0%	0.00007%	0.3%	1.4%
	$\rho = 0.2$	197.44605e−08	0.03077263172	0.72026982842	2.52439498744
		0%	0.03%	0.7%	2.5%

To analyze the behavior of system' remaining characteristics (i.e. immediate establishment probability of retrial elastic session and emptiness probabilities of buffer and its retrial group), we use method not depending on buffer' storage capacity, namely the numerical solution (6). The analysis is focused on average' offered load of incoming elastic sessions dependence. Note that, the use of numerical solution method permits to obtain accurate computation results of investigated characteristics only for small values of incoming' elastic sessions average offered load. However, the graphs provided below are illustrative of the general behavior of the characteristics for the various system' parameters.

As shown by curves in Fig. 8(a), the incoming' elastic sessions immediate establishment probability decreases as their average offered load increases. This is indeed explained by the augmentation of elastic' sessions quantity in buffer with increase of the average' offered load value. Thus, since the elastic sessions awaiting in buffer depart for its retrial group after some exponentially distributed time, it is likely that the buffer might be empty for some time at a certain point. This explains the increase of retrial' elastic sessions immediate establishment probability up to a certain maximum value and its sudden decrease in Fig. 8(b).

Table 3. Absolute error Δ for incoming' elastic session immediate establishment probability calculation, $\theta = 8\,\text{MB}$, $\rho = 0.2$, $b = 5\,\text{MBps}$, $\lim_{n_1 \to \infty} n_1 = 100$.

–		$\beta^{-1} = 1\text{s}$	$\beta^{-1} = 3\text{s}$	$\beta^{-1} = 5\text{s}$
$R_{nc} = 10\text{MBps}$	$\alpha^{-1} = 1\text{s}$	12.980479045e−03 0.13%	98.119119090e−03 0.98%	962.35722294e−03 0.96%
	$\alpha^{-1} = 3\text{s}$	3.7791409654e−03 0.37%	13.207337838e−03 0.13%	27.242831947e−03 0.27%
	$\alpha^{-1} = 5\text{s}$	2.1744519479e−03 0.2%	7.0868379176e−03 0.7%	13.094733275e−03 0.13%
$R_{nc} = 20\,\text{MBps}$	$\alpha^{-1} = 1\text{s}$	574.37975167e−06 0.00057%	4.7753074010e−03 0.47%	998.52931050e−03 0.99%
	$\alpha^{-1} = 3\text{s}$	164.92051128e−06 0.00016%	576.79645625e−06 0.00057%	1.1988472551e−03 0.12%
	$\alpha^{-1} = 5\text{s}$	94.715729012e−06 0.0009%	307.72631725e−06 0.0003%	568.47476749e−06 0.00057%
$R_{nc} = 30\text{MBps}$	$\alpha^{-1} = 1\text{s}$	23.099966729e−06 0.00023%	194.48883387e−06 0.0002%	999.94060287e−03 0.99%
	$\alpha^{-1} = 3\text{s}$	6.6810457051e−06 0.00067%	23.357388823e−06 0.00023%	48.485771984e−06 0.00048%
	$\alpha^{-1} = 5\text{s}$	3.8664118408e−06 0.00038%	12.530227719e−06 0.00012%	23.091889055e−06 0.00023%

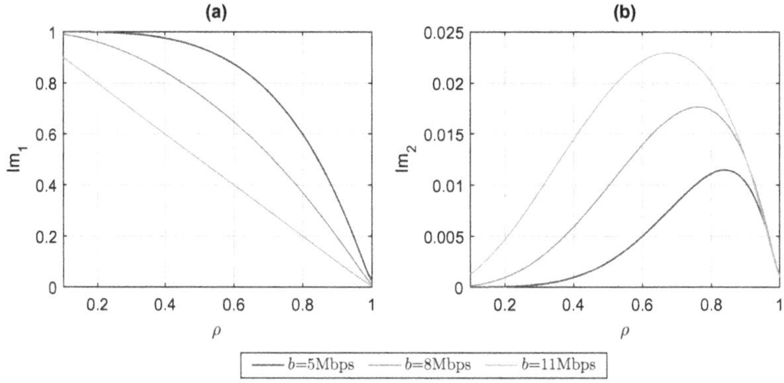

Fig. 8. Immediate establishment probabilities depending on average offered load ρ of incoming elastic sessions for $R_{nc} = 20\text{MBps}$, $\theta = 8\text{MB}$, $\alpha^{-1} = 5\text{s}$, $\beta^{-1} = 3\text{s}$, $\lim_{n_1 \to \infty} n_1 = 100$.

From resulting plots in Fig. 9, one can see that the emptiness probabilities of buffer and its retrial group decreases as the average offered load of incoming elastic sessions tends towards value 1. This is also explained by the fact that,

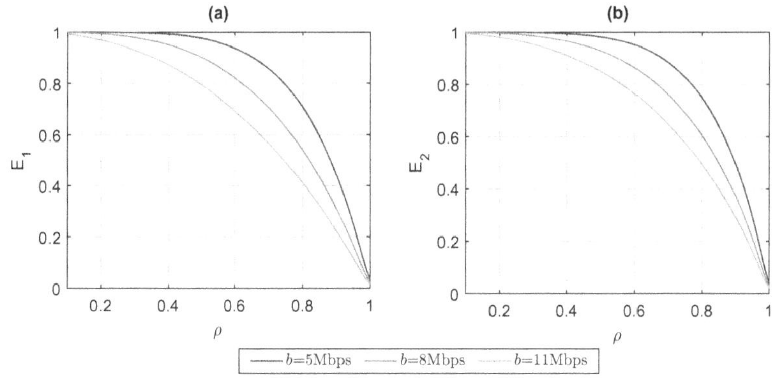

Fig. 9. Emptiness probabilities depending on average offered load ρ of incoming elastic sessions for R_{nc} = 20MBps, θ = 8MB, α^{-1} = 5s, β^{-1} = 3s, $\lim_{n_1\to\infty} n_1 = 100$.

when increasing their average offered load, the incoming' elastic sessions quantity in buffer augments. Thus, the elastic sessions spend more time in the buffer and its retrial group, i.e. buffer and its retrial group are likely to extremely rarely, if not, never be empty.

5 Conclusion

We analyzed the single server queueing system (QS) with retrial queue. Product form solution and numerical solution of the equilibrium equations system, as well as generating function-based approach solution, for computing stationary probability distribution were obtained. A comparative analysis of incoming' elastic requests immediate service probability using all three calculations methods was given. As result, all three methods coincide. Measurement of the elapsed computation time through each of the methods was provided, concluding that computation through generating function-based and product form solutions usually takes less time than numerical solution method.

We continued study on QS model with buffer and its retrial group for modeling network within slicing technology. Product form and numerical solutions for calculating stationary probability distribution were obtained. A comparative analysis of incoming' elastic sessions immediate establishment probability computation using both solutions was given. As result, both product form and numerical solution coincide with absolute error value equaling 0%. An analysis of the remaining performance measures using only numerical solution was given. Generating function-based approach will be implemented in further investigations, and the operation of multiple VNOs renting from one MNO a multi-service wireless network slice will be considered as last stage of this research subject.

References

1. 3GPP: 5G System Session Management Policy Control Service Stage 3. White paper, 3GPP, September 2020
2. Artalejo, J.R.: A classified bibliography of research on retrial queues: progress in 1990–1999. Top **7**(2), 187–211 (1999). https://doi.org/10.1007/BF02564721
3. Bocharov, P.P., D'Apice, C., Pechinkin, A.V.: Queueing Theory. De Gruyter, January 2003. https://doi.org/10.1515/9783110936025
4. China Mobile: Categories and Services Levels of Network Slice. White paper, Huawei and Industry Partners, March 2020
5. Cisco Systems: Cisco Annual Internet Report (2018–2023). White paper, Cisco Systems, Inc. (2020)
6. Dudin, A., Klimenok, V.: A retrial BMAP/SM/1 system with linear repeated requests. Queueing Syst. **34**(1), 47–66 (2000). https://doi.org/10.1023/A:1019196701000
7. Ericsson: Ericsson mobility report. White paper, Telephone Stock Company of LM Ericsson, November 2020
8. Falin, G.I., Templeton, J.G.C.: Retrial Queues, Monographs on Statistics and Applied Probability, vol. 75. Chapman & Hall/CRC, Boca Raton (1997)
9. GSMA: The state of mobile internet connectivity report 2020. White paper, GSM Association, September 2020
10. He, S., Qiu, Y., Xu, J.: Invalid-resource-aware spectrum assignment for advanced-reservation traffic in elastic optical network. Sensors **20**(15), 4190 (2020). https://doi.org/10.3390/s20154190
11. IBM: Telecom's 5G future. White paper, International Business Machines Corporation, February 2020
12. Kim, C.S., Klimenok, V., Lee, S.C., Dudin, A.: The retrial queueing system operating in random environment. J. Stat. Plan. Inference **137**(12), 3904–3916 (2007). https://doi.org/10.1016/j.jspi.2007.04.009
13. Markova, E., Adou, Y., Ivanova, D., Golskaia, A., Samouylov, K.: Queue with retrial group for modeling best effort traffic with minimum bit rate guarantee transmission under network slicing. In: Vishnevskiy, V.M., Samouylov, K.E., Kozyrev, D.V. (eds.) DCCN 2019. LNCS, vol. 11965, pp. 432–442. Springer, Cham (2019). https://doi.org/10.1007/978-3-030-36614-8_33
14. Moiseev, A., Nazarov, A., Paul, S.: Asymptotic diffusion analysis of multi-server retrial queue with hyper-exponential service. Mathematics **8**(4), 531 (2020). https://doi.org/10.3390/math8040531
15. Nazarov, A., Moiseev, A., Phung-Duc, T., Paul, S.: Diffusion limit of multi-server retrial queue with setup time. Mathematics **8**(12) (2020). https://doi.org/10.3390/math8122232
16. Olimid, R., Nencioni, G.: 5G network slicing: a security overview. IEEE Access (2020). https://doi.org/10.1109/ACCESS.2020.2997702
17. Vishnevsky, V., Dudin, A., Klimenok, V.: Stochastic systems with correlated flows. Theory and application in telecommunication networks. Technosphere, Moscow, MSK (2018)
18. Vlaskina, A., Polyakov, N., Gudkova, I.: Modeling and performance analysis of elastic traffic with minimum rate guarantee transmission under network slicing. In: Galinina, O., Andreev, S., Balandin, S., Koucheryavy, Y. (eds.) NEW2AN/ruSMART -2019. LNCS, vol. 11660, pp. 621–634. Springer, Cham (2019). https://doi.org/10.1007/978-3-030-30859-9_54

The Two-Dimensional Output Process of Retrial Queue with Two-Way Communication

Alexey Blaginin and Ivan Lapatin(✉)

Tomsk State University, Lenina pr.36, 634050 Tomsk, Russia
rector@tsu.ru
http://www.tsu.ru

Abstract. In this paper, we review a two-dimensional output process of the system with repeated calls and called applications. In a system with repeated calls, incoming applications, which found serving unit busy, move to the source of repeated calls, where carry out random exponentially distributed delay, after which try to receive serving again. While serving unit is free, it can call applications itself with exponentially distributed intensity, which will serve with their serving time parameter. This feature characterises a system as one with called applications. An asymptotic approximation of the two-dimensional characteristic function is obtained under the condition of a large delay of applications in the orbit. Using integral transformations, the asymptotic two-dimensional distribution of the probabilities of the number of applications of different types that have finished serving in the system is found. A numerical analysis of the values of the correlation coefficient of the components of the considered two-dimensional output is carried out.

Keywords: Output process · Retrial queue · Two-way communication · Asymptotic analysis method · Correlation coefficient

1 Introduction

In this paper we review two-dimensional output process of the queueing system [6,8] with repeated calls [1] and called applications [7]. Such system can be interpreted as a processing node with multiple random access, which in spare time of processing requests can request self-diagnosis or any other procedure, which will continue during the random time. Also considered system can be applied for modelling processing nodes with different types of applications. Applications of one type are not lost and will be served in any case, while applications of another type will be served only with a free resource.

The individual nodes form a communication network model in which the outgoing flow from one node is incoming to another. In the case of applications of different types, after service at a certain node, they leave along their routes. Therefore, the results of the study of the output processes of queuing systems

A. Dudin et al. (Eds.): ITMM 2020, CCIS 1391, pp. 279–290, 2021.
https://doi.org/10.1007/978-3-030-72247-0_21

are widely applicable for the design of real data transmission systems and the analysis of complex processes consisting of several stages. In this regard, for modelling networks, it is important to have information about the presence of a correlation between processes in it. A weak correlation makes it possible to consider processes as independent ones in modelling, which can significantly simplify the model and its research.

In this paper, we consider the influence of the system parameters on the values of the asymptotic correlation coefficient of the components of the two-dimensional output process of different types of applications. To study the system, the method of asymptotic analysis is used to find the form of the limiting two-dimensional distribution of the number of served applications of the input process and the number of served called applications for some time t, provided that there is a large delay of applications in the orbit [10].

2 Mathematical Model

Let's consider the RQ-system, the input of which is supplied with the Poisson process with the intensity of λ. Input stream application, entering the system and finding the device free, takes him, the device, in turn, begins serving for some random time, distributed exponentially with parameter μ_1. If upon entering the system, the application finds the device busy, it instantly goes to the orbit, where carries out a random delay during an exponentially distributed time with parameter σ. In its free time from serving applications from the input process, the device itself calls applications with intensity α and serves them for exponentially distributed time with parameter μ_2.

Let us denote following notations: *i(t)* – the number of application in the system at the moment t, *k(t)* – state of the device: *0* – the device is free, *1* – the device is busy, *2* – the device is busy serving retrial application; $m_1(t)$ – the number of served applications from input process at the moment t, $m_2(t)$ – the number of served called applications at the moment t (Fig. 1).

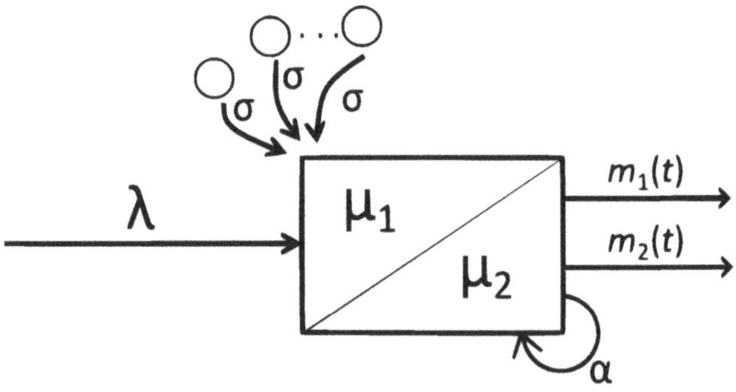

Fig. 1. System model

3 Kolmogorov Equations

Let us consider four-dimensional Markov process

$$\{k(t), i(t), m_1(t), m_2(t)\}$$

Let us set up for probabilities $P\{k(t) = k, i(t) = i, m_1(t) = m_1, m_2(t) = m_2\}$ a system of Kolmogorov differential equations

$$\begin{aligned}
\frac{\partial P_0(i,m_1,m_2,t)}{\partial t} &= -(\lambda + i\sigma + \alpha)P_0(i,m_1,m_2,t) + P_1(i,m_1-1,m_2,t)\mu_1 \\
&+ P_2(i,m_1,m_2-1,t)\mu_2, \\
\frac{\partial P_1(i,m_1,m_2,t)}{\partial t} &= -(\lambda + \mu_1)P_1(i,m_1,m_2,t) + (i+1)\sigma P_0(i+1,m_1,m_2,t) \\
&+ \lambda P_0(i,m_1,m_2,t), \\
\frac{\partial P_2(i,m_1,m_2,t)}{\partial t} &= -(\lambda + \mu_2)P_2(i,m_1,m_2,t) + \lambda P_2(i-1,m_1,m_2,t) \\
&+ \alpha P_0(i,m_1,m_2,t).
\end{aligned} \tag{1}$$

It is not possible to solve provided equations analytically since it is a system of an infinite number of differential finite-difference equations with variable coefficients. In order to pass to a finite number of equations, we introduce the partial characteristic functions, denoting $j = \sqrt{-1}$,

$$H_k(u,u_1,u_2,t) = \sum_{i=0}^{\infty}\sum_{m_1=0}^{\infty}\sum_{m_2=0}^{\infty} e^{jui}e^{ju_1m_1}e^{ju_2m_2}P_k(i,m_1,m_2,t).$$

Then the system (1) takes following form

$$\begin{aligned}
\frac{\partial H_0(u,u_1,u_2,t)}{\partial t} &= -(\lambda+\alpha)H_0(u,u_1,u_2,t) + j\sigma\frac{\partial H_0(u,u_1,u_2,t)}{\partial u} \\
&+ \mu_1 e^{ju_1}H_1(u,u_1,u_2,t) + \mu_2 e^{ju_2}H_2(u,u_1,u_2,t), \\
\frac{\partial H_1(u,u_1,u_2,t)}{\partial t} &= -(\lambda+\mu_1)H_1(u,u_1,u_2,t) - j\sigma e^{-ju}\frac{\partial H_0(u,u_1,u_2,t)}{\partial u} \\
&+ \lambda H_0(u,u_1,u_2,t) + \lambda e^{ju}H_1(u,u_1,u_2,t), \\
\frac{\partial H_2(u,u_1,u_2,t)}{\partial t} &= -(\lambda+\mu_2)H_2(u,u_1,u_2,t) + \lambda e^{ju}H_2(u,u_1,u_2,t) \\
&+ \alpha H_0(u,u_1,u_2,t).
\end{aligned} \tag{2}$$

4 Method of Asymptotic Analysis

The resulting system of differential equations in partial derivatives (2) will be solved by the method of asymptotic analysis in the limit condition of a large delay of applications in the orbit ($\sigma \to 0$).

Let us denote $\epsilon = \sigma, u = \epsilon w, F_k(w, u_1, u_2, t, \epsilon) = H_k(u, u_1, u_2, t)$, then the system will be written as

$$\begin{aligned}
\frac{\partial F_0(w,u_1,u_2,t,\epsilon)}{\partial t} &= -(\lambda+\alpha)F_0(w,u_1,u_2,t,\epsilon) + j\frac{\partial F_0(w,u_1,u_2,t,\epsilon)}{\partial w} \\
&+ \mu_1 e^{ju_1}F_1(w,u_1,u_2,t,\epsilon) + \mu_2 e^{ju_2}F_2(u,u_1,u_2,t,\epsilon), \\
\frac{\partial F_1(w,u_1,u_2,t,\epsilon)}{\partial t} &= -(\lambda+\mu_1)F_1(w,u_1,u_2,t,\epsilon) - je^{-j\epsilon w}\frac{\partial F_0(w,u_1,u_2,t,\epsilon)}{\partial w} \\
&+ \lambda F_0(w,u_1,u_2,t,\epsilon) + \lambda e^{j\epsilon w}F_1(w,u_1,u_2,t,\epsilon), \\
\frac{\partial F_2(w,u_1,u_2,t,\epsilon)}{\partial t} &= -(\lambda+\mu_2)F_2(w,u_1,u_2,t,\epsilon) + \lambda e^{j\epsilon w}F_2(w,u_1,u_2,t,\epsilon) \\
&+ \alpha F_0(w,u_1,u_2,t,\epsilon).
\end{aligned} \tag{3}$$

Let us note that, using the consistency condition for multidimensional distributions, the characteristic function of the processes $m_1(t)$ and $m_1(t)$ is expressed as follows with the introduced functions

$$M\{\exp(ju_1m_1(t))\exp(ju_2m_2(t))\} = \sum_{k=0}^{2} H_k(0,u_1,u_2,t) = \sum_{k=0}^{2} F_k(0,u_1,u_2,t,\epsilon).$$

Theorem 1. *The asymptotic approximation of the two-dimensional characteristic function of the number of served applications of the input process and the number of served called applications for some time t has the form*

$$\begin{aligned}
\boldsymbol{F}(u_1,u_2,t) &= \lim_{\sigma\to 0} M\{\exp(ju_1m_1(t))\exp(ju_2m_2(t))\} = \\
&= \lim_{\epsilon\to 0}\sum_{k=0}^{2} F_k(0,u_1,u_2,t,\epsilon) = \boldsymbol{R}\cdot\exp\{G(u_1,u_2)t\}\cdot\boldsymbol{E}
\end{aligned}$$

where

$$\boldsymbol{G}(u_1,u_2) = \begin{bmatrix} -(\lambda+\alpha+\kappa) & \mu_1 e^{ju_1} & \mu_2 e^{ju_2} \\ \kappa+\lambda & -\mu_1 & 0 \\ \alpha & 0 & -\mu_2 \end{bmatrix}^T,$$

row vector $\boldsymbol{R} = \{R_0, R_1, R_2\}$ *is the stationary probability distribution of the state of the device*

$$\boldsymbol{R} = \{\frac{\mu_2(\mu_1-\lambda)}{\mu_1(\mu_2-\alpha)}, \frac{\lambda}{\mu_1}, \frac{\alpha(\mu_1-\lambda)}{\mu_1(\mu_2+\alpha)}\},$$

κ *is the normalized average number of applications in the orbit*

$$\kappa = \frac{\lambda(\lambda\mu_2+\alpha\mu_1)}{\mu_2(\mu_1-\lambda)},$$

and $\boldsymbol{E}$ *is a unit column vector of the corresponding dimension.*

Proof. Taking the limit value $\lim_{\epsilon \to 0} F_k(w, u_1, u_2, t, \epsilon) = F_k(w, u_1, u_2, t)$ in the resulting system (3) , the system of equations will be written as

$$\begin{aligned} \frac{\partial F_0(w,u_1,u_2,t)}{\partial t} &= -(\lambda+\alpha)F_0(w,u_1,u_2,t) + j\frac{\partial F_0(w,u_1,u_2,t)}{\partial w} \\ &+ \mu_1 e^{ju_1}F_1(w,u_1,u_2,t) + \mu_2 e^{ju_2}F_2(w,u_1,u_2,t), \\ \frac{\partial F_1(w,u_1,u_2,t)}{\partial t} &= -(\lambda+\mu_1)F_1(w,u_1,u_2,t) - j\frac{\partial F_0(w,u_1,u_2,t)}{\partial w} \\ &+ \lambda F_0(w,u_1,u_2,t) + \lambda F_1(w,u_1,u_2,t), \\ \frac{\partial F_2(w,u_1,u_2,t)}{\partial t} &= -(\lambda+\mu_2)F_2(w,u_1,u_2,t) + \lambda F_2(w,u_1,u_2,t) \\ &+ \alpha F_0(w,u_1,u_2,t). \end{aligned} \tag{4}$$

The solution to system (4) will be sought in the following form

$$F_k(w,u_1,u_2,t) = \Phi(w)F_k(u_1,u_2,t). \tag{5}$$

$\Phi(w)$ is an asymptotic approximation of the characteristic function of the number of applications in the orbit under the condition of a large delay.

Substituting (5) into the system (4) and dividing both sides of the equations by $\Phi(w)$ we obtain

$$\begin{aligned} \frac{\partial F_0(u_1,u_2,t)}{\partial t} &= -(\lambda+\alpha)F_0(u_1,u_2,t) + j\frac{\Phi'(w)}{\Phi(w)}F_0(u_1,u_2,t) \\ &+ \mu_1 e^{ju_1}F_1(u_1,u_2,t) + \mu_2 e^{ju_2}F_2(u_1,u_2,t), \\ \frac{\partial F_1(u_1,u_2,t)}{\partial t} &= -(\lambda+\mu_1)F_1(u_1,u_2,t) - j\frac{\Phi'(w)}{\Phi(w)}F_0(u_1,u_2,t) \\ &+ \lambda F_0(u_1,u_2,t) + \lambda F_1(u_1,u_2,t), \\ \frac{\partial F_2(u_1,u_2,t)}{\partial t} &= -(\lambda+\mu_2)F_2(u_1,u_2,t) + \lambda F_2(u_1,u_2,t) \\ &+ \alpha F_0(u_1,u_2,t). \end{aligned} \tag{6}$$

Let us note, that w is only contained in the relation, and the remaining terms and the left side of the equations are independent of w. That means $\Phi(w)$ is an exponential function. Considering that $\Phi(w)$ makes sense of the asymptotic approximation of the characteristic function of the number of applications in the orbit, we can specify the form of this function as

$$\frac{\Phi'(w)}{\Phi(w)} = \frac{e^{j\kappa w}j\kappa}{e^{j\kappa w}},$$

where κ is the normalized average number of applications in the orbit, which was obtained in [10] and has form

$$\kappa = \frac{\lambda(\lambda\mu_2 + \alpha\mu_1)}{\mu_2(\mu_1 - \lambda)}.$$

On this basis, the system (6) takes the following form

$$\begin{aligned}\frac{\partial F_0(u_1,u_2,t)}{\partial t} &= -(\lambda+\alpha+\kappa)F_0(u_1,u_2,t)\\ &+ \mu_1 e^{ju_1}F_1(u_1,u_2,t) + \mu_2 e^{ju_2}F_2(u_1,u_2,t),\\ \frac{\partial F_1(u_1,u_2,t)}{\partial t} &= (\lambda+\kappa)F_0(u_1,u_2,t) - \mu_1 F_1(u_1,u_2,t)\\ &+ 0F_2(u_1,u_2,t),\\ \frac{\partial F_2(u_1,u_2,t)}{\partial t} &= \alpha F_0(u_1,u_2,t) + 0F_1(u_1,u_2,t)\\ &- \mu_2 F_2(u_1,u_2,t).\end{aligned} \tag{7}$$

Let us denote following notations

$$\boldsymbol{F}(u_1,u_2,t) = \{F_0(u_1,u_2,t), F_1(u_1,u_2,t), F_1(u_1,u_2,t)\}$$

$$\boldsymbol{G}(u_1,u_2) = \begin{bmatrix} -(\lambda+\alpha+\kappa) & \mu_1 e^{ju_1} & \mu_2 e^{ju_2} \\ \kappa+\lambda & -\mu_1 & 0 \\ \alpha & 0 & -\mu_2 \end{bmatrix}^T,$$

$\boldsymbol{G}(u_1,u_2)$ is the transposed matrix of coefficients of the system (7). Then we get the following matrix equation

$$\frac{\partial \boldsymbol{F}(u_1,u_2,t)}{\partial t} = \boldsymbol{F}(u_1,u_2,t)\boldsymbol{G}(u_1,u_2),$$

general solution of which has the form

$$\boldsymbol{F}(u_1,u_2,t) = \boldsymbol{C}e^{\boldsymbol{G}(u_1,u_2)t}. \tag{8}$$

In order to obtain the only solution, which corresponds to behaviour of the considered system, let us assume the initial condition

$$\boldsymbol{F}(u_1,u_2,0) = \boldsymbol{R}, \tag{9}$$

where row vector $\boldsymbol{R}$ is the stationary probability distribution of the state of the device, i.e. process $k(t)$, which has form [10]

$$\boldsymbol{R} = \{\frac{\mu_2(\mu_1-\lambda)}{\mu_1(\mu_2-\alpha)}, \frac{\lambda}{\mu_1}, \frac{\alpha(\mu_1-\lambda)}{\mu_1(\mu_2+\alpha)}\}.$$

With the initial condition described, we can move to solve the Cauchy problem (8, 9).

Since we are interested in the probability distribution of the number of applications in the output processes, it is necessary to find the marginal distribution. For this, let us summarize components of the row vector $\boldsymbol{F}(u_1,u_2,t)$ over k by multiplying it by unit column vector $\boldsymbol{E}$. The result is

$$\boldsymbol{F}(u_1,u_2,t)\boldsymbol{E} = \boldsymbol{R}e^{\boldsymbol{G}(u_1,u_2)t}\boldsymbol{E}. \tag{10}$$

This formula allows finding an asymptotic approximation of the characteristic function of the number of called and incoming applications served by the system at some moment t. In other words, formula (10) is the solution for the considered system.

5 Conversion to Explicit Probability Distribution

Obtained characteristic function (10) just like probability distribution fully describes the processes $m_1(t)$ and $m_2(t)$, however it does this in implicit form. Therefore, to use the obtained formula for calculations, it is necessary to obtain explicit probability distribution from it. But first, let us note, that the resulting formula (10) contains matrix exponent, which cannot be evaluated in its initial form. In order to evaluate matrix exponent, let us apply similarity transformation [2], which has the following view

$$\boldsymbol{G}(u_1,u_2) = \boldsymbol{T}(u_1,u_2)\boldsymbol{GJ}(u_1,u_2)\boldsymbol{T}(u_1,u_2)^{-1},$$

where $\boldsymbol{T}(u_1,u_2)$ – matrix of eigenvectors of matrix $\boldsymbol{G}(u_1,u_2)$, and $\boldsymbol{GJ}(u_1,u_2)$ is a diagonal matrix, containing eigenvalues of $\boldsymbol{G}(u_1,u_2)$. This transformation is valid for any power m of some matrix $\boldsymbol{A}^m$, what follows, that it is also valid for matrix exponent

$$e^{\boldsymbol{G}(u_1,u_2)t} = \boldsymbol{T}(u_1,u_2)\cdot \begin{bmatrix} e^{t\Lambda_1(u_1,u_2)} & 0 & 0 \\ 0 & e^{t\Lambda_2(u_1,u_2)} & 0 \\ 0 & 0 & e^{t\Lambda_3(u_1,u_2)} \end{bmatrix} \cdot \boldsymbol{T}(u_1,u_2)^{-1},$$

where Λ_n is eigenvalue of matrix $\boldsymbol{G}(u_1,u_2)$. Then the distribution takes the following form

$$\boldsymbol{F}(u_1,u_2,t) = \boldsymbol{R}\cdot \boldsymbol{T}(u_1,u_2)\cdot \begin{bmatrix} e^{t\Lambda_1(u_1,u_2)} & 0 & 0 \\ 0 & e^{t\Lambda_2(u_1,u_2)} & 0 \\ 0 & 0 & e^{t\Lambda_3(u_1,u_2)} \end{bmatrix} \cdot \boldsymbol{T}(u_1,u_2)^{-1}\cdot \boldsymbol{E}.$$

In order to obtain an explicit distribution of the number of served called and incoming applications, let us use the property of the characteristic function, from which it follows that the distribution is always restorable from the characteristic function. For the restoration of the function, we apply inverse Fourier transform for discrete random variables

$$P(m_1,m_2,t) = \frac{1}{2\pi}\int_{-\pi}^{\pi}\int_{-\pi}^{\pi} e^{-i\cdot u_1\cdot m_1} e^{-i\cdot u_2\cdot m_2}\boldsymbol{F}(u_1,u_2,t)du_2du_2.$$

Obtained distribution characterizes the probability of serving m_1 incoming applications and m_2 called applications at the moment t in the considered system.

6 Correlation Coefficient

The resulting asymptotic approximation of the characteristic function (10) allows us to study in more detail the output processes of the system under consideration, namely, to find the correlation dependence of the random processes $m_1(t)$ and $m_2(t)$. Consider finding the correlation coefficient, which will depend on the parameter t

$$r(t) = \frac{\mathrm{cov}(m_1(t),m_2(t))}{\sqrt{D(m_1(t))}\sqrt{D(m_2(t))}}.$$

Let us use the property of the characteristic function about the existence of its n-th derivative corresponding to the n-th raw moment of the random variable. Then the covariance and variance will be calculated as follows

$$\operatorname{cov}(m_1(t), m_2(t)) = M\{m_1(t)m_2(t)\} - M\{m_1(t)\}M\{m_2(t)\} = \frac{1}{j^2}\frac{\partial^2}{\partial u_1 \partial u_2}\boldsymbol{F}(u_1, u_2, t)\Big|_{\substack{u_1=0\\u_2=0}}$$

$$-\frac{1}{j^2}\frac{\partial}{\partial u_1}\boldsymbol{F}(u_1, u_2, t)\Big|_{\substack{u_1=0\\u_2=0}}\frac{\partial}{\partial u_2}\boldsymbol{F}(u_1, u_2, t)\Big|_{\substack{u_1=0\\u_2=0}},$$

$$D\{m_1(t)\} = M^2\{m_1(t)\} - (M\{m_1(t)\})^2 = \frac{1}{j^2}\frac{\partial^2}{\partial {u_1}^2}\boldsymbol{F}(u_1, u_2, t)\Big|_{\substack{u_1=0\\u_2=0}}$$

$$-\left(\frac{1}{j^2}\frac{\partial}{\partial u_1}\boldsymbol{F}(u_1, u_2, t)\Big|_{\substack{u_1=0\\u_2=0}}\right)^2,$$

$$D\{m_2(t)\} = M^2\{m_2(t)\} - (M\{m_2(t)\})^2 = \frac{1}{j^2}\frac{\partial^2}{\partial {u_2}^2}\boldsymbol{F}(u_1, u_2, t)\Big|_{\substack{u_1=0\\u_2=0}}$$

$$-\left(\frac{1}{j^2}\frac{\partial}{\partial u_2}\boldsymbol{F}(u_1, u_2, t)\Big|_{\substack{u_1=0\\u_2=0}}\right)^2.$$

The resulting formulas allow us to numerically research the behaviour of the system for different parameters.

7 Numerical Examples

Let us present the results of a numerical example showing how the correlation dependence of the random processes $m_1(t)$ and $m_2(t)$ changes for different parameters of the system

Table 1. Surface matrix

μ_1/μ_2	3	4	5	6	7	8	9	10	11	12	13	14	15
3	−0.285	−0.303	−0.307	−0.304	−0.299	−0.293	−0.286	−0.281	−0.275	−0.27	−0.265	−0.261	−0.257
4	−0.288	−0.322	−0.334	−0.336	−0.334	−0.33	−0.325	−0.319	−0.314	−0.309	−0.305	−0.301	−0.297
5	−0.268	−0.31	−0.323	−0.332	−0.331	−0.328	−0.324	−0.319	−0.314	−0.309	−0.305	−0.3	−0.296
6	−0.244	−0.292	−0.312	−0.318	−0.319	−0.316	−0.312	−0.307	−0.303	−0.298	−0.293	−0.289	−0.285
7	−0.222	−0.273	−0.295	−0.303	−0.304	−0.301	−0.298	−0.293	−0.289	−0.284	−0.279	−0.275	−0.271
8	−0.202	−0.256	−0.279	−0.287	−0.289	−0.287	−0.283	−0.279	−0.275	−0.27	−0.266	−0.261	−0.257
9	−0.185	−0.24	−0.264	−0.274	−0.276	−0.274	−0.27	−0.266	−0.262	−0.257	−0.253	−0.249	−0.245
10	−0.171	−0.227	−0.251	−0.261	−0.263	−0.262	−0.259	−0.255	−0.25	−0.246	−0.241	−0.237	−0.233
11	−0.158	−0.215	−0.24	−0.25	−0.253	−0.251	−0.248	−0.244	−0.24	−0.235	−0.231	−0.227	−0.223
12	−0.147	−0.204	−0.23	−0.241	−0.243	−0.242	−0.239	−0.235	−0.231	−0.226	−0.222	−0.218	−0.214
13	−0.137	−0.195	−0.221	−0.232	−0.235	−0.234	−0.231	−0.227	−0.222	−0.218	−0.214	−0.21	−0.206
14	−0.129	−0.187	−0.213	−0.224	−0.227	−0.226	−0.223	−0.219	−0.215	−0.211	−0.207	−0.202	−0.199
15	−0.121	−0.18	−0.206	−0.217	−0.22	−0.219	−0.217	−0.213	−0.208	−0.204	−0.2	−0.196	−0.192

With the given parameters of the system, it is observable on Fig. 2 and Table 1, that the largest absolute value of the correlation coefficient of the processes

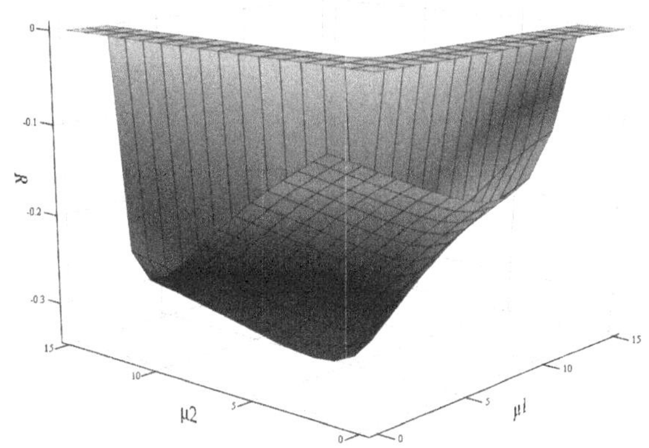

Fig. 2. Change in the correlation coefficient of the processes $m_1(t)$ and $m_2(t)$ depending on the parameters μ_1 and μ_2.

$m_1(t)$ and $m_2(t)$ is achieved at $\mu_1 = 4$ and $\mu_2 = 6$. For provided calculations were used the following system parameters: $\lambda = 1, \alpha = 3, t = 150, \mu_1 \in [0, 15], \mu_2 \in [0, 15]$.

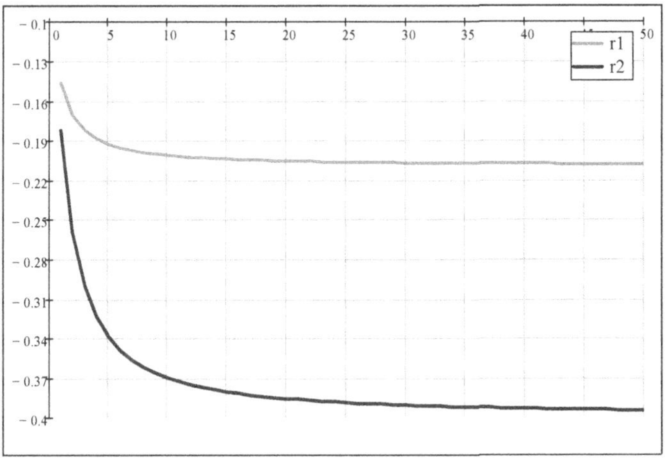

Fig. 3. Change in the correlation coefficient of the processes $m_1(t)$ and $m_2(t)$ depending on the parameter t.

On Fig. 3 it is observable, that as system comes to stationary state during some time t, correlation coefficient also comes to stationary value. We used following system parameters: r_1: $\lambda = 1, \alpha = 2, \mu_1 = 2, \mu_2 = 1, t \in [0, 50]$; r_2: $\lambda = 5, \alpha = 1, \mu_1 = 6, \mu_2 = 1, t \in [0, 50]$. Considering chosen parameters, it is noticeable in this example, that higher serving intensity of incoming applications leads to an increase of correlation dependence of random processes $m_1(t)$

and $m_2(t)$. An absolute value higher then 0.30 is a strong dependence between random processes, which cannot be disregarded.

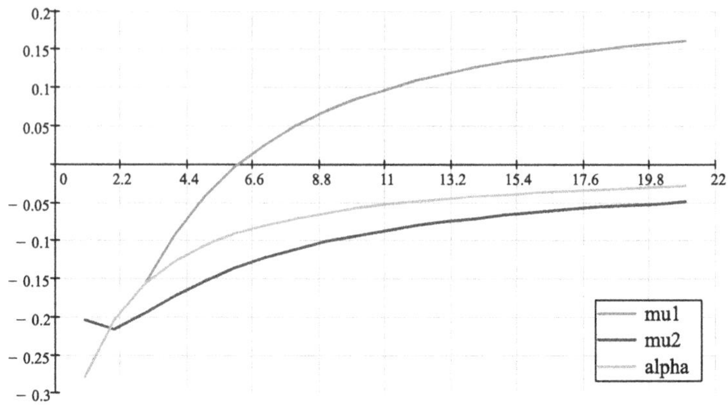

Fig. 4. Change in the correlation coefficient of the processes $m_1(t)$ and $m_2(t)$ depending on different parameters.

System parameters on Fig. 4

$$mu1 : \lambda = 1, \alpha = 2, \mu_2 = 1, t = 15, \mu_1 \in [2, 21]$$
$$mu2 : \lambda = 1, \alpha = 2, \mu_1 = 2, t = 15, \mu_2 \in [1, 21]$$
$$alpha : \lambda = 1, \mu_1 = 2, \mu_2 = 1, t = 15, \alpha \in [1, 21].$$

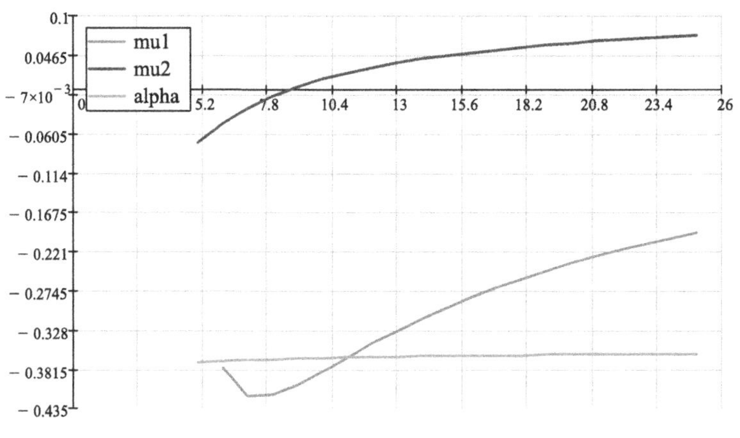

Fig. 5. Change in the correlation coefficient of the processes $m_1(t)$ and $m_2(t)$ depending on different parameters.

System parameters on Fig. 5

$$mu1 : \lambda = 5, \alpha = 1, \mu_2 = 1, t = 15, \mu_1 \in [6, 25];$$
$$mu2 : \lambda = 5, \alpha = 1, \mu_1 = 6, t = 15, \mu_2 \in [5, 25];$$
$$alpha : \lambda = 5, \mu_1 = 6, \mu_2 = 1, t = 15, \alpha \in [5, 25].$$

As can be seen in Fig. 4 and 5, the correlation coefficient, depending on the changed parameters of the system, behaves completely differently, including taking positive values.

8 Conclusion

Thus, we have obtained a formula for finding the asymptotic approximation of the two-dimensional characteristic function of the number of applications from the input process that have finished serving in a Markov queueing system with repeated calls and called applications under the condition of a large delay in the orbit. It was shown that using the inverse Fourier transform it is possible to calculate the numerical values of the probability distribution using the found asymptotic approximation of the characteristic function of the processes $m_1(t)$ and $m_2(t)$. A numerical experiment showing the correlation dependence of the random processes $m_1(t)$ and $m_2(t)$ has been carried out. Depending on the system parameters, the asymptotic correlation coefficient takes on both positive and negative values. And its absolute values can be quite small, which indicates an insignificant correlation dependence. In the future, it is necessary to determine at what ratios of parameters the correlation has values close to zero.

References

1. Artalejo, J.R., Gómez-Corral, A.: Retrial Queueing Systems: A Computational Approach. Springer, Heidelberg (2008). https://doi.org/10.1007/978-3-540-78725-9
2. Bronson, R.: Matrix Methods: An Introduction. Gulf Professional Publishing, Houston (1991)
3. Burke, P.: The output process of a stationary M/M/s queueing system. Ann. Math. Stat. **39**(4), 1144–1152 (1968)
4. Daley, D.: Queueing output processes. Adv. Appl. Probab. **8**(2), 395–415 (1976)
5. Gharbi, N., Dutheillet, C.: An algorithmic approach for analysis of finite-source retrial systems with unreliable servers. Comput. Math. Appl. **62**(6), 2535–2546 (2011)
6. Kendall, D.G.: Stochastic processes occurring in the theory of queues and their analysis by the method of the imbedded Markov chain. Ann. Math. Stat. **24**, 338–354 (1953)
7. Kulkarni, V.G.: On queueing systems with retrials. J. Appl. Probab. **20**, 380–389 (1983)

8. Lapatin, I., Nazarov, A.: Asymptotic analysis of the output process in retrial queue with Markov-modulated poisson input under low rate of retrials condition. In: Vishnevskiy, V.M., Samouylov, K.E., Kozyrev, D.V. (eds.) DCCN 2019. CCIS, vol. 1141, pp. 315–324. Springer, Cham (2019). https://doi.org/10.1007/978-3-030-36625-4_25
9. Mirasol, N.M.: The output of an m/g/∞ queuing system is poisson. Oper. Res. **11**(2), 282–284 (1963)
10. Nazarov, A.A., Paul, S., Gudkova, I., et al.: Asymptotic analysis of Markovian retrial queue with two-way communication under low rate of retrials condition (2017)
11. Paul, S., Phung-Duc, T.: Retrial queueing model with two-way communication, unreliable server and resume of interrupted call for cognitive radio networks. In: Dudin, A., Nazarov, A., Moiseev, A. (eds.) ITMM/WRQ -2018. CCIS, vol. 912, pp. 213–224. Springer, Cham (2018). https://doi.org/10.1007/978-3-319-97595-5_17

Analysis of Retrial Queueing System M/G/1 with Impatient Customers, Collisions and Unreliable Server Using Simulation

János Sztrik[1], Ádám Tóth[1(✉)], Elena Yu. Danilyuk[2], and Svetlana P. Moiseeva[2]

[1] University of Debrecen, 26 Kassai Road, Debrecen, Hungary
{sztrik.janos,toth.adam}@inf.unideb.hu
[2] National Research Tomsk State University, Tomsk, Russian Federation

Abstract. In this paper, we consider a retrial queueing system of $M/G/1$ type with an unreliable server, collisions, and impatient customers. The novelty of our work is to carry out a sensitivity analysis applying different distributions of service time of customers on significant performance measures for example on the probability of abandonment, the mean waiting time of an arbitrary, successfully served, impatient customer, etc. A customer is able to depart from the system in the orbit if it does not get its appropriate service after a definite random waiting time so these will be the so-called impatient customers. In the case of server failure, requests are allowed to enter the system but these will be forwarded immediately towards the orbit. The service, retrial, impatience, operation, and repair times are supposed to be independent of each other. Several graphical illustrations demonstrate the comparisons of the investigated distributions and the interesting phenomena which are obtained by our self-developed simulation program. The achieved results are compared to the results of the [2] to check how the system characteristics changes if we use other distributions of service time and to present the advantages of performing simulations in certain scenarios.

Keywords: Retrial queue · Impatient customers · Collisions · Unreliable server · Simulation · Sensitivity analysis

1 Introduction

With the growing number of users, devices, and networks it is crucial developing and applying new methods and ideas for designing communication systems even

The work of Dr. János Sztrik is supported by the EFOP-3.6.1-16-2016-00022 project. The project is co-financed by the European Union and the European Social Fund. The work of Ádám Tóth is supported by the ÚNKP-20-4 new national excellence program of the ministry for innovation and technology from the source of the national research, development and innovation fund.

A. Dudin et al. (Eds.): ITMM 2020, CCIS 1391, pp. 291–303, 2021.
https://doi.org/10.1007/978-3-030-72247-0_22

in the case of the existing systems. Across industries, more and more companies expand their services using a higher number of devices and cloud networking resulting in big data transmission. Consequently, creating mathematical and simulation models of modern telecommunication systems are necessary because these investigations can lessen the hardship of modifying or creating systems. In real life, in many cases, customers encountering the service units in busy state may make a decision to attempt to be served after some random time remaining in the system. Instead of residing in a queue these customers are located in a virtual waiting room called orbit and can be modeled with retrial queues. Queuing systems with retrial queues are widely used tools modelling emerging problems in major telecommunication systems, such as telephone switching systems or call centres. Many papers dealt with these types of systems which can be viewed in the following works like in [4,9,19].

Models with customers impatience in queues like the process of reneging and balking have been studied by various authors in the past. Most recent results about systems having the impatience property can be found for example in [7,8,16].

In certain scenarios during the transmission of a message, another message may appear in the channel which makes both impossible to decode causing a conflict. This can happen due to the limited number of communication channels and sometimes the launched uncoordinated attempts leading to the loss of the transmission and consequently the necessity for retransmission. In such cases, these requests go into orbit and after a random waiting time other attempts will be initiated in order to reach the service facility again. Investigating and building up efficient procedures for preventing conflicts and corresponding message delays are needed. Of course, there are papers that have studied retrial queues with collisions see for example [11–15].

Seeking in the available literature it is assumed that the components of the system are accessible all the time. In practice, this is quite unrealistic and scientists can not ignore examining the reliability of retrial queueing systems because server breakdowns and repairs have a great influence on the system characteristics and the performance measures. In real-life systems typical problems arise like a power outage, human errors, or in wireless communication packets can suffer transmission failure, interruptions throughout their transfer and unfortunately it can happen at any time. These systems with an unreliable server were analyzed in several papers, for example in [3,6,10,18,20].

In the paper of [2] a retrial queueing system of $M/M/1$ type with Poisson flow of arrivals, impatient customers, collisions, and unreliable service device is presented. In that, an asymptotic analysis method is used to define the stationary distribution of the number of customers in the orbit. We investigate the same model as in [2], but the results are gathered by our simulation program package. With this approach, it is possible to calculate performance measures that can not be determined or almost impossible to give exact formulas using numerical or asymptotic analysis. Various software packages exist which are capable to describe and perform an evaluation of complex systems if all the random variables are exponentially distributed but undoubtedly the usage of simulation has a tremendous advantage: besides exponential, any other distribution can be integrated into the code. The novelty of our work is the inclusion of other distributions of service time in the previously developed models to carry out a sensitivity analysis to see whether the observed curiosities are valid for this model or how this modification alters the performance measures. To do so we use stochastic simulation because using this method it is feasible to calculate the desired measures while obtaining analytical results, which in this case, are a difficult task if at all possible. With the help of this program, we present graphical results revealing interesting phenomena.

2 System Model

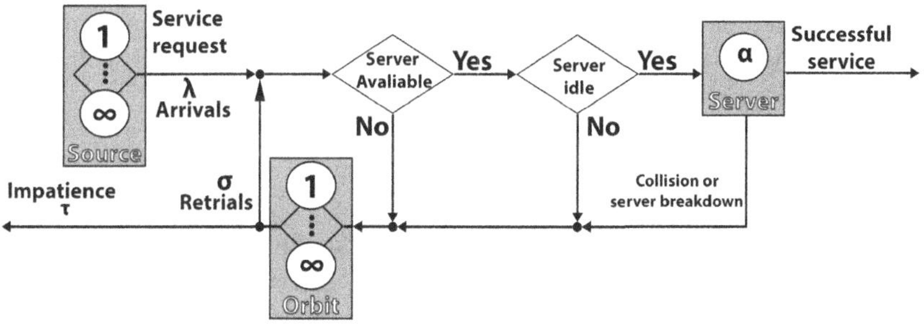

Fig. 1. The considered model.

We consider a queueing system of $M/G/1$ with collisions, impatience of the customers, and an unreliable server which is shown in Fig. 1. The system arrival process is characterized by the Poisson process with a rate of λ. The arriving customer occupies instantly the service unit in idle state and the distribution of its service is according to exponentially, gamma, Pareto, lognormal, hypo-exponentially, and hyper-exponentially distributed random variable with the same mean value and variance but with different parameters. Otherwise, it is forwarded toward the orbit. The retrial time of the requests is assumed to be exponentially distributed with a rate of σ. In the case of a busy server an arriving customer brings about a collision and both requests enter the orbit. It is supposed that the server is unreliable so it breaks down from time to time according to an exponential distribution with parameter γ_0 when the server is idle and with parameter γ_1 when it is busy. In that period generation of new requests continues but each of them is sent to orbit. After a breakdown, it is immediately sent for repair and the recovery process is also an exponential random variable with the rate γ_2. Every customer possesses an "impatience" property meaning that a customer may depart from the system earlier after waiting a random time in the orbit. The distribution of the impatient time follows an exponential distribution with parameter τ. In this unreliable model after interruption or breakdown, it is supposed that requests immediately are placed in orbit. Every service is independent of the other service including the interrupted ones, too.

3 Simulation Results

To obtain the results of our simulation program a statistic package is used that was developed by Andrea Francini in 1994 [5]. With the help of this tool, it is possible to make a quantitative estimation of the mean and variance values of the desired variables using the method of batch means. There are n observations in every batch and the useful run is divided into a predetermined number of batches. In order for the estimation to work correctly, the batches are necessary to be long enough and approximately independent. It is one of the most popular confidence interval techniques for a steady-state mean of a process. The following works contain more detailed information about this method in [1]. The simulations are performed with a confidence level of 99.9%. The relative half-width of the confidence interval required to stop the simulation run is 0.00001.

3.1 First Scenario

The realization of the sensitivity analysis includes four different distributions of service time to compare the performance measures with each other. In every case, the parameters are selected in a way that the mean and variance would be equal. To accomplish that we applied a fitting process that is required to be done and [17] contains detailed information about the whole process describing every used distribution. Two scenarios are developed to investigate the effect of the various distributions. Table 2 shows the chosen parameters of the distribution of service time while Table 1 the values of other parameters. In the first one, the squared coefficient of variation is greater than one and the following distributions are used: hyper-exponential, gamma, Pareto, and lognormal. Results in connection with the second scenario (when the squared coefficient of variation is less than one) were also examined but because of the page limitation, these will be intended to be published in the extended version of the paper.

Table 1. Numerical values of model parameters

σ	γ_0	γ_1	γ_2	τ
0.01; 0.001	0.1	0.2	1	0.02; 0.002

In Figs. 2 and 3 the comparison of steady-state distribution of the number of customers in the orbit can be seen when the distribution of service time of the incoming customers is different. It demonstrates the probability $(P(i))$ of

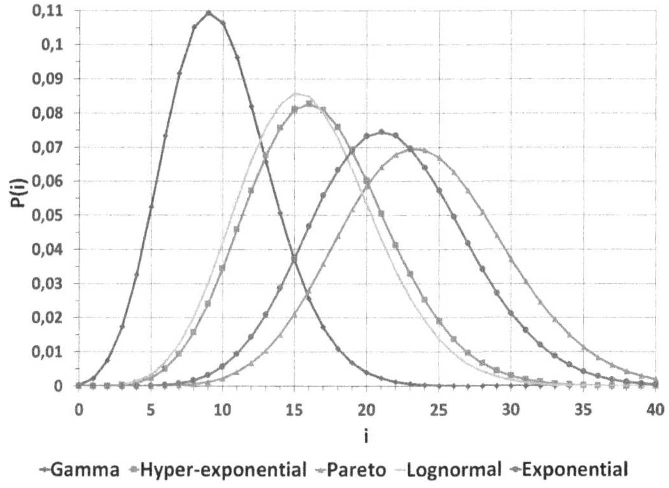

Fig. 2. Distribution of the number of customers in the orbit using various distributions, $\sigma = 0.01$, $\tau = 0.02$, $\lambda = 0.7$.

how many customers (i) residing in the orbit. Taking a closer look at the curves in more detail they coincide with normal distribution regardless of the used parameter setting. The figures also show the case of exponential distribution with the same mean as the other applied distributions. The mean number of customers in the orbit significantly differs from each other, at gamma distribution

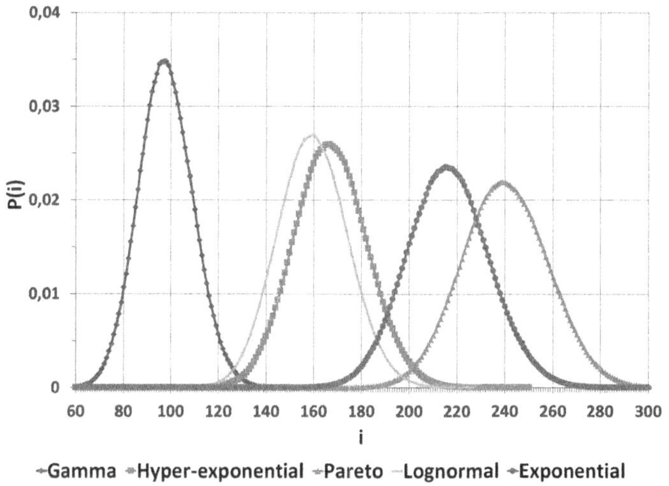

Fig. 3. Distribution of the number of customers in the orbit using various distributions, $\sigma = 0.001$, $\tau = 0.002$, $\lambda = 0.7$.

Table 2. Parameters of service time of incoming customers

Distribution	Gamma	Hyper-exponential	Pareto	Lognormal
Parameters	$\alpha = 0.0816$	$p = 0.4607$	$\alpha = 2.040$	$m = -1.292$
	$\beta = 0.0816$	$\lambda_1 = 0.9214$	$k = 0.5098$	$\sigma = 1.6075$
		$\lambda_2 = 1.0786$		
Mean	1			
Variance	12.25			
Squared coefficient of variation	12.25			

customers spend the fewest at Pareto distribution the highest time for waiting which is quite interesting.

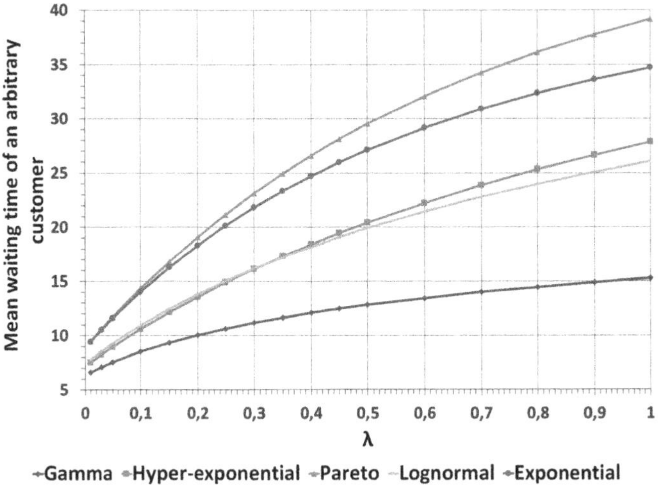

Fig. 4. Mean waiting time of an arbitrary customer vs. arrival intensity using various distributions, $\sigma = 0.01$, $\tau = 0.02$.

The mean waiting time of an arbitrary customer is presented in the function of the arrival intensity of incoming customers in Figs. 4 and 5. Even though the mean and the variance are identical huge gaps develop among the applied distributions. With the increment of the arrival intensity, the mean waiting time of an arbitrary customer increases as well. The same tendency is observable when we use other values of retrial and impatience time. The usage of gamma distribution results in lower mean waiting time compared to the others, especially versus gamma and Pareto distributions.

Figure 6 and 7 demonstrate the development of the probability of abandonment of a customer besides increasing arrival intensity. This measure shows the probability that an arbitrary customer leaves the system throughout the orbit

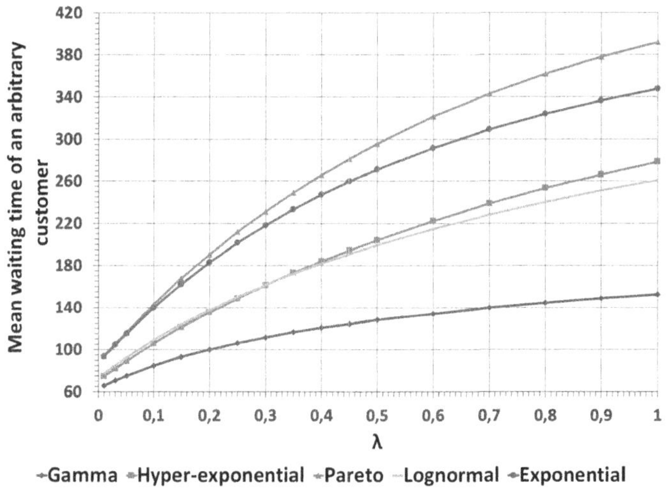

Fig. 5. Mean waiting time of an arbitrary customer vs. arrival intensity using various distributions, $\sigma = 0.001$, $\tau = 0.002$.

which means the request does not get its appropriate service requirement (impatient customers). As λ increases the value of this performance measure raises as well which is true for every used distribution but the difference is quite high among them. At gamma distribution, the tendency of leaving the system earlier is much less than the others especially compared to Pareto and exponential distributions. Taking a closer look at the Fig. 6 and 7 the obtained values of

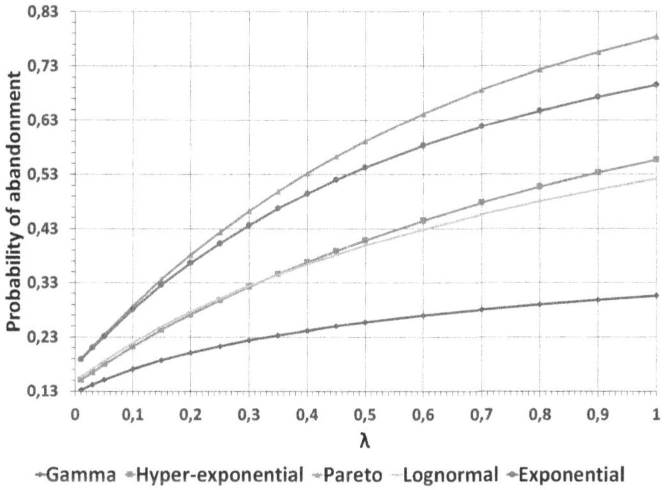

Fig. 6. Comparison of probability of abandonment, $\sigma = 0.01$, $\tau = 0.02$.

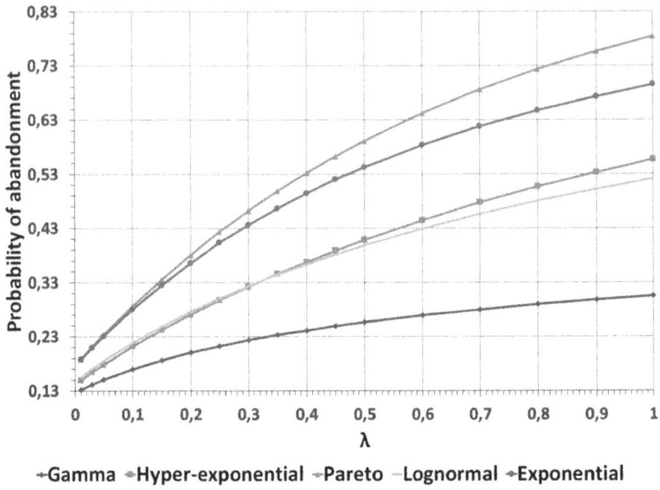

Fig. 7. Comparison of probability of abandonment, $\sigma = 0.001$, $\tau = 0.002$.

this measure are basically identical because the relationship remains the same between σ and τ.

3.2 Second Scenario

After observing the results and the tendencies of the previous section we modified the parameters of service time of incoming customers to see how this new parameter setting affects the performance measures. In this scenario, the squared coefficient of variation is less than one meaning that instead of hyper-exponential we used hypo-exponential distribution. We go over the same figures as in the first scenario but with the new applied parameters of service time which can be viewed in Table 3. All the other parameters remained unchanged (see Table 1).

Table 3. Parameters of service time of incoming customers

Distribution	Gamma	Hypo-exponential	Pareto	Lognormal
Parameters	$\alpha = 1.6$	$\mu_1 = 4$	$\alpha = 2.6125$	$m = -0.2428$
	$\beta = 1.6$	$\mu_2 = 1.3333$	$k = 0.6172$	$\sigma = 0.6968$
Mean	1			
Variance	0.625			
Squared coefficient of variation	0.625			

Figures 8 and 9 display the steady-state distribution of the number of customers in the orbit using various distributions of service time. The obtained curves are much closer to each other with this parameter setting even though

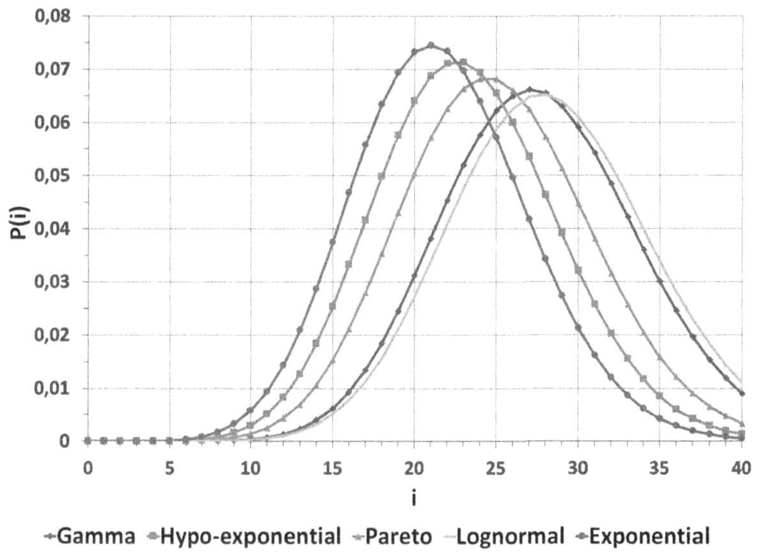

Fig. 8. Distribution of the number of customers in the orbit using various distributions, $\sigma = 0.01$, $\tau = 0.02$, $\lambda = 0.7$.

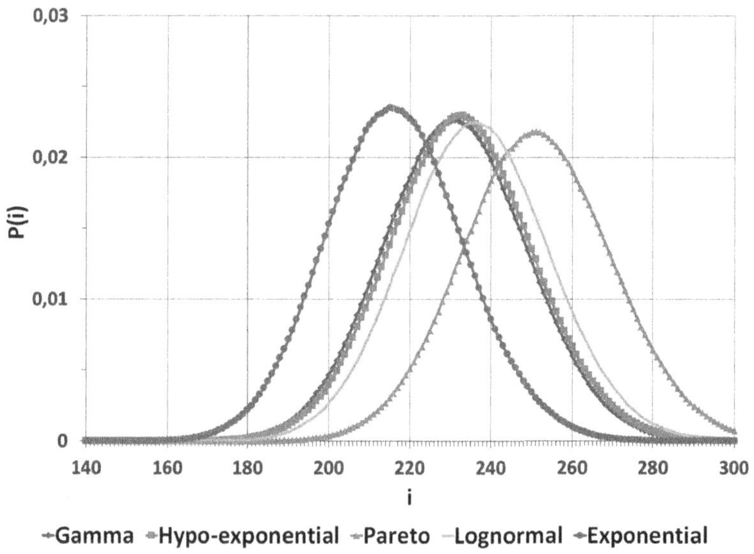

Fig. 9. Distribution of the number of customers in the orbit using various distributions, $\sigma = 0.001$, $\tau = 0.002$, $\lambda = 0.7$.

the difference is still quite significant in Figure 9 and the shape of the curves resemble the normal distribution. The value of the mean number of customers is higher in the case of gamma and lognormal distribution compared to Figures 2 and 3.

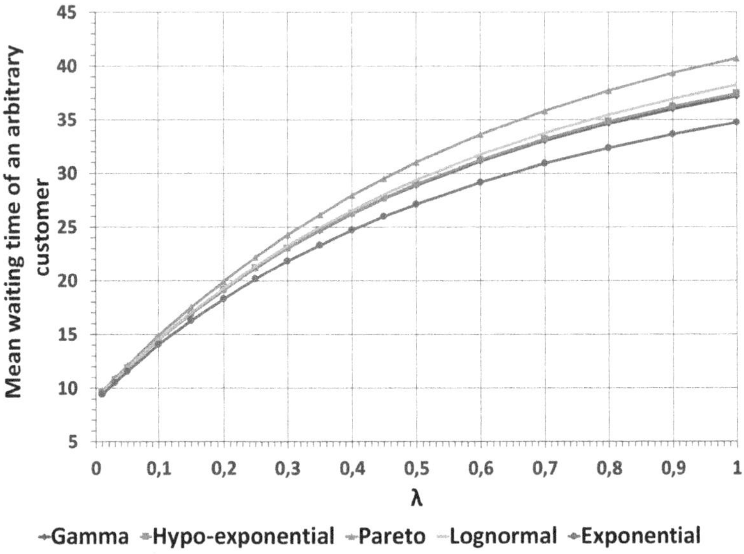

Fig. 10. Mean waiting time of an arbitrary customer vs. arrival intensity using various distributions, $\sigma = 0.01$, $\tau = 0.02$.

The next two figures (Figs. 10 and 11) are related to the mean waiting of an arbitrary customer. Evaluating the results it can be stated that very slight differences occur although in the case of Pareto distribution the values are a little bit higher. Otherwise, they almost overlap each other and the same tendency can be observed in both figures. The mean waiting time increases with the increment of arrival intensity. Obviously, the achieved results indicate that the characteristics of the system are different using these parameters of service time among the applied distributions collated in this scenario with the former one.

Finally, to have a total comparison between the investigated scenarios Fig. 12 exhibits the probability of abandonment in the function of arrival intensity. After examining the two previous figures it is no wonder how this measure develops. The realization of the attained values shows how close the applied distributions with each other and the probability that an arbitrary customer leaves the system from the orbit increases besides higher arrival intensity. The represented values are almost totally identical with the results of Fig. 12 when $\sigma = 0.001$ and $\tau = 0.002$ as in the previous section.

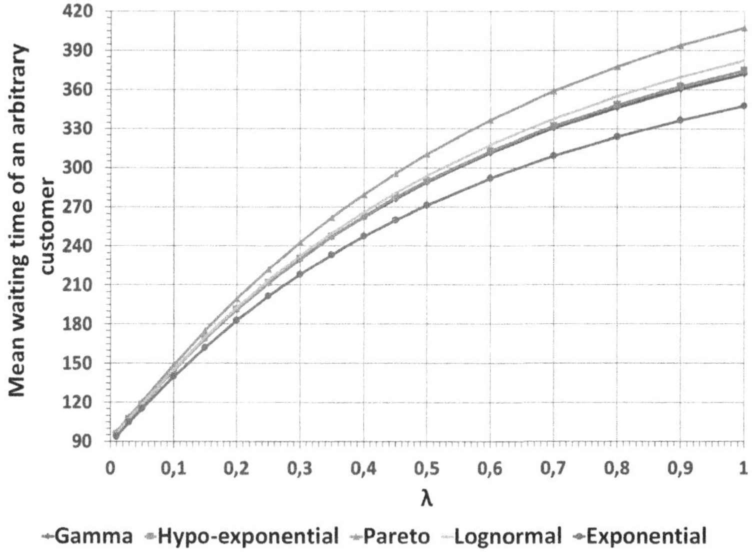

Fig. 11. Mean waiting time of an arbitrary customer vs. arrival intensity using various distributions, $\sigma = 0.001$, $\tau = 0.002$.

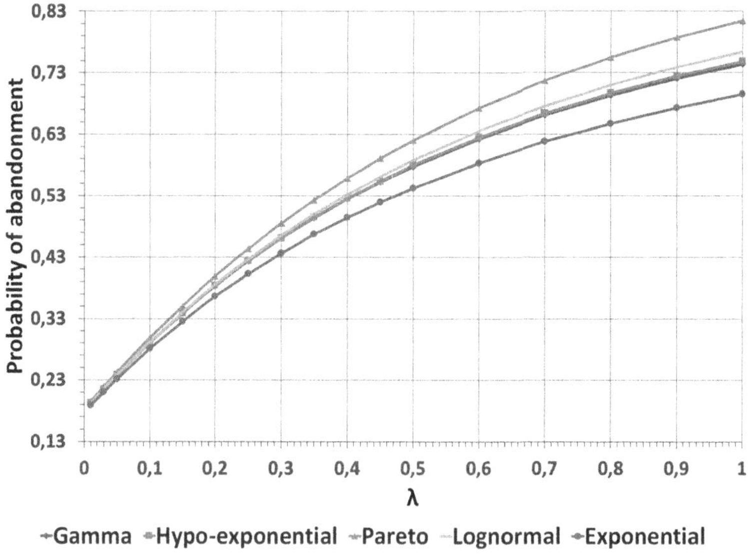

Fig. 12. Comparison of probability of abandonment, $\sigma = 0.01$, $\tau = 0.02$.

4 Conclusion

We studied the development of performance measures like the mean number of customers in the orbit or the mean waiting time of an arbitrary customer in a retrial queueing system of type $M/G/1$ with a non-reliable server and impatient customers in the orbit. Simulation has been carried out, the obtained results demonstrate that the number of customers in the orbit corresponds to the normal distribution in the case of every applied distribution. It is also displayed how the different distributions affect the performance measures despite the equality of mean value and variance when the squared coefficient of variation is more than one. In the case of the other scenario when the squared coefficient of variation is less than one, results clearly illustrated the moderate effect on the performance measures compared to the first scenario. In the future, we would extend this sensitivity analysis including more distributions or expanding the system with other features like two-way communication or other operation modes during a server failure.

References

1. Chen, E.J., Kelton, W.D.: A procedure for generating batch-means confidence intervals for simulation: checking independence and normality. Simulation **83**(10), 683–694 (2007)
2. Danilyuk, E.Y., Moiseev, S.P., Sztrik, J.: Analysis of retrial queueing system $M/M/1$ with impatient customers, collisions and unreliable server. J. Sib. Fed. Univ. Math. Phys. **13**(2), 218–230 (2020)
3. Dragieva, V.I.: Number of retrials in a finite source retrial queue with unreliable server. Asia-Pac. J. Oper. Res. **31**(2), 23 (2014)
4. Fiems, D., Phung-Duc, T.: Light-traffic analysis of random access systems without collisions. Ann. Oper. Res. **277**, 311–327 (2019). https://doi.org/10.1007/s10479-017-2636-7
5. Francini, A., Neri, F.: A comparison of methodologies for the stationary analysis of data gathered in the simulation of telecommunication networks. In: Proceedings of MASCOTS 1996 - 4th International Workshop on Modeling, Analysis and Simulation of Computer and Telecommunication Systems, pp. 116–122, February 1996
6. Gharbi, N., Nemmouchi, B., Mokdad, L., Ben-Othman, J.: The impact of breakdowns disciplines and repeated attempts on performances of small cell networks. J. Comput. Sci. **5**(4), 633–644 (2014)
7. Jouini, O., Koole, G., Roubos, A.: Performance indicators for call centers with impatient customers. IIE Trans. **45**(3), 341–354 (2013). https://doi.org/10.1080/0740817X.2012.712241
8. Kim, C., Dudin, S., Dudin, A., Samouylov, K.: Analysis of a semi-open queuing network with a state dependent marked Markovian arrival process, customers retrials and impatience. Mathematics **7**(8), 715–734 (2019). https://doi.org/10.3390/math7080715
9. Kim, J., Kim, B.: A survey of retrial queueing systems. Ann. Oper. Res. **247**(1), 3–36 (2016). https://doi.org/10.1007/s10479-015-2038-7

10. Krishnamoorthy, A., Pramod, P.K., Chakravarthy, S.R.: Queues with interruptions: a survey. TOP **22**(1), 290–320 (2014). https://doi.org/10.1007/s11750-012-0256-6
11. Kvach, A., Nazarov, A.: Sojourn time analysis of finite source Markov retrial queuing system with collision. In: Dudin, A., Nazarov, A., Yakupov, R. (eds.) ITMM 2015. CCIS, vol. 564, pp. 64–72. Springer, Cham (2015). https://doi.org/10.1007/978-3-319-25861-4_6
12. Kvach, A.: Numerical research of a Markov closed retrial queueing system without collisions and with the collision of the customers. In: Proceedings of Tomsk State University. A series of physics and mathematics. Tomsk. Materials of the II All-Russian Scientific Conference, vol. 295, pp. 105–112. TSU Publishing House (2014). (in Russian)
13. Kvach, A., Nazarov, A.: Numerical research of a closed retrial queueing system M/GI/1//N with collision of the customers. In: Proceedings of Tomsk State University. A series of physics and mathematics. Tomsk. Materials of the III All-Russian Scientific Conference, vol. 297, pp. 65–70. TSU Publishing House (2015). (in Russian)
14. Kvach, A., Nazarov, A.: The research of a closed RQ-system M/GI/1//N with collision of the customers in the condition of an unlimited increasing number of sources. In: Probability Theory, Random Processes, Mathematical Statistics and Applications: Materials of the International Scientific Conference Devoted to the 80th Anniversary of Professor Gennady Medvedev, Doctor of Physical and Mathematical Sciences, pp. 65–70 (2015). (in Russian)
15. Nazarov, A., Kvach, A., Yampolsky, V.: Asymptotic analysis of closed Markov retrial queuing system with collision. In: Dudin, A., Nazarov, A., Yakupov, R., Gortsev, A. (eds.) ITMM 2014. CCIS, vol. 487, pp. 334–341. Springer, Cham (2014). https://doi.org/10.1007/978-3-319-13671-4_38
16. Rakesh, K., Sapana, S.: Transient performance analysis of a single server queuing model with retention of reneging customers. Yugoslav J. Oper. Res. **28**(3), 315–331 (2018). https://doi.org/10.2298/YJOR170415007K
17. Sztrik, J., Tóth, Á., Pintér, Á., Bács, Z.: Simulation of finite-source retrial queues with two-way communications to the orbit. In: Dudin, A., Nazarov, A., Moiseev, A. (eds.) ITMM 2019. CCIS, vol. 1109, pp. 270–284. Springer, Cham (2019). https://doi.org/10.1007/978-3-030-33388-1_22
18. Tóth, Á., Bérczes, T., Sztrik, J., Kvach, A.: Simulation of finite-source retrial queueing systems with collisions and non-reliable server. In: Vishnevskiy, V.M., Samouylov, K.E., Kozyrev, D.V. (eds.) DCCN 2017. CCIS, vol. 700, pp. 146–158. Springer, Cham (2017). https://doi.org/10.1007/978-3-319-66836-9_13
19. Tóth, Á., Sztrik, J.: Simulation of finite-source retrial queuing systems with collisions, non-reliable server and impatient customers in the orbit. In: Proceedings of 11th International Conference on Applied Informatics, pp. 408–419. CEUR-WS. http://ceur-ws.org/Vol-2650/#paper42
20. Zhang, F., Wang, J.: Performance analysis of the retrial queues with finite number of sources and service interruptions. J. Korean Stat. Soc. **42**(1), 117–131 (2013)

A Method for Solving Stationary Equations for Priority Time-Sharing Service Process in Random Environment

Andrei V. Zorine(✉) and Kseniya O. Sizova

Lobachevsky State University of Nizhni Novgorod,
Nuzhni Novgorod 603950, Russian Federation
andrei.zorine@itmm.unn.ru
http://itmm.unn.ru/pi

Abstract. A queueing system with conflicting non-ordinary input flows in considered. The flows are modulated by a random external environment with two states. Serviced customers may return instantly to waiting line, following the general Bernoulli feedback rules. After each service act, a setup-and-control act takes place. A mathematical model for the process is a discrete-time denumerable multivariate Markov chain which includes the server state, numbers in the queues and random environment state. The model extends the class of queueing models studied by G.P.Klimov, M.Yu.Kitaev, V.V.Rykov, M.A. Fedotkin et al. Functional equation for partial probability generating functions are studied for a class of non-preemptive priority service. An algorithm for solving these equations is proposed. Some steps of the algorithm are theoretically justified. Its implementation in an open-source computer algebra language is used to demonstrate practical usefulness of the algorithm, and to investigate inner properties of some important stationary probabilities.

Keywords: Time-sharing queueing system with readjustment times · Random external environment · Non-ordinary input flows · Stationary probability distribution · Stationary probability computation algorithm · Symbolic manipulation program application

1 Introduction

In papers [1–5] a queueing system with conflicting ordinary or non-ordinary Poisson input flows by a time-sharing algorithms was studied. The notion of "time-sharing" was borrowed from computing. It means basically that the total execution time of a task is divided into small slots. When a time quant doesn't result in the process termination, the task is put back into a waiting queue (possibly, switching to another queue for different service type). A class of admissible control algorithms there allowed switching between queues based on observed queues' lengths. The main result in those papers was proving the optimality property for non-preemptive priority policy. The objective function there was the mean sojourn time for all customers per unit of time, or during a working

A. Dudin et al. (Eds.): ITMM 2020, CCIS 1391, pp. 304–318, 2021.
https://doi.org/10.1007/978-3-030-72247-0_23

and setup acts of the server. An important point in the proof was to find explicit formulas for certain stationary probabilities. In particular, these stationary probabilities didn't remained the same for all admissible control policies.

A generalization of [3–5] was investigated in [6, 7]. Input flows of the queueing system were modulated by a random external environment with two states. Only under additional constraints on the system's parameters some explicit formulas for necessary stationary probabilities were found there. The arrival intensities needed to be invariant for all environment states. In general case we failed to analytically deduce these stationary probabilities. So, it is still an open question, whether they depend on a control policy, or not.

In the present study, steady-state equations for non-preemptive priority policy are analyzed. Using a symbolic expressions manipulation software called Maxima [8], a computational procedure to solve for stationary probabilities is built. It allows to compare numerically approximated values for these probabilities for different priority policies.

2 Problem Statement

Let us recall the problem statement and necessary notations from [6, 7]. Flows Π_1, Π_2, ..., Π_m enter the queueing system, $m < \infty$. Customers from flow Π_j join a queue O_j of infinite capacity, $j = 1, 2, \ldots, m$. Set $n = m + 1$. Served customer from queue O_j can be redirected to the queue O_r with probability $p_{j,r}$, or leaves, with the probability

$$p_{j,n} = 1 - \sum_{r=1}^{m} p_{j,r}.$$

After each service act the server performs inner setup and readjustments, and makes controling decision on the next queue to serve. Let a probability distribution $B_j(t)$, $B_j(+0) = 0$ define i.i.d. service act durations for customers from a queue O_j, and let a probability distribution function $\bar{B}_j(t)$, $\bar{B}_j(+0) = 0$ define i.i.d. setup-and-readjustment act durations after a service act for the j-th queue. If at the termination instant of a setup-and-readjustment act the queues are empty the servers switches to an 'idle' state and waits for the first arrival. The first arriving customer get taken for service, the others join appropriate queues. On the other hand, if at the termination instant of an setup-and-readjustment act the queue lengths make a non-zero vector $x = (x_1, x_2, \ldots, x_m)$ then the server switches to a queue O_j with index $j = h(x)$ instantly and begins work there. Here $h(\cdot)$ is a given mapping of an m-dimensional lattice non-negative

$$X = \{0, 1, \ldots,\} \times \{0, 1, \ldots,\} \times \ldots \times \{0, 1, \ldots\}$$

onto a set $\{1, 2, \ldots, n\}$. The mapping $h(\cdot)$ should be such that $h(x) = j$ implies $x_j > 0$ for $j = 1, 2, \ldots, m$, and the only preimage of the number n is the zero vector $\bar{0} = (0, 0, \ldots, 0) \in X$.

Primary customer arrivals in each flow are modulated by a two-stated random environment, its states are denoted by $e^{(1)}$ and $e^{(2)}$. Environment states can alter only at service act termination instants and at readjustment act termination instants. The transition probability from a state $e^{(k)}$ to a state $e^{(l)}$ equals $a_{k,l}$ where k, $l \in \{1,2\}$. On interval on constancy of the environment states, the primary flows Π_1, Π_2, ..., Π_m are conditionally independent. Customers from a flow Π_j in the environment state $e^{(k)}$ arrive in batches, a batch contains b customers with probability $p_j(b;k)$, $b = 1, 2, \ldots$, and batches arrive according to a Poisson process with intensity $\lambda_j^{(k)}$.

Set $\tau_0 = 0$ and let τ_i be either a service act termination instant, or a readjustment act termination instant, at that $\tau_{i+1} > \tau_i$. Let us introduce necessary random variables and random elements. Let $\chi_i \in \{e^{(1)}, e^{(2)}\}$ be the random environment state during the time interval $(\tau_i, \tau_{i+1}]$. Let a random element $\Gamma_i \in \Gamma = \{\Gamma^{(1)}, \Gamma^{(2)}, \ldots, \Gamma^{(n)}\}$ for $i = 1, 2, \ldots$ define the server state during the time interval $(\tau_{i-1}, \tau_i]$. Equality $\Gamma_i = \Gamma^{(s)}$ for $s = 1, 2, \ldots, m$ indicates that a service act for the queue O_s, and for $s = n$ that a readjustment act takes place. A random element Γ_0 from Γ defines the initial server state (at time τ_0). Denote by $\kappa_{j,i}$ the number in the queue O_j at time τ_i after a secondary (redirected) customer joins the queue. Define vectors $\kappa_i = (\kappa_{1,i}, \kappa_{2,i}, \ldots, \kappa_{m,i})$, $i = 0, 1, \ldots$. Under the assumptions above, a multivariate sequence $\{(\Gamma_i, \kappa_i, \chi_i); i = 0, 1, \ldots\}$, given the initial vector $(\Gamma_0, \kappa_0, \chi_0)$, is an irreducible periodic Markov chain with two cyclic classes

$$\{(\Gamma^{(s)}, x, e^{(k)}) \colon s \neq m+1, x \in X, k = 1, 2\},$$
$$\{(\Gamma^{(m+1)}, x, e^{(k)}) \colon x \in X, k = 1, 2\}.$$

Denote by $Q(j, x, k)$ the stationary probability of the Markov chain's state $(\Gamma^{(j)}, x, e^{(k)})$, $j = 1, 2, \ldots, n$, $x \in X$, $k = 1, 2$. For an integer-valued vector $x \in X$ and a vector $v = (v_1, v_2, \ldots, v_m) \in \mathbb{C}^m$ we'll write $v^x = v_1^{x_1} v_2^{x_2} \times \cdots \times v_m^{x_m}$, in particular we set $\bar{0}^{\bar{0}} = 1$. Let us define some partial probability generating functions:

$$f_j^{(k)}(z) = \sum_{b=1}^{\infty} p_j(b;k) z^b, \qquad z \in \mathbb{C},$$
$$R_j(v) = v_j^{-1}\Big(p_{j,n} + \sum_{r=1}^{m} p_{j,r} v_r\Big), \qquad v \in \mathbb{C}^m,$$
$$q_j^{(k)}(v) = \int_0^{\infty} \prod_{r=1}^{m} \exp\{(\lambda_r^{(k)}(f_r^{(k)}(v_r) - 1)t)\} dB_j(t),$$
$$\bar{q}_j^{(k)}(v) = \int_0^{\infty} \prod_{r=1}^{m} \exp\{(\lambda_r^{(k)}(f_r^{(k)}(v_r) - 1)t)\} d\bar{B}_j(t),$$
$$\Psi(v, s, k) = \textstyle\sum_{x \in X} Q(s, x, k) v^x,$$
$$\Phi(v, j, k) = \textstyle\sum_{x \in X_j} Q(n, x, k) v^x$$

and set $\lambda_+^{(k)} = \lambda_1^{(k)} + \lambda_2^{(k)} + \ldots + \lambda_m^{(k)}$ for $k = 1, 2$. Earlier in [6,7] the following equations were given:

$$\Psi(v, j, l) = \sum_{k=1}^{2} a_{k,l} q_j^{(k)}(v) R_j(v) \Big(\Phi(v, j, k) + \frac{\lambda_j^{(k)}}{\lambda_+^{(k)}} f_j^{(k)}(v_j) Q(n, \bar{0}, k)\Big), \quad (1)$$

$j = 1, 2, \ldots, m,$

$$\Psi(v,n,l) = \sum_{k=1}^{2} a_{k,l} \sum_{j=1}^{m} \bar{q}_j(v)\Psi(v,j,k), \tag{2}$$

They hold at least inside a closed polydisk

$$\overline{U}_m = \{(v_1, v_2, \ldots, v_m) \colon |v_j| \leqslant 1, j = 1, 2, \ldots, m\}.$$

The system of Eqs. (1), (2) isn't complete, since it lacks relations for unknown functions $\Phi(v,j,k)$ and probabilities $Q(n,\bar{0},k)$ for $j = 1, 2, \ldots, m$, $k = 1, 2$. It contains only $2n$ relations between unknown functions $\Psi(v,s,k)$, $s = 1, 2, \ldots, n$, $k = 1, 2$, and the above-mentioned functions and probabilities. However, it turned out to be possible, by virtue of putting constraints on environment parameters or input flows' parameters, to obtain explicit formulas for several stationary probabilities. Let us cite these results here. The following notations will be used. Let E denote an identity matrix of size $m \times m$,

$$\begin{aligned}
\beta_r &= \textstyle\int_0^\infty t\, dB_r(t), & \bar{\beta}_r &= \textstyle\int_0^\infty t\, d\bar{B}_r(t),\\
\beta &= (\beta_1, \ldots, \beta_m), & \bar{\beta} &= (\bar{\beta}_1, \ldots, \bar{\beta}_m),\\
\Pi &= (p_{j,r})_{j,r=\overline{1,m}}, & \bar{\lambda}_j^{(k)} &= \lambda_j^{(k)} \textstyle\sum_{b=1}^\infty b p_j(b;k),\\
\bar{\lambda}^{(k)} &= (\bar{\lambda}_1^{(k)}, \ldots, \bar{\lambda}_m^{(k)})^\mathsf{T}, & \Lambda^{(k)} &= (E - \Pi^T)^{-1}\bar{\lambda}^{(k)},\\
\rho^{(k)} &= (\beta + \bar{\beta})\Lambda^{(k)}, & \bar{1} &= (1, \ldots, 1),\\
\Psi_j &= \Psi(\bar{1}, j, 1) + \Psi(\bar{1}, j, 2), & \Psi &= (\Psi_1, \ldots, \Psi_m)^\mathsf{T}
\end{aligned}$$

it is assumed that all quantities defined above are finite, and the matrix $(E - \Pi)$ is invertible. Let us interpret these quantities: β_r is the mean service act duration for a customer from the queue O_r, $\bar{\beta}_r$ is the mean readjustment act duration after a service act for O_r, $\bar{\lambda}_j^{(k)}$ is the overall arrival intensity (i.e. the mean number of customers per unit of time, not batches) from the flow Π_j when the environment state is $e^{(k)}$.

It was discovered earlier [6] that when $a_{1,1} = a_{2,1} = \alpha$ one has *for any* $h(\cdot)$ that

$$\begin{aligned}
Q(\Gamma^{(m+1)}, \bar{0}, k) = \frac{a_{1,k}}{2}(1 - \alpha\rho^{(1)} - (1-\alpha)\rho^{(2)}))\Bigg(\alpha(1-\alpha)\\
\times \Big(\frac{1 + \beta\lambda^{(1)}}{\lambda_+^{(1)}} - \frac{1 + \beta\lambda^{(2)}}{\lambda_+^{(2)}}\Big)(\Lambda_+^{(2)}\rho^{(1)} - \Lambda_+^{(1)}\rho^{(2)})\\
+ \alpha\Lambda_+^{(1)}\frac{1 + \beta\lambda^{(1)}}{\lambda_+^{(1)}} + (1-\alpha)\Lambda_+^{(2)}\frac{1 + \beta\lambda^{(2)}}{\lambda_+^{(2)}}\\
- \Big(\frac{\alpha\beta\lambda^{(1)}}{\lambda_+^{(1)}} + \frac{(1-\alpha)\beta\lambda^{(2)}}{\lambda_+^{(2)}}\Big)(\alpha\Lambda_+^{(1)} + (1-\alpha)\Lambda_+^{(2)})\Bigg)^{-1}
\end{aligned}$$

The second known case assumes $\bar{\lambda}^{(1)} = \bar{\lambda}^{(2)}$, i.e. that arrival intensities remain the same for all states of the external environment. Then, *for any* $h(\cdot)$, one has:

$$\Psi = \frac{\Lambda}{2\Lambda_+}, \qquad \sum_{k=1}^{2} \frac{Q(n, \bar{0}, k)}{\lambda_+^{(k)}} = \frac{1 - \rho^{(1)}}{2\Lambda_+}. \tag{3}$$

However, no separate formulas for the probabilities $Q(n, \bar{0}, 1)$, $Q(n, \bar{0}, 2)$ are known in this case.

3 Solution Algorithm in Case of Non-preemptive Priority Policies

We will narrow the class of admissible control policies in order to fill in missing relations for the partial probability generating functions $\Phi(v, j, k)$. A non-preemptive priority policies is uniquely determined by priority indices for the queues. Up to relabeling the queues, it is enough to assume that queue O_1 has the highest priority index, O_2 is the second in priority, etc. Formally speaking, let $h(x) = \min\{j : x_j \neq 0\}$. Let us use operators $\mathbf{0}^j(\cdot) : \mathbb{C}^m \to \mathbb{C}^m$, $j = 1, 2, \ldots, n$ which were introduced in [9] by means of equations

$$\mathbf{0}^j v = (0, \ldots, 0, v_j, v_{j+1}, \ldots, v_m).$$

Let us remark that $\mathbf{0}^1 v = v$, $\mathbf{0}^n v = \bar{0}$. Then one has

$$\Phi(v, j, k) = \Psi(\mathbf{0}^j v, n, k) - \Psi(\mathbf{0}^{j+1} v, n, k)$$

Now the problem is reduced to finding functions $\Psi(v, n, k)$, $k = 1, 2$, analytic in a polydisk [10]

$$U_m = \{(v_1, v_2, \ldots, v_m) : |v_1| < 1, |v_2| < 1, \ldots, |v_m| < 1\} \in \mathbb{C}^m$$

and satisfying equations

$$\Psi(v, n, k) = \sum_{l_1=1}^{2} a_{l_1,k} \sum_{j=1}^{m} \bar{q}_j^{(l_1)}(v) \sum_{l=1}^{2} a_{l,l_1} q_j^{(l)}(v) R_j(v) \times \left(\Psi(\mathbf{0}^j v, n, l) - \Psi(\mathbf{0}^{j+1} v, n, l) + \Psi(\bar{0}, n, l) f_j^{(l)}(v_j) \frac{\lambda_j^{(l)}}{\lambda_+^{(l)}} \right). \tag{4}$$

Now, to explicate the solution algorithm, let us switch to matrix form of equations. Let

$$\bar{\Psi}(v) = \begin{pmatrix} \Psi(v, n, 1) \\ \Psi(v, n, 2) \end{pmatrix},$$

and for $j = 1, 2, \ldots, m$ let

$$Q_j(v) = \begin{pmatrix} \sum\limits_{l=1}^{2} a_{1,l}a_{l,1}\bar{q}_j^{(l)}(v)q_j^{(1)}(v)R_j(v) & \sum\limits_{l=1}^{2} a_{2,l}a_{l,1}\bar{q}_j^{(l)}(v)q_j^{(2)}(v)R_j(v) \\ \sum\limits_{l=1}^{2} a_{1,l}a_{l,2}\bar{q}_j^{(l)}(v)q_j^{(1)}(v)R_j(v) & \sum\limits_{l=1}^{2} a_{2,l}a_{l,2}\bar{q}_j^{(l)}(v)q_j^{(2)}(v)R_j(v) \end{pmatrix},$$

$$D_j(v_j) = \begin{pmatrix} f_j^{(1)}(v_j)\dfrac{\lambda_j^{(1)}}{\lambda_+^{(1)}} & 0 \\ 0 & f_j^{(2)}(v_j)\dfrac{\lambda_j^{(2)}}{\lambda_+^{(2)}} \end{pmatrix}.$$

Then Eq. (4) takes form

$$\bar{\Psi}(v) = \sum_{j=1}^{m} Q_j(v)\big(\bar{\Psi}(\mathbf{0}^j v) - \bar{\Psi}(\mathbf{0}^{j+1}v) + D_j(v_j)\bar{\Psi}(\bar{0})\big).$$

After rearranging terms we finally get:

$$(E - Q_1(v))\bar{\Psi}(v) = \sum_{j=2}^{m}(Q_j(v) - Q_{j-1}(v))\bar{\Psi}(\mathbf{0}^j v) + \Big(\sum_{j=1}^{m} Q_j(v)D_j(v_j) - Q_m(v)\Big)\bar{\Psi}(\bar{0}). \quad (5)$$

Since the entries of the functional matrix $(v_1 E - v_1 Q_1(v))$ have no poles inside the polydisk U_m, one can solve Eq. (5) for the desired vector of functions $\bar{\Psi}(v)$:

$$\bar{\Psi}(v) = (v_1 E - v_1 Q_1(v))^{-1} \sum_{j=2}^{m} v_1(Q_j(v) - Q_{j-1}(v))\bar{\Psi}(\mathbf{0}^j v) + (v_1 E - v_1 Q_1(v))^{-1}\Big(v_1 \sum_{j=1}^{m} Q_j(v)D_j(v_j) - v_1 Q_m(v)\Big)\bar{\Psi}(\bar{0}). \quad (6)$$

Equality (6) holds at those points of U_m where

$$\det(v_1 E - v_1 Q_1(v)) \neq 0. \quad (7)$$

Since the functions $\Psi(v, n, k)$, $k = 1, 2$ must be holomorphic [10] everywhere in U_m, the yet undefined functions $\bar{\Psi}(\mathbf{0}^j v)$, $j = 2, 3, \ldots, n$ must satisfy additional equations which we now aim to describe.

Let us define functional matrices

$$Q_{1,j}(v_1, v_2, \ldots, v_m) = \big(Q_{1,j;k,l}(v_1, v_2, \ldots, v_m)\big)_{k,l=1,2},$$
$$\tilde{Q}_1(v_1, v_2, \ldots, v_m) = \big(\tilde{Q}_{1;k,l}(v_1, v_2, \ldots, v_m)\big)_{k,l=1,2},$$

through equations

$$Q_{1,1}(v_1, v_2, \ldots, v_m) = v_1(Q_1(v) - E),$$
$$Q_{1,j}(v_1, v_2, \ldots, v_m) = v_1(Q_j(v) - Q_{j-1}(v)),\ j = 2, 3, \ldots, m,$$
$$\tilde{Q}_1(v_1, v_2, \ldots, v_2) = \left(v_1 \sum_{j=1}^{m} Q_j(v) D_j(v_j) - v_1 Q_m(v)\right).$$

Equation (5) becomes

$$\sum_{j=1}^{m} Q_{1,j}(v_1, v_2, \ldots, v_m)\bar{\Psi}(\mathbf{0}^j v) + \tilde{Q}_1(v_1, v_2, \ldots, v_m)\bar{\Psi}(\bar{0}) = 0. \tag{8}$$

Let us multiply Eq. (8) from left by the matrix $Q^*_{1,1}(v_1, v_2, \ldots, v_m)$ adjoint to $Q_{1,1}(v_1, v_2, \ldots, v_m)$ (an asterix marks an adjoint matrix). Then we get for all $v \in U_m$, satisfying the following equation

$$\det Q_{1,1}(v_1, v_2, \ldots, v_m) = 0 \tag{9}$$

(it's the same as Eq. (7)), the following matrix equation should hold:

$$\sum_{j=2}^{m} Q^*_{1,1}(v_1, v_2, \ldots, v_m) Q_{1,j}(v_1, v_2, \ldots, v_m)\bar{\Psi}(\mathbf{0}^j v) \tag{10}$$
$$+ Q^*_{1,1}(v_1, v_2, \ldots, v_m)\tilde{Q}_1(v_1, v_2, \ldots, v_m)\bar{\Psi}(\bar{0}) = 0.$$

Let us assume that Equation (9) implicitly defines two distinct functions

$$v_1 = v_1^{(k)}(v_2, \ldots, v_m), \quad k = 1, 2,$$

holomorphic in a polydisk

$$U_{m-1} = \{(v_2, \ldots, v_m) \colon |v_2| < 1, \ldots, |v_m| < 1\}.$$

Let us remark that vectors $\bar{\Psi}(\mathbf{0}^j v)$ for $j = 2, 3, \ldots, m$, depend only on the variables $v_2, v_3, \ldots, v_m$. So, a substitution $v = (v_1^{(k)}, v_2, \ldots, v_m)$ into matrix Eq. (10) gives a new relation between these vectors:

$$\sum_{j=2}^{m} Q_{2,m}(v_2, \ldots, v_m)\bar{\Psi}(\mathbf{0}^j v) + \tilde{Q}_2(v_2, \ldots, v_m)\bar{\Psi}(\bar{0}) = 0,$$

where matrices $Q_{2,j}(v_2, \ldots, v_m) = \left(Q_{2,j;k,l}(v_2, \ldots, v_m)\right)_{k,l=1,2}$, $j = 2, 3, \ldots, m$ and $\tilde{Q}_2(v_2, \ldots, v_m) = \left(\tilde{Q}_{2;k,l}(v_2, \ldots, v_m)\right)_{k,l=1,2}$ have entries

$$Q_{2,j;1,1}(v_2, \ldots, v_m) = v_2 \cdot \big(Q_{1,1;2,2}(v_1^{(1)}, v_2, \ldots, v_m) Q_{1,j;1,1}(v_1^{(1)}, v_2, \ldots, v_m)$$
$$- Q_{1,1;1,2}(v_1^{(1)}, v_2, \ldots, v_m) Q_{1,j;2,1}(v_1^{(1)}, v_2, \ldots, v_m)\big),$$

$$Q_{2,j;1,2}(v_2,\dots,v_m)=v_2\cdot\big(Q_{1,1;2,2}(v_1^{(1)},v_2,\dots,v_m)Q_{1,j;1,2}(v_1^{(1)},v_2,\dots,v_m)$$
$$-\,Q_{1,1;1,2}(v_1^{(1)},v_2,\dots,v_m)Q_{1,j;2,2}(v_1^{(1)},v_2,\dots,v_m)\big),$$
$$Q_{2,j;2,1}(v_2,\dots,v_m)=v_2\cdot\big(Q_{1,1;1,1}(v_1^{(2)},v_2,\dots,v_m)Q_{1,j;2,1}(v_1^{(2)},v_2,\dots,v_m)$$
$$-\,Q_{1,1;2,1}(v_1^{(2)},v_2,\dots,v_m)Q_{1,j;1,1}(v_1^{(2)},v_2,\dots,v_m)\big),$$
$$Q_{2,j;2,2}(v_2,\dots,v_m)=v_2\cdot\big(Q_{1,1;1,1}(v_1^{(2)},v_2,\dots,v_m)Q_{1,j;2,2}(v_1^{(2)},v_2,\dots,v_m)$$
$$-\,Q_{1,1;2,1}(v_1^{(2)},v_2,\dots,v_m)Q_{1,j;1,2}(v_1^{(2)},v_2,\dots,v_m)\big),$$
$$\tilde{Q}_{2;1,1}(v_2,\dots,v_m)=v_2\cdot\big(Q_{1,1;2,2}(v_1^{(1)},v_2,\dots,v_m)\tilde{Q}_{1;1,1}(v_1^{(1)},v_2,\dots,v_m)$$
$$-\,Q_{1,1;1,2}(v_1^{(1)},v_2,\dots,v_m)\tilde{Q}_{1;2,1}(v_1^{(1)},v_2,\dots,v_m)\big),$$
$$\tilde{Q}_{2;1,2}(v_2,\dots,v_m)=v_2\cdot\big(Q_{1,1;2,2}(v_1^{(1)},v_2,\dots,v_m)\tilde{Q}_{1;1,2}(v_1^{(1)},v_2,\dots,v_m)$$
$$-\,Q_{1,1;1,2}(v_1^{(1)},v_2,\dots,v_m)\tilde{Q}_{1;2,2}(v_1^{(1)},v_2,\dots,v_m)\big),$$
$$\tilde{Q}_{2;2,1}(v_2,\dots,v_m)=v_2\cdot\big(Q_{1,1;1,1}(v_1^{(2)},v_2,\dots,v_m)\tilde{Q}_{1;2,1}(v_1^{(2)},v_2,\dots,v_m)$$
$$-\,Q_{1,1;2,1}(v_1^{(2)},v_2,\dots,v_m)\tilde{Q}_{1;1,1}(v_1^{(2)},v_2,\dots,v_m)\big),$$
$$\tilde{Q}_{2;2,2}(v_2,\dots,v_m)=v_2\cdot\big(Q_{1,1;1,1}(v_1^{(2)},v_2,\dots,v_m)\tilde{Q}_{1;2,2}(v_1^{(2)},v_2,\dots,v_m)$$
$$-\,Q_{1,1;2,1}(v_1^{(2)},v_2,\dots,v_m)\tilde{Q}_{1;1,2}(v_1^{(2)},v_2,\dots,v_m)\big).$$

Let us explain that the k-th row of the matrix $Q_{2,j}(v_2,\dots,v_m)$ is obtained by substituting $v_1=v_1^{(k)}(v_2,\dots,v_m)$ into the k-th row of the matrix

$$v_2\cdot(v_1E-v_1Q_1(v))^*v_1(Q_j(v)-Q_{j-1}(v)),$$

and the k-th row of the functional matrix $\tilde{Q}_2(v_2,\dots,v_m)$ is obtained by substituting $v=(v_1^{(k)},v_2,\dots,v_m)$ into the k-the row of the matrix

$$v_2\cdot(v_1E-v_1Q_1(v))^*\Big(v_1\sum_{j=1}^{m}Q_j(v)D_j(v_j)-v_1Q_m(v)\Big).$$

The structure of the obtained equation is similar to the original Eq. (8), but it contains fewer unkown functions and fewer independent variables. Let us assume that at the s-th step we got the following equation:

$$0=\sum_{j=s}^{m}Q_{s,j}(v_s,\dots,v_m)\bar{\Psi}(\mathbf{0}^jv)+\tilde{Q}_s(v_s,\dots,v_m)\bar{\Psi}(\bar{0}),$$

Next, let us assume that the equation

$$\det Q_{s,s}(v_s,\dots,v_m)=0 \tag{11}$$

implicitly defines two distinct functions

$$v_2=v_2^{(k)}(v_3,\dots,v_m),\quad k=1,2,$$

holomorphic in a polydisk

$$U_{m-2} = \{(v_3, \dots, v_m) \colon |v_3| < 1, \dots, |v_m| < 1\}.$$

Then, repeating the steps as above, we obtain a new equation of the form

$$\sum_{j=s+1}^{m} Q_{s+1,j}(v_{s+1}, \dots, v_m)\bar{\Psi}(\mathbf{0}^j v) + \tilde{Q}_{s+1}(v_{s+1}, \dots, v_m)\bar{\Psi}(\bar{0}) = 0.$$

In particular, after $(m-1)$ steps we finally get

$$Q_{m,m}(v_m)\bar{\Psi}(\mathbf{0}^m v) + \tilde{Q}_m(v_m)\bar{\Psi}(\bar{0}) = 0. \tag{12}$$

If the determinant $\det Q_{m,m}(v_m)$ has two zeros inside the unit disk $|v_m| < 1$ then the described procedure of exclusion of unknown functions leads to two homogeneous linear equations for $\bar{\Psi}(\bar{0})$, its solution is only the zero vector $\bar{0}$. It contradicts to the stationary distribution existence, $(\Gamma^{(n)}, \bar{0}, e^{(1)})$ and $(\Gamma^{(n)}, \bar{0}, e^{(1)})$ are essential states. So, let us assume that the determinant $\det Q_{m,m}(v_m)$ has a single zero $v_m^\dagger$ inside the disk $|v_m| < 1$, the other zero $v_m = 1$ lying on the disk boundary. Using this $v_m^\dagger$, we get a linear homogeneous equation

$$\tilde{Q}_{m+1}\bar{\Psi}(\bar{0}) = 0 \tag{13}$$

where $\tilde{Q}_{m+1}$ is a constant row-vector.

Proposition 1. *The following formulas take place*

$$\Psi(\bar{1}, n, 1) = \frac{a_{2,1}}{2(a_{1,2} + a_{2,1})}, \qquad \Psi(\bar{1}, n, 2) = \frac{a_{1,2}}{2(a_{1,2} + a_{2,1})}. \tag{14}$$

Proof. Set $v = \bar{1}$ in (4) and use the normalization condition

$$\Psi(\bar{1}, n, 1) + \Psi(\bar{1}, n, 2) = 1/2$$

(taking into account the cyclic classes), we get the claimed formulas.

Since a single Eq. (13) binds two unknown quantities $\Psi(\bar{0}, n, 1)$ and $\Psi(\bar{0}, n, 2)$, we need to find more relations for these quantities. The next Proposition suggests useful equations in this respect.

Proposition 2. *The following equality hold:*

$$0 = \sum_{l_1=1}^{2}\sum_{l=1}^{2}\sum_{j=1}^{m} a_{l,l_1}(\bar{\lambda}_g^{(l_1)}\bar{\beta}_{j,1} + \bar{\lambda}_g^{(l)}\beta_{j,1} + p_{j,g} - \delta_{g,j}) \tag{15}$$
$$\times\left(\Psi(\mathbf{0}^j\bar{1}, n, l) - \Psi(\mathbf{0}^{j+1}\bar{1}, n, l) + \Psi(\bar{0}, n, l)\frac{\lambda_j^{(l)}}{\lambda_+^{(l)}}\right) + \sum_{l=1}^{2}\Psi(\bar{0}, n, l)\frac{\bar{\lambda}_g^{(l)}}{\lambda_+^{(l)}}.$$

Proof. Let us sum Eqs. (4) w.r.t. $k = 1, 2$. We get

$$\sum_{k=1}^{2} \Psi(v,n,k) = \sum_{l_1=1}^{2}\sum_{j=1}^{m} \bar{q}_j^{(l_1)}(v) \sum_{l=1}^{2} a_{l,l_1} q_j^{(l)}(v) R_j(v)$$
$$\times \left(\Psi(\mathbf{0}^j v, n, l) - \Psi(\mathbf{0}^{j+1} v, n, l) + \Psi(\bar{0}, n, l) f_j^{(l)}(v_j) \frac{\lambda_j^{(l)}}{\lambda_+^{(l)}} \right).$$

Take derivatives w.r.t. v_g, $g = 1, 2, \ldots, m$ and then set $v = \bar{1}$. We get:

$$\sum_{k=1}^{2} \frac{\partial}{\partial v_g} \Psi(v,n,k)\Big|_{v=\bar{1}} = \sum_{l_1=1}^{2}\sum_{l=1}^{2}\sum_{j=1}^{m} \frac{\partial}{\partial v_g} \left(\bar{q}_j^{(l_1)}(v) a_{l,l_1} q_j^{(l)}(v) R_j(v) \right)\Big|_{v=\bar{1}}$$
$$\times \left(\Psi(\mathbf{0}^j \bar{1}, n, l) - \Psi(\mathbf{0}^{j+1} \bar{1}, n, l) + \Psi(\bar{0}, n, l) \frac{\lambda_j^{(l)}}{\lambda_+^{(l)}} \right)$$
$$+ \sum_{l=1}^{2}\sum_{j=1}^{m} \left(\frac{\partial}{\partial v_g} \left(\Psi(\mathbf{0}^j v, n, l) - \Psi(\mathbf{0}^{j+1} v, n, l) \right)\Big|_{v=\bar{1}} + \delta_{j,g} \Psi(\bar{0}, n, l) \frac{\mu_j^{(l)} \lambda_j^{(l)}}{\lambda_+^{(l)}} \right).$$

After combining similar terms we finally get Eq. (15).

To summarize, Eqs. (13), (14), and (15) provide $(m+3)$ linearly independent linear equations for $(2m+2)$ quantities $\bar{\Psi}(\mathbf{0}^j \bar{1}, n, k)$, $j = 1, 2, \ldots, n$, $k = 1, 2$. Another $(m-1)$ new equations we can have by using that solution $v_s(1, \ldots, 1)$ of Eq. (11) with $v_{s+1} = 1, \ldots, v_m = 1$, which satisfied the inequality $|v_s(1, \ldots, 1)| < 1$ (the other solution is $v_s(1, \ldots, 1) = 1$).

4 Numerical Routine Details and Experiments

At each step of the algorithm presented in the previous section, several assumptions need quantitative verification.

1. Determining the number of zeros of a determinant in an appropriate disk $|v_s| \leqslant 1$.
2. Finding all zeroes in the disk.

The first problem can be solved by applying the famous Cauchy's Theorem (see, e.g., [12]). To count zeros of $W(v_s) = \det Q_{s,s}(v_s, \ldots, v_m)$ inside $|v_s| < 1$ for fixed $v_{m+1}, \ldots, v_m$ use a Couchy integral

$$\frac{1}{2\pi \mathbf{i}} \int_{\partial D} \frac{W'(z)}{W(z)} dz, \quad D = \{|z| \leqslant 1\}.$$

By change of variable it reduces to a Riemann integral over a finite segment and can by computed using standard numerical quadrature routines. We used an adaptive integrator from QUADPACK library [13] distributed with Maxima CAS [8].

For an approximate computation of zeros ζ_l, $l = 1, 2$ in the disk $|z| < 1$ we used an algorithm of Delves and Lyness [11]. First, using the same integrator from QUADPACK we compute sums

$$S_k = \frac{1}{2\pi\mathbf{i}} \int_D z^k \frac{W'(z)}{W(z)}\, dz = \sum_{l=1}^{2} \zeta_l^k, \quad k = 1, 2,$$

and then solve a quadratic algebraic equation

$$z^2 - S_1 z + \frac{S_1^2 - S_2}{2} = 0$$

whose coefficients are symmetric polynomials of powers of the zeros by Vieta's formula.

The next claim is necessary to count zeros of (9). Counting zeros of (11) inside $|v_s| < 1$ for $s \geqslant 2$ is still an open problem.

Proposition 3. *For $|v_2| < 1, \ldots, |v_m| < 1$, Eq. (7) has two zeros (counting their multiplicities) in the disk $|v_1| \leqslant 1$.*

Proof. For a 2×2 matrix, its determinant equals the product of diagonal entries minus the product of off-diagonal entries. We aim to apply the Rouche's theorem to prove that the determinant has as many zeros as the product of the diagonal elements does. Indeed, for $|v_1| = 1$, $|v_2| \leqslant 1$ we have

$$\left| \sum_{l=1}^{2} a_{k,l} a_{l,k} \bar{q}_1^{(l)}(v) q_1^{(k)}(v) v_1 R_1(v) \right| \leqslant \sum_{l=1}^{2} a_{k,l} a_{l,k} < 1 = |v_1|, \quad k = 1, 2.$$

Let us prove that

$$\left| v_1 - \sum_{l=1}^{2} a_{1,l} a_{l,1} \bar{q}_1^{(l)}(v) q_1^{(1)}(v) v_1 R_1(v) \right| \geqslant \left| \sum_{l=1}^{2} a_{1,l} a_{l,2} \bar{q}_1^{(l)}(v) q_1^{(1)}(v) v_1 R_1(v) \right|$$

Let us write the complex-valued quantities of interest in polar form:

$$r_l(t) e^{\mathbf{i} A_l(t)} = \bar{q}_1^{(l)}(v) q_1^{(1)}(v) v_1 R_1(v),$$

where $0 \leqslant r_l(t) \leqslant 1$ and $0 \leqslant A_l(t) \leqslant 2\pi$ for $0 \leqslant t \leqslant 2\pi$, $l = 1, 2$. Then, on a circle $|v_1| = 1$ the left-hand side can be bounded from below as follows:

$$\begin{aligned} \left| e^{\mathbf{i}u} - \sum_{l=1}^{2} a_{1,l} a_{l,1} r_l(u) e^{\mathbf{i} A_l(u)} \right|^2 &= \left(1 - a_{1,1}^2 r_1(u) - a_{1,2} a_{2,1} r_2(u)\right)^2 \\ &+ \sum_{l=1}^{2} 2 a_{1,l} a_{l,1} r_l(u) (1 - \cos(u - A_l(u))) \\ &+ 2 a_{1,1}^2 a_{1,2} a_{2,1} r_1(u) r_2(u) (1 - \cos(A_1(u) - A_2(u))) \\ &\geqslant \left(1 - a_{1,1}^2 r_1(u) - a_{1,2} a_{2,1} r_2(u)\right)^2, \end{aligned}$$

and for the right-hand side,

$$\left|\sum_{l=1}^{2} a_{1,l}a_{l,2}r_l(u)e^{\mathrm{i}A_l(u)}\right|^2 = \big(a_{1,1}(1-a_{1,1})r_1(t)+a_{1,2}(1-a_{2,1})r_2(t)\big)^2$$
$$-2a_{1,1}(1-a_{1,1})a_{1,2}(1-a_{2,1})r_1(u)r_2(u)(1-\cos(A_1(u)-A_2(u)))$$
$$\leqslant \big(1-a_{1,1}^2r_1(u)-a_{1,2}a_{2,1}r_2(u)-1+a_{1,1}r_1(t)+a_{1,2}r_2(u)\big)^2$$
$$\leqslant \big(1-a_{1,1}^2r_1(u)-a_{1,2}a_{2,1}r_2(u)\big)^2,$$

since

$$0 \leqslant a_{1,1}r_1(t)+a_{1,2}r_2(u) \leqslant a_{1,1}+a_{1,2}=1.$$

The proof is completed.

In the rest of the paper we'll study a queueing system with two input flows, $m=2$.

One may ask, are solution $v_1^{(1)}(v_2)$, $v_1^{(2)}(v_2)$ to Eq. (9) actually holomorphic in the open disk $|v_2|<1$? Singular points of these function should satisfy the system

$$F(v_1,v_2) := \det \tilde{Q}_{1,1}(v_1,v_2)=0, \qquad F_1(v_1,v_2) := \frac{\partial}{\partial v_1}\det \tilde{Q}_{1,1}(v_1,v_2)=0. \tag{16}$$

We can prove absence of singular points by computing the number of solutions of equations (16) inside a polydisk $\{(v_1,v_2)\colon |v_1|<1, |v_2|<1\}$. This number of solutions can be found using a multidimensional logarithmic residue formula (see [12,15]). It equals I_1+I_2, where

$$I_k = \int_{[0,1]^3} \frac{\det J_F(v_1,v_2)\overline{\det J_{[k]}(v_1,v_2)}}{(|F(v_1,v_2)|^2+|F_1(v_1,v_2)|^2)^2} e^{2\pi\mathrm{i}\theta_k} r_{\bar{k}}, dr_{\bar{k}}d\theta_1 d\theta_2, \quad k \neq \bar{k} \in \{1,2\},$$

$$J_F = \begin{pmatrix} \frac{\partial F}{\partial v_1} & \frac{\partial F}{\partial v_2} \\ \frac{\partial F_1}{\partial v_1} & \frac{\partial F_1}{\partial v_2} \end{pmatrix}, \quad J_{[1]} = \begin{pmatrix} \frac{\partial F}{\partial v_1} & F \\ \frac{\partial F_1}{\partial v_1} & F_1 \end{pmatrix}, \quad J_{[2]} = \begin{pmatrix} F & \frac{\partial F}{\partial v_2} \\ F_1 & \frac{\partial F_1}{\partial v_2} \end{pmatrix},$$

To evaluate these integrals a routine DCUHRE [16] was used. We'll say in advance that the integral was close to zero in all our experiments.

Let us assume that both service times and readjustment times are exponentially distributed,

$$B_j(t) = \begin{cases} 0 & \text{if } t \leqslant 0, \\ 1-e^{-t/\beta_j} & \text{if } t>0, \end{cases} \quad \text{and} \quad \bar{B}_j(t) = \begin{cases} 0 & \text{if } t \leqslant 0, \\ 1-e^{-t/\bar{\beta}_j} & \text{if } t>0 \end{cases}$$

for $j=1$, 2. Let the input flows be ordinary in the state $e^{(1)}$ ($f_j^{(1)}(z)=z$), and Gnedenko–Kovalenko type [14] in the state $e^{(2)}$ ($f_j^{(2)}(z)=g_jz+(1-g_j)z^2$ with $0<g_j<1$).

For the first experiment, set $a_{1,1} = 0.33$, $a_{2,2} = 0.6$ for the environment, $\lambda_1^{(1)} = 0.625$, $\lambda_2^{(1)} = 0.15$, $\lambda_1^{(2)} = 0.5$, $\lambda_2^{(2)} = 0.1$, $g_1 = 0.75$, $g_2 = 0.5$ for the input flows, $\beta_1 = 0.2$, $\bar{\beta}_1 = 0.05$, $\beta_2 = 0.375$, $\bar{\beta}_2 = 0.125$ for the server, $p_{1,1} = 0.25$, $p_{1,2} = 0$, $p_{2,1} = 0.43$, $p_{2,2} = 0.1$ for the feedback flows. It is easy to check that here $\bar{\lambda}_j^{(1)} = \lambda_j^{(1)} = \bar{\lambda}_j^{(2)} = (2 - g_j)\lambda_j^{(2)}$ for all $j = 1, 2$, so equalities (3) need to hold, we'll use them to assess accuracy of computed probabilities. The results are shown in Table 1 together with rough Monte–Carlo estimates. We will call the case when O_1 has priority over O_2 "order 1", and the opposite case when O_2 has priority over O_1 "order 2". In the last column, correct digits in the control sum are underlined. So, the absolute accuracy seems to be of order 10^{-6}. In the second and third column coinciding digits are underlined, and we find five and six coinciding digits in the probabilities related to different control policies.

For example, for "order 1" the zeros of $\det(v_2\tilde{Q}_{2,2}(v_2)) = 0$ were found to be approximately

$$v_2 = 0.9999999999992426 - 1.866770016832244 \cdot 10^{-14}\,\mathbf{i} \approx 1$$

and

$$v_2 = 0.001675593522702356 - 1.861484450736359 \cdot 10^{-14}\,\mathbf{i}$$

Then, (13) becomes

$$3.385895222324992 \cdot 10^{-7}\,\Psi(\bar{0}, 3, 2) - 5.664247186616641\bar{0}^{-7}\,\Psi(\bar{0}, 3, 1) = 0,$$

Equations (15) turn into

$$\begin{aligned} &-0.071850588988876\,\Psi(\bar{0}, 3, 2) - 0.27114573127955\,\Psi(\bar{0}, 3, 1) \\ &\quad + 1.3362207067866\,\Psi(\mathbf{0}^2\bar{1}, 3, 2) + 1.3362207067866\,\Psi(\mathbf{0}^2\bar{1}, 3, 1) = 0.296875, \\ &0.96875585864266\,\Psi(\bar{0}, 3, 2) + 0.88911857288000\,\Psi(\bar{0}, 3, 1) \\ &\quad - 0.86250703037120\,\Psi(\mathbf{0}^2\bar{1}, 3, 2) - 0.86250703037120\,\Psi(\mathbf{0}^2\bar{1}, 3, 1) = -0.01875; \end{aligned}$$

finally, using the zero $v_1(1)$ with $|v_1(1)| < 1$ we get from (10)

$$\begin{aligned} &2.672611460126473 \cdot 10^{-7}\,\Psi(\bar{0}, 3, 2) - 1.044306811888245 \cdot 10^{-6}\,\Psi(\bar{0}, 3, 1) \\ &-7.911439931036352 \cdot 10^{-4}\Psi(\mathbf{0}^2\bar{1}, 3, 2) + 0.0013244460897152 7\Psi(\mathbf{0}^2\bar{1}, 3, 1) = 0. \end{aligned}$$

For the second example, set $a_{1,1} = 0.3$, $a_{2,2} = 0.1$, $\lambda_1^{(1)} = 0.001$, $\lambda_2^{(1)} = 0.004$, $\lambda_1^{(2)} = 0.001$, $\lambda_2^{(2)} = 0.003$, $g_1 = 0.6$, $g_2 = 0.7$, $\beta_1 = 1$, $\beta_2 = 0.5$, $\bar{\beta}_1 = -0.5$, $\bar{\beta}_2 = 1$, $p_{1,1} = 0.1$, $p_{1,2} = 0.3$, $p_{2,1} = 0.3$, $p_{2,2} = 0.1$. Now the effective input intensities are different, $\bar{\lambda}_j^{(1)} \neq \bar{\lambda}_j^{(2)}$, and we can only rely on the accuracy estimate from Example 1. The computation results are shown in Table 2

Comparing only the underlined digits which we consider to be reliable, we may say that the probabilities are the same for both priority orderings.

Table 1. Computation results from the first experiment

Order	Probabilities		$\sum_{k=1}^{2} \frac{Q(n,\bar{0},k)}{\lambda_{+}^{(k)}}$
	$\Psi(\bar{0},3,1)$	$\Psi(\bar{0},3,2)$	(0.312373225)
Order 1	0.07659145571302693	0.1281294632748519	0.312376790
(simul.)	(0.0765674)	(0.128134)	
Order 2	0.07659030797447868	0.128130351846631	0.312376790
(simul.)	(0.0765644)	(0.128142)	

Table 2. Computation results from the second experiment

Order	Probabilities	
	$\Psi(\bar{0},3,1)$	$\Psi(\bar{0},3,2)$
Order 1	0.15166248480232	0.10904766425247
(simul.)	(0.15160)	(0.10905)
Order 2	0.15166407152574	0.10904654283947
(simul.)	(0.15168)	(0.10905)

5 Conclusion

An algorithm to solve stationary equations for probability distribution functions was proposed and probated on several parameter sets. Comparisons to known predicted combinations of stationary probabilities and rough Monte–Carlo simulation demonstrate that the method is viable. Further research is needed though to improve accuracy.

On the practical side, up to computation errors the stationary probabilities for empty queues seem independent of the server switching policy, which agrees with previous work using computer simulations.

References

1. Klimov, G.P.: Time-sharing service systems. I. Theory Probab. Appl. **19**(3), 532–551 (1975)
2. Kitaev, M.Yu., Rykov, V.V.: On a queuing system with the branching flow of secondary demands. Autom. Remote Control **9**, 52–61 (1980)
3. Fedotkin, M.A.: Optimal control for conflict flows and marked point processes with selected discrete component. I. Liet. Mat. Rinkinys. **4**(28), 783–794 (1988)
4. Fedotkin, M.A.: Optimal control for conflict flows and marked point processes with selected discrete component. II. Liet. Mat. Rinkinys. **1**(29), 148–159 (1989)
5. Fedotkin, M.A., Vysotsky, A.A.: Bartlett flow control in time-sharing systems. In: Twelfth Prague Conference on Information Theory, Statistical Decision Functions and Random Processes. Booklet of abstracts, Prague, pp. 110–122 (1994)

6. Fedotkin, M.A., Zorine, A.V.: Optimization of control of doubly stochastic nonordinary flows in time-sharing systems. Autom. Remote Control **7**(66), 1115–1124 (2005). https://doi.org/10.1007/s10513-005-0152-8
7. Fedotkin, M.A., Zorine, A.V.: Optimization of control of conflict flows with repeated service. J. Math. Sci. **4**(191), 492–505 (2013)
8. Maxima, a Computer Algebra System. Version 5.44.0. http://maxima.sourceforge.net/. Accessed 20 Dec 2020
9. Klimov, G.P.: Stochastic Queueing Systems. Nauka, Moscow (1996). (in Russian)
10. Shabat, B.V.: Introduction to complex analysis [Translated from Russian by J.S. Joel], vol. 2. AMS (1992)
11. Delves, L.M., Lyness, J.N.: A numerical method for locating the zeros of an analytic function. Math. Comput. **100**(21), 543–560 (1967)
12. Kravanja, P., Van Barel, M.: Computing the Zeros of Analytic Functions. LNM, vol. 1727. Springer, Heidelberg (2000). https://doi.org/10.1007/BFb0103927
13. Piessens, R., de Doncker-Kapenga, E., Uberhuber, C.W., Kahaner, D.K.: QUADPACK: A Subroutine Package for Automatic Integration. Springer Series in Computational Mathematics, vol. 1, 1st edn. Springer, Berlin (1983). https://doi.org/10.1007/978-3-642-61786-7
14. Gnedenko, B.V., Kvalenko, I.N.: Introduction to Queueing Theory. Birkhäuser, Boston, Basel, Berlin (1989)
15. Aizenberg, L.A., Yuzhakov, A.P.: Integral Representations and Residues in Multidimensional Complex Analysis. Nauka, Novosibirsk (1979). English Translation: Translations of Mathematical Monographs, vol. 58, Providence (1983)
16. Berntsen, J., Espelid, T.O., Genz, A.: Algorithm 698: DCUHRE - an adaptive multidimensional integration routine for a vector of integrals. ACM Trans. Math. Softw. **4**(17), 452–456 (1991)

Virtual Waiting Time in Single-Server Queueing Model M|G|1 with Unreliable Server and Catastrophes

Ruben Kerobyan[1] and Khanik Kerobyan[2(✉)]

[1] University of California San Diego, San Diego, CA, USA
[2] California State University Northridge, Northridge, CA, USA
khanik.kerobyan@csun.edu

Abstract. In the present paper, the single-server queue model M|G|1|∞ with unreliable server subject to catastrophes is considered. The transient and stationary distributions of virtual waiting time, busy period and idle state probability for two basic models with reliable and unreliable server are obtained. Different generalizations of basic models are considered: model with batch arrival of customers, model with non-homogeneous streams of customers and catastrophes, model with k types of customers, model with k types of priority customers. For those models, the virtual waiting time distribution and idle state probability are found.

Keywords: Virtual waiting time · Single-server queue · Catastrophes · Unreliable server · Idle state probability

1 Introduction

Over the last two decades, an increasing interest in queueing models with "negative" customers has been observed (see reviews [1–3]). Queueing networks with "negative" customers have been introduced and investigated in [4]. Different mechanisms of interaction between "regular" and "negative" customers have been considered. For example, the arriving "negative" customers can destroy, transfer or trigger some number of "regular" customers which are waiting in the model or are being served in the server. If an arriving "negative" customer can destroy all system workload, i.e. remove all "regular" customers which have been waiting and being served in the model, this type of "negative" customer is called catastrophe or disaster. Stochastic systems with clearing mechanism are considered in [5] using methods of renewal processes. Different queueing models with catastrophes and disasters and their applications are considered in [6–18]. Later in this paper we will use customers and catastrophes instead of "regular" and "negative" customers.

In queuing models with catastrophes, subjects of interest are three main characteristics: queue size, busy period, and virtual waiting time or workload of the model. One of first queuing models with catastrophes was considered in [6]. The product form solution for queue size distribution in steady state was obtained. The queuing model $M|G|1$ with catastrophes was investigated in [7]. The stationary workload process is

A. Dudin et al. (Eds.): ITMM 2020, CCIS 1391, pp. 319–336, 2021.
https://doi.org/10.1007/978-3-030-72247-0_24

considered and the analog of Pollaczeck-Khinchin formula for the model was found. The single server model $M|G|1$ with "stochastic clearing" mechanism-disaster and with server repair after the disaster was considered in [8]. The distribution of the number of customers in the model and sojourn time distribution in steady state were obtained. The models M|G|1, $M_x|G|1$ with recovery time after the catastrophes were investigated in [9, 10]. The queue size and busy period distributions in steady state are obtained. A queueing model M|G|1 with Poisson arrival of "negative" customers is considered in [11, 12]. It is showed that the workload distribution in this model equals the waiting time distribution in an equivalent GI/G/1 queue with "regular" customers only. The queue model M|G|1 with unreliable server and catastrophes was considered in [13] using supplementary variables and supplementary event methods. The transient and steady state distributions of queue size and busy period are investigated. Single server queue MAP|G|1 with Markov Arrival Process input of customers and catastrophes was considered in [14] using the supplementary variable method. The transient and stationary behavior of queue length, busy period, and virtual waiting time of the model are investigated. The model BMAP|SM|1 with Batch MAP input of customers and catastrophes was considered in [15] using Semi Markov (SM) processes and embedded Markov chain methods. The stationary distributions of EMC and SMP, distribution of actual and virtual waiting times of customers were found. The queue MAP|G|1 with MAP arrival of k types of customers and catastrophes is considered in [16]. The queue size and workload distributions in steady state are obtained. The queue model M|M|1 with catastrophes and two preemptive priority customers were introduced in [17] for the purpose of analyzing distributed database systems that undergo site failure. The waiting time distributions for different priority customers are found. The queue model $M_r|G_r|1$ with preemptive and non-preemptive customers and catastrophes is considered in [18]. The transient and stationary queue size and busy period distributions of the model are obtained.

In this paper, we consider transient and stationary distributions of virtual waiting time of the models M|G|1 with catastrophes and unreliable server. We suppose that service time of customers and recovery time of server have general distribution and arrival of customers, occurrence of disasters have Poisson distribution. This model is extended the results of [6–11, 18]. As an application of obtained results we consider the queueing model $M_r|G_r|1|\infty$ with priority service of customers and catastrophes.

2 Model Description

We consider a single server queueing model $M|G|1|\infty$ with unreliable server and catastrophes The arrival of customers and occurrence of catastrophes happen according to a Poisson distribution with parameters λ and ν, respectively. Catastrophes cannot be served and accumulated in the model. They can just remove all the customers being in the model at the moment of their occurrence. Service times of customers are independent and identically distributed (i.i.d.) random variables (r.v.) β with general distribution function (PDF) $B(x) = P(\beta < x)$, and finite mean value b_1 The server is unreliable: it can fail and be repaired only when the model is free of customers. Time intervals between two consecutive failures of server have exponential PDF with parameter α, and the repair times have general PDF $C(x)$ with mean value c_1.

The catastrophes act in the following manner: if the model is empty and the server is reliable, then they disappear without any influence on the model. If the server is busy by serving customers or is in repair, then the catastrophes remove all customers in the model, including the one in service, and interrupt the server repair. After the catastrophes, the model continues its work from an empty and reliable server state.

All customers serve in the model according to FCFS (first come – first serve) discipline. The model has unlimited waiting space. At the initial moment $t = 0$, the model is empty, the server is reliable and ready to serve the customers.

We will consider two cases of the model: first, model with reliable server, and second, model with unreliable server when the model is empty.

3 Virtual Waiting Time

Let's introduce following notations and definitions: $\{\xi(t),\ t \geq 0\}$ - is a stochastic process (SP) which describes workload or virtual waiting time of the model at the moment t and takes values in the state space $[0, \infty)$. For the considering model, SP $\{\xi(t),\ t \geq 0\}$ is a homogeneous Markov process with respect to time. For the farther convenience of analysis we will consider two different states of the SP $\xi(t)$: 0 - state, where the model is free of customers and $(0, \infty)$ - where the model is busy either serving the customers or repairing the server.

$W(x,t) = P\{\xi(t) < x\}$ - is the probability distribution function for SP $\{\xi(t),\ t \geq 0\}$,

$p_0(t) = W(0+,t) = P\{\xi(t) = 0\}$ - is an idle state probability of the model, i.e. model is free from customers and the server is reliable, at the moment t.

$\hat{W}(x,t) = W(x,t) - p_0(t) = P\{0 < \xi(t) < x\}$ - is a busy state probability of the model, i.e. the model is busy, at the moment t and total workload (virtual waiting time) of the model is less than x [19–21].

Obviously, between $W(x,t)$, $p_0(t)$, and $\hat{W}(x,t)$ the following relations take place:

$$W(x,t) = \begin{cases} 0, & \text{if } x \leq 0, \\ \hat{W}(x,t) + p_0(t), & \text{if } x > 0, \end{cases}$$

$$W(\infty,t) = \hat{W}(\infty,t) + p_0(t) = 1.$$

First of all, let's consider a model M|G|1 with reliable server and catastrophes. Using standard probabilistic arguments for $p_0(t)$ and $\hat{W}(x,t)$ we derive the following equations:

Case 1

$$\frac{d}{dt}p_0(t) = -\lambda p_0(t) + \frac{\partial}{\partial x}\hat{W}(x,t)|_{x=0} + v\hat{W}(\infty,t), \tag{1}$$

$$\frac{\partial}{\partial t}\hat{W}(x,t) - \frac{\partial}{\partial x}\hat{W}(x,t) = -(\lambda+v)\hat{W}(x,t) - \frac{\partial}{\partial x}\hat{W}(x,t)|_{x=0}$$
$$+\lambda\int_0^x \hat{W}(x-y,t)dB(y) + \lambda B(x)p_0(t),$$

Case 2

$$\frac{d}{dt}p_0(t) = -(\lambda+\alpha)p_0(t) + \frac{\partial}{\partial x}\hat{W}(x,t)|_{x=0} + v\hat{W}(\infty), \tag{2}$$

$$\frac{\partial}{\partial t}\hat{W}(x,t) - \frac{\partial}{\partial x}\hat{W}(x,t) = -(\lambda+v)\hat{W}(x,t) - \frac{\partial}{\partial x}\hat{W}(x,t)|_{x=0}$$
$$+\lambda\int_0^x \hat{W}(x-y,t)dB(y) + [\alpha C(x) + \lambda B(x)]p_0(t),$$

with initial conditions: $W(\infty,t) = \hat{W}(\infty,t) + p_0(t) = 1$, and $W(0+,0) = p_0(0) = 1$.

First let's analyze the steady state solution of the model,

$$\hat{W}(x) = \lim_{t\to\infty}\hat{W}(x,t), \; p_0 = \lim_{t\to\infty}p_0(t).$$

where p_0 - is a steady state probability of emptiness of the model, and $\hat{W}(x)$ is a steady state PDF of workload of the model, respectively.

From (1) and (2) for p_0 and $\hat{W}(x)$ we obtain

Case 1

$$0 = -\lambda p_0 + \frac{\partial}{\partial x}\hat{W}(x)|_{x=0} + v\hat{W}(\infty), \tag{3}$$

$$\frac{\partial}{\partial x}\hat{W}(x) = (\lambda+v)\hat{W}(x) + \frac{\partial}{\partial x}\hat{W}(x)|_{x=0} - \lambda\int_0^x \hat{W}(x-y)dB(y) - \lambda B(x)p_0,$$

Case 2

$$0 = -(\lambda+\alpha)p_0 + \frac{\partial}{\partial x}\hat{W}(x)|_{x=0} + v\hat{W}(\infty), \tag{4}$$

$$\frac{\partial}{\partial x}\hat{W}(x) = (\lambda+v)\hat{W}(x) + \frac{\partial}{\partial x}\hat{W}(x)|_{x=0} - \lambda\int_0^x \hat{W}(x-y)dB(y) - [\alpha C(x) + \lambda B(x)]p_0.$$

Now taking into account the relations between p_0, $\hat{W}(x)$ and $W(x)$ for PDF of virtual waiting time $W(x)$ we derive
Case 1

$$\frac{\partial}{\partial x}W(x) = (\lambda + v)W(x) - \lambda\int_0^x W(x-y)dB(y) - v. \tag{5}$$

Case 2

$$\frac{\partial}{\partial x}W(x) = (\lambda + v)W(x) - \lambda\int_0^x W(x-y)dB(y) - v + \alpha[1 - C(x)]p_0.$$

Let $\tilde{W}(s)$ be the Laplace - Stieltjes transformation (LST) of the function $W(x)$

$$\tilde{W}(s) = \int_0^\infty e^{-sx}dW(x).$$

Then from (5) and we get
Case 1

$$\tilde{W}(s) = \frac{sp_0 - v}{s - v - \lambda(1 - \tilde{B}(s))}, \tag{6}$$

Case 2

$$\tilde{W}(s) = \frac{[s + \alpha(1 - \tilde{C}(s))]p_0 - v}{s - v - \lambda(1 - \tilde{B}(s))}.$$

From (6) we can define the LST of distribution $\tilde{\hat{W}}(s)$,
Case 1

$$\tilde{\hat{W}}(s) = \tilde{W}(s) - p_0 = \frac{p_0\lambda\tilde{\bar{B}}(s) - v(1 - p_0)}{s - v - \lambda\tilde{\bar{B}}(s)} = \frac{p_0\lambda\tilde{\bar{B}}(s) - v\hat{W}(\infty)}{s - v - \lambda\tilde{\bar{B}}(s))},$$

Case 2

$$\tilde{\hat{W}}(s) = \frac{p_0[\alpha\tilde{\bar{C}}(s) + \lambda\tilde{\bar{B}}(s)] - v(1 - p_0)}{s - v - \lambda\tilde{\bar{B}}(s)} = \frac{p_0[\alpha\tilde{\bar{C}}(s) + \lambda\tilde{\bar{B}}(s)] - v\hat{W}(\infty)}{s - v - \lambda\tilde{\bar{B}}(s)},$$

where $\bar{A}(x) = 1 - A(x)$, and $\hat{W}(\infty) = 1 - p_0$.

$\hat{W}(x)$ can be interpreted as a workload which can be destroyed by the catastrophe at the moment of its occurrence, whereas $\hat{W}(\infty)$ - can be interpreted as a probability that the model is busy.

To find the value of unknown factor p_0 we can use different approaches. For instance, the factor p_0 can be found by using the fact that LST of virtual waiting time distribution is an analytic function (see [20–22]). We can also use the regenerative processes arguments [21, 23, 24], relation between characteristics of two M|G|1 models with and without catastrophes [25–29], and collective marks method [25–33]. In the next section we shall discuss the usage of these three methods to find the idle state transient and steady state probabilities $p_0(t)$, p_0 of the model.

4 The Idle State Probabilities

Let's use alternating renewal processes to define the idle state probability p_0 of the model. According to this approach the model can be described by the renewal process $\psi(t)$ which has two alternative periods: idle period when the model is free of customers and the server is reliable, and busy period - when the model serves customers or the server repairs after the failure. These two periods we denote farther by (0) and (1), respectively. The length of idle period is a r.v. ζ which has a Poisson distribution with parameter λ for case 1 and $\lambda+\alpha$ for case 2. The length of busy period is a r.v. $\hat{\eta}$ which has a PDF $\hat{\pi}(t) = P(\hat{\eta} \leq t)$, and mean value $\hat{\bar{\pi}}_1$.

Let p_0 and p_1 are stationary probabilities of states (0) and (1) at moment t. Then, according to [23], for the steady-state probabilities p_0 and p_1 we obtain

Case 1

$$p_0 = \frac{1}{1+\lambda\hat{\bar{\pi}}_1}, \; p_1 = \frac{\lambda\hat{\bar{\pi}}_1}{1+\lambda\hat{\bar{\pi}}_1}. \tag{7}$$

Case 2

$$p_0 = \frac{1}{1+\hat{\bar{\pi}}_1(\lambda+\alpha)}, \; p_1 = \frac{\hat{\bar{\pi}}_1(\lambda+\alpha)}{1+\hat{\bar{\pi}}_1(\lambda+\alpha)}.$$

To find the mean value of the busy period for the model with catastrophes we can use well known result for the busy period of the standard model M|G|1 without catastrophes, as shown in [25]. Let the r.v θ is the length of interval between two successive catastrophes, and the r.v. η_i, $i=1,2$ is the length of busy period of the standard M|G|1 model without catastrophes, where r.v. η_1 correspond to model with reliable server and r.v. η_2 correspond to model with unreliable server. Then the length of busy period $\hat{\eta}$ of the model with catastrophes can be find from $\hat{\eta}_i = \min\{\theta, \eta_i\}$. Let $F(t) = P(\theta < t)$ and $\pi_i(t) = P(\eta_i < t)$ are the PDFs of r.v. η_i and θ, and $\bar{\pi}_{i1}$ and $\bar{\theta}_1$ are their mean values, respectively. As it is well known (see for example [30, 31]), the LST of busy period for standard M|G|1 model $\tilde{\pi}_1(s)$ with reliable server is a smallest positive real root of the functional equation $\tilde{\pi}_1(s) = \tilde{B}(s+\lambda(1-\tilde{\pi}_1(s)))$, $\text{Re}s > 0$.

For the model with unreliable server the LST of busy period $\tilde{\pi}_2(s)$ can be found from functional equations (see for example [22, 30, 31, 34])

$$\tilde{\pi}_2(s) = \frac{\lambda}{\lambda+\alpha}\tilde{B}(s+(\lambda+\alpha)(1-\tilde{\pi}_2(s)) + \frac{\alpha}{\lambda+\alpha}\tilde{C}(s+(\lambda+\alpha)(1-\tilde{\pi}_2(s)) \quad \mathrm{Re}s > 0.$$

Thus, for the distribution of busy period of the model with catastrophes $\hat{\pi}(t)$ and its mean value $\hat{\bar{\pi}}_1$ we obtain

$$\hat{\pi}(t) = \int_0^t (1-\pi(x))dF(x) + \int_0^t (1-F(x))d\pi(x), \; \hat{\bar{\pi}}_1 = \int_0^\infty (1-\hat{\pi}(t))dt.$$

If catastrophes occur according to Poisson process with parameter v, $\bar{\theta}_1 = v^{-1}$, then for LST of PDF $\tilde{\tilde{\pi}}_i(s)$, and its mean value $\hat{\bar{\pi}}_{1i}$, $i = 1, 2$, (see [17, 23]) we derive

$$\tilde{\tilde{\pi}}_i(s) = \tilde{\pi}_i(s+v) + \frac{v}{s+v}[1-\tilde{\pi}_i(s+v)] = \frac{v+s\tilde{\pi}_i(s+v)}{s+v}, \; \hat{\bar{\pi}}_{1i} = \frac{1-\tilde{\pi}_i(v)}{v}, \; i = 1, 2. \tag{8}$$

The steady-state probabilities p_0 and p_1 can be obtained from (7)
Case 1

$$p_0 = \frac{v}{v+\lambda[1-\tilde{\pi}_1(v)]}, \; p_1 = \frac{\lambda[1-\tilde{\pi}_1(v)]}{v+\lambda[1-\tilde{\pi}_1(v)]}, \tag{9}$$

Case 2

$$p_0 = \frac{v}{v+[1-\tilde{\pi}_2(v)](\lambda+\alpha)}, \; p_1 = \frac{[1-\tilde{\pi}_2(v)](\lambda+\alpha)}{v+[1-\tilde{\pi}_2(v)](\lambda+\alpha)}.$$

After substitution of p_0 into (6) we find the LST of virtual waiting time distribution $\tilde{W}(s)$ for the corresponding models M|G|1 with catastrophes
Case 1

$$\tilde{W}(s) = \frac{v[s-v-a(1-\hat{\tilde{\pi}}(v))]}{[s-v-\lambda\bar{B}(s)][v+a(1-\hat{\tilde{\pi}}(v))]}. \tag{10}$$

Case 2

$$\tilde{W}(s) = \frac{v[s-v(1-\tilde{C}(s))]\{1-\tilde{C}(s)[v+\lambda[1-\tilde{\pi}(v)](1+\alpha v)]\}}{[s-v-\lambda(1-\tilde{B}(s))][v+\lambda[1-\tilde{\pi}(v)](1+\alpha v)]}.$$

Let $\bar{\omega}_1$ be a mean value of virtual waiting time, then from (10) for the case 1 we derive

$$\bar{\omega}_1 = \lim_{s \to 0}[-W'(s)] = \frac{p_0 - [1 - \lambda b_1]}{v} = \frac{p_0 - [1 - \rho]}{v}, \tag{11}$$

where ρ is a workload of the model, $\rho = \lambda b_1$.

Obviously when the arrival rate of catastrophes approaches zero $v \to 0$, from (11) and (9) we obtain the mean value of virtual waiting time $\hat{\bar{\omega}}_1$ and idle state stationary probability for the model without catastrophes $\hat{\bar{\omega}}_1 = \frac{\lambda b_2}{2(1-\rho)}$ and $\hat{p}_0 = 1 - \rho$.

Here b_2 is the second moment of the service time distribution.

To evaluate the mean value of the virtual waiting time $\bar{\omega}_1$ and idle state probability of the model p_0 for small value of v first we define the expansion of LST of a busy period distribution $\tilde{\pi}(v)$

$$\tilde{\pi}(v) = \int_0^\infty e^{-vt} d\hat{\pi}(t) = 1 - v\hat{\pi}_1 + \frac{v^2\hat{\pi}_2}{2} - \frac{v^3\hat{\pi}_3}{3!} + O(v),$$

where $\hat{\pi}_1 = \frac{b_1}{1-\rho}$, $\hat{\pi}_2 = \frac{b_2}{(1-\rho)^3}$, $\hat{\pi}_3 = \frac{b_3}{(1-\rho)^4} + \frac{3\lambda b_2^2}{(1-\rho)^5}$.

Then for p_0 and $\bar{\omega}_1$ we obtain

$$p_0 = (1-\rho) + \frac{v}{2}\frac{\lambda b_2}{(1-\rho)} + v^2\left(\frac{1}{6}\frac{\lambda b_3}{(1-\rho)^2} + \frac{3}{4}\frac{(\lambda b_2)^2}{(1-\rho)^3}\right) + O(v^3), \tag{12}$$

$$\bar{\omega}_1 = \frac{p_0 - (1-\rho)}{v} \cong \frac{1}{2}\frac{\lambda b_2}{(1-\rho)} + v\left(\frac{1}{6}\frac{\lambda b_3}{(1-\rho)^2} + \frac{3}{4}\frac{(\lambda b_2)^2}{(1-\rho)^3}\right) + O(v^2). \tag{13}$$

The (12) and (13) can be rewritten in more convenient form

$$p_0 = \hat{p}_0 + \frac{v}{2}\frac{\lambda b_2}{(1-\rho)} + v^2\left(\frac{1}{6}\frac{\lambda b_3}{(1-\rho)^2} + \frac{3}{4}\frac{(\lambda b_2)^2}{(1-\rho)^3}\right) + O(v^3),$$

$$\bar{\omega}_1 = \hat{\bar{\omega}}_1 + v\left(\frac{1}{6}\frac{\lambda b_3}{(1-\rho)^2} + \frac{3}{4}\frac{(\lambda b_2)^2}{(1-\rho)^3}\right) + O(v^2).$$

Where $\hat{\bar{\omega}}_1$ and $\hat{p}_0$ are defined above.

5 Transient Virtual Waiting Time of the Model M|G|1

In this section we consider the transient virtual waiting time distribution $W(x,t)$ of the models M|G|1 with catastrophes. To solve the Eqs. (1) and (2) for $\hat{W}(x,t)$ and $p_0(t)$ with initial conditions $p_0(0) = 0$, $\hat{W}(x,0)$, $x \geq 0$, $\hat{W}(x,\infty) + p_0(\infty) = 1$ we will use LST method.

Let $\tilde{W}(s,t)$ be a LST of $W(x,t)$ with respect to x

$$\tilde{W}(s,t) = \int_0^\infty e^{-sx} dW(x,t), \ \mathrm{Re}(s) > 0.$$

Taking into account the relation $W(x,t) = p_0(t) + \hat{W}(x,t)$, initial condition $W(x,0)$, and normalization condition $W(\infty,t) = 1, \quad t \geq 0$, then, for $\tilde{W}(s,t)$ we derive the following differential equations:

Case 1

$$\frac{\partial}{\partial t}\tilde{W}(s,t) - \tilde{W}(s,t)[s - v - \lambda(1 - \tilde{B}(s))] = v - sp_0(t). \tag{14}$$

Case 2

$$\frac{\partial}{\partial t}\tilde{W}(s,t) - \tilde{W}(s,t)[s - v - \lambda(1 - \tilde{B}(s))] = v - p_0(t)[s + \alpha\tilde{\bar{C}}(s)].$$

The solutions for (14) are

Case 1

$$\tilde{W}(s,t) = e^{\varphi(s,v)t}\left\{\tilde{W}(s,0) - s\int_0^t e^{-\varphi(s,v)u} p_0(u)du + v\int_0^t e^{-\varphi(s,v)u} du\right\}, \tag{15}$$

Case 2

$$\tilde{W}(s,t) = e^{\varphi(s,v)t}\left\{\tilde{W}(s,0) - [s + \alpha(1 - \tilde{C}(s))]\int_0^t e^{-\varphi(s,v)u} p_0(u)du + v\int_0^t e^{-\varphi(s,v)u} du\right\},$$

where $\varphi(s,v) = s - v - \lambda(1 - \tilde{B}(s))$, $\tilde{W}(s,0) = \tilde{\tilde{W}}(s,0)$.

The expressions for unknown function $p_0(t)$ we will derive by supplementary event method [22, 30–34].

Let's suppose that an event A can happen independently from the considering model. The intervals between two consecutive events A have exponential distribution with parameter s.

Then $s\tilde{p}_0(s) = \int\limits_0^\infty p_0(t)dt(1 - e^{-st})$ can be interpreted as a probability of an event {an event A occurs when the model is idle and the server is reliable}.

But this event can be realized in three mutually independent ways:

- {an event A appears earlier than a customer arrives or a server breaks down}. Recall, that at initial moment $t = 0$ the model is in idle (0) state and the server is reliable. In this state there are three competitive events: occurrence of an event A with the rate s, arrival of a customer with the rate λ, and a server breaking down with the rate α. The probability of this event is

$$\int_0^\infty e^{-(\lambda+\alpha)t} d(1-e^{-st}) = \frac{s}{s+\lambda+\alpha}.$$

- {among those three events (an arriving of a customer, the breaking down of a server, and the occurrence of an event A), the first is the customer's arrival. The probability of the first part of this event is

$$\int_0^\infty e^{-(s+\alpha)t} d(1-e^{-\lambda t}) = \frac{\lambda}{s+\lambda+\alpha}.$$

- {after arrival of customer the busy period of the model immediately begins. During this busy period an event A does not occur (with probability $\tilde{\pi}(s)$), it occurs during the following idle period (with probability $s\tilde{p}_0(s)$}. The probability of this event is

$$\frac{\lambda}{s+\lambda+\alpha}\tilde{\pi}(s)s\tilde{p}_0(s).$$

- {among those, the server breaks down first ($\frac{\alpha}{s+\lambda+\alpha}$) and immediately its renewal period begins. During this period, an event A does not occur ($\tilde{c}(s)$). It occurs during the following idle period ($s\tilde{p}_0(s)$)}. The probability of this event is

$$\frac{\alpha}{s+\lambda+\alpha}\tilde{c}(s)s\tilde{p}_0(s).$$

Now by using the full probability rule for $s\tilde{p}_0(s)$ we obtain

$$s\tilde{p}_0(s) = \frac{s}{s+\lambda+\alpha} + \frac{\lambda}{s+\lambda+\alpha}\tilde{\pi}(s)s\tilde{p}_0(s) + \frac{\alpha}{s+\lambda+\alpha}\tilde{c}(s+\lambda(1-\tilde{\tilde{\pi}}(s)))s\tilde{p}_0(s). \quad (16)$$

Hence from (16) for $\tilde{p}_0(s)$ we derive

$$\tilde{p}_0(s) = \frac{1}{s+\lambda(1-\tilde{\pi}(s))+\alpha[1-\tilde{C}(s+\lambda(1-\tilde{\tilde{\pi}}(s)))]}. \quad (17)$$

Now to present the probability $\hat{\tilde{p}}_0(s)$ by the parameters of the model without catastrophes $\tilde{p}_0(s)$ we use the relation [26, 27]:

$$\hat{p}_0(s) = \frac{s+v}{s} p_0(s+v). \tag{18}$$

From (17) and (18) for the $\tilde{p}_0(s)$ of the model M|G|1 with catastrophes we derive

$$\tilde{p}_0(s) = \frac{s+v}{s\{s+v+\lambda(1-\tilde{\tilde{\pi}}(s+v))+\alpha[1-\tilde{C}(s+v+\lambda(1-\tilde{\tilde{\pi}}(s+v)))]\}}. \tag{19}$$

From (19) when $\alpha = 0$ for the model M|G|1 with reliable server and catastrophes we obtain

$$\tilde{p}_0(s) = \frac{s+v}{s\{s+v+\lambda(1-\tilde{\tilde{\pi}}(s+v))\}}. \tag{20}$$

The corresponding steady-state probabilities p_0 for the model M|G|1 with catastrophes we obtain

Case 1

$$p_0 = \lim_{t\to\infty} p_0(t) = \lim_{s\to 0} s\tilde{p}_0(s) = \frac{v}{v+\lambda(1-\tilde{\tilde{\pi}}(v))}, \tag{21}$$

Case 2

$$p_0 = \frac{v}{v+\lambda(1-\tilde{\tilde{\pi}}(v))+\alpha[1-\tilde{C}(v+\lambda(1-\tilde{\tilde{\pi}}(v)))]}.$$

Remark 1. Let's suppose that customers arrive and catastrophes occur according to non-homogeneous Poisson processes with rates $\lambda(t)$ and $v(t)$, respectively. In this case the corresponding differential equations for $\tilde{W}(s,t)$ are

Case 1

$$\frac{\partial}{\partial t}\tilde{W}(s,t) - \tilde{W}(s,t)[s-v(t)-\lambda(t)\tilde{\tilde{B}}(s)] = v(t) - sp_0(t), \tag{22}$$

Case 2

$$\frac{\partial}{\partial t}\tilde{W}(s,t) - \tilde{W}(s,t)[s-v(t)-\lambda(t)\tilde{\tilde{B}}(s)] = v(t) - p_0(t)[s+\alpha\tilde{\tilde{C}}(s)],$$

and their solutions are

Case 1

$$\tilde{W}(s,t) = e^{\omega(t)} \left\{ \tilde{W}(s,0) - s \int_0^t e^{-\omega(u)} p_0(u) du + \int_0^t e^{-\omega(u)} v(u) du \right\}, \tag{23}$$

Case 2

$$\tilde{W}(s,t) = e^{\omega(t)} \left\{ \tilde{W}(s,0) - [s + \alpha(1 - \tilde{C}(s))] \int_0^t e^{-\omega(u)} p_0(u) du + \int_0^t e^{-\omega(u)} v(u) du \right\},$$

where $\varphi(s,v,t) = s - v(t) - \lambda(t)(1 - \tilde{B}(s))$, $\omega(u) = \int_0^u \varphi(s,v,y)dy$.

To solve the differential Eq. (14) we use LT method. Let $\tilde{\tilde{W}}(s,\theta)$ be the LT of PDF $\tilde{W}(s,t)$ with respect to t, namely

From (14), after taking LT, for $\tilde{\tilde{W}}(s,\theta)$ we get the following equations:
Case 1

$$\tilde{\tilde{W}}(s,\theta) = \frac{\tilde{W}(s,0) + \frac{v}{\theta} - s\tilde{p}_0(\theta)}{\theta - s + v + \lambda(1 - \tilde{B}(s))}. \tag{24}$$

Case 2

$$\tilde{\tilde{W}}(s,\theta) = \frac{\tilde{W}(s,0) + \frac{v}{\theta} - [s + \alpha(1 - \tilde{C}(s))]\tilde{p}_0(\theta)}{\theta - s + v + \lambda(1 - \tilde{B}(s))}.$$

As shown in [21], the unknown function $\tilde{p}_0(\theta)$ can be derived by using analytic properties of a function $\tilde{\tilde{W}}(s,\theta)$ with respect to arguments s and θ. Let s be a zero of the denominator of $\tilde{\tilde{W}}(s,\theta)$

$$s = \theta + v + \lambda(1 - \tilde{\pi}(\theta)).$$

Hence, from (24) for $\tilde{p}_0(\theta)$ we obtain
Case 1

$$\tilde{p}_0(\theta) = \frac{\tilde{W}(\theta + v + \lambda(1 - \tilde{\pi}(\theta)), 0) + \frac{v}{\theta}}{\theta + v + \lambda(1 - \tilde{\pi}(\theta))}, \tag{25}$$

Case 2

$$\tilde{p}_0(\theta) = \frac{\tilde{W}(s,0) + \frac{v}{\theta}}{[\alpha(1 - \tilde{C}(\theta + v + \lambda(1 - \tilde{\pi}(\theta)))) + \theta + v + \lambda(1 - \tilde{\pi}(\theta))]}.$$

Remark 2. Let's consider the model BM|G|1 with batch arrival of customers, catastrophes, "negative" customers and unreliable server. The batches arrive according to Poisson distribution with parameter λ. With probability q_r arriving batch contains r regular customers. Let $Q(z)$ be the probability generating function (PGF) of batch size and $\bar{q}$ is its mean value.

$$Q(z) = \sum_{r=0}^{\infty} q_r z^r.$$

"Negative" customers, catastrophes and server failures occur according to Poisson distribution with parameters λ^-, v, and α, respectively. The "negative" customers act in the following manner: if the model is empty or server is in repair station, then "negative" customer disappears without any influences on the model. If the model is busy by serving regular customers, then occurring "negative" customer removes a regular customer in service and with probability p initiates the second type recovery of server or with probability $1-p$ starts serving next regular customer. After the recovery, server continues serving remaining regular customers in the model. Second type recovery time has general PDF $G(t)$ with LST $\tilde{G}(s)$ and mean value $\bar{g}_1$.

For the LST of virtual waiting time distribution of the model we derive

$$\tilde{W}(s,t) = e^{\varphi(s,v)t}\left\{\tilde{W}(s,0) - [\alpha(1 - \tilde{C}_1(s)) + s]\int_0^t e^{\varphi(s,v)u} p_0(u)du + \int_0^t e^{\varphi(s,v)u} v du\right\},$$

where $\varphi(s,v) = s - v - \lambda Q(\tilde{B}_0(s))$, $\tilde{B}_0(s) = \tilde{B}(s+\lambda^-) + \frac{\lambda^-}{s+\lambda^-}\bar{\bar{B}}(s+\lambda^-)[1 - p\tilde{G}(s)]$.

Remark 3. The results for $\tilde{p}_0(s)$, p_0 and virtual waiting time distribution can be generalized in case of the model $M_k|G_k|1$ with k types of customers, unreliable server and catastrophes. Let's suppose that different types of customers arrive according to Poisson processes with the rates $\lambda_1, \lambda_2, \ldots, \lambda_k$ and their service times are i.i.d. r.v. with PDFs $B_1(t), B_2(t), \ldots, B_k(t)$ and finite mean values $\bar{b}_{11}, \bar{b}_{12}, \ldots, \bar{b}_{1k}$. For the model $M_k|G_k|1$ with conservative service discipline and catastrophes for LT of transient and steady state probabilities $\tilde{p}_0(s)$, p_0 we derive

$$\tilde{p}_0(s) = \frac{s+v}{s\{s+v+\sigma\tilde{\tilde{\pi}}(s+v) + \alpha[1 - \tilde{C}(s+v+\sigma(1 - \tilde{\tilde{\pi}}(s+v)))]\}},$$

$$p_0 = \frac{v}{v+\sigma\tilde{\tilde{\pi}}(v) + \alpha[1 - \tilde{C}(v+\sigma(1 - \tilde{\tilde{\pi}}(v)))]}.$$

Where $\tilde{\tilde{\pi}}(s)$ is a unique solution of the functional equation

$$\sigma\tilde{\tilde{\pi}}(s) = \sum_{i=1}^{k} \lambda_i \tilde{B}_i(s + \lambda(1 - \tilde{\tilde{\pi}}(s))), \ \sigma = \sum_{i=1}^{k} \lambda_i.$$

For the LST of steady state virtual waiting time PDF $\tilde{W}(s)$ of the model $M_k|G_k|1$ with k types of customers, catastrophes and unreliable server we derive:

$$\tilde{W}(s) = \frac{[s + \alpha(1 - \tilde{C}(s))]p_0 - v}{s - v - \sum_{i=1}^{k} \lambda_i (1 - \tilde{B}_i(s + \lambda(1 - \tilde{\tilde{\pi}}(s))))}. \tag{26}$$

Remark 4. Let's consider queuing model M|G|1 with catastrophes when arrival rate and PDF of service time of first customer which opens the busy period are λ_0 and $B_0(x)$; the arrival rate and PDF of other customers (during one busy period of the model) are λ and $B(x)$, respectively. As noted in [21, 31] this model can be used for modeling queuing systems with unreliable server, vacations and set up time. We suppose that both PDFs have finite mean values.

The corresponding differential equations for $W(x,t)$, LST $\tilde{W}(s,t)$ and its solution are

$$\begin{aligned}\frac{\partial}{\partial t}W(x,t) - \frac{\partial}{\partial x}W(x,t) &= -(\lambda + v)W(x,t) + \lambda \int_0^x W(x-y,t)dB(y) \\ &\quad + [\lambda_0 B_0(x) - \lambda B(x) - \alpha(1 - C(x))]p_0(t),\end{aligned}$$

$$\begin{aligned}\frac{\partial}{\partial t}\tilde{W}(s,t) - \tilde{W}(s,t)[s - v - \lambda(1 - \tilde{B}(s))] &= v - p_0(t)[s + \alpha(1 - \tilde{C}(s)) \\ &\quad - \lambda_0 \tilde{B}_0(s) + \lambda \tilde{B}(s)],\end{aligned}$$

$$\tilde{W}(s,t) = e^{\varphi(s,v)t}\left\{\tilde{W}(s,0) - [s + \alpha\bar{\tilde{C}}(s) - \lambda_0 \tilde{B}_0(s) + \lambda \tilde{B}(s)]\int_0^t e^{-\varphi(s,v)u}p_0(u)du + v\int_0^t e^{-\varphi(s,v)u}du\right\}.$$

Remark 5. As an application of obtained results let define virtual waiting time distribution for $M_k|G_k|1$ queueing model with priority service of customers and catastrophes. We consider queuing models $M_k|G_k|1$ with non-preemptive and preemptive priorities. Suppose that customers of ith priority (simply i^{th} type) arrive according to Poisson process with the rate λ_i, $i = 1, 2, \ldots, k$ and their service times are i.i.d. r.v. β_i with general PDF $B_i(x)$ and finite mean values b_{1i}. Suppose that i^{th} type customers have higher priority than j^{th} type customers if $i < j$. We will consider the steady state solution.

Let $\tilde{\pi}_n(s)$ be the LST of the busy period of the model with customers of priorities n and higher, and $\tilde{H}_n(s)$ be the LST of PDF of time interval starting with service of nth

type customer up to moment when the model is free of nth type customers and higher priority customers. Then for $\tilde{\pi}_n(s)$ and $\tilde{H}_n(s)$ we get [22],

$$\sigma_n\tilde{\pi}_n(s) = \sum_{j=1}^{n} \lambda_j \tilde{B}_j(s + \sigma_n(1 - \tilde{\pi}_n(s))), \ \sigma_n = \sum_{i=1}^{n} \lambda_i, \ n = 1, 2, \ldots, k.$$

$$\tilde{H}_n(s) = \tilde{B}_n(s + \sigma_{n-1}\bar{\tilde{\pi}}_{n-1}(s)).$$

To define the LST of virtual waiting time for n^{th} priority customers for the model with non-preemptive priorities and preemptive resume priorities we have to do the following substitutions into the results of the model M|G|1 with unreliable server and catastrophes: $\lambda = \lambda_n, \quad B(x) = H_n(x), \quad \alpha = \sigma_{n-1}, \quad \tilde{C}(s) = \tilde{\pi}_{n-1}(s), \quad v = v_n.$

Using the results obtained in [22] for priority queueing models we can define virtual waiting time distribution for different modifications of preemptive priorities, as well as for the models with alternative priorities, vocations and set-up times.

Remark 6. Now let's consider a single server model M|G|1 with recovery time after catastrophes. Catastrophes occur according to Poisson distribution with parameter v. The occurring catastrophe removes all customers in the model including the one in service, and the repair period of the server starts immediately. The customers which arrive during repair period, stay in the model, and get served after the repair of server. Suppose that catastrophes can occur during repair period of server as well and can remove all the customers waiting in the model. Server repair times have general PDF $C_2(x)$, with LST $\tilde{C}_2(s)$, and finite mean value c_2.

By using standard probabilistic arguments for $p_0(t)$ and $\hat{W}(x,t)$ of the model we derive the following differential equations and their transient and steady state solutions

$$\frac{d}{dt}p_0(t) = -\lambda p_0(t) + \frac{\partial}{\partial x}\hat{W}(x,t)|_{x=0}, \tag{27}$$

$$\frac{\partial}{\partial t}\hat{W}(x,t) - \frac{\partial}{\partial x}\hat{W}(x,t) = -(\lambda + v)\hat{W}(x,t) - \frac{\partial}{\partial x}\hat{W}(x,t)|_{x=0} + v\hat{W}(\infty)C_2(x)$$
$$+ \lambda\int_0^x \hat{W}(x-y,t)dB(y) + \lambda B(x)p_0(t),$$

$$\frac{\partial}{\partial t}\tilde{W}(s,t) - \tilde{W}(s,t)[s - v - \lambda(1 - \tilde{B}(s))] = v\tilde{C}_2(s) - p_0(t)[s - v(1 - \tilde{C}_2(s))], \tag{28}$$

and their transient and steady state solutions

$$\tilde{W}(s,t) = e^{\varphi(s,v)t}\left\{\tilde{W}(s,0) - [s - v\tilde{\tilde{C}}_2(s)]\int_0^t e^{-\varphi(s,v)u}p_0(u)du + v\tilde{C}_2(s)\int_0^t e^{-\varphi(s,v)u}du\right\},$$

$$\tilde{\tilde{W}}(s,\theta) = \frac{\tilde{W}(s,0) + \frac{v\tilde{C}_2(s)}{\theta} - \tilde{p}_0(\theta)[s - v(1 - \tilde{C}_2(s))]}{\theta - s + v + \lambda(1 - \tilde{B}(s))}, \tag{29}$$

$$\tilde{p}_0(\theta) = \frac{\tilde{W}(\theta + v + \lambda(1 - \tilde{\pi}(\theta)), 0) + \frac{v}{\theta}\tilde{C}_2(\theta + v + \lambda(1 - \tilde{\pi}(\theta)))}{\theta + v + \lambda(1 - \tilde{\pi}(\theta)) - v(1 - \tilde{C}_2(\theta + v + \lambda(1 - \tilde{\pi}(\theta))))},$$

$$\tilde{W}(s) = \frac{[s - v(1 - \tilde{C}_2(s))]p_0 - v\tilde{C}_2(s))}{s - v - \lambda(1 - \tilde{B}(s))}, \tag{30}$$

where p_0 is defined by $p_0 = \frac{v\tilde{C}_2(\delta))}{\delta - v(1 - \tilde{C}_2(\delta))}$.

Here δ is a unique positive zero of the function $f(s) = s - v - \lambda(1 - \tilde{B}(s))$ in the unite disk $|s| < 1$ of complex plane [30, 31].

6 Conclusion

In the present paper we consider two basic types of queue models M|G|1 with catastrophes, reliable and unreliable server. We also consider some generalizations of these models for queues with priorities, non-homogeneous arrivals of customers and occurrence of catastrophes. To define the virtual waiting time distribution and idle state probability of the model we show possibility of using different methods: for example, collective marks method, integro-differential equations of Takacs, or renewal theory. In the future research we plan to generalize queueing models with catastrophes by considering Marked Markov Arrival Processes stream of customers and occurrence of catastrophes, and develop the supplementary event method for these types of queueing models.

References

1. Artalejo, J.R.: G-networks: a versatile approach for work removal in queueing networks. Eur. J. Oper. Res. **126**(2), 233–249 (2000)
2. Bocharov, P.P., Vishnevskii, V.M.: G-networks: development of the theory of multiplicative networks. Autom. Remote Control **64**(5), 714–739 (2003)
3. Van Do, T.: Bibliography on G-networks, negative customers and applications. Math. Comput. Model. **53**(1–2), 205–212 (2011)
4. Gelenbe, E., Glynn, P., Sigman, K.: Queues with negative arrivals. J Appl. Prob. **28**, 245–250 (1991)
5. Stidham, Sh.: Stochastic clearing systems. Stochast. Process. Their Appl. **2**, 85–113 (1974)

6. Chao, X.: A queueing network model with catastrophes and product form solution. Oper. Res. Lett. **18**, 75–79 (1995)
7. Jain, G., Sigman, K.: A Pollaczek-Khintchine formula for M/G/1 queues with disasters. J. Appl. Probab. **33**(4), 1191–1200 (1996)
8. Yang, W.S., Kim, J.D., Chae, K.C.: Analysis of M/G/1 stochastic clearing systems. Stochast. Anal. Appl. **20**, 1083–1100 (2002)
9. Kim, B.K., Lee, D.H.: The M/G/1 queue with disasters and working breakdowns. Appl. Math. Model. **38**, 1788–1798 (2014)
10. Afthab, M.I., Begum, P., Bama, S.: Mx/G/1 queue with disasters and working breakdowns. Int. J. Sci. Res. Publ. **6**(4), 275–284 (2006)
11. Boucherie, R.J., Boxma, O.J., Sigman, K.: A note on negative customers, GI/G/1 workload, and risk processes. Prob. Eng. Inf. Sci. **11**(03), 305–311 (1997)
12. Boucherie, R.J., Boxma, O.J.: The workload in the M/G/1 queue with work removal. Prob. Eng. Inf. Sci. **10**(2), 261–277 (1996)
13. Kerobyan, K.: The model M|G|1|∞ with unreliable server and "negative" customers. Proc. Yerevan State Univ. Armenia Nat. Sci. Math. **3**, 11–19 (2007)
14. Li, Q.L., Zhao, Y.Q.: A MAP|G|1 queue with negative customers. Queueing Syst. **47**(1), 5–43 (2004)
15. Dudin, A.N., Nishimura, S.: A BMAP|SM|1 queueing system with Markovian arrival input of disasters. J. Appl. Probab. **36**, 868–881 (1999)
16. Inoue, Y., Takine, T.: The workload distribution in a MAP|G|1 queue with disasters. In: The 7th International Conference on Queueing Theory and Network Applications (QTNA2012), Kyoto, Japan, 1–3 August 2012
17. Towsley, D., Tripathi, S.K.: A single server priority queue with server failures and queue flushing. Oper. Res. Lett. **10**, 353–362 (1991)
18. Kerobyan, K.: The Mr|Gr|1|∞ queuing model with the priorities and "negative" customers. Bull. RAU Ser.: Phis.-Math. Nat. Sci. **2**, 6– 18 (2007)
19. Cox, D.R., Isham, V.: The virtual waiting-time and related processes. Adv. Appl. Probab. **18** (2), 558–573 (1986)
20. Takacs, L.: Investigation of waiting time problems by reduction to Markov processes. Acta Math. Acad. Sci. Ilungar. **6**, 101–129 (1955)
21. Gnedenko, B.V., Kovalenko, I.N.: Introduction to Queuing Theory. Nauka, Moscow (1966)
22. Gnedenko, B.V.: Service Systems with Priorities. Moscow University, Moscow (1973)
23. Cox, D.R., Miller, H.D.: The Theory of Stochastic Processes. Chapman & Hall\CRC, London (1996)
24. Cox, D.R: Renewal Theory. Methuen and Company, Ltd./Wiley, London/New York (1962)
25. Kerobyan, K., Covington, R., Kerobyan, R., Enakoutsa, K.: An infinite-server queueing $MMAP_k|G_k|\infty$ model in semi-Markov random environment subject to catastrophes. In: Dudin, A., Nazarov, A., Moiseev, A. (eds.) ITMM/WRQ -2018. CCIS, vol. 912, pp. 195–212. Springer, Cham (2018). https://doi.org/10.1007/978-3-319-97595-5_16
26. Kerobyan, K., Kerobyan, R.: Infinite-server queueing MAPk|Gk|1 model with k markov arrival streams, random volume of customers in random environment subject to catastrophes. In: Dudin, A., Nazarov, A., Moiseev, A. (eds.) ITMM/WRQ -2018. CCIS, vol. 912, pp. 305–321. Springer, Cham (2018)
27. Böhm, W.: A note on queueing systems exposed to disasters. Research report series. Department of Statistics and Mathematics, 79. WU Vienna University of Economics and Business, Vienna (2008)
28. Economou, A., Fakinos, D.: Alternative approaches for the transient analysis of Markov chains with catastrophes. J. Stat. Theory Pract. **2**(2), 183–197 (2008)

29. Pakes, A.G.: Killing and resurrection of Markov processes. Comm. Stat. – Stochast. Models **13**, 255–269 (1997)
30. Klimov, G.P.: Stochastic Service Ssystems. Nauka, Moscow (2018)
31. Matveev, V.F., Ushakov, V.G.: Queueing Systems. MSU, Moscow (1984)
32. Runnenburg, J.Th.: Probabilistic interpretation of some formulae in queueing theory. Bull. Inst. Int. Statist. **37**, 405–414 (1958)
33. Runnenburg, J.Th.: On the use of the method of collective marks in queueing theory. In: Smith, W.L, Wilkinson, W.E. (eds.) Congestion Theory, Chapel Hill, pp. 399–438 (1965)
34. Kleinrock, L.: Queueing Systems Vol I: Theory. Wiley, New York (1975)

A Queueing System with Probabilistic Joining Strategy for Priority Customers

Dhanya Babu(✉), Varghese C. Joshua, and Achyutha Krishnamoorthy

Department of Mathematics, CMS College, Kottayam 686001, Kerala, India
{dhanyababu,vcjoshua,krishnamoorthy}@cmscollege.ac.in

Abstract. We consider a single-server queueing system with two parallel queues of which one is a finite buffer for priority customers and the other infinite for ordinary customers. Two types of customers arrive according to the Marked Markovian Arrival Process (MMAP). Service times are assumed to follow phase-type distributions. A customer gets priority either by paying a cost or by any other means. Priority customers receive service on the basis of a token system which works according to the following rule: $K-1$ lower priority customers are served consecutively, and the Kth one is from the priority queue, if there is any. Priority customers have the right to take the strategic decision in choosing the queue on arrival, if such a customer joins an ordinary queue then he loses the special benefit that he would have got otherwise. We introduce the joining strategy for the priority customers and call it 'K-policy' along with probabilistic decision whether to give up their additional benefit (reward). Steady-state analysis of the model is done. Some system characteristics are evaluated, a social optimization problem is discussed and numerical illustrations are provided.

Keywords: Token · Joining strategy · Marked Markovian arrival process · Net benefits

1 Introduction

Classical priority queueing systems consider two or more parallel queues, formed by customers of distinct priorities $1(\text{highest}), 2, ..., m(\text{least})$, respectively. White and Christie were the first to consider priority queues. In [13], they considered the simple 2-priority $(M, M)/M/1$ queue and analyzed both non-preemptive and preemptive cases of services to derive the system state probability distribution, thereby deriving all-important system performance measures. Miller [10] employed the matrix geometric method to analyze the same system. Sapna and Stanford extended Miller's results to n-priority system in [4], with service time having phase-type distribution. Krishnamoorthy and Manjunath in [7] considered a priority system where only one queue is formed externally. In a subsequent paper [6], Krishnamoorthy and Manjunath extended the procedure of generating low priority queues internally to feedback queues. Analysis of various priority queues is described in [12].

A. Dudin et al. (Eds.): ITMM 2020, CCIS 1391, pp. 337–351, 2021.
https://doi.org/10.1007/978-3-030-72247-0_25

Multiple queues with simultaneous services are analyzed in [1]. MMAP model is described in [3] and [9]. Steady-state probabilities are computed using Matrix Geometric methods in [11]. The rate matrix is computed using Ramaswami's Logarithmic Reduction Algorithm by [8]. In [2], Deepak T.G. et al. considered queues with postponed work. The present work is an extension of [5], a single server queueing model with an optimal joining strategy for priority customers.

In this paper, we introduce an entirely different priority queue, two types of priority waiting lines are formed at a counter in a single server queue. One is called ordinary customers queue (OC) with infinite capacity waiting for space and the second one is referred to as queue of priority customers (PC) with finite capacity waiting for space.

The problems addressed in this paper are what is the optimal joining strategy of a priority customer upon arrival, whether to join the PC queue or OC queue to reduce his waiting time assuming there is no additional benefit (reward) for joining the PC queue, except that he is offered service as per the K-policy, and the other is what is the optimal K value in the sense of reducing the waiting of PC customers and minimizing their loss due to finite capacity of the PC queue? Along with the joining strategy, we introduced a probabilistic joining strategy for PC customers to join in PC queue or in OC queue by assuming additional benefit (reward). The case of $K = 1$ leads to a classical 2-priority queue; $K = 2$ results in the alternating queue and K infinity leads to exhaustive service discipline of OC queue before the PC queue is attended.

The rest of the article is organized as follows. Section 2 provides the mathematical modeling of the problem under study. Steady-state analysis of the model and some system performance measures are presented in Sect. 3.

2 Model Description

We consider a queueing system with two types of priority waiting for lines formed in front of a single server. One is called the ordinary customers (OC) queue and the second one is referred to as the priority customers (PC) queue. The ordinary queue is of infinite size and the priority queue is of finite size M. Customer priority may be either by paying a cost or by any other means.

2.1 Service Policy

Service policy is on the basis of a token system. A token is circulating from $1, 2, ..., K-1, K$. The system assigns priority to priority customers as follows. After serving continuously $K-1$ ordinary customers a priority customer, if present, will be taken for service. In the absence of a priority customer, ordinary customers, if present, are again served. Each time, a cycle of $K-1$ ordinary customers complete service continuously (one after the other), the server goes to PC queue to serve the head of the queue. If the PC queue is empty, he returns to OC queue. If OC queue gets empty before the counter reaches K, the server visits PC queue and the counter restarts. Also, the priority customer not only observes the token number but also the occupancy of both queues as well. The token number of the ordinary customers being served $1, 2, ..., K-1$, and K standing for a priority customer currently in service is displayed so that a priority customer can, upon arrival decide which queue he should join to minimize his waiting time. We assume that such an option is given to the priority customers alone. Every time the server serves a priority customer, he checks the OC queue to serve customers there, if there is any, else he attends the next priority customer. We call this process of assigning priority to priority customers "K-policy of priority".

2.2 Joining Strategy of Priority Customer

We introduce a joining strategy of an arriving priority customer as follows. An arriving priority customer observes the number of ordinary customers waiting, the type of customer being served and the number of priority customers in the queue. This information will help him to decide his joining strategy.

Suppose there are n ordinary customers in waiting, service in progress is that of an ordinary customer with his token number j, m priority customers are waiting and the K-policy is followed.

Thus if the tagged priority customer joins the PC queue:

Waiting time of PC = The service time of m PC's joined ahead of him + The residual service time of the ordinary customer in-service + Service time of $(K-1) - j$ ordinary customers in the present cycle + Service time of a maximum of $[(K-1)m]$ ordinary customers (future arrivals also to be taken into consideration).

Suppose instead when $n+1$ customers in OC queue get served before his turn comes in PC queue, if he were to join the PC queue, then he should join the OC queue(provided there is no extra benefit for joining it).

Let $S_1 = \lfloor \frac{i+r-(K+1)}{K-1} \rfloor (K-1)$, $S_2 = (K-1)j$, and $S_1^* = \lfloor \frac{i}{K-1} \rfloor (K-1)$.

In effect with K-policy for taking for service, the PC customer arriving at the station decide to join PC or OC queue according to the following conditions:

- If an OC customer is in service,
  An arriving PC customer decide to join PC queue if $S_1 > S_2$.
  An arriving PC customer decide to join OC queue if $S_1 \leq S_2$.
- If a PC customer is in service,
  An arriving PC customer decide to join PC queue if $S_1^* > S_2$.
  An arriving PC customer decide to join OC queue if $S_1^* \leq S_2$.

The arrival process is governed by a continuous-time Markov chain $\{A(t), t \geq 0\}$ with state-space $\{1, 2, ...n\}$. Even if the priority customer has the opportunity to get the chance to enter into service immediately as an ordinary customer than to stand in the priority queue, we assume that some of the priority customers take the probabilistic decision whether to give up their additional benefit (reward) or not with a probability. Instead of joining in the ordinary queue according to the optimal joining strategy, let the priority customer join in PC queue itself with probability p, where $0 \leq p \leq 1$. The sojourn time in the state i^1 is exponentially distributed with a positive λ_i^1, when the sojourn time in the state i^1

expires, the process jumps to the state j^1 without generation of a customer with probability $\lambda_i^1 p_{i^1 j^1}(0) = d_{i^1 j^1}(0)$ where i^1, j^1 $=\{1, 2, ..., n\}$. The process $J_2(t)$ jumps from state i^1 to j^1 with the generation of ordinary or priority customer with the probability $d_{i^1 j^1}(1)$ and $d_{i^1 j^1}(2)$, respectively, $\lambda_i^1 p_{i^1 j^1}(1) = d_{i^1 j^1}(1)$ and $\lambda_i^1 p_{i^1 j^1}(2) = d_{i^1 j^1}(2)$.

Let $D_0 = (d_{i^1 j^1}(0))$, $D_1 = (d_{i^1 j^1}(1))$ and $D_2 = (d_{i^1 j^1}(2))$. The matrix $D = D_0 + D_1 + D_2$ represents the generator of the process $\{J(t), t \geq 0\}$. The average total arrival intensity λ is defined by $\lambda = \boldsymbol{\theta}(D_1 + D_2)\mathbf{e}$, where $\boldsymbol{\theta}$ is an invariant vector of the stationary distribution of the Markov chain $\{A(t), t \geq 0\}$. The vector $\boldsymbol{\theta}$ is the unique solution to the system $\boldsymbol{\theta} D = \mathbf{0}$ and $\boldsymbol{\theta}\mathbf{e} = 1$, where $\mathbf{e}$ is a column-vector consisting of ones and $\mathbf{0}$ is a zero row-vector. The average arrival intensities of ordinary and priority customers are given by $\lambda^{(1)} = \boldsymbol{\theta} D_1 \mathbf{e}$ and $\lambda^{(2)} = \boldsymbol{\theta} D_2 \mathbf{e}$. The two types of customer's arrive according to a Marked Markovian Arrival Process (MMAP) with representation (D_0, D_1, D_2) with order n. Service time of an ordinary customer follows phase type distribution with representation $PH(\alpha, T)$ of order s_1 and that of priority customer follows phase type distribution with representation $PH(\beta, S)$ of order s_2.

Notations

Let

- $N_1(t)$ be the number of customers at time t in ordinary queue;
- $N_2(t)$ be the number of customers at time t in priority queue;
- $R(t)$ be the token numbers from $1, 2, ..., K-1, K$;
- $A(t)$ be the arriving phase of customer;
- $S(t)$ be the service phase of customer in service.

Let $\{(N_1(t), N_2(t), R(t), S(t), A(t)); t \geq 0\}$ be the Markov Process on the state space:

$\Omega = l^* \cup (\cup_{i=1}^{\infty} l(i))$, where

$l^* = \{(0, 0, 0, 0, v); v = 1, 2, ..l\}$, for $i \geq 1$,
$l(i) = \{i, j, r, u, v); i \geq 1, 0 \leq j \leq M, 1 \leq r \leq K, 1 \leq u \leq m, 1 \leq v \leq n\}$, depending on the joining strategy, where $m = \delta_{[r:1<K]} s_1 + \delta_{[r=K]} s_2$,

$$\delta_{[condition]} = \begin{cases} 1, & \text{if condition is true.} \\ 0, & \text{otherwise.} \end{cases}$$

3 Steady-State Analysis

The infinitesimal generator of the Markov chain is

$$Q = \begin{pmatrix} A_{10} & A_{00} & & & & & \\ A_{21} & A_{11} & A_{01} & & & & \\ & A_{22} & A_{12} & A_{02} & & & \\ & & \ddots & \ddots & \ddots & & \\ & & & A_{2\,N} & A_{1\,N} & A_{0\,N} & \\ & & & & A_2 & A_1 & A_0 \\ & & & & & \ddots & \ddots & \ddots \end{pmatrix},$$

where $N = M(K-1)+1$.

This model is a level independent quasi birth and death process (LIQBD). This can be conveniently and efficiently solved by the classical matrix analytic method.

The matrices A_0, A_1 and A_2 are as follows:

$$A_0 = diag\{E, \ldots, E_M\},$$

where

$$E = diag\{I_{s_1} \otimes D_1, \ldots, I_{s_2} \otimes D_1\},$$

$$E_M = diag\{I_{s_1} \otimes (D_1 + D_2), \ldots, I_{s_2} \otimes (D_1 + D_2)\}.$$

The matrix A_1 is

$$A_1 = \begin{pmatrix} F_1 & F_0 & & & & \\ F_2 & F_1 & F_0 & & & \\ & F_2 & F_1 & F_0 & & \\ & & \ddots & \ddots & \ddots & \\ & & & \ddots & \ddots & \ddots \\ & & & F_2 & F_1 & F_0 \\ & & & & F_2 & F_1 \end{pmatrix},$$

where $F_1 = diag\{T \oplus D_0, \ldots, S \oplus D_0\}, F_0 = diag\{I_{s_1} \otimes D_2, \ldots, I_{s_2} \otimes D_2\}$,

$$F_2 = \begin{pmatrix} O & O & \ldots & O \\ O & O & \ldots & O \\ \vdots & \vdots & \vdots & \vdots \\ O & O & \ldots & (T^0 \otimes \beta) \otimes I_l \\ O & O & \ldots & O \end{pmatrix}.$$

$$A_2 = diag\{G_0, G, \ldots, G\},$$

where

$$G_0 = \begin{pmatrix} & (T^0 \otimes \alpha) \otimes I_n & & \\ & & \ddots & \\ & & & (T^0 \otimes \alpha) \otimes I_n \\ (S^0 \otimes \alpha) \otimes I_n & & & \\ (T^0 \otimes \alpha) \otimes I_n & & & \end{pmatrix},$$

$$G = \begin{pmatrix} & (T^0 \otimes \alpha) \otimes I_n & & \\ & & \ddots & \\ & & & (T^0 \otimes \alpha) \otimes I_n \\ (T^0 \otimes \alpha) \otimes I_n & & & \\ (S^0 \otimes \alpha) \otimes I_n & & & \end{pmatrix}.$$

The matrix $A = A_0 + A_1 + A_2$ can be written as

$$A = \begin{pmatrix} C_1 & C_0 & & & & \\ C_2 & C_1 & C_0 & & & \\ & C_2 & C_1 & C_0 & & \\ & & \ddots & \ddots & \ddots & \\ & & & \ddots & \ddots & \ddots \\ & & & C_2 & C_1 & C_0 \\ & & & & C_2^1 & C_1^M \end{pmatrix},$$

where the matrices C_0, C_1, C_2, C_1^M are of same orders:

$$C_1 = \begin{pmatrix} T \oplus (D_0 + D_1) & (T^0 \otimes \alpha) \otimes I_l & & & \\ \ddots & \ddots & & & \\ & \ddots & \ddots & & \\ O & \dots & \dots & T \oplus (D_0 + D_1) & (T^0 \otimes \alpha) \otimes I_l \\ (S^0 \otimes \alpha) \otimes I_l & O & \dots & \dots & S \oplus (D_0 + D_1) \end{pmatrix},$$

$$C_2 = \begin{pmatrix} O & O \dots\dots & O \\ \vdots & \vdots \;\; \vdots \;\; \vdots & \vdots \\ T \oplus (D_0 + D_1) & O \dots O & (T^0 \otimes \beta) \otimes I_l \\ O & O \dots O & O \end{pmatrix}, \quad C_2^1 = \begin{pmatrix} O \; O \dots\dots & O \\ \vdots \;\; \vdots \;\; \vdots \;\; \vdots & \vdots \\ O \; O \dots O & (T^0 \otimes \beta) \otimes I_l \\ O \; O \dots O & O \end{pmatrix},$$

$$C_1^M = \begin{pmatrix} T \oplus D & (T^0 \otimes \alpha) \otimes I_l & & & \\ \ddots & \ddots & & & \\ & \ddots & \ddots & & \\ O & \dots & \dots & T \oplus D & (T^0 \otimes \alpha) \otimes I_l \\ (S^0 \otimes \alpha) \otimes I_l & O & \dots & \dots & S \oplus D \end{pmatrix},$$

$$C_0 = diag\{I_{s_1} \otimes D_2, \dots, I_{s_1} \otimes D_2\}.$$

3.1 Stability Condition

We see that A is an irreducible infinitesimal generator matrix and so there exists the stationary vector π of A such that $\pi A = 0$ and $\pi \mathbf{e} = 1$:

$$\pi_i = \pi_M \prod_{j=0}^{M-1-i} H_{M-1-j} \tag{1}$$

for $i = 0, 1, \ldots\ldots M-1$.

The sequence of matrices H_i are defined as $H_i = -C_2[H_{i-1}C_0 + C_1]^{-1}$ for $i = 1, 2, \ldots\ldots, M-1$ and $H_0 = -C_2[C_1^0]^{-1}$.

The vector π_M is obtained from the equation $\pi\mathbf{e} = 1$, where

$$\pi_M \left[\sum_{i=0}^{M-1} \prod_{j=0}^{M-1-i} H_{M-1-j} + I \right] \mathbf{e} = 1. \tag{2}$$

The stability condition is given by

$$\pi A_0 \mathbf{e} < \pi A_2 \mathbf{e},$$

where

$$\pi A_0 \mathbf{e} = [\pi_0 + \pi_1 + \pi_2 + \ldots\pi_{M-1}]\, E\mathbf{e} + \pi_M E_M \mathbf{e} = E\mathbf{e} + \pi_M(E_M - E)\mathbf{e},$$

$$\pi A_2 \mathbf{e} = [\pi_0 + \pi_1 + \pi_2 + \ldots\pi_{M-1}]\, G\mathbf{e} + \pi_M G_M \mathbf{e} = G\mathbf{e} + \pi_M(G_M - G)\mathbf{e}.$$

3.2 Computation of the Steady-State Vector

The Quasi-birth-death processes can be conveniently and efficiently solved using the Matrix Analytic Method.

The stationary distribution of the Markov process under consideration is obtained by solving the set of equations $\mathbf{x}Q = 0$, $\mathbf{x}\mathbf{e} = 1$.

Let $\mathbf{x}$ be the steady-state probability vector of Q. Partition this vector as: $\mathbf{x} = (\mathbf{x}_0, \mathbf{x}_1, \mathbf{x}_2, \ldots)$, where $\mathbf{x}_i = (\mathbf{x}_{i0}, \mathbf{x}_{i1}, \ldots \mathbf{x}_{iM})$, $\mathbf{x}_{ij} = (\mathbf{x}_{ij0}, \mathbf{x}_{ij1}, \mathbf{x}_{ij2}, \mathbf{x}_{ij3}, \ldots \mathbf{x}_{ijK})$ for $j = 0, 1, 2, \ldots, M$, whereas for r=1, 2 ...K, the vectors

$$\mathbf{x}_{ijr} = (\mathbf{x}_{ijr1}, \mathbf{x}_{ijr2}, \ldots, \mathbf{x}_{ijrm}),$$

where $\mathbf{x}_{ijru}$ is the probability of being in state (i, j, r, u) for $r = 1, 2, \ldots K, i \geq 0$, $j = 0, 1, 2 \ldots M, u = 1, 2, \ldots m$, the vectors $\mathbf{x}_{ijru} = (\mathbf{x}_{ijru1}, \mathbf{x}_{ijru2}, \ldots, \mathbf{x}_{ijrun})$, where $\mathbf{x}_{ijruv}$ is the probability of being in state (i, j, r, u, v) for $r = 1, 2, \ldots N$, $i \geq 0, j = 0, 1, 2 \ldots M, u = 1, 2, \ldots\ldots m, v = 1, 2, \ldots\ldots n$.

The steady-state vector $\mathbf{x}_i$ is obtained as

$$\mathbf{x}_i = \mathbf{x}_{i-1} L_i,$$

where $L_i = -A_{0(i-1)}[A_{1i} + L_{i+1}A_{2(i+1)}]^{-1}, 1 \leq i \leq N$,

$$\mathbf{x}_{M(K-1)+i} = \mathbf{x}_{M(K-1)} R^i, i \geq 1, \tag{3}$$

where the matrix R is the minimal non negative solution to the matrix quadratic equation

$$R^2 A_2 + R A_1 + A_0 = 0 \tag{4}$$

and R can be obtained by successive substitution procedure $R_0 = 0$ and $R_{k+1} = -V - R_k^2 W$, where $V = A_2 A_1^{-1}$, $W = A_0 A_1^{-1}$ by *Logarithmic Reduction Algorithm* developed by Latouche and Ramaswamy in [8].

The vectors $\mathbf{x}_0, ...\mathbf{x}_{N-1}$ are obtained by solving

$$\begin{aligned} \mathbf{x}_0 A_{10} + \mathbf{x}_1 A_{21} &= 0, \\ \mathbf{x}_{i-1} A_{0(i-1)} + \mathbf{x}_i A_{1i} + \mathbf{x}_{i+1} A_2 &= 0, 1 \leq i \leq N-1, \\ \mathbf{x}_{N-1} A_{0\,N-1} + \mathbf{x}_N A_1 + \mathbf{x}_{N+1} A_2 &= 0, \\ \mathbf{x}_{i-1} A_0 + \mathbf{x}_i A_1 + \mathbf{x}_{i+1} A_2 &= 0, i \geq N+1, \end{aligned} \tag{5}$$

subject to the normalizing condition

$$\mathbf{x}_0 \left[I + \sum_{j=2}^{M} (K-1) \prod_{i=1}^{j} L_i + \prod_{j=1}^{M(K-1)} L_j (I-R)^{-1} \right] \mathbf{e} = 1.$$

3.3 Waiting Time Distribution of an Ordinary Customer

For deriving the waiting time distribution of a tagged ordinary customer who joins the queue as the r^{*th} customer, we consider a Markov process. Let the Markov process be

$$\mathbf{X(t)} = \{(N(t), N_2(t), R(t), J_1(t), J_2(t)); t \geq 0\},$$

where

- $N(t)$ denotes the rank of the customer,
- $N_2(t)$ denotes the number of priority customers,
- $R(t)$ denotes the token numbers,
- $J_1(t)$ denotes the phase of the service process,
- $J_2(t)$ denotes the phase of the arrival process.

The state space of $\mathbf{X(t)}$ is $\{r^*, r^*-1, r^*-2, ...2, 1\} \times \{0, 1, ...M\} \times \{1, 2, ...K\} \times \{1, 2, ...m\} \times \{1, 2, ...n\} \cup 0^*$ denotes the absorbing state that a tagged customer is entered into service. The infinitesimal generator of waiting time distribution of tagged ordinary customer is

$$Q^* = \begin{pmatrix} T^* & T^{*0} \\ O & O \end{pmatrix},$$

where

$$T^* = \begin{pmatrix} W_1 & W_0 & & & \\ & W_1 & W_0 & & \\ & & W_1 & W_0 & \\ & & & \ddots & \\ & & & & W_1 \end{pmatrix}, \quad T^{*0} = \begin{pmatrix} O \\ O \\ O \\ \vdots \\ W^* \end{pmatrix},$$

where the matrix W_1 is

$$W_1 = \begin{pmatrix} X_1 & X_0 & & & & \\ X_2 & X_1 & X_0 & & & \\ & X_2 & X_1 & X_0 & & \\ & & \cdots & \cdots & \cdots & \\ & & & \cdots & \cdots & \cdots \\ & & & X_2 & X_1 & X_0 \\ & & & & X_2 & X_1 \end{pmatrix},$$

where

$$X_1 = diag\{T \oplus (D_0 + D_1), \ldots, S \oplus (D_0 + D_1)\},$$

$$X_0 = diag\{I_{s_1} \otimes D_2, \ldots, I_{s_2} \otimes D_2\},$$

$$X_2 = \begin{pmatrix} O & O & \ldots & O \\ \vdots & \vdots & \vdots & \vdots \\ O & O & \ldots & (T^0 \otimes \beta) \otimes I_n \end{pmatrix}, \quad W^* = \begin{pmatrix} Y_1 \\ \vdots \\ \vdots \\ Y_1 \end{pmatrix}, \quad Y_1 = \begin{pmatrix} (T^0 \otimes \alpha) \otimes I_n \\ (T^0 \otimes \alpha) \otimes I_n \\ \vdots \\ (S^0 \otimes \alpha) \otimes I_n \end{pmatrix},$$

$$W_0 = diag\{W_0^1, W_0^{11}, \ldots, W_0^{11}\},$$

$$W_0^1 = \begin{pmatrix} (T^0 \otimes \alpha) \otimes I_n & & & \\ & \ddots & & \\ & & (T^0 \otimes \alpha) \otimes I_n & \\ & & & (T^0 \otimes \alpha) \otimes I_n \\ (T^0 \otimes \alpha) \otimes I_n & & & \\ (S^0 \otimes \alpha) \otimes I_n & & & \end{pmatrix},$$

$$W_0^{11} = \begin{pmatrix} (T^0 \otimes \alpha) \otimes I_n & & \\ & \ddots & \\ & & (T^0 \otimes \alpha) \otimes I_n \\ (S^0 \otimes \alpha) \otimes I_n & & \end{pmatrix}.$$

Expected waiting time of an ordinary customer who joins the queue as r^{*th} customer is

$$W_{Q_1} = -(W_1)^{-1}(I - (W_0 W_1)^{-1})^{r^*}(I - W_0 W_1)^{-1}.$$

3.4 Waiting Time Distribution of a Priority Customer

For deriving the waiting time distribution of a tagged ordinary customer who joins the queue as the s^{*th} customer, we consider a Markov process. Let the Markov process be the Markov process

$$X^*(t) = \{(N(t), N_1(t), R(t), J_1(t), J_2(t)); t \geq 0\},$$

where

- $N(t)$ denotes the rank of the customer,
- $N_1(t)$ denotes the number of ordinary customers which is chosen such as Pr{Number of customers in the ordinary queue $\geq N\}< \epsilon$, for sufficiently small ϵ,
- $R(t)$ denotes the token numbers,
- $J_1(t)$ denotes the phase of the service process,
- $J_2(t)$ denotes the phase of the arrival process.

The state space of $X(t)$ is $\{s^*, s^*-1, s^*-2, ...2, 1\} \times \{0, 1, ...N\} \times \{1, 2, ...K\} \times \{1, 2, ...m\} \times \{1, 2, ...n\} \cup 0^*$, where 0^* denotes the absorbing state that the tagged customer is entered into service.

The infinitesimal generator of waiting time distribution of tagged ordinary customer is

$$Q^* = \begin{pmatrix} T^{1^*} & T^{1^{*0}} \\ O & O \end{pmatrix},$$

where

$$T^{1^*} = \begin{pmatrix} W^1{}_1 & W^1{}_0 & & & \\ & W^1{}_1 & W^1{}_0 & & \\ & & W^1{}_1 & W^1{}_0 & \\ & & & \ddots & \\ & & & & W^1{}_1 \end{pmatrix}, \quad T^{1^{*0}} = \begin{pmatrix} O \\ O \\ O \\ \vdots \\ W^{1*} \end{pmatrix},$$

where the matrix $W^1{}_1$ is

$$W^1{}_1 = \begin{pmatrix} X^1{}_1 & X^1{}_0 & & & & \\ X^1{}_2 & X^1{}_1 & X^1{}_0 & & & \\ & X^1{}_2 & X^1{}_1 & X^1{}_0 & & \\ & & \cdots & \cdots & \cdots & \\ & & & \cdots & \cdots & \cdots \\ & & & X^1{}_2 & X^1{}_1 & X^1{}_0 \\ & & & & X^1{}_2 & X^1{}_1 \end{pmatrix},$$

$$X^1{}_1 = diag\{T \oplus (D_0 + D_2), \ldots, S \oplus (D_0 + D_2)\},$$
$$X^1{}_0 = diag\{I_{s_1} \otimes D_1, \ldots, I_{s_2} \otimes D_1\},$$

$$X^1{}_2 = \begin{pmatrix} (T^0 \otimes \beta) \otimes I_n \end{pmatrix}, \quad W^{1^*} = \begin{pmatrix} Y^1{}_1 \\ \vdots \\ \vdots \\ Y^1{}_1 \end{pmatrix}, \text{ where } Y^1{}_1 = \begin{pmatrix} O \\ \vdots \\ \vdots \\ (T^0 \otimes \alpha) \otimes I_n \\ O \end{pmatrix},$$

$$W^1{}_0 = \begin{pmatrix} W^{1}{}^{1}_{0} & & \\ & W^{1}{}^{11}_{0} & \\ & & \ddots \\ & & & W^{1}{}^{11}_{0} \end{pmatrix}, \quad W^{1}{}^{1}_{0} = \begin{pmatrix} O & O & \ldots & (T^0 \otimes \alpha) \otimes I_n \\ \vdots & \vdots & \vdots & \vdots \\ O & O & \ldots & (T^0 \otimes \alpha) \otimes I_n \\ O & O & \ldots & (S^0 \otimes \alpha) \otimes I_n \end{pmatrix},$$

$$W^{1}{}^{11}_{0} = \begin{pmatrix} O & O & \ldots & O \\ \vdots & \vdots & \vdots & \vdots \\ O & O & \ldots & (T^0 \otimes \alpha) \otimes I_n \\ O & O & \ldots & O \end{pmatrix}.$$

Expected waiting time of an priority customer who joins the queue as s^{*th} customer is

$$W_{Q_2} = -(W^1{}_1)^{-1}(I - (W^1{}_0 W^1{}_1)^{-1})^{s^*}(I - W^1{}_0 W^1{}_1)^{-1}.$$

3.5 Performance Measures

1. Expected number of ordinary customers in the system:

$$E[N_1] = \sum_{i=0}^{\infty} i\mathbf{x}_i\mathbf{e}.$$

2. Expected Number of priority customers in the system:

$$E[N_2] = \sum_{j=1}^{M} j\mathbf{x}_{ij}\mathbf{e}.$$

3. Expected number of customers in the system:

$$E[N] = E[N_1] + E[N_2].$$

4. The probability that the server is idle:

$$t_0 = x_{00}.$$

5. The probability that the server is busy with an ordinary customer:

$$t_1 = \sum_{i=0}^{\infty}\sum_{j=0}^{M}\sum_{r=1}^{K-1} \mathbf{x}_{ijr}\mathbf{e}.$$

6. The probability that the server is busy with a priority customer:

$$t_2 = \sum_{i=0}^{\infty}\sum_{j=0}^{M} \mathbf{x}_{ijK}\mathbf{e}.$$

7. The probability that a priority customer is blocked from entering the system upon arrival:

$$P_b = \sum_{i=0}^{\infty}\sum_{r=1}^{K} \mathbf{x}_{iMr}\mathbf{e}.$$

3.6 Social Optimization Problem

In this section we propose a social optimization problem:

- a reward or benefit R_1 monetary units for a priority customer joining in ordinary queue;
- a reward or benefit R_2 monetary units for a priority customer joining in priority queue;
- a waiting cost h_1 monetary units for each unit of time that a priority customer waiting in ordinary queue;

- a waiting cost h_2 monetary units for each unit of time that a priority customer waiting in priority queue;
- a cost of c_1 monetary units for each priority customer being served in ordinary queue;
- a cost of c_2 monetary units for each priority customer being served in priority queue;
- W_{Q_1} be the expected waiting time of an ordinary customer;
- W_{Q_2} be the expected waiting time of priority customer.

Net Benefit for a priority customer if he joins in the ordinary queue Q_1 according to joining strategy is

$$B_1 = R_1 - (1-p)h_1 W_{Q_1} - c_1.$$

Net Benefit for a priority customer if he joins in the priority queue Q_2

$$B_2 = R_2 - ph_2 W_{Q_2} - c_2.$$

- If $B_1 > 0$ and $B_1 > B_2$, then the priority customer join in ordinary queue Q_1.
- If $B_1 > 0$ and $B_1 < B_2$, then the priority customer join in priority queue Q_2.
- If $B_1 < 0$ and $B_2 > 0$ then the priority customer join in priority queue Q_2.
- $B_1 < 0$ and $B_2 < B_1$, then the priority customer join in ordinary queue Q_1.
- If $B_1 = 0$ and $B_2 = 0$, then the priority customer either join in ordinary queue Q_1 or in priority queue Q_2.
- If $B_1 = 0$ and $B_1 > B_2$, then the priority customer join in ordinary queue Q_1.
- If $B_1 = 0$ and $B_1 < B_2$, then the priority customer join in priority queue Q_2. Similarly,
- If $B_2 > 0$ and $B_2 > B_1$, then the priority customer join in priority queue Q_2.
- If $B_2 > 0$ and $B_2 < B_1$, then the priority customer join in ordinary queue Q_1.
- If $B_2 < 0$ and $B_1 > 0$ then the priority customer join in ordinary queue Q_1.
- $B_2 < 0$ and $B_1 < B_2$, then the priority customer join in priority queue Q_2.
- If $B_2 = 0$ and $B_1 < B_2$, then the priority customer join in priority queue Q_2.
- If $B_2 = 0$ and $B_1 > B_2$, then the priority customer join in ordinary queue Q_1.

Example 1. In this example we look at the effect of varying p on some measures.

For the arrival process we consider Markovian arrival processes with representation D_0, D_1 and D_2 given by

$$D_0 = \begin{pmatrix} -2.0044 & 2.0044 & 0.0000 \\ 0.0000 & -2.0044 & 0.0000 \\ 0.0000 & 0.0000 & -451.50000 \end{pmatrix}, D_1 = \begin{pmatrix} 0.0000 & 0.0000 & 0.0000 \\ 1.1906 & 0.0000 & 0.0120 \\ 2.7090 & 0.0000 & 268.1910 \end{pmatrix},$$

$$D_2 = \begin{pmatrix} 0.0000 & 0.0000 & 0.0000 \\ 0.7938 & 0.0000 & 0.0080 \\ 1.8060 & 0.0000 & 178.7940 \end{pmatrix}.$$

This MAP process is normalized so as to have an arrival rate $\lambda = 2$, where $\lambda_1 = 1.2$, $\lambda_2 = 0.8$ and has correlated arrivals with positive correlation between successive inter-arrival times given by 0.4886.

For the service time distribution of ordinary and priority customers we consider the following phase type distributions $\alpha = (0, 1)$, $T = \begin{pmatrix} -5 & 5 \\ 0 & -5 \end{pmatrix}$, $\beta = (0, 1)$, and $S = \begin{pmatrix} -6 & 6 \\ 0 & -6 \end{pmatrix}$.

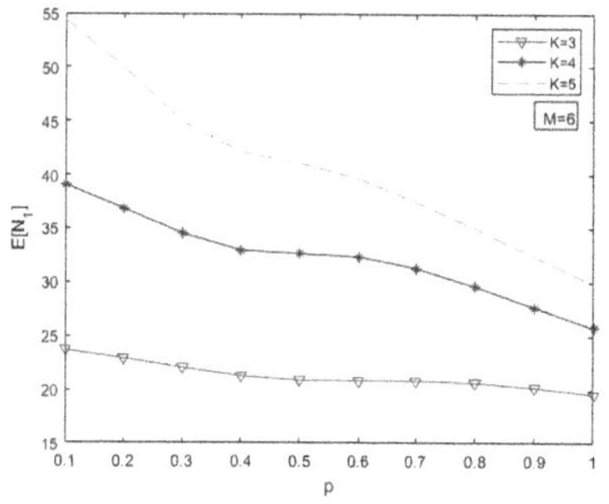

Fig. 1. Effect of p on $E[N_1]$ for different K

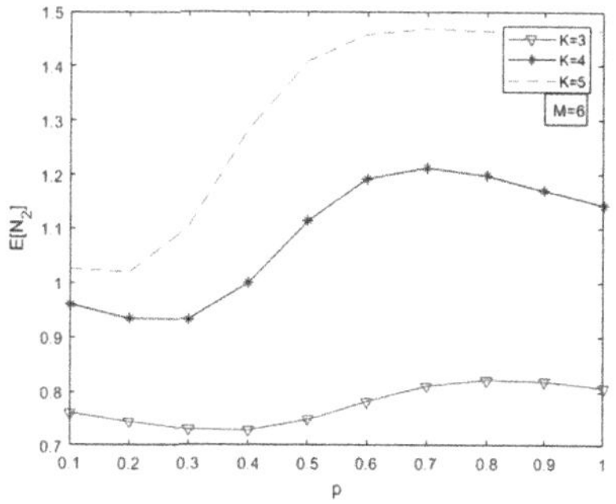

Fig. 2. Effect of p on $E[N_2]$ for different K

- From the Fig. 1, we can say that the expected number of customers in ordinary queue is monotonically decreasing as p increases. Figure 2 shows expected number of priority queue increases as p increases.This is because as p increases more customers join in priority queue than in ordinary queue.

Example 2. In this example we look at the effect of varying p on net benefits.

For the arrival process we consider Markovian arrival processes with representation D_0 and D_1:

$$D_0 = \begin{pmatrix} -1.0049 & 1.0049 & 0.0000 \\ 0.0000 & -1.0049 & 0.0000 \\ 0.0000 & 0.0000 & -226.3485 \end{pmatrix}, D_1 = \begin{pmatrix} 0.0000 & 0.0000 & 0.0000 \\ 0.9924 & 0.0000 & 0.0900 \\ 2.0430 & 0.0000 & 223.9818 \end{pmatrix},$$

$$D_2 = \begin{pmatrix} 0.0000 & 0.0000 & 0.0000 \\ 0.0024 & 0.0000 & 0.0010 \\ 2.2205 & 0.0000 & 0.1033 \end{pmatrix}.$$

This MAP process is normalized so as to have an arrival rate $\lambda = 1$, where $\lambda_1 = 0.9$, $\lambda_2 = 0.1$ and has correlated arrivals with positive correlation between successive inter-arrival times given by 0.4886.

For the service time distribution of ordinary and priority customers we consider the following phase type distributions $\alpha = (0.2, 0.8)$, $T = \begin{pmatrix} -10 & 10 \\ 0 & -10 \end{pmatrix}$, $\beta = (0.2, 0.8)$, and $S = \begin{pmatrix} -5 & 5 \\ 0 & -5 \end{pmatrix}$. Here we illustrate the individual strategy of priority customers with some examples.

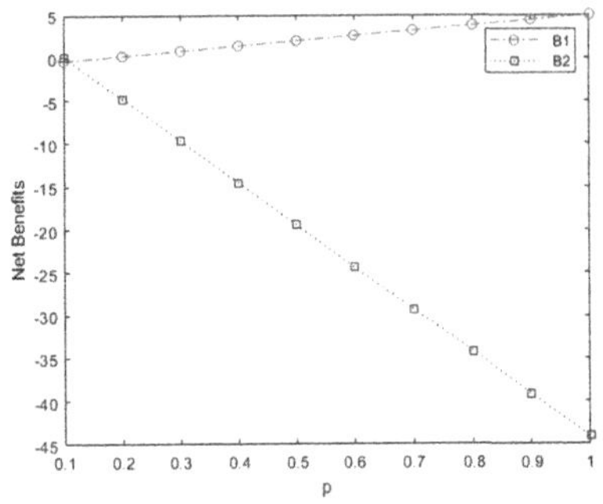

Fig. 3. Effect of p on B_1 and B_2

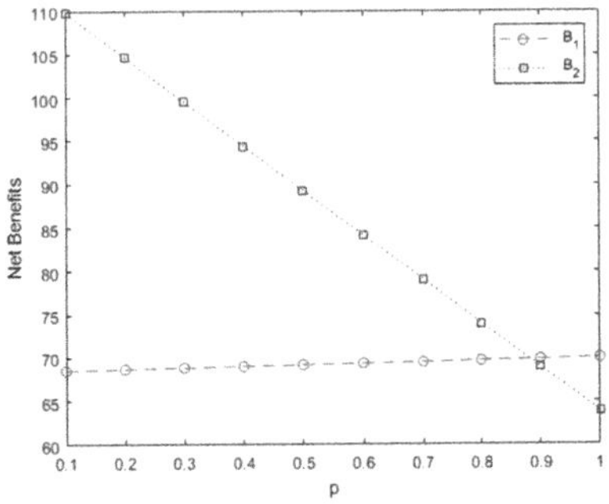

Fig. 4. Effect of p on B_1 and B_2

- For $K = 2$ and $M = 3$, $R_1 = 35$, $R_2 = 40$, $h_1 = 5$, $h_2 = 10$, $c_1 = 30$, $c_2 = 35$. Figure 3 shows that as p increases, the net benefit for a priority customer to join in ordinary queue increases and is positive so priority customer join in Q_1.
- For $K = 5$ and $M = 10$, $R_1 = 100$, $R_2 = 150$, $h_1 = 5$, $h_2 = 10$, $c_1 = 30$, $c_2 = 35$, Fig. 4 shows that the net benefit for a priority customer to join in priority queue decreases as p increases and is positive and is greater than B_1, so priority customer join in Q_2.

Conclusion

In this paper, we considered a queueing system with non-preemptive priority and a joining strategy for priority customers. Even if an optimal joining strategy exists for a priority customer upon arrival whether to join the priority queue or ordinary queue to reduce his waiting time, priority customers take probabilistic decision whether to give up their additional benefit (reward) for joining the priority queue. From the social optimisation problem, we conclude that priority customers take decision depending on the values of net benefits.

Acknowledgments. The work of first author was supported by the Maulana Azad National fellowship F1-17.1/2015-16/MANF-2015-17-KER-65493 of University Grants commission, India.

References

1. Chakravarthy, S., Thiagarajan, S.: Two parallel finite queues with simultaneous services and markovian arrivals. J. Appl. Math. Stochast. Anal. **10**(4) (1997). Article ID 394369. https://doi.org/10.1155/S1048953397000439
2. Deepak, T.G., Joshua, V.C., Krishnamoorthy, A.: Queues with postponed work. Top **12**, 375–398 (2004). https://doi.org/10.1007/BF02578967
3. Dudin, S., Kim, C., Dudina, O.: MMAP/M/N queueing system with impatient heterogeneous customers as a model of a contact center. Comput. Oper. Res. **40**(7), 1790–1803 (2013). https://doi.org/10.1016/j.cor.2013.01.023
4. Isotupa, S., Stanford, D.: An infinite-phase quasi-birth-and-death model for the non-preemptive priority M/PH/1 queue. Stoch. Model. **18**(3), 387–424 (2007). https://doi.org/10.1081/STM-120014219
5. Krishnamoorthy, A., Joshua, V., Babu, D.: A token based parallel processing queueing system with priority. Commun. Comput. Inf. Sci. **700**, 231–239 (2017). https://doi.org/10.1007/978-3-319-66836-9_19
6. Krishnamoorthy, A., Manjunath, A.: On queues with priority determined by feedback. Calcutta Stat. Assoc. Bull. **70**, 33–56 (2018). https://doi.org/10.1177/0008068318767271
7. Krishnamoorthy, A., Manjunath, A.S.: On priority queues generated through customer induced service interruption. Neural Parallel and Sci. Comput. **23**, 459–486 (2015)
8. Latouche, G., Ramaswami, V.: Introduction to matrix analytic methods in stochastic modeling. Soc. Ind. Appl. Math. (1999). https://doi.org/10.1137/1.9780898719734
9. Mathew, A., Krishnamoorthy, A., Joshua, V.: A retrial queueing system with orbital search of customers lost from an offer zone. Commun. Comput. Inf. Sci. **912**, 39–54 (2018). https://doi.org/10.1007/978-3-319-97595-5_4
10. Miller, D.R.: Computation of steady-state probabilities for M/M/1 priority queues. Oper. Res. **29**(5), 945–958 (1981). https://doi.org/10.1287/opre.29.5.945
11. Neuts, M.F.: Matrix-Geometric Solutions in Stochastic Models: An Algorithmic Approach. Dover Publications (1995)
12. Takagi, H.: Queueing Analysis: A Foundation of Performance Evaluation, vol. 1. North-Holland (1991). https://doi.org/10.1145/122564.1045501
13. White, H., Christie, L.S.: Queuing with preemptive priorities or with breakdown. Oper. Res. **6**(1), 76–96 (1958). https://doi.org/10.1287/opre.6.1.79

Analysis of Closed Unreliable Queueing Networks with Batch Service

Elena Stankevich(✉), Igor Tananko, and Oleg Osipov

Saratov State University, 83 Astrakhanskaya Street, Saratov 410012, Russia

Abstract. In this article, we study closed queuing networks with batch services of customers. Each node in the queueing network is an infinite capacity single-server queueing system under a RANDOM discipline. Customers move among the nodes following a routing matrix. We assume queueing systems in the network operate under the general batch service rule. The lower and upper bounds for the batch size are given. The batch service time is exponentially distributed. We presents an analysis of the queueing network using a Markov chain with continuous time. The qenerator matrix is constructed for the underlying Markov chain. We obtain expressions for the performance measures. In addition, we consider an unreliable case and propose an approximation. Some numerical examples are provided. The results can be used for the performance analysis manufacturing systems, production lines, trucking, ship locks.

Keywords: Queueing networks · Unreliable server · Batch service

1 Introduction

Queueing models with batch arrivals and services [1–5] are widely used for system performance evaluation and prediction for different kinds of real systems (production lines, buses, trucking, ship locks). In article [6] Chaudhry M. L. and Templeton J. G. C. present an overview of the main results for queueing systems with batch arrivals and batch services. There are a lot of results for both queueing systems and queueing networks [3,7,8].

Chao X. et al. [2] consider an open queueing network with single arrivals and batch services at each node. At a service completion the entire batch coalesces into a single unit, and it either leaves the system or goes to another node according to given routing probabilities. If the number of units present at a service completion epoch is less than the required number of units, then all the units coalesce into an incomplete batch which leaves the system. Main result is that this network possesses a geometric product form solution with a special type of traffic equations which depend on the batch size distribution at each node.

This work was supported by the Ministry of science and education of the Russian Federation in the framework of the basic part of the scientific research state task, project FSRR-2020-0006.

A. Dudin et al. (Eds.): ITMM 2020, CCIS 1391, pp. 352–362, 2021.
https://doi.org/10.1007/978-3-030-72247-0_26

In contrast [2], Economou A. [8] studied an open queueing network without incomplete batches. It is shown that the stationary distribution of the queueing systems has a nearly geometric form. Using quasi-reversibility arguments a modified network was constructed. This network has a tractable product-form stationary distribution which can be used as a bound or as an approximation for the stationary distribution of the original network under suitable conditions.

Miyazawa M. and Taylor P. G. [9] consider a continuous-time open network with batch movements and introduce the assembly-transfer batch service discipline. In the network a requested number of customers is simultaneously served at a node, and transferred to another node as, possibly, a batch of different size, if there are sufficient customers there; the node is emptied otherwise.

Articles [10,11] extend the works of Miyazawa M. and Taylor P. G. [9] and present queueing networks with triggered concurrent batch arrival and concurrent batch departure processes. It is shown that if an additional arrival process is introduced, the network has a product form which depends on a class of nonlinear traffic equations.

Economou A. [12] obtained a stochastic lower bound for the queueing network [9] by defining an additional departure process at server which tends to remove all the customers present in it.

Article [3] considers a single chain open queueing networks with a fixed batch size. A decomposition method based on a $GI^X/G^{(b,b)}/c$ queue approximation is proposed.

Paper [1] considers a closed queueing network with batch services and consequently batch arrivals. The network consists of $M/M^x/1$ and $M^x/M/1$ queueing systems. A method based on MVA [13] is presented.

The study of unreliable queueing systems with batch movements is one of the new fields within queueing theory. For example, this kind of queueing networks arise naturally in the performance analysis of manufacturing systems where stations are unreliable. In article [14], an unreliable closed queueing network with batch movements is considered. Each node in the network contains the same number of servers as the number of customers in the network, the service times of customers are exponential. The stationary distribution and various performance measures are obtained.

Optimization of queueing networks with batch services is considered in [4,15].

Mitici M. et al. [4] show that a tandem network of queues with batch services has a geometric product-form steady-state distribution and determine the service allocation that minimizes the waiting time in the system.

The problem of dynamic allocation of a single server with batch processing capability to a set of parallel queues is considered in [15]. Customers from different classes cannot be processed together in the same batch. It is shown that for the case of infinite buffers, allocating the server to the longest queue, stochastically maximizes the aggregate throughput of the system. For the case of equal-size finite buffers the same policy stochastically minimizes the loss of customers due to buffer overflows.

Paper [16] considers a closed queuing network with batch service and movements of customers in continuous time. Each node in the queueing network is an infinite capacity single server queueing system. Customers are served in batches of a fixed size. If a number of customers in a node is less than the size, the server of the system is idle until the required number of customers arrive at the node. An arriving at a node customer is placed in the queue if the server is busy. The batch service time is exponentially distributed. After a batch finishes its execution at a node, each customer of the batch, regardless of other customers of the batch, immediately moves to another node in accordance with the routing probability. The main performance measures are obtained.

This paper extends the queueing model presented in [16]. We assume queueing systems in the network operate under the general batch service rule, there are a minimum number and a maximum one for batch sizes. We obtain the main performance measures for the network. In addition, we consider an unreliable case and propose an approximation.

The rest of the paper is organized as follows. Section 2 describes the closed queueing network with batch services and batch movements. In Sect. 3, we consider the unreliable queueing networks. In Sect. 4 we obtain the main performance measures for the queueing networks. Section 5 provides various numerical examples. Finally, a section of conclusions commenting the main research contributions of this paper is presented.

2 The Model

Consider a closed queueing network $\mathcal{N}$ consisting of L nodes S_i, $i \in I = \{1, \dots, L\}$. There are H customers in the network.

Each node S_i, $i = 1, \dots, L$, operates like an infinite capacity single-server queueing system under a RANDOM discipline.

Arriving customers are placed in the queue if the node server is busy. Customers are served in batches, servers operate under the general batch service rule. The server at a node S_i may take in a batch of customers with minimum number x_i of customers and with maximum number y_i of customers. The service of a batch is started immediately after there are at least x_i customers in the queue.

Thus if the server finds q customers in the queue, there are the following possibilities

- $0 \leq q < x_i$, then the server waits till the queue size grows to x_i,
- $x_i \leq q \leq y_i$, then the server takes a batch of size q for service,
- $q > y_i$, then server takes a batch of size y_i for service.

The service times of batches at node S_i are exponentially distributed with parameter μ_i, $i = 1, \dots, L$. Therefore, the extended Kendall/Gnedenko notation for node S_i is $G^{a_i}/M^{(x_i,y_i)}/1$, $i = 1, \dots, L$, where a_i is the random variable and denotes the arriving batch size, its probability distribution will be defined below.

After a batch finishes its execution at a node, the customers leave the node. Transitions of single customers between nodes are defined by the routing matrix $\Theta = (\theta_{ij})$, $i, j = 1, \ldots, L$. Thus, a customer completing service at node S_i will go to node S_j with probability θ_{ij}, $i, j = 1, \ldots, L$.

Let s_i denote the number of customers at node S_i, $i = 1, \ldots, L$. Define a queueing network state as the vector $s = (s_1, \ldots, s_L)$.

Denote by $X = \left\{ s : s_i \geq 0, \sum_{i=1}^{L} s_i = H \right\}$ the state space of the queueing network. By V_i denote the set $V_i = \{j \in I\colon \theta_{ij} > 0\}$, $i = 1, \ldots, L$, the set defines nodes, which are reachable from node S_i.

Each transition is associated with an event in the system, consider a transition from state $s \in X$ to state $s' \in X$, $s \neq s'$:

1. Let a batch complete its service at node S_i, $i \in I$, and thus g_i customers leave node S_i. Denote by $d = (d_1, \ldots, d_L)$ a vector representing departing customers, all components of the vector equal to 0, except the ith, which is g_i. Let D be the set of the departing vectors.
2. Each of d_i customers in the batch, leaving from S_i, is routed independently to nodes according to probability θ_{ij}, $j \in V_i$. The customers join in batches of random sizes. Thus a batch of a_j customers goes to node S_j with probability $\theta_{ij}^{a_j}$.
3. A batch of a_j customers arrives at node S_j, then $d_i = \sum_{j \in V_i} a_j$. Thus vector $a = (a_1, \ldots, a_L)$ represents the arriving customers. Denote by A the set of the arriving vectors.
4. Thus for the state s' we have $s' = s - d + a$.

It is easy to see, the process $\{s(t), t > 0\}$ is a continuous time Markov chain on the state space X. As shown in [17], the transition rate $q(s, s')$ from state s to state s' is defined as:

$$q(s, s') = \sum_{\substack{s' \in X, \\ s' = s - d + a}} u(s) \rho(d, a), \quad d \in D, \quad a \in A, \quad s \in X, \tag{1}$$

$u(s)$ is a function associated with the service rates of the network, $\rho(d, a)$ is a function associated with the routing probabilities of the network.

According to the service policy we have

$$u(s) = \sum_{i=1}^{L} \mu_i \mathbf{1}(s_i \geq x_i), \tag{2}$$

where $\mathbf{1}(s_i \geq x_i) = 1$, if $s_i \geq x_i$, and $\mathbf{1}(s_i \geq x_i) = 0$ otherwise.

Let a batch of customers finish its service at node S_i, customers leave the node and independently arrive at nodes according to vector a.

Denote by ζ_j the random variable representing the size of the batch arriving at node S_j. As customers go between nodes of the network independently on each other, the random variables are independent random variables with the multinomial distribution. Thus we can write

$$P_{d_i}(\zeta_1 = a_1, \ldots, \zeta_L = a_L) = \binom{d_i}{a_1, \ldots, a_L} \prod_{j=1}^{L} \theta_{ij}^{a_j}.$$

Let us define now the probability distribution $\rho(d, a)$ for sizes of arriving batches, we can write

$$\rho(d, a) = \sum_{i=1}^{L} \binom{d_i}{a_1, \ldots, a_L} \prod_{j=1}^{L} \theta_{ij}^{a_j}, \quad d_i = \sum_{j \in V_i} a_j, \quad d \in D, \quad a \in A. \tag{3}$$

Substituting expressions for $u(s)$ and $\rho(d, a)$ in (1) into (2) and (3), we have

$$q(s, s') = \sum_{\substack{s' \in X, \\ s' = s-d+a}} \sum_{i=1}^{L} \mu_i \mathbf{1}(s_i \geq x_i) \binom{d_i}{a_1, \ldots, a_L} \prod_{j=1}^{L} \theta_{ij}^{a_j},$$

$s \in X$, $d_i = \sum_{j \in V_i} a_j$, $d \in D$, $a \in A$.

3 Unreliable Networks

Let $\bar{\mathcal{N}}$ be a network with the same parameters as network $\mathcal{N}$, but nodes in the network are subject to breakdown and repair when active.

Assume, at any arbitrary point in time, there are three possible states for each server: idle, active or inactive. The server at node S_i is idle when there are no customers to be served, active when a batch is being served with the corresponding rate μ_i, and inactive when the server is down and under repair.

The failure and repair times for each server at node S_i have exponential distributions with the corresponding parameters α_i and β_i. Thus, the total service time for the unreliable network can be exactly represented with a two-stage Coxian distribution.

Let $\frac{1}{\alpha_i} >> \frac{1}{\beta_i}$, $i = 1, \ldots, L$, then we model [18] the total service time as the exponential random variable with the expected value

$$\bar{r}_i = \frac{1}{\mu_i} \frac{\alpha_i + \beta_i}{\beta_i}, \quad i = 1, \ldots, L. \tag{4}$$

For the unreliable queueing network, the correspondinng Markov chain transition rates have the following form

$$q(s, s') = \sum_{\substack{s' \in X, \\ s' = s-d+a}} \sum_{i=1}^{L} \frac{\mathbf{1}(s_i \geq x_i)}{\bar{r}_i} \binom{d_i}{a_1, \ldots, a_L} \prod_{j=1}^{L} \theta_{ij}^{a_j}, \tag{5}$$

$s \in X$, $d_i = \sum_{j \in V_i} a_j$, $d \in D$, $a \in A$.

4 Performance Measures

The stationary distribution $\pi = (\pi(s))$, $s \in X$, for the queueing model is defined as a solution of the equation

$$\pi Q = 0, \qquad \sum_{s \in X} \pi(s) = 1,$$

here Q is the generator matrix, $Q = (q(s, s'))$, $s, s' \in X$.

Once the stationary distribution is computed, a variety of other performance measures may be obtained.

The average number $\bar{s}_i$ of customers at the node S_i

$$\bar{s}_i = \sum_{k=1}^{H} k \sum_{\substack{s \in X, \\ s_i = k}} \pi(s), \quad i = 1, \dots, L,$$

the arrival rate λ_i to node S_i

$$\lambda_i = \frac{1}{\bar{r}_i} \sum_{k=x_i}^{y_i} \Big(1 - \sum_{s \in X} \sum_{s_i=0}^{k-1} \pi(s)\Big), \quad i = 1, \dots, L,$$

the average response time $\bar{u}_i$ for node S_i

$$\bar{u}_i = \frac{\bar{s}_i}{\lambda_i}, \quad i = 1, \dots, L,$$

the average idle time $\bar{v}_i$ for node S_i

$$\bar{v}_i = \frac{\sum_{k=0}^{x_i-1} (x_i - k) \sum_{\substack{s \in X, \\ s_i = k}} \pi(s)}{\lambda_i \sum_{k=0}^{x_i-1} \sum_{\substack{s \in X, \\ s_i = k}} \pi(s)}, \quad i = 1, \dots, L,$$

the average waiting time $\bar{w}_i$ for node S_i

$$\bar{w}_i = \bar{u}_i - \frac{1}{\mu_i}, \quad i = 1, \dots, L,$$

the average number $\bar{b}_i$ of customers in the queue for node S_i

$$\bar{b}_i = \bar{w}_i \lambda_i, \quad i = 1, \dots, L.$$

Note, if $\alpha_i = 0$, $i = 1, ..., L$, then the performance measures for $\mathcal{N}$ and $\bar{\mathcal{N}}$ are the same.

5 Numerical Examples

Consider a queueing network which consists of $L = 7$ nodes with service rates $\mu = (0.8, 0.8, 0.9, 0.6, 1.0, 0.7, 0.7)$ and routing matrix Θ, where

$$\Theta = \begin{bmatrix} 0 & 0.3 & 0 & 0.7 & 0 & 0 & 0 \\ 0 & 0 & 0.5 & 0 & 0.5 & 0 & 0 \\ 0 & 0 & 0 & 0 & 0.2 & 0.8 & 0 \\ 0 & 0 & 0 & 0 & 0 & 0 & 1 \\ 0.2 & 0.2 & 0 & 0.4 & 0.2 & 0 & 0 \\ 0 & 0.6 & 0 & 0 & 0.4 & 0 & 0 \\ 0.4 & 0 & 0.4 & 0 & 0 & 0.2 & 0 \end{bmatrix}.$$

There are $H = 20$ customers in the network, the threshold values for queueing systems are $x = (2, 2, 2, 4, x_5, 2, 2)$, $y = (5, 5, 5, 7, y_5, 5, 5)$. Let us consider five examples, in each example we change the parameters only for node S_5.

The first example considers the reliable queuing network where $x_5 = y_5 = 1, \ldots, 10$. For this example, Figs. 1, 2, 3 illustrate the arrival rates, the average number of customers and the average idle times for all nodes. It may be seen that as x_5 and y_5 increase from 1 to 4, the arrival rates for all nodes increase too. There is a slight decrease for $x_5 = y_5 = 5, \ldots, 10$. The lower batch size thresholds influence both the arrival rates and the average idle times for all nodes, this is especially appreciable for Fig. 3 where depicted the average numbers of customers for node S_5.

In the second and third experiments, we examine the same performance measures, in the second experiment, x_5 increases from 1 to 10 for fixed $y_5 = 10$ (Figs. 4, 5 and 6), and in the third one, we assume $x_5 = 1$, and y_5 increases from 1 to 10 (Fig. 7, 8 and 9).

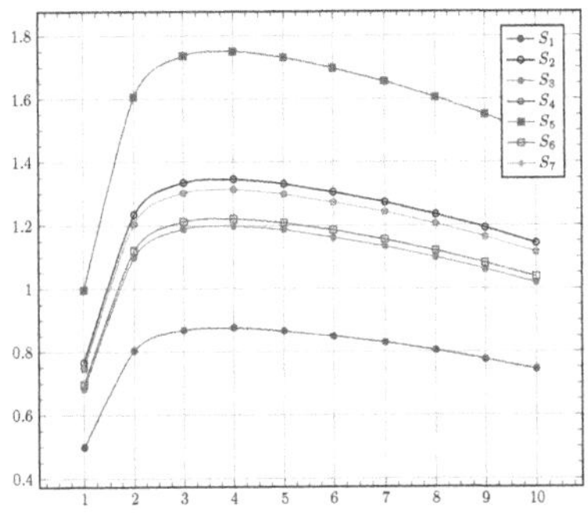

Fig. 1. Example 1: The arrival rates

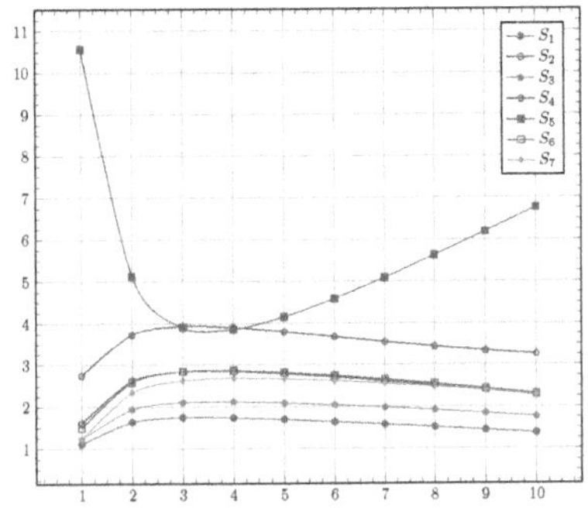

Fig. 2. Example 1: The average number of customers

The results of the second example show that the increasing of the lower threshold batch size at node S_5 (for a sufficiently large the upper threshold

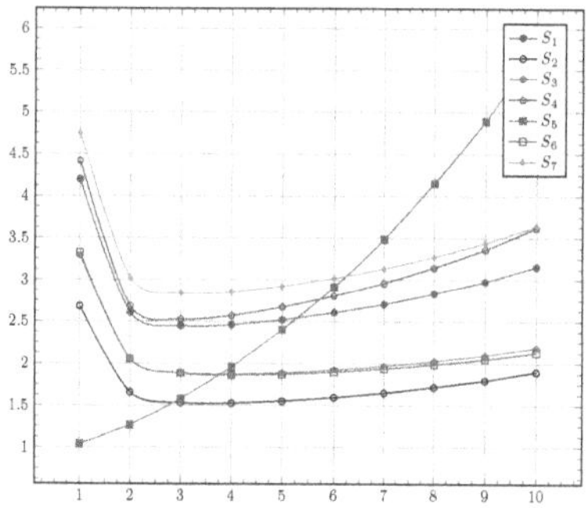

Fig. 3. Example 1: The average server idle times

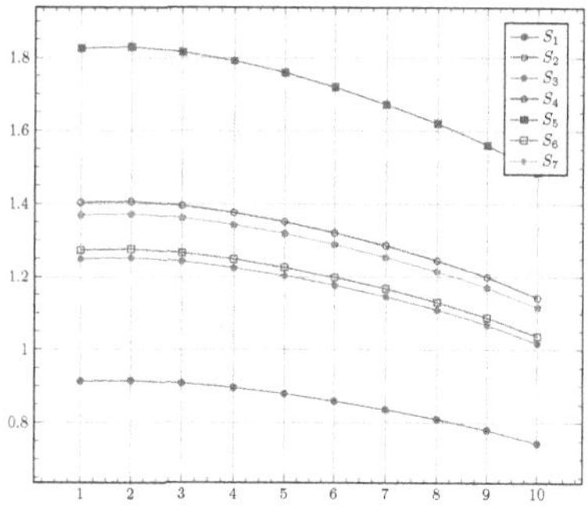

Fig. 4. Example 2: The arrival rates

value) leads to a decrease in the arrival rates in all nodes (Fig. 4). At the same time, the average idle times are decreased for all nodes (Fig. 6). Since we consider a closed queueing network, a significant increase in the average number of customers in node S_5 leads to a decrease in the number of customers for other nodes in the network (Fig. 5).

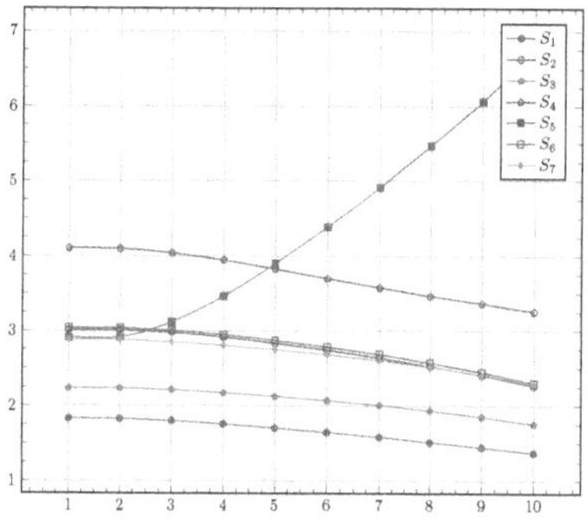

Fig. 5. Example 2: The average number of customers

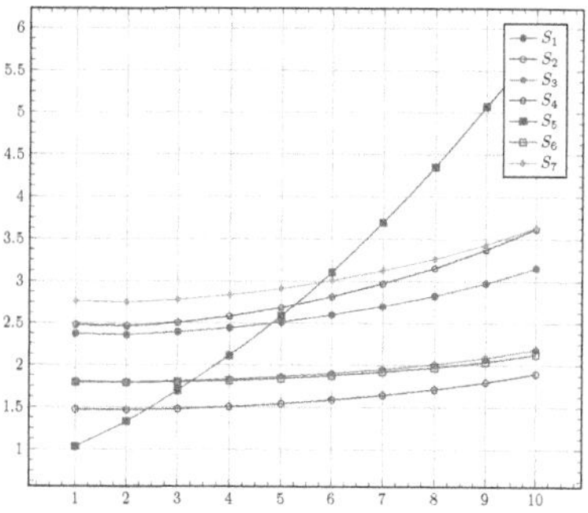

Fig. 6. Example 2: The average server idle times

In the third experiment, we investigate influence for the upper threshold batch size in node S_5. It leads to increases of the arrival rates for all nodes (Fig. 7) and decreases for the average idle times (Fig. 9). The average number of customers in node S_5 increases at first, and then tends to a limit value (Fig. 8).

In the fourth and fifth examples, we assume that node S_5 is unreliable. In the fourth example, we compare the performance measures obtained using our approximation and simulation approach. In this example, we assume $\alpha_5 = 0.1$, $\beta_5 = 1.5$. Tables 1, 2 present the average number of customers and the average idle times for different values of $x_5 = 1, \ldots, 6$, we fixed $y_5 = 6$.

All simulation results were obtained with confidence probability 0.95. A comparison of the results of analytical and simulation modeling (Tables 1, 2) shows

Table 1. Average number of customers at node S_5

x_5	1	2	3	4	5	6
Approximation approach	1.216	1.183	1.378	1.737	2.225	2.890
Simulation approach	1.270	1.237	1.433	1.788	2.279	2.942

Table 2. Average idle time for node S_5

x_5	1	2	3	4	5	6
Approximation approach	1.035	1.332	1.695	2.098	2.524	2.902
Simulation approach	1.040	1.335	1.700	2.105	2.530	2.906

the accuracy of the proposed method for analyzing networks is acceptable for real applications.

The fifth example contains the performance measures for different values of $\beta_5 = 1, \ldots, 4$. The following parameters were used: $x_5 = 2$, $y_5 = 5$, $\alpha_5 = 0.1$ (Fig. 10, 11 and 12).

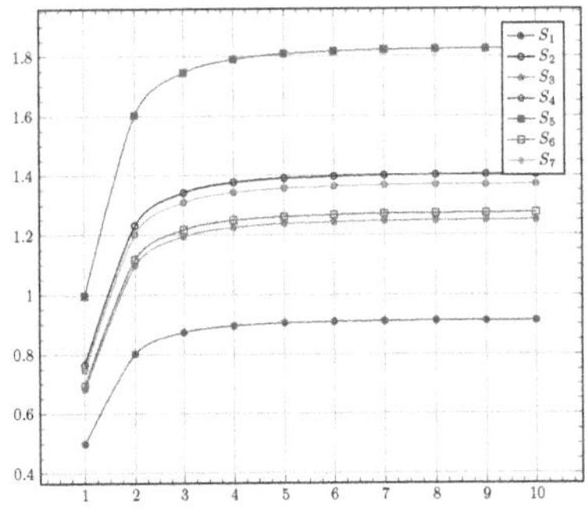

Fig. 7. Example 3: The arrival rates

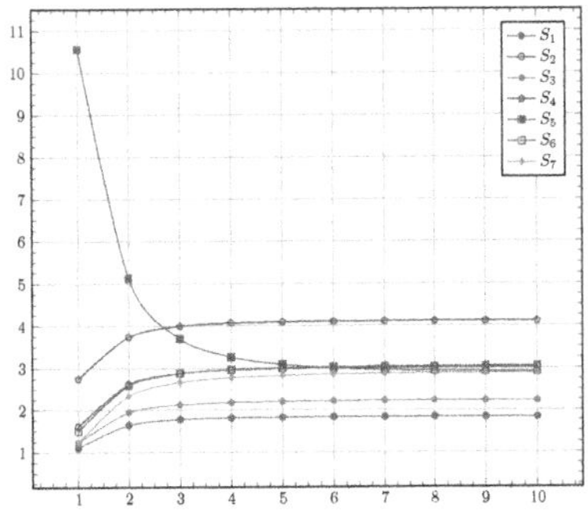

Fig. 8. Example 3: The average number of customers

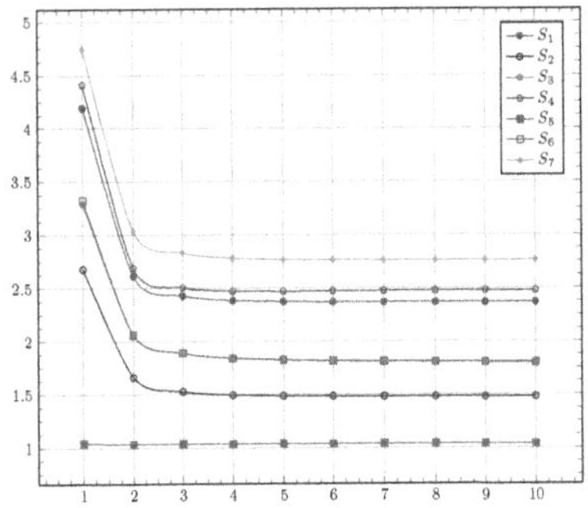

Fig. 9. Example 3: The average server idle times

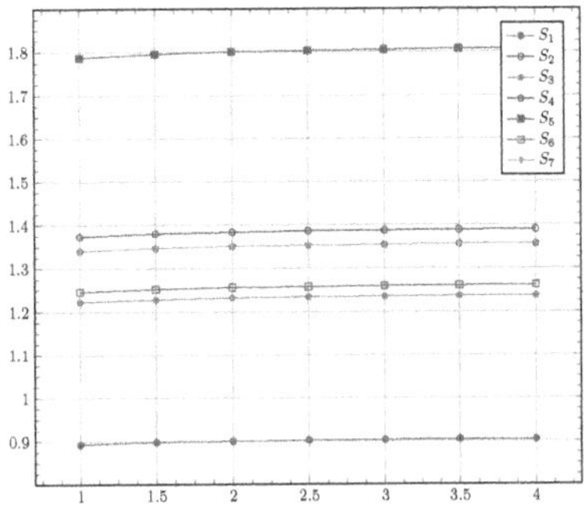

Fig. 10. Example 5: The arrival rates

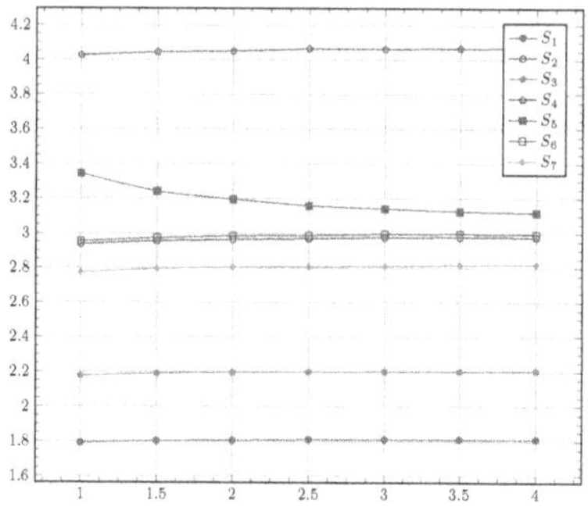

Fig. 11. Example 5: The average number of customers

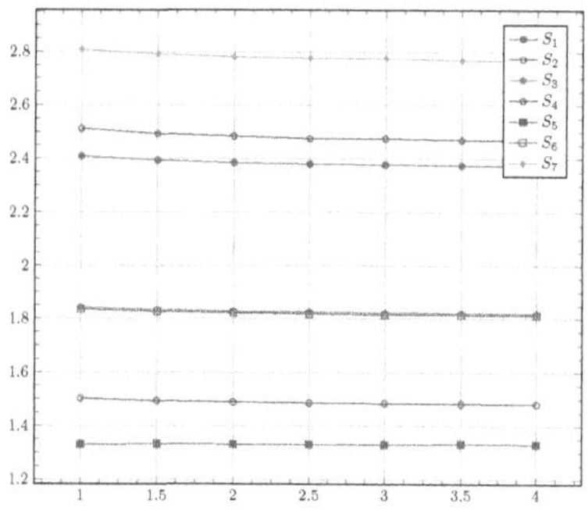

Fig. 12. Example 5: The average server idle times

As it can be seen (Fig. 10, 11 and 12), the greatest changes were made to the minimum number of customers in node S_5 (Fig. 11). The maximum and minimum values differ by about 7%. On all other figures, there is a slight change in the performance measures of queueing systems. This can be explained by the fact that increasing the recovery rate of β_5 by 4 times increases the service rate μ_5 of batches by only 7%.

6 Conclusion

This paper analyzed a closed unreliable queueing network with batch services. The queueing systems in the network operate under the general batch service rule. For the reliable network we obtain the stationary distribution of the network and its performance measures, for the unreliable case we present an approximation. At the end, some examples are presented. The results can be used for the performance analysis of transport vehicles, telecommunication systems and manufacturing systems.

References

1. Evequoz, C., Tropper, C.: Approximate analysis of bulk closed queueing networks. INFOR: Inf. Syst. Oper. Res. **33**(3), 179–204 (1995). https://doi.org/10.1080/03155986.1995.11732280
2. Chao, X., Pinedo, M., Shaw, D.: Networks of queues with batch services and customer coalescence. J. Appl. Probab. **33**(3), 858–869 (1996). https://doi.org/10.2307/3215364
3. Hanschke, T., Zisgen, H.: Queueing networks with batch service. Eur. J. Ind. Eng. **5**(3), 313–326 (2011). https://doi.org/10.1504/EJIE.2011.041619
4. Mitici, M., Goseling, J., van Ommeren, J.-K., de Graaf, M., Boucherie, R.J.: On a tandem queue with batch service and its applications in wireless sensor networks. Queueing Syst. **87**(1), 81–93 (2017). https://doi.org/10.1007/s11134-017-9534-1

5. Klünder, W.: Decomposition of open queueing networks with batch service. Oper. Res. Proc. **2016**, 575–581 (2017). https://doi.org/10.1007/978-3-319-55702-1_76
6. Chaudhry, M.L., Templeton, J.G.C.: A First Course in Bulk Queues. John Wiley, New York (1983)
7. Krishnamoorthy, A., Ushakumari, P.A.: A queueing system with single arrival bulk service and single departure. Math. Comput. Model. **31**(2–3), 99–108 (2000). https://doi.org/10.1016/S0895-7177(99)00226-5
8. Economou, A.: An alternative model for queueing systems with single arrivals, batch services and customer coalescence. Queueing Syst. **40**(4), 407–432 (2002). https://doi.org/10.1023/a:1015089518876
9. Miyazawa, M., Taylor, P.G.: A geometric product-form distribution for a queueing network with non-standar batch arrivals and batch transfers. Adv. Appl. Probab. **29**(2), 523–544 (1997). https://doi.org/10.2307/1428015
10. Chao, X.: Partial balances in batch arrival batch service and assemble-transfer queueing networks. J. Appl. Probab. **34**(3), 745–752 (1997). https://doi.org/10.2307/3215099
11. Chao, X., Zheng, S.: Triggered concurrent batch arrivals and batch departures in queueing networks. Disc. Event Dyn. Syst.: Theory Appl. **10**, 115–129 (2000). https://doi.org/10.1023/A:1008339216447
12. Economou, A.: A stochastic lower bound for assemble-transfer batch service queueing networks. J. Appl. Probab. **37**(3), 881–889 (2000). https://doi.org/10.1017/s0021900200016065
13. Reiser, M., Lavenberg, S.S.: Mean-value analysis of closed multichain queueing networks. J. ACM **27**(2), 313–322 (1980)
14. Tananko, I.E., Fokina, N.P.: Analysis of closed unreliable queueing networks with batch movements of customers. Izv. Saratov Univ. (N. S.) Ser. Math. Mech. Inform. **13**(2–1), 111–117 (2013). https://doi.org/10.18500/1816-9791-2013-13-2-1-111-117. (in Russian)
15. Xia, C.H., Michailidis, G., Bambos, N., Glynn, P.W.: Optimal control of parallel queues with batch service. Probab. Eng. Inf. Sci. **16**(3), 289–307 (2002). https://doi.org/10.1017/S0269964802163029
16. Stankevich, E.P., Tananko, I.E., Dolgov, V.I.: Analysis of closed queueing networks with batch service. Izv. Saratov Univ. (N. S.) Ser. Math. Mech. Inform. **20**(4), 527–533 (2020). https://doi.org/10.18500/1816-9791-2020-20-4-527-533
17. Boucherie, R.J., Dijk, N.M.: Product forms for queueing networks with state-dependent multiple job transitions. Adv. Appl. Probab. **23**(1), 152–187 (1991). https://doi.org/10.2307/1427516
18. Vinod, B., Altiok, T.: Approximating unreliable queueing networks under the assumption of exponentiality. J. Oper. Res. Soc. **37**(3), 309–316 (1986). https://doi.org/10.1057/jors.1986.49

On an MMAP/(PH, PH)/1/(∞, N) Queueing-Inventory System

Nisha Mathew[1](✉), Varghese C. Joshua[2], and Achyutha Krishnamoorthy[2]

[1] Department of Mathematics, B.K College Amalagiri, Kottayam, India
nishamathew@cmscollege.ac.in

[2] Department of Mathematics, CMS College, Kottayam, India
{vcjoshua,krishnamoorthy}@cmscollege.ac.in

Abstract. In this paper, we consider a single-server queueing inventory model with two types of customers, say, type-1 and type-2. The queue formed by the type-1 customer is infinite. The queue formed by the type-2 customer can accommodate a maximum of N customers. Even though the same server provides service to both types of customers, type-1 customers are served one by one, while type-2 customers are served in batches of varying sizes. The service is initiated only when inventory is available and the service time is assumed to be positive. If at least one item is available in the inventory and the server is idle, an arriving type-1 customer can directly enter into the service. The service of the type-2 customer is initiated either upon realization of a random clock that started ticking with the arrival of the first type-2 customer or by the accumulation of N type-2 customers, whichever occurs first. The arrivals follow a Marked Markovian process. Service time distribution of both type-1 and type-2 customers follow two different phase-type distribution. Replenishment of inventory follows the (s, S) policy having a positive lead time. It is assumed that N is less than s. Steady-state analysis, as well as evaluation of some performance measures, have been done. The model is analyzed numerically and graphically.

Keywords: Lead time · Positive service time · MMAP · Phase type distribution

1 Introduction

In [1], Sigman and Simchi-Levi introduced inventory models with positive service time. In [2], Krishnamoorthy et al. give a survey of inventory with positive service time. Dudin et al. in [3] consider a multi-server queueing system with an infinite buffer. In [3], there are two types of customers whose arrival are according to a Marked Markovian arrival process. A MAP/PH/1 queueing model with server vacations is given by Sreenivasan et al. in [4]. In [5], Nisha et al. consider a single-server queueing inventory model with two channels of service. A single server queueing model with batch service is considered by Chakravarthy

A. Dudin et al. (Eds.): ITMM 2020, CCIS 1391, pp. 363–377, 2021.
https://doi.org/10.1007/978-3-030-72247-0_27

et al. in [6]. In [7], Anbazhagan et al. consider inventory system with (s, S) replenishment policy and two types of services. In [8], Krishnamoorthy et al. consider a retrial queueing model in which service time distribution is of phase-type distribution. [9–11] consider queueing inventory models in which lead time is assumed to be exponential.

The present-day retail system operates on multiple platforms which are either online or offline. The physical presence of a customer is no longer a mandatory element in present-day shopping. The customers prefer to shop online through the various virtual platforms that are available. However, in certain cases, customers go out shopping. In order to boost sales, the sellers use various methods of marketing. Specific algorithms have to be designed in order to determine the inventory required by the sellers so that there is no delay or failure in meeting the customer demands.

Our model is motivated by two types of customer demands that arrive at a shop: one is an online customer and the other is the customer that directly comes to the shop. The customers that are physically present are served by the system on a FIFO basis. The system follows a different algorithm while addressing the service request of an online customer. The demands of online customers are addressed by the system only when the random clock realizes or when the number of such demands reaches a prefixed number N, whichever occurs first.

The remaining sections of the paper are organized as follows: The model is described in Sect. 2. Section 3 formulates the model mathematically. It also includes the steady-state analysis of the model. Evaluation of some performance measures is done in Sect. 4. Numerical and graphical illustrations are provided in Sect. 5. We conclude the paper in Sect. 6.

2 Model Description

In the present model, we consider the problem of selling out a single product via two different platforms, say, physical and virtual. Physical platform refers to a shop, where customers can directly come to purchase commodities. Such customers are named type-1 customers. Virtual platform refers to some online facilities for booking items. Customers can order items through these online facilities and such customers are named type-2 customers. Both the customers are served by a single server. The arrival of both types of customers is according to a stationary Marked Markovian Arrival Process (MMAP) having m phases. Let (D_0, D_1, D_2) denote the matrix representation of the MMAP guiding the arrival process. The service time is assumed to be positive. Replenishment follows (s, S) policy. Lead time is assumed to be exponential with parameter γ. It is assumed that each customer demands one unit of item. Inventory is required for service. The distribution of service time of type-1 customer follows phase-type with irreducible representation PH$(\boldsymbol{\alpha}, T)$ with m_1 phases and that of a batch of type-2 customers follow phase-type with irreducible representation PH$(\boldsymbol{\beta}, U)$ with m_2 phases. The vectors $\boldsymbol{T^0}$ and $\boldsymbol{U^0}$ are given by $\boldsymbol{T^0} = -T\boldsymbol{e}$ and $\boldsymbol{U^0} = -U\boldsymbol{e}$.

The queue of the type-1 customer is of infinite capacity. The type-2 customer joins a finite buffer of size N. Service of customers is done as per the following rule: when a type-1 customer arrives, if the server is idle and there is at least one item in inventory, that customer can directly enter the service. Otherwise, they wait in their respective queue. For controlling the long waiting time of the type-2 customer, a random clock is set. The clock starts ticking with the first arrival of the type-2 customer in every cycle. It works for a duration of time, which is exponentially distributed with parameter θ. The entry of the type-2 customer to the buffer is permitted only up to the time where the number of type-2 customers reaches N or when the random clock expires, whichever occurs first. Only after these customers enter service, new type-2 customers are allowed to join. When the number of type-2 customers reaches N or when the random clock expires, if the server is idle and required inventory is available, type-2 customers can enter service immediately. Otherwise, they are served immediately after the current service. Service of type-2 customers occurs as batches of a size that varies from 1 to N. It is assumed that N is less than s.

At the beginning of service of a type-1 customer, the level of inventory drops by one unit. When the service of a batch of type-2 begins, the level of inventory drops by n_2 units, where n_2 is the size of that batch.

3 Mathematical Formulation

We define the necessary random variables in the model as follows.
$N_1(t)$ – the number of type-1 customers in the queue at time t.
$N_2(t)$ – the number of type-2 customers in the finite buffer at time t.
$B(t)$ – the server status at time t,

$$B(t) = \begin{cases} 0, & \text{if the server is idle.} \\ 1, & \text{if the server is busy with a type-1 customer.} \\ 2, & \text{if the server is busy with a batch of type-2 customer.} \end{cases}$$

$C(t)$ – the clock status at time t.

$$C(t) = \begin{cases} 0, \text{if the clock is off.} \\ 1, \text{if the clock is on.} \end{cases}$$

$I(t)$ – the number of items in the inventory at time t.
$J(t)$ – the phase of the service process at time t.
$A(t)$ – the phase of the arrival process at time t.

Then $\{(N_1(t), N_2(t), B(t), C(t), I(t), J(t), A(t)); t \geq 0\}$ is a continuous-time Markov chain on the state space to be described below. This model can be considered as a Level Independent Quasi-Birth-Death (LIQBD) process and a solution is obtained by Matrix-Analytic Method. We define the state space of the QBD under consideration and analyze the structure of its infinitesimal generator.

The state space $\Omega = \Omega_1 \bigcup \Omega_2 \bigcup \Omega_3 \bigcup \Omega_4 \bigcup \Omega_5$, where
$\Omega_1 = \{(0, 0, 0, 0, i, 0^*, a)/0 \leq i \leq S; a = 1, 2, \ldots, m\}$,
$\Omega_2 = \{(n_1, 0, 0, 0, 0, 0^*, a) \bigcup (n_1, n_2, 0, 1, 0, 0^*, a)/n_1 \geq 1; 1 \leq n_2 \leq N-1;$

$a = 1, 2, \ldots, m\}$,
$\Omega_3 = \{(n_1, n_2, 0, 0, i, 0^*, a)/n_1 \geq 0; 1 \leq n_2 \leq N; 0 \leq i \leq n_2 - 1; a = 1, 2, \ldots, m\}$,
$\Omega_4 = \{(n_1, n_2, 1, c, i, j, a)/n_1 \geq 0; 0 \leq n_2 \leq N; c = 0, 1; 0 \leq i \leq S;$
$j = 1, 2, \ldots, m_1; a = 1, 2, \ldots, m\}$,
$\Omega_5 = \{(n_1, n_2, 2, 1, i, j, a) \bigcup (n_1, 0, 2, 0, i, j, a) \bigcup (n_1, N, 2, 0, i, j, a)/n_1 \geq 0;$
$0 \leq n_2 \leq N - 1; 0 \leq i \leq S; j = 1, 2, \ldots, m_2; a = 1, 2, \ldots, m\}$,
0^* represents phase of an idle server.

The rate of transitions are given in the Tables 1–2 below.

Table 1. Transition table

From	To	Rate	Description
$(0, 0, 0, 0, i, 0^*, a)$	$(0, 0, 1, 0, i-1, j, a')$	$d_{aa'}(1)\alpha_j$	$i \geq 1, j = 1, 2, \cdots m_1$, $a, a' = 1, 2, \ldots, m$
$(0, n_2, 0, 1, i, 0^*, a)$	$(0, n_2, 1, 1, i-1, j, a')$	$d_{aa'}(1)\alpha_j$	$i \geq 1, j = 1, 2, \cdots m_1$, $1 \leq n_2 \leq N - 1$, $a, a' = 1, 2, \ldots, m$
$(n_1, n_2, 1, 1, i, j, a)$	$(n_1 + 1, n_2, 1, 1, i, j, a')$	$d_{aa'}(1)$	$n_1 \geq 0, j = 1, 2, \cdots m_1$, $1 \leq n_2 \leq N - 1$, $a, a' = 1, 2, \ldots, m$
$(n_1, n_2, 1, 0, i, j, a)$	$(n_1 + 1, n_2, 1, 0, i, j, a')$	$d_{aa'}(1)$	$n_1 \geq 0, j = 1, 2, \cdots m_1$, $0 \leq n_2 \leq N$, $a, a' = 1, 2, \ldots, m$
$(n_1, 0, 1, 0, i, j, a)$	$(n_1, 1, 1, 1, i, j, a')$	$d_{aa'}(2)$	$n_1 \geq 0, j = 1, 2, \cdots m_1$, $a, a' = 1, 2, \ldots, m$
$(0, 0, 0, 0, i, 0^*, a)$	$(0, 1, 0, 1, i, 0^*, a')$	$d_{aa'}(2)$	$i \geq 0, a, a' = 1, 2, \ldots, m$
$(n_1, n_2, 1, 1, i, j, a)$	$(n_1, n_2 + 1, 1, 1, i, j, a')$	$d_{aa'}(2)$	$n_1 \geq 0, 1 \leq n_2 \leq N - 2$, $j = 1, 2, \cdots m_1$, $a, a' = 1, 2, \ldots, m$
$(0, n_2, 0, 1, i, 0^*, a)$	$(0, n_2 + 1, 0, 1, i, 0^*, a')$	$d_{aa'}(2)$	$1 \leq n_2 \leq N - 2$, $a, a' = 1, 2, \ldots, m$
$(n_1, N - 1, 1, 1, i, j, a)$	$(n_1, N, 1, 0, i, j, a')$	$d_{aa'}(2)$	$n_1 \geq 0, j = 1, 2, \cdots m_1$ $a, a' = 1, 2, \ldots, m$
$(n_1, 0, 2, 0, i, j, a)$	$(n_1, 1, 2, 1, i, j, a')$	$d_{aa'}(2)$	$n_1 \geq 0, j = 1, 2, \cdots m_2$ $a, a' = 1, 2, \ldots, m$
$(n_1, n_2, 2, 1, i, j, a)$	$(n_1, n_2 + 1, 2, 1, i, j, a')$	$d_{aa'}(2)$	$n_1 \geq 0, 1 \leq n_2 \leq N - 2$, $j = 1, 2, \cdots m_2$, $a, a' = 1, 2, \ldots, m$
$(n_1, N - 1, 2, 1, i, j, a)$	$(n_1, N, 2, 0, i, j, a')$	$d_{aa'}(2)$	$n_1 \geq 0, j = 1, 2, \cdots m_2$ $a, a' = 1, 2, \ldots, m$

The infinitesimal generator Q of the LIQBD describing the above single server queueing inventory system is of the form

$$\begin{pmatrix} B_{00} & B_{01} & O & \cdots\cdots & & & \\ B_{10} & A_1 & A_0 & O & \cdots\cdots & & \\ O & A_2 & A_1 & A_0 & O & \cdots\cdots & \\ O & O & A_2 & A_1 & A_0 & O & \cdots \\ & \ddots & \ddots & \ddots & \ddots & \ddots & \ddots \\ & & \ddots & \ddots & \ddots & \ddots & \ddots \end{pmatrix},$$

where B_{00}, A_0, A_1, A_2 are all square matrices of appropriate order. The structure of the matrices A_0 and A_2 are as follows:

$$A_0 = I_K \otimes D_1,$$

where $K = (S+1)(2m_1N + m_2(N+1)) + (N/2)(N+3)$;

$$A_2 = \begin{pmatrix} H_1^0 & & & & \\ & H_1^1 & & & \\ & & \ddots & & \\ & & & H_1^{N-1} & \\ & & & & H_1^N \end{pmatrix},$$

where $H_1^0 = \begin{pmatrix} O & Y & O \\ O & Z & O \end{pmatrix}$, $H_1^j = \begin{pmatrix} O & O & O \\ O & Y & O \\ O & O & O \\ O & Z & O \end{pmatrix}$ for $j = 1$ to $N-1$, H_1^0 is a square matrix of order $m[1+(S+1)(m_1+m_2)]$ and H_1^j are square matrices of order $m[(j+1)+(S+1)(2m_1+m_2)]$, H_1^N is a zero square matrix of order $m[N+(S+1)(m_1+m_2)]$;

$Y = \begin{pmatrix} O & \gamma\alpha \otimes I_m & O \end{pmatrix}$ is a matrix of order $m \times [(S+1)(m_1 m)]$;

$Z = \begin{pmatrix} O & O \\ I_S \otimes T^0 \otimes \alpha \otimes I_m & O \\ O & O \\ I_S \otimes U^0 \otimes \alpha \otimes I_m & O \end{pmatrix}$ is a matrix of order $m[(S+1)(m_1+m_2)] \times [(S+1)(m_1 m)]$.

3.1 Stability Condition

The Markov chain with generator Q is positive recurrent if and only if

$$\boldsymbol{\pi}[I_K \otimes D_1]\mathbf{e} < \sum_{\mathbf{j}=0}^{\mathbf{N}} \boldsymbol{\pi}_{\mathbf{j}} \mathbf{H}_1^{\mathbf{j}} \mathbf{e}, \tag{1}$$

Table 2. Transition table

From	To	Rate	Description
$(n_1, n_2, b, c, i, j, a)$	$(n_1, n_2, b, c, i, j, a^{'})$	$d_{aa^{'}}(0)$	$n_1 \geq 0, n_2 \geq 0, b = 0, 1, 2,$ $c = 0, 1, 0 \leq i \leq s,$ $a = 1, 2, \ldots, m$
$(n_1, n_2, 1, 1, i, j, a)$	$(n_1, n_2, 1, 0, i, j, a)$	θ	$n_1 \geq 0, j = 1, 2, \cdots m_1$ $1 \leq n_2 \leq N-1,$ $a = 1, 2, \ldots, m$
$(0, n_2, 0, 1, i, 0^*, a)$	$(0, 0, 2, 0, i-n_2, j, a)$	$\theta\beta_j$	$i \geq n_2, j = 1, 2, \cdots m_2,$ $a = 1, 2, \ldots, m$
$(n_1, n_2, b, c, i, j, a)$	$(n_1, n_2, b, c, S, j, a)$	γ	$n_1 \geq 0, n_2 \geq 0, b = 0, 1, 2,$ $c = 0, 1, 0 \leq i \leq s,$ $a = 1, 2, \ldots, m$
$(n_1, n_2, 0, 1, 0, 0^*, a)$	$(n_1-1, n_2, 1, 1, S-1, j, a)$	$\gamma\alpha_j$	$n_1 \geq 1, j = 1, 2, \cdots m_1$ $a = 1, 2, \ldots, m$
$(n_1, n_2, 0, 0, i, 0^*, a)$	$(n_1, 0, 2, 0, S-n_2, j, a)$	$\gamma\beta_j$	$n_1 \geq 0, 1 \leq n_2 \leq N, i < n_2,$ $a = 1, 2, \ldots, m,$ $j = 1, 2, \cdots m_2$
$(0, n_2, 1, 1, i, j, a)$	$(0, n_2, 0, 1, i, 0^*, a)$	T_j^0	$n_2 \geq 1, j = 1, 2, \cdots m_1,$ $a = 1, 2, \ldots, m$
$(n_1, n_2, 1, 1, 0, j, a)$	$(n_1, n_2, 0, 1, 0, 0^*, a)$	T_j^0	$n_2 \geq 1, j = 1, 2, \cdots m_1$ $a = 1, 2, \ldots, m$
$(n_1, n_2, 1, 1, i, j, a)$	$(n_1-1, n_2, 1, 1, i-1, k, a)$	$T_j^0\alpha_k$	$n_1, n_2 \geq 1, a = 1, 2, \ldots, m,$ $i \geq 1, j, k = 1, 2, \cdots m_1$
$(n_1, n_2, 1, 0, i, j, a)$	$(n_1, 0, 2, 0, i-n_2, k, a)$	$T_j^0\beta_k$	$i \geq n_2, a = 1, 2, \ldots, m,$ $j = 1, 2, \cdots m_1,$ $k = 1, 2, \cdots m_2$
$(n_1, 0, 1, 0, i, j, a)$	$(n_1, 0, 1, 0, i, k, a)$	T_{jk}	$n_1 \geq 0, i \geq 0,$ $j, k = 1, 2, \cdots m_1$
$(n_1, n_2, 1, 1, i, j, a)$	$(n_1, n_2, 1, 1, i, k, a)$	T_{jk}	$n_1 \geq 0, j, k = 1, 2, \cdots m_1,$ $a = 1, 2, \ldots, m$
$(0, 0, 2, 0, i, j, a)$	$(0, 0, 0, 0, i, 0^*, a)$	U_j^0	$j = 1, 2, \cdots m_2,$ $a = 1, 2, \ldots, m$
$(n_1, 0, 2, 0, 0, j, a)$	$(n_1, 0, 0, 0, 0, 0^*, a)$	U_j^0	$j = 1, 2, \cdots m_2,$ $a = 1, 2, \ldots, m$
$(n_1, n_2, 2, 1, 0, j, a)$	$(n_1, n_2, 0, 1, 0, 0^*, a)$	U_j^0	$n_1 \geq 1, 1 \leq n_2 \leq N-1,$ $a = 1, 2, \ldots, m, j = 1, 2, \cdots m_2$
$(n_1, N, 2, 0, i, j, a)$	$(n_1, N, 0, 0, i, 0^*, a)$	U_j^0	$n_1 \geq 0, i \leq N-1,$ $a = 1, 2, \ldots, m, j = 1, 2, \cdots m_1$
$(n_1, 0, 2, 0, i, j, a)$	$(n_1-1, 0, 1, 0, i-1, k, a)$	$U_j^0\alpha_k$	$n_1 \geq 1, i \geq 1, a = 1, 2, \ldots, m$ $j = 1, 2, \cdots m_2, k = 1, 2, \cdots m_1$ $a = 1, 2, \ldots, m$
$(n_1, n_2, 2, 1, i, j, a)$	$(n_1-1, n_2, 1, 1, i-1, k, a)$	$U_j^0\alpha_k$	$n_1 \geq 1, 1 \leq n_2 \leq N-1, i \geq 1,$ $j = 1, 2, \cdots m_2, k = 1, 2, \cdots m_1$ $a = 1, 2, \ldots, m$
$(n_1, N, 2, 0, i, j, a)$	$(n_1, 0, 2, 0, i-N, k, a)$	$U_j^0\beta_k$	$n_1 \geq 0, i \geq N, a = 1, 2, \ldots, m$ $j, k = 1, 2, \cdots m_1$
$(n_1, 0, 2, 0, i, j, a)$	$(n_1, 0, 2, 0, i, k, a)$	U_{jk}	$n_1 \geq 0, j, k = 1, 2, \cdots m_2$ $a = 1, 2, \ldots, m$
$(n_1, n_2, 2, 1, i, j, a)$	$(n_1, n_2, 2, 1, i, k, a)$	U_{jk}	$n_1 \geq 0, j, k = 1, 2, \cdots m_2$ $a = 1, 2, \ldots, m$

where the stationary vector $\boldsymbol{\pi}$ of A is obtained by solving

$$\boldsymbol{\pi} A = 0; \boldsymbol{\pi}\mathbf{e} = 1, \tag{2}$$

where the matrix A be defined as $A = A_0 + A_1 + A_2$.

3.2 Stationary Distribution

The stationary distribution of the Markov process under consideration is obtained by solving the set of equations:

$$\mathbf{x}Q = 0; \mathbf{x}\mathbf{e} = \mathbf{1}. \tag{3}$$

Let $\mathbf{x}$ be decomposed in conformity with Q. Then

$$\mathbf{x} = (\mathbf{x_0}, \mathbf{x_1}, \mathbf{x_2}, \ldots),$$

where $\mathbf{x_i} = (\mathbf{x_{i0}}, \mathbf{x_{i1}}, \ldots\ldots \mathbf{x_{iN}})$, $\mathbf{x_{ij}} = (\mathbf{x_{ij0}}, \mathbf{x_{ij1}}, \mathbf{x_{ij2}})$ for $j = 1, 2, \ldots, N$, whereas for $k = 0, 1, 2$, the vectors

$$\mathbf{x_{ijk}} = (\mathbf{x_{ijk0}}, \mathbf{x_{ijk1}}),$$

$\mathbf{x_{ijkl}} = (\mathbf{x_{ijkl1}}, \mathbf{x_{ijkl2}}, \ldots\ldots \mathbf{x_{ijklS}})$ for $l = 0, 1$,

$$\mathbf{x_{ijklr}} = (\mathbf{x_{ijklr1}}, \mathbf{x_{ijklr2}}, \ldots\ldots \mathbf{x_{ijklrt}}),$$

for $k = 1, 2\text{m}$, where $t = m_k$

$$\mathbf{x_{ijklru}} = (x_{ijklru1}, x_{ijklru2}, \ldots\ldots x_{ijklrum}),$$

where $x_{ijklrua}$ is the probability of being in state (i, j, k, l, r, u, a) for $i \geq 0 : j = 1, 2, \ldots, N; k = 1, 2; l = 0, 1; 0 \leq r \leq S; u = 1, 2, \ldots, m_k, a = 1, 2, \ldots, m$ and x_{ij0lr0^*a} is the probability of being in state $(i, j, 0, l, r, 0^*, a)$.

From $\mathbf{x}Q = 0$, we get the following equations:

$$\mathbf{x_0}B_{00} + \mathbf{x_1}B_{10} = 0, \tag{4}$$

$$\mathbf{x_0}B_{01} + \mathbf{x_1}A_1 + \mathbf{x_2}A_2 = 0, \tag{5}$$

$$\mathbf{x_1}A_0 + \mathbf{x_2}A_1 + \mathbf{x_3}A_2 = 0, \tag{6}$$

$$\mathbf{x_{i-1}}A_0 + \mathbf{x_i}A_1 + \mathbf{x_{i+1}}A_2 = 0, i = 2, 3, .. \tag{7}$$

It may be shown that there exists a constant matrix R such that

$$\mathbf{x_i} = \mathbf{x_{i-1}}R, i = 2, 3, ... \tag{8}$$

The sub vectors $\mathbf{x_i}$ are geometrically related by the equation

$$\mathbf{x_i} = \mathbf{x_1}R^{i-1}, i = 2, 3, ... \tag{9}$$

R can be obtained from the matrix quadratic equation

$$R^2 A_2 + RA_1 + A_0 = 0. \tag{10}$$

We can find $\mathbf{x_0}$ and $\mathbf{x_1}$ by solving Eqs. (4) and (5). Then we normalize $\mathbf{x_0}$ and $\mathbf{x_1}$ by using the normalizing condition $\mathbf{x_0}\mathbf{e} + \mathbf{x_1}(1 - R)^{-1}\mathbf{e} = 1$.

4 Performance Measures

In this section we evaluate some performance measures of the system.

1. Expected number of type-1 customers in the system:

$$E[N_1] = \sum_{i=0}^{\infty} i\mathbf{x_i}\mathbf{e}. \tag{11}$$

2. Expected number of type-2 customers in the system:

$$E[N_2] = \sum_{i=0}^{\infty}\sum_{j=0}^{N} j\mathbf{x_{ij}}\mathbf{e}. \tag{12}$$

3. Expected number of items in the inventory:

$$E[I] = \sum_{i=0}^{\infty}\sum_{j=0}^{N}\sum_{k=0}^{2}\sum_{l=0}^{1}\sum_{r=0}^{S} r\mathbf{x_{ijklr}}\mathbf{e}. \tag{13}$$

4. Expected number of customers waiting in the system due to lack of inventory:

$$E[W] = \sum_{i=0}^{\infty}\sum_{j=1}^{N-1} i\mathbf{x_{ij010}}\mathbf{e} + \sum_{i=0}^{\infty}\sum_{j=1}^{N}\sum_{r=0}^{j-1} j\mathbf{x_{ij00r}}\mathbf{e}. \tag{14}$$

5. The probability that the server is idle:

$$b_0 = \sum_{i=0}^{\infty}\sum_{j=0}^{N} \mathbf{x_{ij0}}\mathbf{e}. \tag{15}$$

6. The probability that the server is busy with the type-1 customer:

$$b_1 = \sum_{i=0}^{\infty}\sum_{j=0}^{N} \mathbf{x_{ij1}}\mathbf{e}. \tag{16}$$

7. The probability that the server is busy with the type-2 customer:

$$b_2 = \sum_{i=0}^{\infty}\sum_{j=0}^{N} \mathbf{x_{ij2}}\mathbf{e}. \tag{17}$$

8. The probability that the clock is on:

$$c_1 = \sum_{i=0}^{\infty}\sum_{j=1}^{N}\sum_{k=0}^{2} \mathbf{x_{ijk1}}\mathbf{e}. \tag{18}$$

9. The expected rate at which replenishment of inventory occurs:

$$E_R = \sum_{i=0}^{\infty}\sum_{j=0}^{N}\sum_{k=0}^{2}\sum_{l=0}^{1}\sum_{r=0}^{s} \gamma \mathbf{x_{ijklr}e}. \tag{19}$$

10. The probability that the type-2 customer is blocked from entering the system:

$$p_b = \sum_{i=0}^{\infty}\sum_{j=1}^{N-1}\sum_{k=1}^{2} \mathbf{x_{ijko}e} + \sum_{i=0}^{\infty} \mathbf{x_{iN}e}. \tag{20}$$

5 Numerical Examples

In this section, we provide a numerical illustration of the system performance measures with variation in values of underlying parameters. We consider a MMAP in which the arrivals of the type-1 and type-2 customers are described by matrices (D_0, D_1, D_2). In this example the following values are kept fixed:

$$\alpha = \begin{pmatrix}0.3 & 0.7\end{pmatrix}; \quad T = \begin{pmatrix}-10 & 5 \\ 4 & -12\end{pmatrix}; \quad T_0 = \begin{pmatrix}5 \\ 8\end{pmatrix};$$

$$\beta = \begin{pmatrix}0.5 & 0.4 & 0.1\end{pmatrix}; U = \begin{pmatrix}-6.0 & 3.0 & 2.0 \\ 1.0 & -6.0 & 2.0 \\ 2.5 & 1.5 & -8.0\end{pmatrix}; U_0 = \begin{pmatrix}1 \\ 3 \\ 4\end{pmatrix};$$

$$D_0 = \begin{pmatrix}-9.3 & 0.5 \\ 0.8 & -5.1\end{pmatrix}; D_1 = \begin{pmatrix}2.5 & 0.6 \\ 1.1 & 0.2\end{pmatrix}; D_2 = \begin{pmatrix}3.2 & 2.5 \\ 1.9 & 1.1\end{pmatrix}.$$

5.1 Effect of Parameter θ on Performance Measures

We fix $m = 2$; $m_1 = 2$; $m_2 = 3$; $s = 6$; $S = 13$; $\gamma = 5$; $N = 5$. Table 3 indicates the variation in the system performance measures with variation of θ.

Table 3. Effect of θ on various performance measures

θ	$E[N_1]$	$E[N_2]$	$E[W]$	b_0	b_1	b_2	c_1	p_b
2	2.0447	1.8146	0.0131	0.2106	0.3314	0.458	0.6	0.1508
3	2.4734	1.7082	0.0125	0.1718	0.3314	0.4968	0.5592	0.1665
4	2.8917	1.6367	0.012	0.1442	0.3314	0.5243	0.5287	0.1787
5	3.2991	1.5865	0.0116	0.124	0.3314	0.5446	0.5054	0.1882
6	3.6958	1.5499	0.0113	0.1087	0.3314	0.5598	0.487	0.1958
7	4.0819	1.5223	0.0111	0.0969	0.3314	0.5716	0.4723	0.202
8	4.4576	1.5008	0.0109	0.0876	0.3314	0.581	0.4603	0.207
9	4.8231	1.4837	0.0108	0.08	0.3314	0.5885	0.4503	0.2112
10	5.1787	1.4698	0.0107	0.0738	0.3314	0.5948	0.4419	0.2147

As θ increases, the chance of getting service for a batch of type-2 customers increases. So as θ increases expected number of type-1 customers in the system increases while the expected number of type-2 customers in the system decreases. Also, as θ increases the probability that the server is idle decreases, while the probability that the server is busy with a batch of type-2 customers increases. Also, as θ increases the probability that the clock is on decreases and the probability that the type-2 customers are blocked from entering the system increases. This is because as θ increases, the chance of service of type-2 customers increases and so the probability that the clock is on decreases. As type-2 customers are blocked from entering the system when the clock expires, the probability that the type-2 customers are blocked from entering the system increases as θ increases (Figs. 1, 2 and 3).

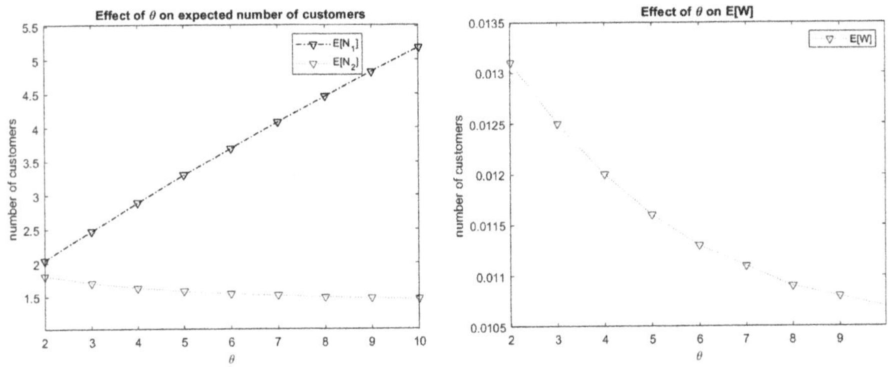

Fig. 1. Effect of θ on the expected number of customers and expected number of customers waiting in the system due to lack of inventory

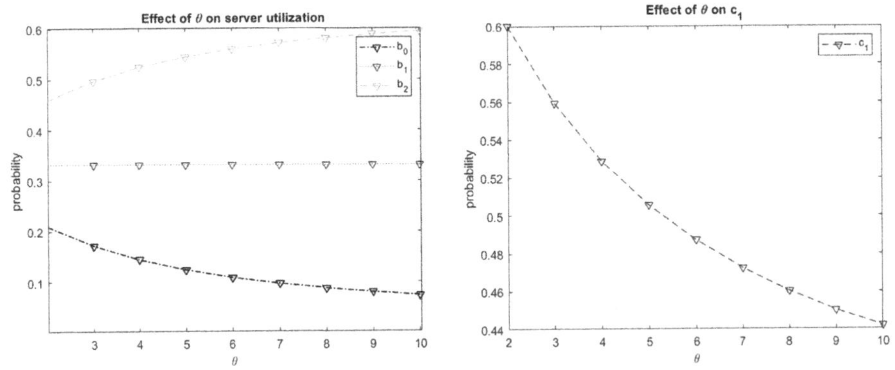

Fig. 2. Effect of θ on server utilization and the probability that the clock is on

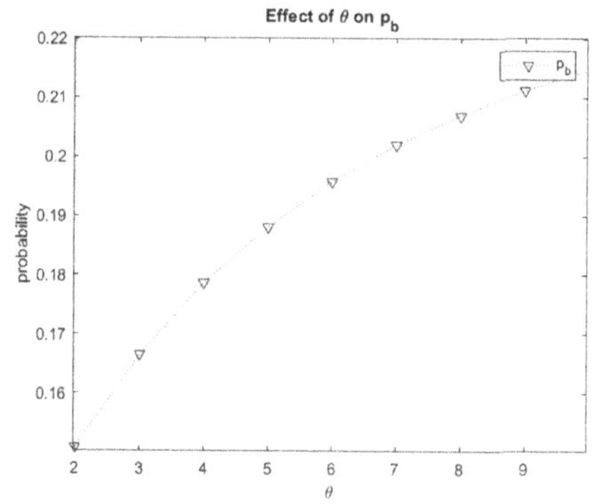

Fig. 3. Effect of θ on the probability that the type-2 customer is blocked from entering the system

5.2 Effect of s on Various Performance Measures

We fix $m = 2$; $m_1 = 2$; $m_2 = 3$; $\theta = 6$; $S = 13$; $\gamma = 5$; $N = 5$. Table 4 indicates the variation in the system performance measures with variation of s.

Table 4. Effect of s on various performance measures

s	$E[N_1]$	$E[N_2]$	$E[W]$	b_0	b_1	b_2	c_1	p_b
6	3.6659	1.5494	0.007	0.1082	0.3314	0.5604	0.4875	0.1918
7	3.6457	1.5485	0.0042	0.1078	0.3314	0.5607	0.4878	0.1916
8	3.6339	1.5479	0.0025	0.1076	0.3314	0.561	0.488	0.1915
9	3.6273	1.5474	0.0015	0.1075	0.3314	0.5611	0.4881	0.1905
10	3.6234	1.5471	0.0009	0.1074	0.3314	0.5612	0.4882	0.1888
11	3.6211	1.5467	0.0006	0.1074	0.3314	0.5612	0.4882	0.1879
12	3.6197	1.5464	0.0004	0.1073	0.3314	0.5613	0.4883	0.1873
13	3.6187	1.546	0.0002	0.1073	0.3314	0.5613	0.4883	0.1868
14	3.6183	1.5455	0.0001	0.1073	0.3314	0.5613	0.4883	0.1865

As s increases, expected a number of customers waiting in the system due to lack of inventory decreases, because as safety stock s increases, the chance that inventory is not available decreases. As s increases, the probability that the server is idle decreases, because as safety stock s increases the probability that the server is idle due to lack of inventory decreases.

5.3 Combined Effect of s and θ on Sever Being Idle

We fix $m = 2$; $m_1 = 2$; $m_2 = 3$; $S = 13$; $\gamma = 5$; $N = 5$. Table 5 indicates the combined effect of s and θ on sever being idle (Figs. 4 and 5).

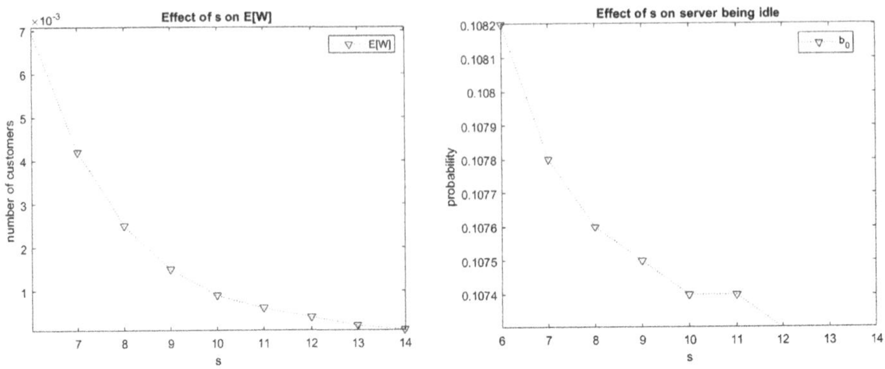

Fig. 4. Effect of s on the expected number of customers waiting in the system due to lack of inventory and probability that the server is idle

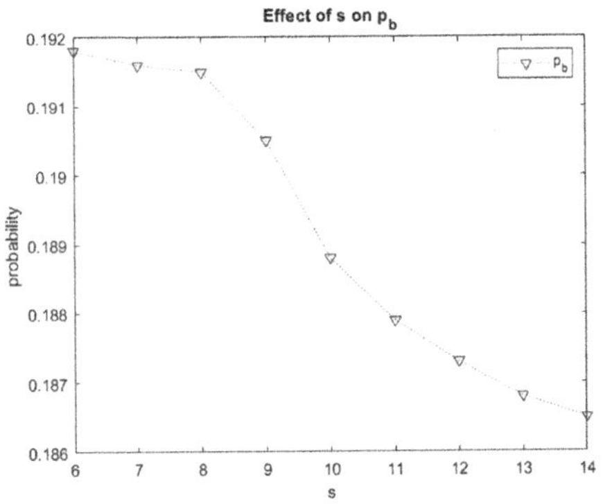

Fig. 5. Effect of s on the probability that the type-2 customer is blocked from entering the system

Table 5. Combined effect of s and θ on sever being idle

s	$\theta = 2$	$\theta = 4$	$\theta = 6$	$\theta = 8$	$\theta = 10$	$\theta = 12$	$\theta = 14$	$\theta = 16$
6	0.2101	0.1437	0.1082	0.087	0.0733	0.0637	0.0567	0.0514
7	0.2098	0.1434	0.1078	0.0867	0.0729	0.0633	0.0563	0.051
8	0.2097	0.1432	0.1076	0.0864	0.0726	0.0631	0.0561	0.0508
9	0.2096	0.143	0.1075	0.0863	0.0725	0.0629	0.0559	0.0506
10	0.2095	0.143	0.1074	0.0862	0.0724	0.0629	0.0559	0.0505
11	0.2094	0.1429	0.1074	0.0862	0.0724	0.0628	0.0558	0.0505
12	0.2094	0.1429	0.1073	0.0861	0.0724	0.0628	0.0558	0.0505
13	0.2094	0.1429	0.1073	0.0861	0.0723	0.0628	0.0558	0.0505
14	0.2094	0.1429	0.1073	0.0861	0.0723	0.0628	0.0558	0.0504

As s and θ increases, the probability that the server is idle decreases. This is because as safety stock s increases the probability that the server is idle due to lack of inventory decreases and as θ increases, the chance of service of type-2 customers increases. Both these reduce the probability of the server is idle. For a fixed value of s, the probability that the server is idle decreases with increasing θ. For a fixed value of θ the probability that the server is idle decreases with increasing s. It is clear from the table that when s becomes 11, there is not much change in the probability (Figs. 6 and 7).

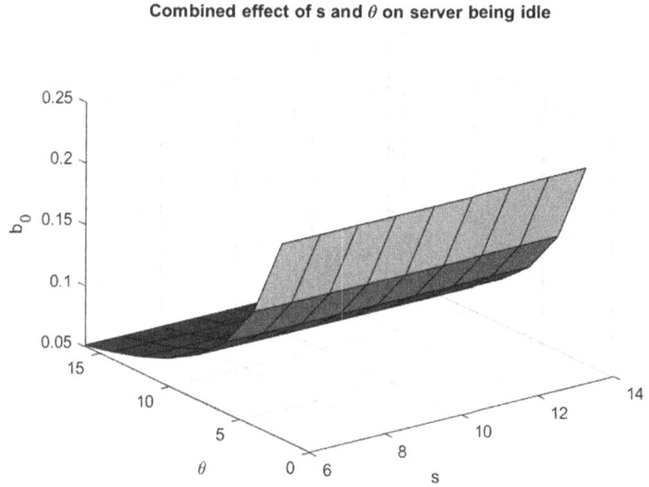

Fig. 6. The combined effect of s and θ on sever being idle

5.4 Combined Effect of s and θ on Sever Busy with a Batch of Type-2 Customers

We fix $m = 2$; $m_1 = 2$; $m_2 = 3$; $S = 13$; $\gamma = 5$; $N = 5$. Table 6 indicates the combined effect of s and θ on sever busy with a batch of type-2 customers.

As s and θ increases, the probability that the server is busy with a batch of type-2 customers increases. This is because as safety stock s increases the probability that the server is idle due to lack of inventory decreases and as θ increases, the chance of service of type-2 customers increases. Both these increase the probability of the server being busy with a batch of type-2 customers. It is clear from the table that when s becomes 11, there is not much change in the probability of the server being busy with a batch of type-2 customers.

Table 6. Combined effect of s and θ on sever busy with a batch of type-2 customers.

s	$\theta = 2$	$\theta = 4$	$\theta = 6$	$\theta = 8$	$\theta = 10$	$\theta = 12$	$\theta = 14$	$\theta = 16$
6	0.4585	0.5249	0.5604	0.5816	0.5953	0.6049	0.6119	0.6172
7	0.4587	0.5252	0.5607	0.5819	0.5957	0.6053	0.6123	0.6176
8	0.4589	0.5254	0.561	0.5822	0.5959	0.6055	0.6125	0.6178
9	0.459	0.5255	0.5611	0.5823	0.5961	0.6056	0.6126	0.618
10	0.4591	0.5256	0.5612	0.5824	0.5961	0.6057	0.6127	0.618
11	0.4591	0.5257	0.5612	0.5824	0.5962	0.6058	0.6128	0.6181
12	0.4592	0.5257	0.5613	0.5824	0.5962	0.6058	0.6128	0.6181
13	0.4592	0.5257	0.5613	0.5825	0.5962	0.6058	0.6128	0.6181
14	0.4592	0.5257	0.5613	0.5825	0.5962	0.6058	0.6128	0.6181

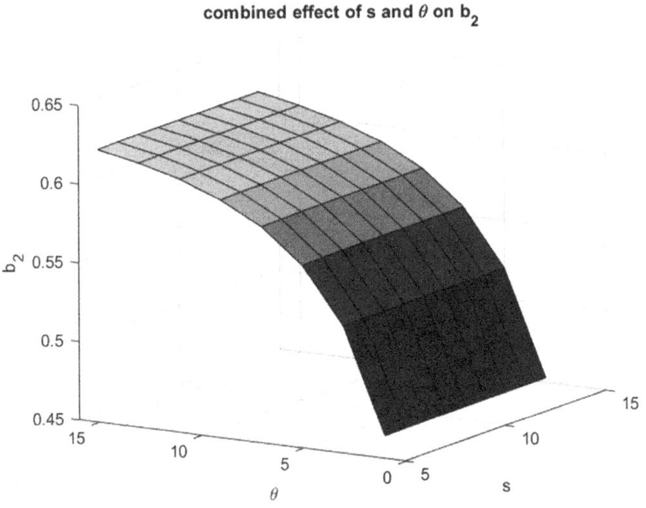

Fig. 7. The combined effect of s and θ on sever busy with a batch of type-2 customers

6 Conclusion

In the model considered here, customers opting for two different platforms of service are present. We analyze the model for the conditions of stability and evaluated the stationary distribution. We plan to extend this paper by analyzing it for cost-effectiveness. We plan to find the optimal values of the maximum buffer size N.

References

1. Sigman, K., Simchi-Levi, D.: Light traffic heutrestic for an M/G/1 queue with limited inventory. Ann. Oper. Res. **40**, 371–380 (1992)

2. Krishnamoorthy, A., Shajin, D., Narayanan, V.C.: Inventory with positive service time: a survey. OPSEARCH **48**(2), 153–169 (2011). Advanced Trends in Queueing Theory: Series of Books "Mathematics and Statistics", Sciences. ISTE & Wiley, London (2019)
3. Dudin, A., Dudin, S., Dudina, O.: Analysis of an MMAP/PH1, PH2/N/∞ queueing system operating in a random environment. Int. J. Appl. Math. Comput. Sci. **24**, 485–501 (2014)
4. Sreenivasan, C., Chakravarthy, S.R., Krishnamoorthy, A.: MAP/PH/1 queue with working vacations, vacation interruptions and N policy. Appl. Math. Model. **37**(6), 3879–3893 (2013)
5. Mathew, N., Joshua, V.C., Krishnamoorthy, A.: A queueing inventory system with two channels of service. In: Vishnevskiy, V.M., Samouylov, K.E., Kozyrev, D.V. (eds.) DCCN 2020. LNCS, vol. 12563, pp. 604–616. Springer, Cham (2020). https://doi.org/10.1007/978-3-030-66471-8_46
6. Chakravarthy, S.R., Maity, A., Gupta, U.C.: An '(s, S)' inventory in a queueing system with batch service facility. Ann. Oper. Res. **258**(2), 263–283 (2015). https://doi.org/10.1007/s10479-015-2041-z
7. Anbazhagan, N., Vigneshwaran, B., Jeganathan, K.: Stochastic Inventory system with two types of services. Int. J. Adv. Appl. Math. Mech. **2**(1), 120–127 (2014)
8. Krishnamoorthy, A., Joshua, V.C., Ambily, P.M.: A retrial queueing system with abandonment and search for priority customers. Commun. Comput. Inf. Sci. **700**, 98–107 (2017)
9. Krishnamoorthy, A., Benny, B., Shajin, D.: A revisit to queueing-inventory system with reservation, cancellation and common life time. OPSEARCH **54**(2), 336–350 (2016). https://doi.org/10.1007/s12597-016-0278-1
10. Shajin, D., Lakshmy, B., Manikandan, R.: On a two stage queueing-inventory system with rejection of customers. Neural Parallel Sci. Comput. **23**, 111–128 (2015)
11. Krishnamoorthy, A., Manikandan, R., Lakshmy, B.: A revisit to queueing inventory system with positive service time. Ann. Oper. Res. **233**(1), 221–236 (2013)

Resource Sharing Model with Minimum Allocation for the Performance Analysis of Network Slicing

Kirill Ageev[1(✉)], Eduard Sopin[1,2], and Konstantin Samouylov[1,2]

[1] Peoples Friendship University of Russia (RUDN University), 6 Miklukho-Maklaya Street, 117198 Moscow, Russian Federation
{ageev-ka,sopin-es,samuylov-ke}@rudn.ru

[2] Institute of Informatics Problems, FRC CSC RAS, 44-2 Vavilova Street, 119333 Moscow, Russian Federation

Abstract. In modern infrastructure, the mobile network operator, the owner of the equipment, leases part of the resources of the base stations. Network slicing allows several virtual network operators to use the resources of one base station. This allows operators and resource owners to provide and manage multiple dedicated logical networks with specific functionality running on top of a shared infrastructure. The key task is to manage resource allocation between slices. In the current work, we develop resource queuing model for the analysis of the network slicing technology. Besides, algorithms are being proposed to optimize the operation of slicing mechanisms. The paper provides a brief overview of resource slicing strategies, describes a simplified mathematical model of network slicing, and analyzes it using an effective recurrent algorithm.

Keywords: Network slicing · Queueing systems · Resource systems · Limited resources · Resource sharing

1 Introduction

Network slicing is one of the top features of modern network systems that provides an opportunity to a single network to simultaneously support a wide range of application scenarios (e.g., automotive, utilities, smart cities, high-tech manufacturing) and business models that impose a wide variety of requirements on network functions and expected performance. This allows operators to create and manage multiple dedicated logical networks with specific functionality running on top of the overall infrastructure. Each of these logical networks is called a network slice and can be adapted to provide specific system behavior to best support specific service/application domains [1].

This paper has been supported by the RUDN University Strategic Academic Leadership Program (recipient Samouylov K., methodology). The reported study was funded by RFBR, projects number 19-07-00933 (recipient Sopin E., mathematical model) and 19-37-90147 (recipient Ageev K., numerical analysis).

A. Dudin et al. (Eds.): ITMM 2020, CCIS 1391, pp. 378–389, 2021.
https://doi.org/10.1007/978-3-030-72247-0_28

Network slicing allows a mobile network operator to provide a part of the radio re-source to virtual network operators in the form of network slices. A network slice can be allocated for specific types of services to several virtual network operators that provide similar services, or separately for each virtual operator [2].

The list of services which provided in wireless networks by VNOs can be different. However, for the correct interaction of the network and VRRM, a list of parameters is defined that must be set for each service before it is put into interaction: class of service, service priority, violation priority, maximum rate, minimum rate, maximum waiting time for the start of service [3]. Table 1 summarizes the SLA types and their requirements [1].

Table 1. Types of service.

SLA type	Serving type	Service example
Guaranteed Bitrate (GB)	Conversational	Voice over IP (VoIP); real-time games
Guaranteed Bitrate (GB)	Streaming	Buffered video; video streaming
Best effort with minimum Guaranteed (BG)	Interactive	Web browsing, multimedia data transmission
Best Effort (BE)	Background	Email, File Transfer Protocol (FTP)

GB has a high priority among other types of SLA. VNO must provide a persistent connection with necessary bitrate. BG has a high priority too: connection must be stable, but bitrate can change between maximum and minimum borders. BE doesn't need in persistent connection at all. Services can wait some period before being done.

A model of radio resources sharing with a dynamic distribution of resources be-tween classes was described in [4], the characteristics were calculated and compared with the results of simulations. In [3,5], authors considered the theoretical basis for Multi-Operator scheduling (MOS). By dynamically adapting to the channel and load, the centralized approach maximizes spectral efficiency for multiple operators with full control over sharing guarantees.

The very important criteria of network slicing are efficient resource usage, fairness and performance isolation of slices. In [6] proposed a flexible and customizable resource slicing model which satisfied these criteria. This model is suitable QoS-aware, service-oriented slicing, where each slice is homogeneous with respect to traffic characteristics and QoS requirements.

Network slicing models can be analyzed with resource queuing systems (RQS). In [7,8] studied basis and review on research about RQS.

In this article we considered a simplified model for sharing radio resources as RQS. The model assumes a slicing of radio resources with GB SLA type. The total resource volume is divided into $K+1$ blocks. $1,..,K$ blocks are isolated from each other, each of them serves its own flow of customers with corresponding service intensity and resource distribution. There is also the share-able block 0, which accepts customers, that are blocked in $1,..,K$. We considered each block as a type of customers. Each block serv one type of customers.

2 Resource Sharing Strategies

A key feature of the slicing is sharing resources. It can be completed by different method and strategies. On figure below (see Fig. 1) present strategies how the resources can be sliced. Kleinrock and Kamoun applied these strategies for memory sharing in [9].

The first (and the simplest) is the complete partitioning scheme where actually no sharing is provided, but where the entire finite resources are permanently partitioned among the several blocks. The second scheme is complete sharing, which is such that an arriving customer is accepted if any necessary resources are available, independent of the block to which directed. With normal traffic conditions and for balanced input systems complete sharing is better in achieving a better performance. In strongly asymmetrical customer input rates and equal service rates this strategy tends to heavily favor block with higher input rates. In this case other customer flows have a high dropping probability.

Fails of first and second strategies suggest that contention for resources must be limited in some way. In order to avoid the possible utilization of the entire resources by any particular output channel, authors in [9] impose a limit on the number of resources to be allocated at any time to any blocks. This idea is incorporated in their third scheme: sharing with maximum number of resources (SMNR). SMNR still does not guarantee a full utilization of the resources under heavy traffic conditions. This deficiency motivates the fourth scheme: sharing with minimum allocation (SMA) scheme. With SMA, a minimum number of resources is always reserved for each server and, in addition, a common pool of resources is to be shared among all servers. With SMA, the shared area tends to be unfairly utilized as mentioned earlier; hence, authors offered the fifth scheme: sharing with a maximum number of resources and minimum allocation (SMRMA).

In this article we considered the sharing strategy similar to the fourth one.

3 Network Slicing Model

3.1 Model Description

K Poisson flows of customers arrive into multiserver queuing system with intensities $\lambda_k, k = 1..K$. For each type of customers, the required number of resources

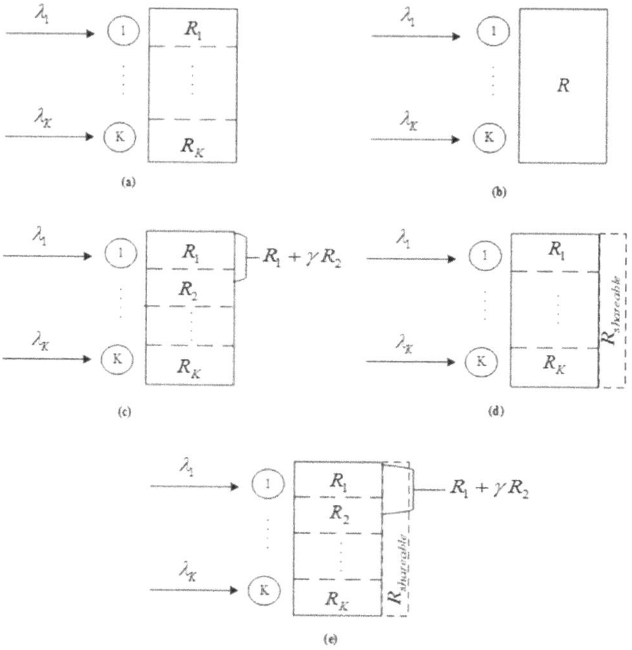

Fig. 1. (a) - CP. (b) - CS. (c) - SMNR. (d) - SMA. (e) - SMRMA.

is defined by the prob-ability distribution $\{p_{k,r}, r \geq 1\}$. Service times are exponentially distributed with intensity $\mu_k, k = 1..K$.

Customers of type k firstly arrive to the block k, which has $R_k, k = 1..K$ resources. As the simplification, we assume that the resource volumes $R_k, k = 1..K$ are fixed for all blocks and do not change in time. Maximum number of customers in block k is $N_k = R_k$.

Besides, there is also a shareable block with R_0 resources. If a k-type customer is blocked due to insufficient resources at block k, then it is redirected to the shareable block.

Total volume of resources is the following:

$$R = \sum_{k=0}^{K} R_k \tag{1}$$

Let $n_k(t)$ be the number of customers in block k at moment $t, (t > 0)$. and $\gamma_k(t) = (\gamma_{k,1}(t), ..., \gamma_{k,n_k}(t))$ - the vector of occupied resources by each customer in block. To simplify the description of the set of states we applied the aggregation of occupied volume of resources. $r_k(t)$ total occupied volume of resources in block k [4]. Therefore, the set of states can be expressed as:

$$X_k = \{(n_k, r_k) : 0 \leq n_k \leq N_k, 0 \leq r_k \leq R_k\}, k = 1, ..., K \tag{2}$$

Assume that a k-type customer arrives to the system and it requires r resources. Then, if there are not enough resources in block k to meet the resource requirements, $r_k + r \geq R_k$, the customer is redirected to the common block R_0. If there are not enough resources for the customer in the common block, $r_0 + r \geq R_0$, the customer is lost.

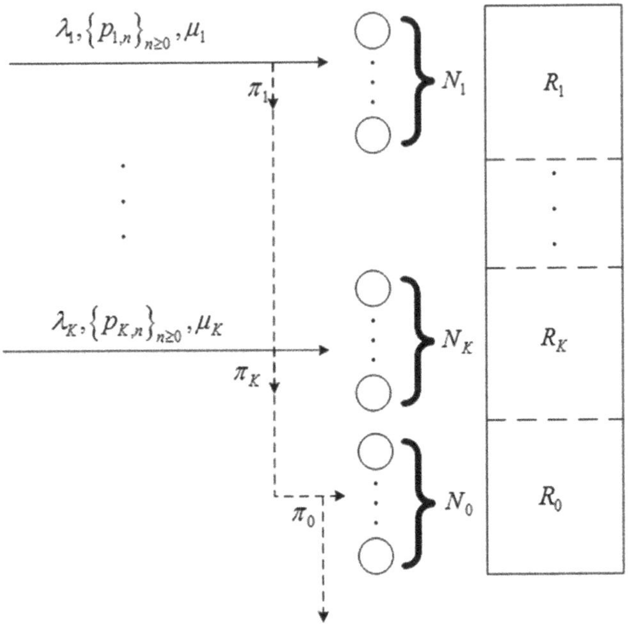

Fig. 2. The scheme of the model

Denote the stationary distribution of process X_k:

$$q_{k,0} = \left(1 + \sum_{n=0}^{N_k} P_k^{(n)} \frac{\rho_k^n}{n!}\right)^{-1} \tag{3}$$

$$q_{k,n}(r) = q_{k,0} \frac{\rho_k^n}{n!} P_{k,r}^{(n)} \tag{4}$$

where $P_{k,r}^{(n)}$ is the n-fold convolution of resource requirements function, and $\rho_k = \frac{\lambda_k}{\mu_k}$ - is the offered load. $q_{k,n}(r)$ - is a probability that there are n customers occupying r resources at block k. Distribution of resource requirements for the flow of rejected customers in block k can be calculated according to the formula (5).

$$\tilde{p}_{k,r} = \frac{1}{\pi_k} p_r \left(\sum_{1 \le n \le N_k - 1, R_{k-r+1} \le j \le R_k} q_{k,n}(j) + \sum_{1 \le j \le R_k} q_{k,N_k}(j) \right) \quad (5)$$

Probability of redirecting from k-block to shareable block can be calculated according to formula (6) and the average volume of occupied resources according to formula (7).

$$\pi_k = 1 - q_{k,0} \sum_{n=0}^{N_k-1} \frac{\rho_k^n}{n!} P_{k,r}^{(n)} \quad (6)$$

$$b_k = q_{k,0} \sum_{n=1}^{N_k} \frac{\rho_k^n}{n!} \sum_{r=1}^{R_k} r P_{k,r}^{(n)} \quad (7)$$

The scheme of the described model is presented on figure (see Fig. 2).

To define the resource requirements distribution of the customers that arrive to the shareable resource block, we combine distributions of resource requirements for each block k in formula (9). The offered load at the shareable block can be calculated ac-cording to formula (8).

$$\rho_0 = \sum_{k=1}^{K} \rho_k \pi_k \quad (8)$$

$$P_{0,r} = \sum_{k=1}^{K} \frac{\rho_k \pi_k}{\rho_0} \tilde{p}_{k,r} \quad (9)$$

3.2 Analysis of the Performance Measures

Denote

$$G_k(n, r) = \sum_{i=0}^{n} \frac{\rho_k^i}{i!} \sum_{0 \le j \le r} p_{k,j}^{(I)},$$

where $p_{k,j}^{(i)}$ is the i-fold convolution of the distribution $\{p_{k,j}\}$. According to [10], the stationary probability distributions at block k can be evaluated using the functions $G_k(n, r)$ in the following way:

$$G_k(N_k, R_k)^{-1} G_k(n, r) = \sum_{0 \le i \le n, 0 \le j \le r} q_{k,i}(j) \quad (10)$$

Based on recurrent algorithm, presented in [11], we can find G by the recurrent relation (11).

$$G(n, r) = \frac{\rho_k}{n_k} \sum_{0 \le j \le r} p_{k,j} (G(n-1, r-j) - G(n-2, r-j)) + G(n-1, r), 2 \le n \le N_k \quad (11)$$

The initial values are as follows:

$$G(1,r) = 1 + \rho_k \sum_{0 \le j \le r} p_{k,j}, 0 \le r \le R_k \tag{12}$$

$$G(0,r) = 1, 0 \le r \le R_k \tag{13}$$

We calculated the probability of redirecting customers from block k to the shareable block 0 by formula (14).

$$\pi_k = 1 - G_k^{-1}(N_k, R_k) \sum_{0 \le j \le R_k} p_{k,j} G_k(N_k - 1, R_k - j) \tag{14}$$

We used formula (15) to calculate the average volume of occupied resources in block k.

$$b_k = R_k - G_k^{-1}(N_k, R_k) \sum_{r=1}^{R_k} G_k(N_k, R_k - r) \tag{15}$$

Proposition 1. The resource requirements distribution of the redirected customers at block 0 can be also obtained with the help of $G_k(n,r)$, by substituting (10) into (7):

$$\begin{aligned} \tilde{p}_{k,r} = p_{k,r} G_k(N_k, R_k)^{-1} (G_k(N_k - 1, R_k) - G_k(N_k - 1, R_k - r) \\ + G_k(N_k, R_k) - G_k(N_k - 1, R_k)) \end{aligned} \tag{16}$$

$$\tilde{p}_{k,r} = p_{k,r} G_k(N_k, R_k)^{-1} (G_k(N_k, R_k) - G_k(N_k - 1, R_k - r)) \tag{17}$$

Dropping probability for k-type customers derived by formula (18).

$$B_k = \pi_k \pi_0 \tag{18}$$

4 Case Study

4.1 Equal Resources for K Blocks

For the experiment we took the number of blocks $K = 3$. The distribution of resource requirements was taken from [12] and is assumed the same for each block. Note that the average resource requirement is 2,99. For blocks 1–3 we set the offered load $\rho_1 = 200$, $\rho_2 = 60$, $\rho_3 = 32$.

Blocks 1–3 are assumed to have the same number of resources. The results were obtained for different capacity of the shareable block R_0 with fixed total number of resources R. The range of the resource volumes in blocks 1–3 is $R_1 = R_2 = R_3 = [60, 100]$. The range for block 0 is $R_0 = [50, 170]$. Total volumes of servers and resources are $R = N = 350$.

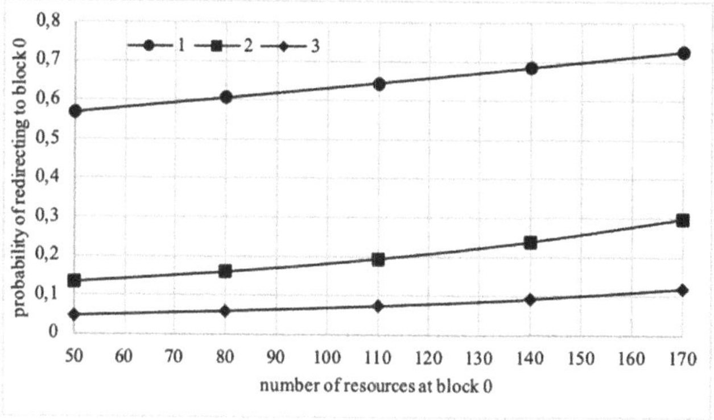

Fig. 3. Probability of redirecting to the shareable block as a function of its capacity.

The diagram (see Fig. 3) shows the dependence of the probability of redirecting customers from blocks 1–3 to the shareable block on its capacity. Note that the probability is increasing, which is correct, because we decrease the resources for blocks 1–3.

The next diagram (see Fig. 4) shows the dependence of dropping probability of each customer types on the capacity of the shareable block.

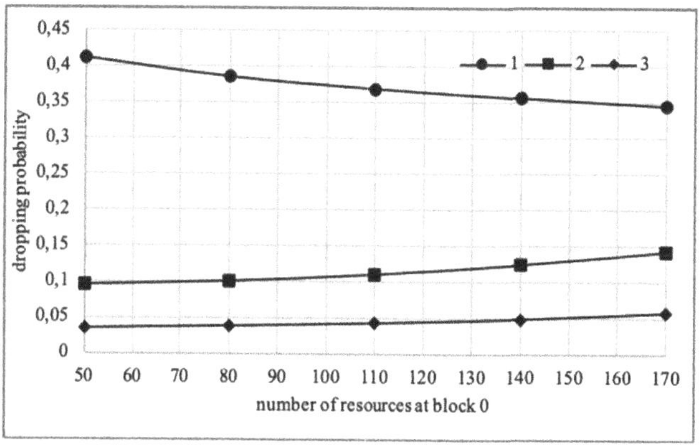

Fig. 4. Dropping probability depending on distribution of resources.

On Fig. 4 we can see, that for customers in blocks 2 and 3 dropping probability increase and stay lower than for block 1. While for customers from block 1, the prob-ability of drop decreases. This behavior is explained by the fact that an equal number of resources is allocated for blocks 1–3, and the offered load for block 1 is significantly higher than for blocks 2 and 3. Thus, when the volume of resources for blocks 1–3 decreases, the intensity of redirections from block 1 to block 0 increases. That allows customers from block 1 to occupy more resources in block 0, thus reducing the drop-ping probability.

The next diagram (see Fig. 5) shows the dependence of utilization ratio of resources on the capacity of the shareable block.

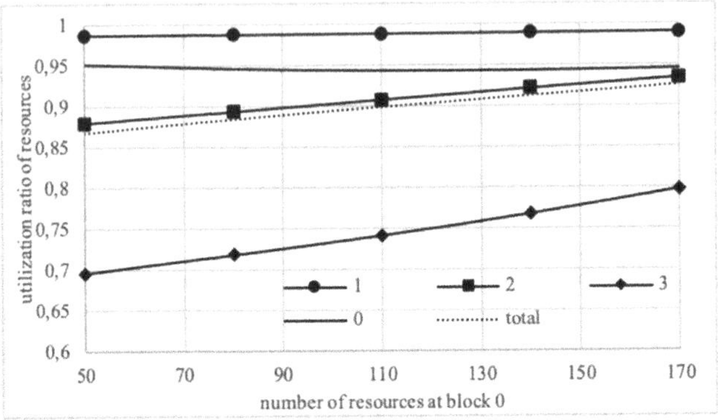

Fig. 5. Utilization ratio of resources depending on distribution of resources.

Utilization ratios for blocks 1–3 depend on offered load and number of resources. That's why it rises with decreasing the number of resources. We can oversee that total utilization ratio increase too. In case, when for blocks 1–3 is 100, for block 0 – 50 – there are many free resources in block 2 and 3. But when the resources for block 0 comes to 170, system can redirect from block 1 to block 0 more customers, that's why the total utilization ration increased.

4.2 Equal Offered Loads for K Blocks

For the next experiment we took the number of blocks $K = 3$. The distribution of resource requirements was taken from [12] and is assumed the same for each block. Note that the average resource requirement is 2,99. For blocks 1–3 we set the offered load $\rho_1 = \rho_2 = \rho_3 = 125$.

We considered five cases with numbers of resources for blocks pointed in Table 2. Total volumes of servers and resources are always $R = N = 350$.

The diagram (see Fig. 6) shows the dependence of the probability of redirecting customers from blocks 1–3 to the shareable block on its capacity. Note that the probability is increasing, which is correct, because we decrease the resources for blocks 1–3. Behavior of lines is similar, because we considered the same offer loads and smoothly decreased numbers of resources for each block.

The next diagram (see Fig. 7) shows the dependence of dropping probability of each customer types on the capacity of the shareable block. Notice that dropping probability is increasing, which is correct too. Behavior on this diagram can be explained by increasing utilization ratio in system.

The diagram (see Fig. 8) shows the dependence of utilization ratio of resources on the capacity of the shareable block.

Table 2. Numbers of resources for each block.

Block	1	2	3	0
Case 1	60	65	70	155
Case 2	70	75	80	125
Case 3	80	85	90	95
Case 4	90	95	100	65
Case 5	100	105	110	35

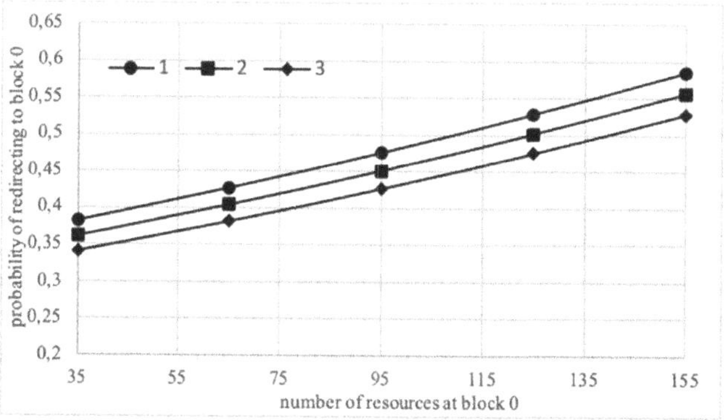

Fig. 6. Probability of redirecting to the shareable block as a function of its capacity.

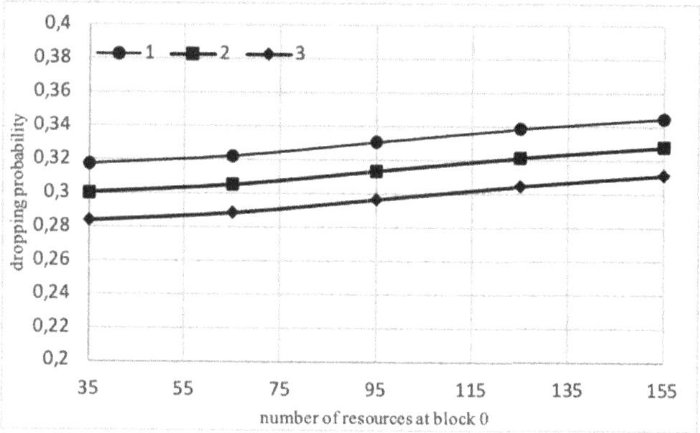

Fig. 7. Dropping probability depending on distribution of resources.

Utilization ratios for blocks 1–3 depend on offered load and number of resources. That's why it rises with decreasing the number of resources. We can oversee that total utilization ratio increase too. In this case we also point out that with increasing re-sources for shareable block the utilization ratio increases too.

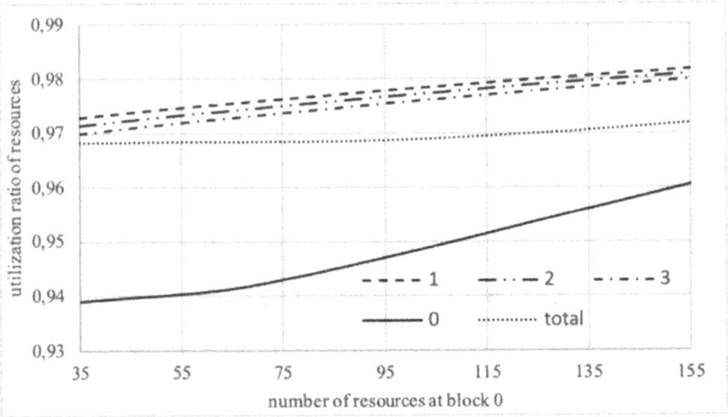

Fig. 8. Utilization ratio of resources depending on distribution of resources.

5 Conclusion

Network slicing is one of the main features of modern wireless technologies. Despite the fact that research has been conducted in this area for several years, a wide range of applications and variations of use allows you to create, describe and analyze new queuing models.

In this paper, we have made a brief overview of possible strategies for sharing network resources and discussed one of them in more detail. Also, a simplified model of radio resource slicing is studied. The system is described in terms of resource queuing systems. The simplification is that we were looking at a system where the resource was divided and committed between slices. At the same time, considered a common block that could receive customers redirected from any other block in moments of a lack of resources.

The stationary characteristics of the model such a blocking probability, dropping probability and the utilization ratio are calculated using a recurrent algorithm with normalization function. This approach worked well in early studies, when direct computing led to high time and resource costs.

Numerical experiments were performed for three blocks. Based on the results, it can be noted that the proposed algorithms and formulas can be applied to the analysis of the systems that use radio resource slicing mechanisms. We can notice that dropping probabilities is rather high. In this case customers can be redirected to close standing systems, which work on unlicensed frequencies. For example to LTE-U (Long-Term Evolution Unlicensed) or LAA (License assisted Access) for LTE, or to NR-U (New Radio Unlicensed) for LTE. As a result the obtained data and developments can contribute to the analysis of services provided through such channels.

The area of future research is wide. At the next stage, we plan to analysis different numerical results for a new practical cases. Then we will extend the simulation tool and make it more flexible for different types of services and traffic. We plan to set and solve the optimization problem for network slicing with different SLA types and initial parameters. Also we will pay more attention to the practical side of the issue and consider the use of network slicing for IoT (Internet of Things) services and road traffic monitoring.

References

1. Ageev, K., et al.: Modelling of virtual radio resources slicing in 5G networks. In: Dudin, A., Nazarov, A., Moiseev, A. (eds.) ITMM 2019. CCIS, vol. 1109, pp. 150–161. Springer, Cham (2019). https://doi.org/10.1007/978-3-030-33388-1_13
2. 3GPP TS 23.501 V15.4.0 - System architecture for the 5G System. http://www.3gpp.org/ftp//Specs/archive/23_series/23.501/23501-f40.zip 3GPP TR 28.801. Study on management and orchestration of network slicing for next generation network. Accessed 20 Jan 2021
3. Malanchini, I., Valentin, S., Aydin, O.: An analysis of generalized resource sharing for multiple operators in cellular networks. In: 2014 IEEE 25th Annual International Symposium on Personal, Indoor, and Mobile Radio Communication (PIMRC), pp. 1157–1162. https://doi.org/10.1109/PIMRC.2014.7136342
4. Ageev, K.A., Sopin, E.S., Yarkina, N.V., Samouylov, K.E., Shorgin, S.Ya.: Analysis of the network slicing mechanisms with guaranteed allocated resources for various traffic types. Informatika i Ee Primeneniya **14**(3), 94–100 (2020). https://doi.org/10.14357/19922264200314
5. Malanchini, I., Valentin, S., Aydin, O.: Wireless resource sharing for multiple operators: generalization, fairness, and the value of prediction. Comput. Netw. **14**, 110–123 (2016). https://doi.org/10.1016/j.comnet.2016.02.014
6. Yarkina, N., Gaidamaka, Y., Correia, L.M., Samouylov, K.: An analytical model for 5G network resource sharing with flexible SLA-oriented slice isolation. Mathematics **8**(7), 1177 (2020). https://doi.org/10.3390/math8071177
7. Gorbunova, A.V., Naumov, V.A., Gaidamaka, Y.V., Samouylov, K.E.: Resource queuing systems as models of wireless communication systems. Informatika i Ee Primeneniya **12**(3), 48–55 (2018). https://doi.org/10.14357/19922264180307
8. Gorbunova, A.V., Naumov, V.A., Gaidamaka, Y.V., Samouylov, K.E.: Resource queuing systems with general service discipline. Informatika i Ee Primeneniya **13**(1), 99–107 (2019). https://doi.org/10.14357/19922264190114
9. Kamoun, F., Kleinrock, L.: Analysis of shared finite storage in a computer network node environment under general traffic conditions. IEEE Trans. Commun. **28**(7), 992–1003 (1980). https://doi.org/10.1109/TCOM.1980.1094756
10. Naumov, V.A., Samuilov, K.E., Samuilov, A.K.: On the total amount of resources occupied by serviced customers. Autom. Remote Control **77**(8), 1419–1427 (2016). https://doi.org/10.1134/S0005117916080087
11. Sopin, E.S., Ageev, K.A., Markova, E.V., Vikhrova, O.G., Gaidamaka, Y.V.: Performance analysis of M2M traffic in LTE network using queuing systems with random resource requirements. Autom. Control Comput. Sci. **52**(5), 345–353 (2018). https://doi.org/10.3103/S0146411618050127
12. Begishev, V., Moltchanov, D., Sopin, E., Samuylov, A., Andreev, S., Koucheryavy, Y., Samouylov, K.: Quantifying the impact of guard capacity on session continuity in 3GPP new radio systems. IEEE Trans. Veh. Technol. **68**(12), 12345–12359 (2019). https://doi.org/10.1109/TVT.2019.2948702

Analysis of a MAP/PH(1), PH(2)/2 Production Inventory System Under the Bernoulli Vacation Scheme

Pathari Beena[1] and K. P. Jose[2(✉)]

[1] Sree Neelakanta Government Sanskrit College, Pattambi, Palakkad 679306, Kerala, India
[2] Department of Mathematics, St. Peter's College, Kolenchery 682311, Kerala, India

Abstract. This article presents an inventory system of production and multiple servers, each of which takes multiple vacations, and the vacations are subject to the Bernoulli vacation policy. The arrival of customers constitutes a Markovian Arrival Process (MAP) and the servers provide phase-type service. Once production starts, the time for making an item, and the vacation of each server are exponentially distributed. The manufacturing process begins when the level reaches a prefixed point and stops production when the inventory level reaches a maximum value S. Matrix Analytic Method (MAM) is used to obtain the algorithmic solution to the model. A suitable cost function based on performance measures is developed and numerical experiments are conducted under various combinations of arrival and service processes.

Keywords: Markovian arrival process · Multiple servers · Multiple vacations · Cost analysis · Bernoulli vacation policy

1 Introduction

The multiple server production inventory models with server vacations have become an intensive area of research in recent years and researchers studied this area extensively due to the various applications in production/inventory systems, communication systems, computer networks, and data switching systems. Vacation in a queuing system means that the servers are not convenient enough to provide the service in a short period. Studies related to vacation in queuing models be seen to have begun in the early 1970s. Doshi [5] and Thegam [15] proposed outstanding survey papers on vacation models. The analysis of the Bernoulli vacation model (BER) was started by Keilson and Servi [8]. Ayyappan and Gouthami [1] analyzed a Bernoulli schedule vacation model with customer reneging and Bernoulli feedback. Banik [2] studied a BMAP/G/1/N system with a p-limited service schedule and vacation to the server depends on the length of the queue. The stationary distribution of the number of customers in the system at different epochs is calculated. Suganya [14] discussed a retrial

A. Dudin et al. (Eds.): ITMM 2020, CCIS 1391, pp. 390–403, 2021.
https://doi.org/10.1007/978-3-030-72247-0_29

inventory system with finite orbit size and multiple servers in which the major contribution was the use of the idle period of the servers. Krishnamoorthy and Viswanath Narayanan [9] analyzed a production inventory system in which the manufacturing process follows a Markovian Production Scheme. Krishnakumar et al. [11] discussed the waiting time distribution and mean waiting time of a multiple server queueing system. Krishnakumar and Pavai Madheswari [10] considered multiple servers queueing model in which each server takes the Bernoulli vacation service schedule. Jose and Beena [6] analyzed an inventory system with production, multiple servers, and customer retrial. Jose and Salini [7] compared two production inventory models with different production rates and with the retrial of customers. Beena and Jose [3] discussed a multiple server production inventory system in which the production rate depends on the stock level. Chakravarthy [4] analyzed a single server vacation model where primary users could choose an additional service with certain probabilities or exit the system with its complementary probabilities.

The following situation in a manufacturing company that produces cattle feed can be considered as an example of the model under consideration. There are two staff- one technician and the other, a worker to assist the technician. Service rates of technician and worker are different. Once the stock level of cattle feed reaches S, the production will stop. If the level of stock drops to s, production will begin immediately until the stock level reaches back to S. The presence of a control parameter is an important factor behind the use of the Bernoulli service strategy, and we can control system congestion by modifying the value of that parameter.

This article is organized as follows. Description of the model and system stability are given in Sects. 2 and 3. Sections 4 and 5 consider the measures of effectiveness and cost analysis. Numerical illustrations are presented in Sect. 6. The conclusion is given in Sect. 7.

2 Description of the Model

In this article, we studied a production inventory system with multiple servers that provide different rates of service to customers. We assume that the arrivals of customers are according to a MAP with representation $(D_0, D_1)_l$. Service rates of server 1 and server 2 are phase type distributed having representations $(\alpha, S)_m$ and $(\beta, T)_n$ respectively. Vacation duration of servers 1 and 2 are exponentially distributed with parameters θ_1 and θ_2. If the system has no customers or the inventory level is zero or both, the servers will always take a vacation. When the service completes, the servers can choose a vacation with probability $p_i, i = 1, 2$ or restart the service with its complementary probability $q_i = 1 - p_i, i = 1, 2,$ if there are positive inventory levels and customers in the waiting area. When the servers return after a vacation, they return to the vacation, if the system is empty or the inventory level is zero, or both. The servers will continue to do this until they find the system nonempty with a positive inventory level. Items are only available after a certain period and are distributed exponentially with parameter $\gamma(> 0)$.

The notations used in this model are

- N(t) indicates the number of customers in the system at time t.
- I(t) describes the level of stock at time t.
- C(t) denotes the status of servers 1 and 2 where

$$C(t):\begin{cases}0, & \text{if both the servers are on vacation}\\ 1, & \text{if server 1 is busy and server 2 is on vacation}\\ 2, & \text{if server 1 is on vacation and server 2 is busy}\\ 3, & \text{if both servers are busy}\end{cases}$$

- J(t) denotes the production status where

$$J(t):\begin{cases}0, \text{ if the production process gets switched OFF}\\ 1, \text{ if the production process gets switched ON}\end{cases}$$

- $J_0(t)$ indicates the phase of the arrival process.
- $J_1(t)$ denotes the phase of the service process of server 1.
- $J_2(t)$ indicates the phase of the service process of server 2.
- l, m, n denote the number of arrival phases, service phases of server 1, and 2 respectively.
- e_p denotes a column vector of dimension p with each of its entries are 1.
- $e_1 = e_{l(2S-s)+lm(2S-s-1)+ln(2S-s-1)+lmn(2S-s-2)}$.
- $e_2 = e_{l(2S-s)+lm(2S-s-1)+ln(2S-s-1)}$ and $e_3 = e_{l(2S-s)}$.

2.1 Steady State Analysis

Then $\{X(t) = (N(t), C(t), J(t), I(t), J_0(t), J_1(t), J_2(t)), t \geq 0\}$ is a continuous time Markov chain on the state space $\Omega = l(0) \cup l(1) \cup l(2) \cup l(3)$ where
$l(0) = (i, 0, 0, k, j_0)|s+1 \leq k \leq S, \cup (i, 0, 1, k, j_0)|0 \leq k \leq S-1; i \geq 0$
$l(1) = (i, 1, 0, k, j_0, j_1)|s+1 \leq k \leq S, \cup (i, 1, 1, k, j_0, j_1)|1 \leq k \leq S-1; i \geq 1$
$l(2) = (i, 2, 0, k, j_0, j_2)|s+1 \leq k \leq S, \cup (i, 2, 1, k, j_0, j_2)|1 \leq k \leq S-1; i \geq 1$
$l(3) = (i, 3, 0, k, j_0, j_1, j_2)|s+1 \leq k \leq S, \cup (i, 3, 1, k, j_0, j_1, j_2)|2 \leq k \leq S-1; i \geq 2$

Generator Q of this process is of the form

$$Q = \begin{bmatrix} B_{00} & B_{01} & 0 & 0 & 0 & \cdots \\ B_{10} & B_{11} & B_{12} & 0 & 0 & \cdots \\ 0 & B_{21} & A_1 & A_0 & 0 & \cdots \\ 0 & 0 & A_2 & A_1 & A_0 & \cdots \\ \vdots & \vdots & \vdots & \vdots & \vdots & \ddots \end{bmatrix}$$

where A_0, A_1, A_2 are square matrices of order $l(2S-s) + lm(2S-s-1) + ln(2S-s-1) + lmn(2S-s-2)$.

$$A_1 = \begin{bmatrix} C_1^{(11)} & C_1^{(12)} & C_1^{(13)} & 0 \\ 0 & C_1^{(22)} & 0 & C_1^{(24)} \\ 0 & 0 & C_1^{(33)} & C_1^{(34)} \\ 0 & 0 & 0 & C_1^{(44)} \end{bmatrix}, \; A_2 = \begin{bmatrix} 0 & 0 & 0 & 0 \\ C_2^{(21)} & C_2^{(22)} & 0 & 0 \\ C_2^{(31)} & 0 & C_2^{(33)} & 0 \\ 0 & C_2^{(42)} & C_2^{(43)} & C_2^{(44)} \end{bmatrix},$$

$$B_{11} = \begin{bmatrix} C_1^{(11)} & C_1^{(12)} & C_1^{(13)} \\ 0 & C_{11}^{(22)} & 0 \\ 0 & 0 & C_{11}^{(33)} \end{bmatrix}, \; B_{21} = \begin{bmatrix} 0 & 0 & 0 \\ C_2^{(21)} & C_2^{(22)} & 0 \\ C_2^{(31)} & 0 & C_2^{(33)} \\ 0 & C_{21}^{(42)} & C_2^{(43)} \end{bmatrix}$$

$$B_{00} = \begin{cases} D_0, \text{ if } 1 \le u \le S-s, v=u \\ D_0 - \gamma I_l, \quad S-s+1 \le u \le 2S-s, v=u \\ \gamma I_l, \quad S-s+1 \le u \le 2S-s-1, v=u+1 \\ \gamma I_l, \quad u=2S-s, v=S-s \\ 0, \quad \text{otherwise} \end{cases}$$

$$B_{01} = \begin{cases} D_1, \quad 1 \le u \le 2S-s, v=u \\ 0, \quad \text{otherwise} \end{cases}$$

$$B_{10} = \begin{cases} I_l \otimes S^0, \quad u=2S-s+1, v=S+1 \\ I_l \otimes S^0, \quad 2S-s+2 \le u \le 3S-2s, 1 \le v \le S-s-1 \\ I_l \otimes S^0, \quad 3S-2s+1 \le u \le 4S-2s-1, S-s+1 \le v \le 2S-s-1 \\ I_l \otimes T^0, \quad u=4S-2s, v=S+1 \\ I_l \otimes T^0, \quad 4S-2s+1 \le u \le 5S-3s-1, 1 \le v \le S-s-1 \\ I_l \otimes T^0, \quad 5S-3s \le u \le 6S-3s-2, S-s+1 \le v \le 2S-s-1 \\ 0, \quad \text{otherwise} \end{cases}$$

$$B_{12} = \begin{cases} D_1, \quad 1 \le u \le 2S-s, v=u \\ D_1 \otimes I_m, \quad 2S-s+1 \le u \le 4S-2s-1, v=u \\ D_1 \otimes I_n, \quad 4S-2s \le u \le 6S-3s-2, v=u \\ 0, \quad \text{otherwise} \end{cases}$$

$$A_0 = \begin{cases} D_1, \quad 1 \le u \le 2S-s, v=u \\ D_1 \otimes I_m, \quad 2S-s+1 \le u \le 4S-2s-1, v=u \\ D_1 \otimes I_n, \quad 4S-2s \le u \le 6S-3s-2, v=u \\ D_1 \otimes I_m \otimes I_n, \quad 6S-3s-1 \le u \le 8S-4s-4, v=u \\ 0, \quad \text{otherwise} \end{cases}$$

$$C_{11}^{(22)} = \begin{cases} (D_0 \oplus S), \quad 1 \le u \le S-s, v=u \\ (D_0 \oplus S) - \gamma I_{lm}, \quad S-s+1 \le u \le 2S-s-1, v=u \\ \gamma I_{lm}, \quad S-s+1 \le u \le 2S-s-2, v=u+1 \\ \gamma I_{lm}, \quad u=2S-s-1, v=S-s \\ 0, \text{ otherwise} \end{cases}$$

$$C_{11}^{(33)} = \begin{cases} (D_0 \oplus T), & 1 \leq u \leq S-s-1, v=u \\ (D_0 \oplus T) - \gamma I_{ln}, & \text{if } S-s+1 \leq u \leq 2S-s-1, v=u \\ \gamma I_{ln}, & S-s+1 \leq u \leq 2S-s-2, v=u+1 \\ \gamma I_{ln}, & u=2S-s-1, v=S-s \\ 0, & \text{otherwise} \end{cases}$$

$$C_{2}^{(21)} = \begin{cases} I_l \otimes p_1 S^0, & u=1, v=S+1 \\ I_l \otimes p_1 S^0, & 2 \leq u \leq S-s, v=u-1 \\ I_l \otimes p_1 S^0, & S-s+1 \leq u \leq 2S-s-1, v=u \\ 0, & \text{otherwise} \end{cases}$$

$$C_{2}^{(22)} = \begin{cases} I_l \otimes q_1\, S^0\alpha, & u=1, v=S \\ I_l \otimes q_1\, S^0\alpha, & 2 \leq u \leq S-s, v=u-1 \\ I_l \otimes q_1\, S^0\alpha, & S-s+2 \leq u \leq 2S-s-1, v=u-1 \\ 0, & \text{otherwise} \end{cases}$$

$$C_{2}^{(31)} = \begin{cases} I_l \otimes p_2 T^0, & u=1\,S, v=S+1 \\ I_l \otimes p_2 T^0, & 2 \leq u \leq S-s, v=u-1 \\ I_l \otimes p_2 T^0, & S-s+1 \leq u \leq 2S-s-1, v=u \\ 0, & \text{otherwise} \end{cases}$$

$$C_{2}^{(33)} = \begin{cases} I_l \otimes q_2 T^0\beta & , u=1, v=S \\ I_l \otimes q_2 T^0\beta, & 2 \leq u \leq S-s, v=u-1 \\ I_l \otimes q_2 T^0\beta, & S-s+2 \leq u \leq 2S-s-1, v=u-1 \\ 0, & \text{otherwise} \end{cases}$$

$$C_{2}^{(42)} = \begin{cases} I_l \otimes (I_m \otimes p_2 T^0), & u=1, v=S \\ I_l \otimes (I_m \otimes p_2 T^0), & 2 \leq u \leq S-s, v=u-1 \\ I_l \otimes (I_m \otimes p_2 T^0), & S-s+1 \leq u \leq 2S-s-2, v=u \\ 0, & \text{otherwise} \end{cases}$$

$$C_{2}^{(43)} = \begin{cases} I_l \otimes (p_1 S^0 \otimes I_n), & u=1, v=S \\ I_l \otimes (p_1 S^0 \otimes I_n), & 2 \leq u \leq S-s, v=u-1 \\ I_l \otimes (p_1 S^0 \otimes I_n), & S-s+1 \leq u \leq 2S-s-2, v=u \\ 0, & \text{otherwise} \end{cases}$$

$$C_{2}^{(44)} = \begin{cases} I_l \otimes (q_1 S^0\alpha \oplus q_2 T^0\beta), & u=1, v=S \\ I_l \otimes (q_1 S^0\alpha \oplus q_2 T^0\beta), & 2 \leq u \leq S-s, v=u-1 \\ I_l \otimes (q_1 S^0\alpha \oplus q_2 T^0\beta), & S-s+2 \leq u \leq 2S-s-2, v=u-1 \\ 0, & \text{otherwise} \end{cases}$$

$$C_{21}^{(42)} = \begin{cases} I_l \otimes (I_m \otimes T^0), & u=1, v=S \\ I_l \otimes (I_m \otimes p_2 T^0), & 2 \leq u \leq S-s, v=u-1 \\ I_l \otimes (I_m \otimes p_2 T^0), & S-s+1 \leq u \leq 2S-s-2, v=u \\ 0, & \text{otherwise} \end{cases}$$

$$C_2^{(43)} = \begin{cases} I_l \otimes (S^0 \otimes I_n), & u = 1, v = S \\ I_l \otimes (S^0 \otimes I_n), & 2 \le u \le S - s, v = u - 1 \\ I_l \otimes (S^0 \otimes I_n), & S - s + 1 \le u \le 2S - s - 2, v = u \\ 0, & \text{otherwise} \end{cases}$$

$$C_1^{(11)} = \begin{cases} D_0 - (\theta_1 + \theta_2) I_l, & 1 \le u \le S - s, v = u \\ D_0 - \gamma I_l, & u = S - s + 1, v = u \\ D_0 - (\theta_1 + \theta_2 + \gamma) I_l, & S - s + 2 \le u \le 2S - s, v = u \\ \gamma I_l, & S - s + 1 \le u \le 2S - s - 1, v = u + 1 \\ \gamma I_l, & u = 2S - s, v = S - s \\ 0, & \text{otherwise} \end{cases}$$

$$C_1^{(12)} = \begin{cases} I_l \otimes \theta_1 \alpha, & `1 \le u \le S - s, v = u \\ I_l \otimes \theta_1 \alpha, & S - s + 2 \le u \le 2S - s, v = u - 1 \\ 0, & \text{otherwise} \end{cases}$$

$$C_1^{(13)} = \begin{cases} I_l \otimes \theta_2 \beta, & 1 \le u \le S - s, v = u \\ I_l \otimes \theta_2 \beta, & S - s + 2 \le u \le 2S - s, v = u - 1 \\ 0, & \text{otherwise} \end{cases}$$

$$C_1^{(22)} = \begin{cases} (D_0 \oplus (S - \theta_2 I_m)), & 1 \le u \le S - s, v = u \\ (D_0 \oplus p_1 S) - \gamma I_{lm} & u = S - s + 1, v = u \\ D_0 \oplus (S - \theta_2 I_m) - \gamma I_{lm}, & S - s + 2 \le u \le 2S - s - 1, v = u \\ \gamma I_{lm}, & S - s + 1 \le u \le 2S - s - 2, v = u + 1 \\ \gamma I_{lm}, & u = 2S - s - 1, v = S - s \\ 0, & \text{otherwise} \end{cases}$$

$$C_1^{(24)} = \begin{cases} I_l \otimes \theta_2 \beta, & 1 \le u \le S - s, v = u \\ I_l \otimes \theta_2 \beta, & S - s + 2 \le u \le 2S - s - 1, v = u - 1 \\ 0, & \text{otherwise} \end{cases}$$

$$C_1^{(33)} = \begin{cases} D_0 \oplus (T - \theta_1 I_n), 1 \le u \le S - s, v = u \\ (D_0 \oplus p_2 T) - \gamma I_{ln}, & u = S - s + 1, v = u \\ D_0 \oplus (T - \theta_1 I_n) - \gamma I_{ln}, & S - s + 2 \le u \le 2S - s - 1, v = u \\ \gamma I_{ln}, & S - s + 1 \le u \le 2S - s - 2, v = u + 1 \\ \gamma I_{ln}, & u = 2S - s - 1, v = S - s \\ 0, & \text{otherwise} \end{cases}$$

$$C_1^{(34)} = \begin{cases} I_l \otimes (\theta_1 \alpha \otimes I_n), & 1 \le u \le S - s, v = u \\ I_l \otimes (\theta_1 \alpha \otimes I_n), & S - s + 2 \le u \le 2S - s - 1, v = u - 1 \\ 0, & \text{otherwise} \end{cases}$$

$$C_1^{(44)} = \begin{cases} D_0 \otimes I_m \otimes I_n + I_l \otimes (S \oplus T), 1 \le u \le S-s, v=u \\ D_0 \otimes I_m \otimes I_n + I_l \otimes (p_1 S \oplus p_2 T) - \gamma I_{lmn}, u = S-s+1, v=u \\ D_0 \otimes I_m \otimes I_n + I_l \otimes (S \oplus T) - \gamma I_{lmn}, \quad S-s+2 \le u \le 2S-s-2, v=u \\ \gamma I_{lmn}, \quad S-s+1 \le u \le 2S-s-3, v=u+1 \\ \gamma I_{lmn}, \quad u=2S-s-2, v=S-s \\ 0, \quad \text{otherwise} \end{cases}$$

3 System Stability

To prove the system stability, we define transition rate matrix $A = A_0 + A_1 + A_2$ of order $l(2S-s) + lm(2S-s-1) + ln(2S-s-1) + lmn(2S-s-2)$. A is irreducible so there exists a stationary probability vector Π satisfying $\Pi A = 0$ and $\Pi \boldsymbol{e} = 1$. Π can be partitioned as $\Pi = (\pi^{[i]}, i = 0,1,2,3)$ where each $\pi^{[i]} = \{(\pi^{[i,0]}, \pi^{[i,1]}), i = 0,1,2,3\}$

$$\pi^{[i,0]} = \begin{cases} (\pi^{[i,0,s+1,l]}, \ldots, \pi^{[i,0,S,l]}), i=0 \\ (\pi^{[i,0,s+1,l,m]}, \ldots, \pi^{[i,0,S,l,m]}), i=1 \\ (\pi^{[i,0,s+1,l,n]}, \ldots, \pi^{[i,0,S,l,n]}), i=2 \\ (\pi^{[i,0,s+1,l,m,n]}, \ldots, \pi^{[i,0,S,l,m,n]}), i=3 \end{cases}$$

$$\pi^{[i,1]} = \begin{cases} (\pi^{[i,1,0,l]}, \ldots, \pi^{[i,1,S-1,l]}), i=0 \\ (\pi^{[i,1,1,l,m]}, \ldots, \pi^{[i,1,S-1,l,m]}), i=1 \\ (\pi^{[i,1,1,l,n]}, \ldots, \pi^{[i,1,S-1,l,n]}), i=2 \\ (\pi^{[i,1,2,l,m,n]}, \ldots, \pi^{[i,1,S-1,l,m,n]}), i=3 \end{cases}$$

From the renowned result of the standard drift condition of Nuets [13], $\Pi A_0 e < \Pi A_2 e$ is a necessary and sufficient condition for the stability of the QBD process. Using the structure of the matrices A_0, A_2 and Q , the stability condition can be found as

$$\Pi A_0 e = \Big[\pi^{[0,0]}[D_1 I_{S-s}]e_{l(S-s)} + \pi^{[0,1]}[D_1 I_S]e_{lS} + \pi^{[1,0]}[D_1 I_{S-s} e_{l(S-s)}] \otimes e_m + \pi^{[1,1]}[D_1 I_{S-1} e_{l(S-1)}] \otimes e_m + \pi^{[2,0]}[D_1 I_{S-s} e_{l(S-s)}] \otimes e_n + \pi^{[2,1]}[D_1 I_{S-1} e_{l(S-1)}] \otimes e_n + \pi^{[3,0]}[D_1 I_{S-s} e_{l(S-s)}] \otimes e_{mn} + \pi^{[3,1]}[D_1 I_{S-2} e_{l(S-2)}] \otimes e_{mn}\Big]$$

and

$$\Pi A_2 \boldsymbol{e} = \pi^{[1,0]}[(e_l \otimes S^0) \otimes e_{S-s}] + \Big[\pi^{[1,1,1]}[e_l \otimes p_1\, S^0] + \pi^{[1,1,2]}[e_l \otimes S^0] + \dots\dots\dots + \pi^{[1,1,S-1]}[e_l \otimes S^0]\Big] + \pi^{[2,0]}[(e_l \otimes T^0) \otimes e_{S-s}] + \Big[\pi^{[2,1,1]}[e_l \otimes p_2 T^0] + \pi^{[2,1,2]}[e_l \otimes T^0] + \dots + \pi^{[2,1,S-1]}[e_l \otimes T^0]\Big] + \pi^{[3,0]}[e_l \otimes (S^0 \oplus T^0) \otimes e_{S-s}] + \Big[\pi^{[3,1,2]}[e_l \otimes (p_1\, S^0 \oplus p_2 T^0)] + \pi^{[3,1,3]}[e_l \otimes (S^0 \oplus T^0)] + \dots\dots + \pi^{[3,1,S-1]}[e_l \otimes (S^0 \oplus T^0)]\Big]$$

3.1 Steady State Probability Vector

Under the stability condition of the system, there exists a steady state probability vector $\mathbf{x} = (\boldsymbol{x}_0, \boldsymbol{x}_1, \dots)$, satisfying $\mathbf{xQ} = \mathbf{0}, \mathbf{xe} = \mathbf{1}$ and the sub-vectors can be obtained as

$$\boldsymbol{x_0} B_{00} + \boldsymbol{x_1} B_{10} = 0 \tag{1}$$

$$\boldsymbol{x_0} B_{01} + \boldsymbol{x_1} B_{11} + \boldsymbol{x_2} B_{21} = 0 \tag{2}$$

$$\boldsymbol{x_1} B_{12} + \boldsymbol{x_2}[A_1 + RA_2] = 0 \tag{3}$$

$$\boldsymbol{x}_i = \boldsymbol{x_2} \boldsymbol{R^{i-1}}, i = 3, 4, 5... \tag{4}$$

subject to the normalizing condition

$$\boldsymbol{x_0 e_3} + \boldsymbol{x_1 e_2} + \boldsymbol{x_2}(I - R)^{-1}\boldsymbol{e_1} = 1 \tag{5}$$

where $\boldsymbol{x_0} = (y_{0,0,0,s+1}, \dots y_{0,0,0,S}, y_{0,0,1,0}, \dots\dots, y_{0,0,1,S-1})$

$$\boldsymbol{x_1} = \begin{cases} (y_{1,0,0,s+1}, \cdots, y_{1,0,0,S}, y_{1,0,1,0}, \cdots, y_{1,0,1,S-1}, y_{1,1,0,s+1}, \cdots, y_{1,1,0,S}) \\ (y_{1,1,1,1}, \dots, y_{1,1,1,S-1}, y_{1,2,0,s+1}, \dots\dots, y_{1,2,0,S}, y_{1,2,1,1}, \dots\dots, y_{1,2,1,S-1}) \end{cases}$$

for $i \geq 2$

$$\boldsymbol{x_i} = \begin{cases} (y_{i,0,0,s+1}, \cdots, y_{i,0,0,S}, y_{i,0,1,0}, \cdots, y_{i,0,1,S-1}, y_{i,1,0,s+1}, \cdots, y_{i,1,0,S}) \\ (y_{i,1,1,1}, \dots, y_{i,1,1,S-1}, y_{i,2,0,s+1}, \dots\dots, y_{i,2,0,S}, y_{i,2,1,1}, \dots\dots, y_{i,2,1,S-1}), \\ y_{i,3,0,s+1}, \cdots, y_{i,3,0,S}, y_{i,3,1,2}, \dots\dots, y_{i,3,1,S-1} \end{cases}$$

R is the minimal nonnegative solution of the matrix quadratic equation $R^2 A_2 + RA_1 + A_0 = 0$, where the spectral radius of R is less than one. The rate matrix R is computed from $R = -A_0(A_1)^{-1} - R^2 A_2(A_1)^{-1}$ and is approximated by the continuous approximation procedure developed by Neuts [12] namely $R_0 = 0, R_{n+1} = -A_0(A_1)^{-1} - R_n^2 A_2(A_1)^{-1}$, $n = 0, 1, 2, \dots$. The process continues until the difference in corresponding values in R, which is obtained in successive iteration falls below a certain tolerance level.

4 Measures of Effectiveness

(i) Expected number of customers in the system:

$$E_{EC} = \boldsymbol{x_1 e_2} + \boldsymbol{x_2}[2(I-R)^{-1} + R(I-R)^{-2}]\boldsymbol{e_1}$$

(ii) Expected switching rate:

$$E_{SWR} = \sum_{i=1}^{\infty} y_{(i,1,0,s+1)}[I_l \otimes S^0]\boldsymbol{e} + \sum_{i=1}^{\infty} y_{(i,2,0,s+1)}[I_l \otimes T^0]\boldsymbol{e} + \sum_{i=2}^{\infty} y_{(i,3,0,s+1)}[I_l \otimes (S^0 \oplus T^0)]\boldsymbol{e}$$

(iii) Expected number of departures after completing service:

$$E_{EDS} = \sum_{i=1}^{\infty}\sum_{k=s+1}^{S} y_{i,1,0,k}(I_l \otimes S^0)\boldsymbol{e} + \sum_{i=1}^{\infty}\sum_{k=1}^{S-1} y_{i,1,1,k}(I_l \otimes S^0)\boldsymbol{e} + \sum_{i=1}^{\infty}\sum_{k=s+1}^{S} y_{i,2,0,k}(I_l \otimes T^0)\boldsymbol{e} + \sum_{i=1}^{\infty}\sum_{k=1}^{S-1} y_{i,2,1,k}(I_l \otimes T^0)\boldsymbol{e} + \sum_{i=2}^{\infty}\sum_{k=s+1}^{S} y_{i,3,0,k}(I_l \otimes (S^0 \oplus T^0))\boldsymbol{e} + \sum_{i=2}^{\infty}\sum_{k=2}^{S-1} y_{i,3,1,k}(I_l \otimes (S^0 \oplus T^0))\boldsymbol{e}$$

(iv) Mean Production rate:

$$E_{EPR} = \gamma\left[\sum_{i=0}^{\infty}\sum_{k=0}^{S-1} y_{i,0,1,k} + \sum_{i=1}^{\infty}\sum_{k=1}^{S-1} y_{i,1,1,k} + \sum_{i=1}^{\infty}\sum_{k=1}^{S-1} y_{i,2,1,k} + \sum_{i=2}^{\infty}\sum_{k=2}^{S-1} y_{i,3,1,k}\right]$$

(v) Expected number of crossovers in one cycle:

$$E_{ECC} = \gamma[\sum_{i=1}^{\infty} y_{(i,1,1,s-1)} + \sum_{i=1}^{\infty} y_{(i,2,1,s-1)} + \sum_{i=2}^{\infty} y_{(i,3,1,s-1)}] + \sum_{i=1}^{\infty} y_{(i,1,0,s+1)}[I_l \otimes S^0]\boldsymbol{e} + \sum_{i=1}^{\infty} y_{(i,2,0,s+1)}[I_l \otimes T^0]\boldsymbol{e} + \sum_{i=2}^{\infty} y_{(i,3,0,s+1)}[I_l \otimes (S^0 \oplus T^0)]\boldsymbol{e}$$

(vi) Expected inventory level:

$$E_{EI} = \sum_{i=0}^{\infty} \sum_{k=s+1}^{S} \sum_{j_0=1}^{l} k y_{(i,0,0,k,j_0)} + \sum_{i=0}^{\infty} \sum_{k=1}^{S-1} \sum_{j_0=1}^{l} k y_{(i,0,1,k,j_0)}$$
$$+ \sum_{i=1}^{\infty} \sum_{k=s+1}^{S} \sum_{j_0=1}^{l} \sum_{j_1=1}^{m} k y_{(i,1,0,k,j_0,j_1)} + \sum_{i=1}^{\infty} \sum_{k=1}^{S-1} \sum_{j_0=1}^{l} \sum_{j_1=1}^{m} k y_{(i,1,1,k,j_0,j_1)}$$
$$+ \sum_{i=1}^{\infty} \sum_{k=s+1}^{S} \sum_{j_0=1}^{l} \sum_{j_2=1}^{n} k y_{(i,2,0,k,j_0,j_2)} + \sum_{i=1}^{\infty} \sum_{k=1}^{S-1} \sum_{j_0=1}^{l} \sum_{j_2=1}^{n} k y_{(i,2,1,k,j_0,j_2)}$$
$$+ \sum_{i=2}^{\infty} \sum_{k=s+1}^{S} \sum_{j_0=1}^{l} \sum_{j_1=1}^{m} \sum_{j_2=1}^{n} k y_{(i,3,0,k,j_0,j_1,j_2)} + \sum_{i=2}^{\infty} \sum_{k=2}^{S-1} \sum_{j_0=1}^{l} \sum_{j_1=1}^{m} \sum_{j_2=1}^{n} k y_{(i,3,1,k,j_0,j_1,j_2)}$$

5 Cost Analysis

For the construction of cost function, we define the following costs c_1: the procurement cost per unit per unit time, c_2: the holding cost of inventory per unit per unit time, c_3: the holding cost of customers per unit per unit time, and c_4: the cost due to service per unit per unit time. Then the expected total cost (T_{cost}) of the system per unit per unit time is given by

$$T_{cost} = c_1 E_{SWR} + c_2 E_{EI} + c_3 E_{EC} + c_4 E_{EDS}$$

6 Numerical Experiments

The model is numerically analyzed by assigning two sets of different values for D_0 and D_1 for the arrival process. The arrival processes marked as MAP(C−) and MAP(C+) respectively, have negative and positive correlations with values −0.2 and 0.2.

a) Map with negative correlation-MAP(C−):

$$D_0 = \begin{bmatrix} -5.307 & 0.001 \\ 0.001 & -0.903 \end{bmatrix}, D_1 = \begin{bmatrix} 0.006 & 5.3 \\ 0.9 & 0.002 \end{bmatrix}$$

b) Map with positive correlation-MAP(C+)

$$D_0 = \begin{bmatrix} -0.9030 & 0.001 \\ 0.0002 & -3.2880 \end{bmatrix}, D_1 = \begin{bmatrix} 0.9 & 0.002 \\ 0.008 & 3.28 \end{bmatrix}$$

Three different PH- distributions are considered for each server's service time distribution, and by normalizing these processes, servers 1 and 2 will have specific service rates.

i) Erlang distribution (ER):

$$\alpha = \begin{bmatrix}1 \; 0\end{bmatrix}, \beta = \begin{bmatrix}1 \; 0\end{bmatrix}, S = \begin{bmatrix}-5.1613 & 5.1613 \\ 0 & -5.1613\end{bmatrix}, T = \begin{bmatrix}-5.0746 & 5.0746 \\ 0 & -5.0746\end{bmatrix}$$

ii) Hyper exponential (HEX):

$$\alpha = \begin{bmatrix}0.8 \; 0.2\end{bmatrix}, \beta = \begin{bmatrix}0.7 \; 0.3\end{bmatrix}, S = \begin{bmatrix}-4.2665 & 0 \\ 0 & -1\end{bmatrix}, T = \begin{bmatrix}-7.437 & 0 \\ 0 & -1\end{bmatrix}$$

iii) Exponential distribution (EX):

$$\alpha = [1], \beta = [1], S = [-2.5806], T = [-2.5373]$$

We consider the long-term expected cost behavior based on variations in the values of different parameters in different service distributions for the positive and negative correlated arrivals. Due to the complex terminology of the cost function that we have acquired, its qualitative nature cannot be studied through analytical methods. Therefore, we find the "local" optimal values of the expected cost function by assigning a small set of integer values to the parameters using a simple numerical procedure.

Tables 1 and 2 provide the details of long term expected costs incurred on different values of θ_1 and θ_2 under Bernoulli vacation schedule and for different service distributions in MAP(C+) and MAP(C−).

Table 1. T_{cost} vs θ_1 ($S = 20, s = 5, \gamma = 5, p_1 = 0.5, p_2 = 0.6, c_1 = 350, c_2 = 4, c_3 = 100, c_4 = 1, \theta_2 = 50$)

θ_1	MAP(C+) and PH service								
	ER			HEX			EX		
	EX	ER	HEX	EX	ER	HEX	EX	ER	HEX
5	48.7974	48.6640	**47.8424**	48.9408	49.6771	48.9502	48.8617	48.6802	49.1114
6	48.7202	48.5888	**47.8110**	48.8649	49.5815	48.8874	48.7867	48.6166	49.0031
7	48.6657	48.5361	**47.7905**	48.8114	49.5127	48.8430	48.7341	48.5723	48.9281
8	48.6251	48.4971	**47.7765**	48.7715	49.4602	48.8099	48.6952	48.5396	48.8733
9	48.5934	48.4670	**47.7667**	48.7405	49.4184	48.7843	48.6652	48.5146	48.8315
10	48.5680	48.4430	**47.7597**	48.7156	49.3841	48.7637	48.6414	48.4949	48.7987
θ_1	MAP(C−) and PH service								
	ER			HEX			EX		
	EX	ER	HEX	EX	ER	HEX	EX	ER	HEX
5	99.4215	116.0370	**76.9550**	94.9408	116.1167	82.7191	96.4769	110.9148	85.9030
6	98.6746	114.7734	**76.4954**	93.8580	114.4247	82.0321	95.4981	109.3975	84.9615
7	98.1389	113.8000	**76.2179**	93.0204	113.0242	81.5396	94.7557	108.1943	84.2918
8	97.7409	113.0222	**76.0537**	92.3464	111.8297	81.1711	94.1700	107.2054	83.7957
9	97.4375	112.3829	**75.9640**	91.7875	110.7886	80.8862	93.6937	106.37073	83.4172
10	97.2019	111.8458	**75.9253**	91.3130	109.8664	80.6605	93.2972	105.6514	83.1218

Table 2. T_{cost} vs θ_2 ($S = 20, s = 8, \gamma = 5, p_1 = 0.5, p_2 = 0.6, c_1 = 350, c_2 = 4, c_3 = 100, c_4 = 1, \theta_1 = 10$)

θ_2	MAP(C+) and PH service								
	ER			HEX			EX		
	EX	ER	HEX	EX	HEX	ER	EX	ER	HEX
10	49.1261	49.0090	**48.3963**	49.3174	49.4216	50.0706	49.3049	49.1662	49.4262
11	49.1010	48.9863	**48.3724**	49.2862	49.3858	50.0276	49.2765	49.1400	49.4000
12	49.0807	48.9679	**48.3521**	49.2607	49.3564	49.9930	49.2529	49.1182	49.3782
13	49.0640	48.9529	**48.3346**	49.2396	49.3318	49.9649	49.2330	49.0998	49.3598
14	49.0501	48.9404	**48.3194**	49.2219	49.3110	49.9417	49.2159	49.0841	49.3439
15	49.0383	48.9299	**48.3060**	49.2069	49.2932	49.9224	49.2012	49.0705	49.3302
θ_2	MAP(C−) and PH service								
	ER			HEX			EX		
	EX	ER	HEX	EX	HEX	ER	EX	ER	HEX
10	70.4356	73.7856	**61.6830**	68.1710	67.6271	76.6408	67.8018	70.4495	67.3274
11	69.9205	73.3525	**61.1545**	67.8403	67.1971	76.4800	67.4532	70.1746	66.9206
12	69.4768	72.9805	**60.7003**	67.5574	66.8256	76.3542	67.1527	69.9405	66.5678
13	69.0899	72.6570	**60.3051**	67.3122	66.5010	76.2547	66.8907	69.7385	66.2583
14	68.7491	72.3728	**59.9579**	67.0974	66.2144	76.1752	66.6597	69.5623	65.9843
15	68.4465	72.1207	**59.6502**	67.9074	65.9593	76.1113	66.4544	69.4069	65.7397

As the server vacation rates increase, the average duration of the server vacations decreases, and hence the average service time increases. So, the overall expected cost decreases as the vacation rate increases. From the table values, one can see that MAP with positive correlation can achieve the lowest possible cost and it is obtained during the positive correlated inter-arrival time when Erlang distribution is considered for the service time of the first server and hyper-exponential for the second server.

6.1 Optimum (s, S) Pair

We present the optimum (s, S) pair for different service distributions under MAP(C+). Assume $c_1 = 2000, c_2 = 0.5, c_3 = 10, c_4 = 10$ and $p_1 = p_2 = 0.5$. Optimum (s, S)pair in ER/HEX service distribution is (8, 19) and the optimum value of the expected cost is 8.1815 (Fig. 1). In ER/EX service distribution $(S^*, s^*) = (4, 20)$ and the optimum value of the expected cost is found to be 8.7775 (Fig. 2).

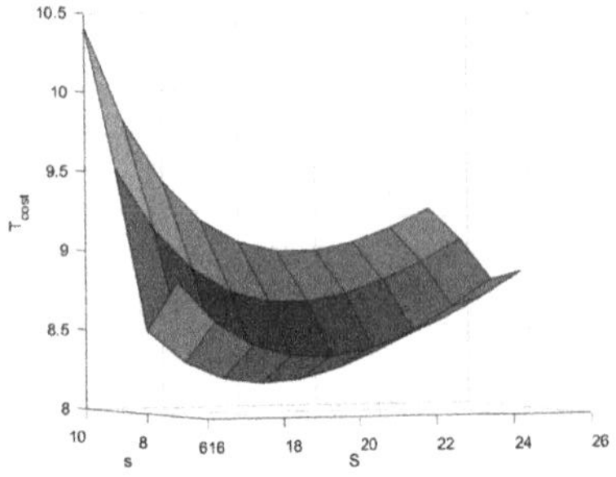

Fig. 1. Three dimensional plot for the convexity of T_{cost} in ER/HEX for $\theta_1 = 60, \theta_2 = 22, \gamma = 10$

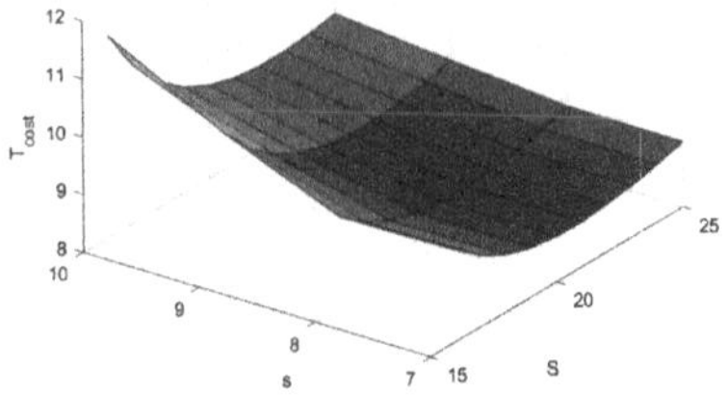

Fig. 2. Three dimensional plot for the convexity of T_{cost} in ER/EX for $\theta_1 = 60, \theta_2 = 22, \gamma = 15$

7 Concluding Remarks

In this article, we considered a production inventory system with MAP arrivals and multiple servers under the Bernoulli vacation schedule. The stability and measures of effectiveness of the system are obtained. We discussed the behaviour of the expected total cost based on variations in the values of distinct parameters in different service distributions for the positive and negative correlated arrivals. Finally, the optimum (s, S) pair for different service time distributions under MAP(C+) is computed. This work can further be extended by considering the number of servers greater than two and the distribution of server relaxation time as phase type.

References

1. Ayyappan, G., Gowthami, R.: Analysis of MAP/PH (1), PH (2)/2 queue with Bernoulli schedule vacation, Bernoulli feedback and renege of customers. Int. J. Appl. Comput. Math. **5**(6), 159 (2019)
2. Banik, A.: Analysis of queue-length dependent vacations and P-limited service in BMAP/G/1/N systems: stationary distributions and optimal control. Int. J. Stochastic Anal. **2013**, 1–15 (2013). Article ID 196372
3. Beena, P., Jose, K.P.: A MAP/PH(1), PH(2)/2 production inventory model with inventory dependent production rate and multiple servers. In: AIP Conference Proceedings, vol. 2261, no. 1, p. 030052 (2020). https://doi.org/10.1063/5.0017008
4. Chakravarthy, S.R., Ozkar, S.: MAP/PH/1 queueing model with working vacation and crowdsourcing. Mathematica Applicanda **44**(2), 263 (2016)
5. Doshi, B.T.: Queueing systems with vacations-a survey. Queueing Syst. **1**(1), 29–66 (1986)
6. Jose, K.P., Beena, P.: On a retrial production inventory system with vacation and multiple servers. Int. J. Appl. Comput. Math. **6**(4), 1–17 (2020)
7. Jose, K.P., Nair, Salini S.: Analysis of two production inventory systems with buffer, retrials and different production rates. J. Ind. Eng. Int. **13**(3), 369–380 (2017). https://doi.org/10.1007/s40092-017-0191-0

8. Keilson, J., Servi, L.: Oscillating random walk models for GI/G/1 vacation systems with Bernoulli schedules. J. Appl. Probab. **23**(3), 790–802 (1986)
9. Krishnamoorthy, A., Narayanan, V.C.: Production inventory with service time and vacation to the server. IMA J. Manage. Math. **22**(1), 33–45 (2011)
10. Kumar, B.K., Madheswari, S.P., et al.: Analysis of an M/N/N queue with Bernoulli service schedule. Int. J. Oper. Res. **5**(1), 48–72 (2009)
11. Kumar, B.K., Rukmani, R., Thangaraj, V.: Analysis of MAP/PH (1), PH (2)/2 queue with Bernoulli vacations. Int. J. Stochastic Anal. **2008**, 1–21 (2008). Article ID 396871
12. Neuts, M.F.: Matrix-geometric solutions to stochastic models. In: DGOR, p. 425. Springer (1984). https://doi.org/10.1007/978-3-642-69546-9_91
13. Neuts, M.F., Rao, B.: Numerical investigation of a multiserver retrial model. Queueing Syst. **7**(2), 169–189 (1990)
14. Suganya, C., Sivakumar, B.: MAP/PH (1), PH (2)/2 finite retrial inventory system with service facility, multiple vacations for servers. Int. J. Math. Oper. Res. **15**(3), 265–295 (2019)
15. Teghem Jr., J.: Control of the service process in a queueing system. Eur. J. Oper. Res. **23**(2), 141–158 (1986)

Spectral Analysis of a Discrete-Time Queueing Model with N -Policy on an Accelerated Service

M. P. Anilkumar[1] and K. P. Jose[2(✉)]

[1] Department of Mathematics, T. M. Govt. College, Tirur 676 502, Kerala, India
[2] Department of Mathematics, St. Peter's College, Kolenchery 682 311, Kerala, India

Abstract. This paper analyses a discrete-time queueing model with two modes of service and N-policy. The arrival of customers constitutes a Bernoulli process. There are two types of service; mode 1 and mode 2 of which service times are geometrically distributed with parameters q_1 and q_2 respectively, where $q_1 < q_2$, so that the mode 2 is an accelerated service mode. Initially, the service starts with mode 1 and when the number of customers reaches N, the server tries to change the type of the service to mode 2 with probability θ. Once the type of the service is changed from mode 1 to mode 2, it will resume the reduced rate when either of two cases happens; i) the number of customers in the system is either less than N and ii) the number of customers is reduced to zero. These two cases are studied in Model I and Model II respectively. The spectral value (or eigenvalue) approach is used to analyze Model I and consequently obtain the rate matrix of the model. Using the rate matrix of Model I, we analyze Model II. On the basis of a suitable cost function, numerical experiments are conducted for the models and obtained the optimum value of N.

Keywords: Discrete-time queue · Bernoulli process · Geometric distribution · Eigenvalue · Rate matrix

1 Introduction

The discrete-time queue is mainly introduced by Meisling [19] in which the author analyzed a single-server queueing model with Bernoulli arrivals and geometrically distributed service time. The author obtained the expressions for expected queue length and expected waiting time and hence the expected measures for the analogue continuous-time model as a limiting case. Later, Hunter [14] discussed the tools such as the generating function method, matrix theory for the discrete-time queues. An extensive study in discrete-time queues using Matrix-Analytic Method (MAM) is carried out by Alfa [1]. Recently, Anilkumar and Jose [3,4] analyzed discrete-time priority queues by applying MAM. The optimum value of queue length for starting the service was the interest of researchers since the work of Yadin and Naor [27]. The authors optimized

A. Dudin et al. (Eds.): ITMM 2020, CCIS 1391, pp. 404–416, 2021.
https://doi.org/10.1007/978-3-030-72247-0_30

the queue size for turning the server on, assuming that the server is turned off when the queue is empty. Heyman [13] studied the economic behavior of an M/G/1 queue by considering start-up cost, server shutdown cost, holding cost of customer and cost per unit time when the service is going on. The author proved that there is a stationary optimal value for the number of customers to be present to start the service. The exact rate of cost as a function of N and a closed-form expression for the optimum value of N is derived. Krishnamoorthy and Deepak [15] studied a modified N- policy model in continuous-time in which the service starts with a batch of size N and subsequent arrivals are served one after the other. As soon as the queue is empty, the server will begin the service only after accumulating N customers. Moreno [23] analyzed the discrete version of the model. Anilkumar and Jose [2] analyzed the model in which served customer provides an item from the inventory. Wang et al. [26] discussed a discrete-time Geo/G/1 queue that operates under (p, N)-policy in which the server is deactivated until N messages are accumulated in the queue. If the number of messages in the queue is accumulated to N, the server is activated for service with probability p and deactivated with probability $(1 - p)$. Based on the relevant system characteristics, an average cost function per unit time is analytically developed and the optimal values of p and N that minimize the cost function are determined.

In working vacation models, the server provides service on the ongoing service at a lower rate during vacation. Due to this reason, the idea of variation in the service rate is used in working vacation models. The multiple vacations with N-policy in continuous time are analyzed using the Matrix-Analytic Method by Zhang and Xu [28]. The authors obtained additional queue length as well as an additional delay for the service due to the reduced rate of service during vacation. Li and Tian [17] analyzed Geo/Geo/1 queue having a single working vacation and formulated an expected regular busy period and cycle. Yong and Jun [24] analyzed a Geo/Geo/1 queue with N-policy on starting time with negative customers who do not accept service but, remove the ordinary customers. Balking of an arriving customer after joining the system with a preassigned probability in an infinite buffer discrete-time working vacation queue is carried out by Goswami [8]. Ma et al. [18] studied a repairable Geo/Geo/1 discrete-time queueing system with pseudo-fault, setup time, N policy and multiple working vacations. Chandrasekaran et al. [5] analysed the developments of investigations in working vacation models in the survey paper. Recently, Lang and Tang [16] explicitly formulated the distribution of transient queue length and recursively derived the distribution of queue length in steady-state. In discrete-time queueing models, the mode of changing the service rate with a preassigned probability when the number of pending accumulated services is greater than N has not been considered till the time. In this paper, we analyze the change of service rate in a discrete-time queueing model under (θ, N)-policy. In this policy, when the number of customers in the queue is greater than or equal to N, the service rate is increased with preassigned probability θ or continued at the same rate. Once the service rate is increased and the number of customers became a number less than N, the service rate continues at the increased rate or it changes to the reduced rate. These two cases are studied separately. By defining a suitable cost function, an optimum value of N that minimizes the cost function is obtained.

In order to obtain the steady steady-state probability distribution of the number of customers present in the system and to decide the mode of service, we use the generalized eigenvalue approach. For the details of the theory of generalized eigenvalue, one can refer to Gohberg et al. [7]. The method of the generalized eigenvalue is successfully applied in queueing theory by Mitrani and Chakka [20] and Haverkort and Ost [12]. The other notable works carried out using generalized eigenvalue approach by Grassmann and Drekic [10] in tandem queues, Grassmann and Tavokali [11] in tandem queues with a movable server and Drekic and Grassmann [6] in priority queues. Recently, Grassmann and Tavakoli [9] demonstrated the fastest way to find the distribution of the queue length in a discrete GI/G/1 queue with bounded support from the waiting time distribution.

The rest of the paper is organized as follows. Section 2 provides mathematical modeling of Model I and calculation of a steady-state probability vector using the eigenvalue approach. The relationship with the matrix analytic method and computation of the rate matrix is discussed in Sect. 3. To reduce the starting cost of mode 2 service per unit time, a modified model (Model II) is discussed in Sect. 4. Finally, Optimization and numerical illustration are discussed in Sect. 5.

2 Model I

In this model, we consider a single-server queueing system in which the arrival of customers follows a Bernoulli process with a parameter p. There are two modes of service; mode 1 and mode 2 of which the service times are geometrically distributed with parameters q_1 and q_2 respectively. If the number of customers in the system is less than N, then the service is in mode 1 and if the number of customers in the system is greater than or equal N, the mode of service may change from mode 1 to mode 2 with a probability θ in each time slot. Once the mode is changed to mode 2, it will continue until the number of customers in the system is less than N. We assume that in a time slot, arrival takes place at the beginning of a slot which is followed by the change in the mode of service and service takes place at the end of a slot.

Notations

$$N(n) : \text{Number of customers in queue at an epoch } n.$$

$$J(n) : \begin{cases} 1, & \text{when server is in mode 1} \\ 2, & \text{when server is in mode 2} \end{cases}$$

$$\bar{x} : 1 - x, \text{ for any real } x.$$

Then $\{(N(n), J(n)) ; n = 0, 1, 2, 3, ..\}$ is a Quasi Birth Death process with state space

$$\{i; 0 \leq i < N\} \cup \{(i, j); i \geq N, j = 1, 2\}$$

We order the state space of the model as dictionary order. Then, the transition probability matrix of the process is given by,

$$P = \begin{bmatrix} \bar{p} & p & & & & & & \\ \bar{p}q_1 & t_1 & p\bar{q}_1 & & & & & \\ & \ddots & \ddots & \ddots & & & & \\ & & \bar{p}q_1 & t_1 & B_0 & & & \\ & & & B_2 & A_1 & A_0 & & \\ & & & & A_2 & A_1 & A_0 & \\ & & & & & \ddots & \ddots & \ddots \end{bmatrix}$$

with $t_i = pq_i + \bar{p}\bar{q}_i$, $B_0 = \begin{bmatrix} p\bar{q}_1\bar{\theta} \; p\bar{q}_1\theta \end{bmatrix}$, $B_2 = \begin{bmatrix} \bar{p}q_1 \\ \bar{p}q_2 \end{bmatrix}$ $A_1 = \begin{bmatrix} t_1\bar{\theta} & t_1\theta \\ 0 & t_2 \end{bmatrix}$
$A_0 = \begin{bmatrix} pq_1\bar{\theta} & pq_1\theta \\ 0 & p\bar{q}_2 \end{bmatrix}$, $A_2 = \begin{bmatrix} \bar{p}q_1\bar{\theta} & \bar{p}q_1\theta \\ 0 & \bar{p}q_2 \end{bmatrix}$

Let $\boldsymbol{\pi} = (\pi_0, \pi_1, \pi_2, \ldots\ldots)$ be the steady state probability vector of P, where $\pi_0, \pi_1, \ldots \pi_{N-1}$ are scalars and $\pi_i = [\pi_{i1}, \pi_{i2}]$, for $i \geq N$
Then

$$\begin{aligned} \pi_0\bar{p} + \pi_1\bar{p}q_1 &= \pi_0 \\ \pi_0 p + \pi_1 t_1 + \pi_2\bar{p}q_1 &= \pi_1 \\ \pi_{i-1}p\bar{q}_1 + \pi_i t_1 + \pi_{i+1}\bar{p}q_1 &= \pi_i, \text{ for } 2 \leq i \leq N-2 \\ \pi_{N-2}p\bar{q}_1 + \pi_{N-1}t_1 + \pi_N B_2 &= \pi_{N-1} \\ \pi_{N-1}B_0 + \pi_N A_1 + \pi_{N+1}A_2 &= \pi_N \\ \pi_{i-1}A_0 + \pi_i A_1 + \pi_{i+1}A_2 &= \pi_i \text{ for } i > N. \end{aligned} \tag{1}$$

From the above set of equations, we have

$$\pi_i = \frac{p}{\bar{p}q_1}\left(\frac{p\bar{q}_1}{\bar{p}q_1}\right)^{i-1}\pi_0 \text{ for } 1 \leq i \leq N-1$$

According to Morse [21] and Mitrani and chakka [20] the Eq. (1) has solution of the form,

$$\pi_n = \boldsymbol{g}x^{n-N} \text{ for } n > N \tag{2}$$

where $\boldsymbol{g} = [g_1, g_2]$, which is different from zero and x is a scalar. There may have different expressions of the form (2), satisfying the Eq. (1). Out of these, we choose only that expression which is consistent with the convergence property $|x| < 1$.
On substituting (2) in (1), we get

$$\boldsymbol{g}x^{n-2}A_0 + \boldsymbol{g}x^{n-1}A_1 + \boldsymbol{g}x^n A_2 = \boldsymbol{g}x^{n-1}$$

This implies

$$\boldsymbol{g}x^{n-2}(A_0 + x(A_1 - I) + x^2 A_2) = 0$$

That is

$$\boldsymbol{g}P(x)=0,\ \text{where } P(x)=A_0+(A_1-I)x+A_2x^2$$

For non zero $\boldsymbol{g}$, this is possible only if, x is a generalized eigen value of $P(x)$ and $\boldsymbol{g}$ is the corresponding eigen vector (Ghoberg [7]).
The equation $det P(x)=0$ implies

$$det\begin{bmatrix} pq_1\bar{\theta}+(t_1\bar{\theta}-1)x+\bar{p}q_1\bar{\theta}x^2 & (pq_1+t_1x+\bar{p}q_1x^2)\theta \\ 0 & pq_2+t_2x+\bar{p}q_2x^2 \end{bmatrix}=0$$

This implies that either $pq_2+t_2x+\bar{p}q_2x^2=0$

or $pq_1\bar{\theta}+(t_1\bar{\theta}-1)x+\bar{p}q_1\bar{\theta}x^2=0$

From the first equation $x=1$ or $\frac{p\bar{q}_2}{\bar{p}q_2}$ Since $|x|<1$ is necessary for the convergence, we can only consider the value $x_2=\frac{p\bar{q}_2}{\bar{p}q_2}$
The second equation leads to

$$\begin{aligned} x &= \frac{\theta+\bar{\theta}t_1\pm\sqrt{(\bar{\theta}t_1+\theta)^2-4\bar{\theta}^2p\bar{p}q_1\bar{q}_1}}{2\bar{\theta}\bar{p}q_1} \\ &= \frac{\theta(\bar{\theta})^{-1}+p\bar{q}_1+\bar{p}q_1\pm\sqrt{(\theta(\bar{\theta})^{-1}+p\bar{q}_1+\bar{p}q_1)^2-4p\bar{p}q_1\bar{q}_1}}{2\bar{p}q_1} \end{aligned}$$

We have $(\theta(\bar{\theta})^{-1}+p\bar{q}_1+\bar{p}q_1)^2-4p\bar{p}q_1\bar{q}_1>-p\bar{q}_1+\bar{p}q_1-\theta(\bar{\theta})^{-1}$

$$\begin{aligned} \text{Hence, } & \frac{\theta(\bar{\theta})^{-1}+p\bar{q}_1+\bar{p}q_1+\sqrt{(\theta(\bar{\theta})^{-1}+p\bar{q}_1+\bar{p}q_1)^2-4p\bar{p}q_1\bar{q}_1}}{2\bar{p}q_1} \\ & > \frac{\theta(\bar{\theta})^{-1}+p\bar{q}_1+\bar{p}q_1+(-p\bar{q}_1+\bar{p}q_1-\theta(\bar{\theta})^{-1})}{2\bar{p}q_1}=1 \end{aligned}$$

Therefore, $x=\frac{\theta(\bar{\theta})^{-1}+p\bar{q}_1+\bar{p}q_1+\sqrt{(\theta(\bar{\theta})^{-1}+p\bar{q}_1+\bar{p}q_1)^2-4p\bar{p}q_1\bar{q}_1}}{2\bar{p}q_1}$ is not admissible.
Since $pq_1\bar{\theta}+(t_1\bar{\theta}-1)x+\bar{p}q_1\bar{\theta}x^2<0$ for x = 1 and >0 for $x=0$,

we have the other root
$x_1=\frac{\theta(\bar{\theta})^{-1}+p\bar{q}_1+\bar{p}q_1-\sqrt{(\theta(\bar{\theta})^{-1}+p\bar{q}_1+\bar{p}q_1)^2-4p\bar{p}q_1\bar{q}_1}}{2\bar{p}q_1}$ which lies between 0 and 1.
Now we will find the corresponding eigen vectors. In order to simplifying the notations, we assume that

$$\begin{aligned} q_1(x) &= pq_1\bar{\theta}+(t_1\bar{\theta}-1)x+\bar{p}q_1\bar{\theta}x^2 \\ q_2(x) &= pq_2+t_2x+\bar{p}q_2x^2 \\ h(x) &= (pq_1+t_1x+\bar{p}q_1x^2)\theta \end{aligned}$$

Let $\boldsymbol{g}^{(i)} = [g_1^{(i)}, g_2^{(i)}]$ be eigen vector corresponding to the eigen value x_i for i = 1,2

Therefore $\boldsymbol{g}^{(1)} P(x_1) = 0$. This leads to

$$g_1^{(1)} q_1(x_1) = 0$$
$$g_1^{(1)} h(x_1) + g_2^{(1)} q_2(x_1) = 0$$

Since $q_1(x_1) = 0$ and $q_2(x_1) \neq 0$ (for, $q_2(x)$ has only one root between 0 and 1), we have $g_1^{(1)} \neq 1$ (otherwise $\boldsymbol{g}^{(1)} = 0$, a contradiction).

Hence without loss of generality, we assume that $g_1^{(1)} = 1$

Then $g_2^{(1)} = -\dfrac{h(x_1)}{q_2(x_1)}$.

Hence $\boldsymbol{g}^{(1)} = \left[1, -\dfrac{h(x_1)}{q_2(x_1)}\right]$

To find $\boldsymbol{g}^{(2)}$, we have $\boldsymbol{g}^{(2)} P(x_2) = \mathbf{0}$.

$$g_1^{(2)} q_1(x_2) = 0$$
$$g_1^{(2)} h(x_2) + g_2^{(2)} q_2(x_2) = 0$$

Since $q_1(x_2) \neq 0$ and $q_2(x_2) = 0$, this is possible only if $g_1^{(2)} = 0$ and $g_2^{(2)} \neq 0$.

Therefore, without loss of generality we take $\boldsymbol{g}^{(2)} = [0, 1]$

Hence $\pi_n = \boldsymbol{g}^{(i)} x_i^{n-1}$ is a solution of (1)

Therefore

$$\pi_n = c_1 \boldsymbol{g}^{(1)} x_1^{n-1} + c_2 \boldsymbol{g}^{(2)} x_2^{n-1} \tag{3}$$

is the general solution of (1).

Let $\mathbf{c} = [c_1\ c_2]$, $G = \begin{bmatrix} 1 & -\dfrac{h(x_1)}{q_2(x_1)} \\ 0 & 1 \end{bmatrix}$ and $\Lambda = \begin{bmatrix} x_1 & \\ & x_2 \end{bmatrix}$, then we have

$$\pi_n = \mathbf{c}\Lambda^{n-1} G \text{ for } n > N$$

where $\mathbf{c}$ is obtained by substituting $\pi_N = \mathbf{c}G$ in the boundary equation

$$\pi_{N-1} B_0 + \pi_N A_1 + \pi_{N+1} A_2 = \pi_N$$

That is, $\pi_{N-1} B_0 + \mathbf{c}GA_1 + \mathbf{c}\Lambda G A_2 = \mathbf{c}G$

This leads to $\mathbf{c}\,(G - GA_1 - \Lambda G A_2) = \pi_{N-1} B_0$

Therefore,

$$\mathbf{c} = \frac{p}{\bar{p}q_1} \left(\frac{p\bar{q}_1}{\bar{p}q_1}\right)^{N-2} B_0 \left(G - GA_1 - \Lambda G A_2\right)^{-1} \pi_0$$

Now, π_0 is obtained by normalizing condition $\sum_{i=0}^{\infty} \pi_i = 1$

$$\pi_0 + \pi_1 + \cdots + \pi_{N-1} + \mathbf{c}(I + \Lambda + \Lambda^2 + \ldots\ldots)Ge = 1$$

$$\frac{p}{(q_1 - p)}\left(1 - (\frac{p\bar{q}_1}{\bar{p}q_1})^{(N-1)}\right)\pi_0 + \frac{1}{(1-x_1)(1-x_2)}c\left[\frac{(1-x_2)\left(1 + \frac{h(x_1)}{q_2(x_1)}\right)}{(1-x_1)}\right] = 1$$

3 Relationship with Matrix-Analytic Method

By analyzing the transition probability matrix, according to Nuets [22], we have

$$\pi_n = \pi_N R^{n-N}, \text{ for } n \geq N$$

where R is the minimal non negative solution of $A_0 + RA_1 + R^2A_2 = R$.
Let $C = \begin{bmatrix} c_1 \ c_2 \end{bmatrix}$, $\Lambda = \begin{bmatrix} x_1 & \\ & x_2 \end{bmatrix}$ and $G = \begin{bmatrix} g_1^{(1)} & g_2^{(1)} \\ g_1^{(2)} & g_2^{(2)} \end{bmatrix}$
From (3), $\pi_n = C\Lambda^{n-N}G$
Therefore

$$\pi_N R^{n-N} = C\Lambda^{n-N}G$$
$$R^{n-N} = G^{-1}\Lambda^{n-N}G$$

Hence

$$R = G^{-1}\Lambda G$$

On simplification, we get

$$R = \begin{bmatrix} x_1 & \dfrac{\theta x_1}{\bar{\theta}\bar{p}q_2(1-x_1)} \\ 0 & x_2 \end{bmatrix}$$

This is similar to the R matrix obtained in [25].

4 Model II

As in the above model, when the number of customer reaches N, the service mode may change from mode 1 to mode 2 with a probability θ. In this model, we assume that the service mode is changed to mode 2. it will continue the same until the number of customers reaches zero. Let $N(n)$ and $J(n)$ be the notations defined as before. Then $(N(n), J(n))$, n = 1,2,... is quasi birth death process with stste space $\{0\} \cup \{(i,j); i \geq 1, j = 1,2\}$. The transition probability matrix is given by,

$$P^* = \begin{bmatrix} \bar{p} & B_0^* & & & & & \\ B_2 & C_1 & C_0 & & & & \\ & C_2 & C_1 & C_0 & & & \\ & & \ddots & \ddots & \ddots & & \\ & & & C_2 & A_1 & A_0 & \\ & & & & A_2 & A_1 & A_0 \\ & & & & & \ddots & \ddots & \ddots \end{bmatrix}$$

with $B_1^* = [p\ 0]$,

$$B_2 = \begin{bmatrix} \bar{p}q_1 \\ \bar{p}q_2 \end{bmatrix}, C_1 = \begin{bmatrix} t_1 & 0 \\ 0 & t_2 \end{bmatrix}, C_0 = \begin{bmatrix} p\bar{q}_1 & 0 \\ 0 & p\bar{q}_2 \end{bmatrix}, C_2 = \begin{bmatrix} \bar{p}q_1 & 0 \\ 0 & \bar{p}q_2 \end{bmatrix}$$

As in the previous case, the system is stable if and only if $p < q_2$ Let $(\pi_1^*, \pi_2^*, \ldots\ldots)$ be the steady state probability vector of P^*. By looking through the structure of P^*, we have $\pi_{i+1}^* = \pi_i^* R$, where R is a 2×2 mtrix having the expression same as before. The boundary probability vectors $\pi_0^*, \pi_1^*, \ldots \pi_{N-1}^*$ are calculated using the equtions,

$$\begin{aligned}
\pi_0^* \bar{p} + \pi_1^* B_1 &= \pi_0^* \\
\pi_0^* B_0 + \pi_1^* C_1^* + \pi_2^* C_2 &= \pi_1^* \\
\pi_{i-1}^* C_0 + \pi_i^* C_1 + \pi_2^* C_2 &= \pi_i^*, \text{ for } 1 \le i \le N-2 \\
\pi_{N-2}^* C_0 + \pi_{N-1}^* (C_1 + RC_2) &= \pi_{N-1}^*
\end{aligned}$$

Assume $\pi_i^* = (\pi_1^*, \pi_2^*)$ for $i \ge 1$. Then on simplifying, we get

$$\begin{aligned}
\pi_{11}^* &= \frac{p}{\bar{p}q_1} \frac{(1-t_1^{N-3})p\bar{q}_1 - (1-t_1^{N-2})(1-t_1-x_1\bar{p}q_1)}{(1-t_1^{N-2})p\bar{q}_1 - (1-t_1^{N-1})(1-t_1-x_1\bar{p}q_1)} \pi_0^* \\
\pi_{12}^* &= \frac{p}{\bar{p}q_2} \frac{p\bar{q}_1(t_1^{N-3} - t_1^{N-2})(1-t_1^{N-1}) - (1-t_1-x_1\bar{p}q_1)(t_1^{N-2} - t_1^{N-3})}{(1-t_1^{N-2})p\bar{q}_1 - (1-t_1^{N-1})(1-t_1-x_1\bar{p}q_1)} \pi_0^* \\
\pi_{i1}^* &= \frac{1-t_1^i}{1-t_1} \pi_{11}^* - \frac{p}{\bar{p}q_1} \frac{1-t_1^{i-1}}{1-t_1} \pi_0^* \text{ for } 2 \le i \le N-1 \\
\pi_{i2}^* &= \frac{1-t_2^i}{1-t_2} \pi_{12}^* \text{ for } 2 \le i \le N-1
\end{aligned}$$

where π_0 is obtained from the normalizing condition

$$\pi_0^* + (\pi_1^* + \pi_2^* + \cdots + \pi_{N-1}(I-R)^{-1})e = 1$$

5 Optimization

We determine the value of N which minimizes a suitably defined cost function depending on some system performances measures. Without loss generality we may assume that $(\psi_0, \psi_1, \psi_2, \ldots, \psi_{N-1}, \psi_N, \ldots)$ be the steady state probability vector for these models with $\psi_i = (\psi_{i1}, \psi_{i2})$ for $i \ge 1$ with the assumption that $\psi_{i2} = 0$ for $1 \le i \le N-1$ for the first model. The following relevent performance measures are considered for defining cost function.

- Expected number of customers in the system, EC, is given by

$$EC = \sum_{i=1}^{\infty} i\psi_i$$

- Probability that the server is busy with mode 1 service is given by

$$PBq_1 = \sum_{i=1}^{\infty} \psi_{i1}$$

- Probability that the server is busy with mode 2 service is given by

$$EBq_2 = \sum_{i=N}^{\infty} \psi_{i2}$$

- Probability of starting a mode 2 service in an epoch is

$$Pstartq_2 = \theta\left(\sum_{i=N}^{\infty}\psi_{i1} + p\psi_{(N-1)1}\right)$$

- Expected departure after completing the service is given by

$$EDS = q_1\sum_{i=1}^{\infty}\psi_{i1} + q_2\sum_{i=1}^{\infty}\pi_{i2}$$

Cost Analysis

On the basis of the above performance measures, we define Expected Total Cost per unit time as

$$ETC = C_1EC + C_2PBq_1 + C_3PBq_2 + C_4Estartq_2 + C_5EDS$$

where the individual costs C_1, C_2, C_3, C_4 and C_5 are given by

C_1 : holding cost of customers/unit/unit time
C_2 : running cost of mode 1 service/unit time
C_3 : running cost of mode 2 service/unit time
C_4 : starting cost mode 2 service/unit time
C_5 : service cost of customers/unit/unit time

Illustrations

Table 1, 2 and 3 illustrates the variations of system performance measures and expected value of total cost per unit time with respect to the parameters q_1 and N for model I with $(p, q_2, \theta, C_1, C_2, C_3, C_4, C_5) = (0.55, 0.7, 0.6, 0.01, 1, 2, 1, 5)$. We assume that the cost associated with the starting of mode 2 service and the working cost for mode 2 service is greater than that of mode 1 service. From the table the value of N at which ETC attains minimum increases with q_1.

Figure (1) compares the expected total cost for both model. It is evident from the figure that for each value of n, the expected total cost per unit time for Model II is less than that of Model I. This is due to the low switching rate of Model II compared to Model I. From the figure, it is also evident that the value N at which ETC is minimum for Model II is greater than that of Model I.

Concluding Remarks

This paper analyzed an eigenvalue approach for finding the optimum value of the number of customers in the system at which the service rate is to be changed. We obtained the rate matrix using the eigenvalue approach and a closed-form solution to the model. To minimize the starting cost of changing the mode of service, a modified model is also discussed. Numerical illustrations are incorporated to compare the models based on the optimum value of the cost function defined. This paper can be extended by assuming the arrival process as a discrete Markovian arrival process and service time as discrete phase-type distribution. One can also be extended the paper by considering the inventory of items in addition to the assumptions in the model.

Table 1. *ETC* vs. N, $q_1 = 0.4$

N	EC	PBq_1	PBq_2	$Estartq_2$	EDS	ETC
2	1.714	0.4216	0.4634	0.1581	0.5390	4.2187
3	2.449	0.5383	0.4067	0.1388	0.5220	4.1249
4	3.268	0.5906	0.3813	0.1301	0.5144	4.0877
5	4.145	0.6165	0.3687	0.1258	0.5106	4.0741
6	5.064	0.6299	0.3622	0.1236	0.5087	**4.0716**
7	6.011	0.6370	0.3587	0.1224	0.5076	4.0749
8	6.978	0.6408	0.3568	0.1218	0.5071	4.0813
9	7.957	0.6429	0.3558	0.1214	0.5068	4.0893
10	8.944	0.6440	0.3553	0.1212	0.5066	4.0982
11	9.936	0.6446	0.3550	0.1211	0.5065	4.1076
12	10.931	0.6450	0.3548	0.1211	0.5065	4.1172

Table 2. *ETC* vs. N, $q_1 = 0.48$

N	EC	PBq_1	PBq_2	$Estartq_2$	EDS	ETC
2	1.573	0.4411	0.4108	0.1629	0.5704	4.2931
3	2.170	0.5814	0.3290	0.1304	0.5524	4.1535
4	2.828	0.6552	0.2860	0.1134	0.5429	4.0835
5	3.533	0.6993	0.2603	0.1032	0.5373	4.0448
6	4.280	0.7276	0.2438	0.0966	0.5336	4.0228
7	5.064	0.7468	0.2326	0.0922	0.5312	4.0107
8	5.879	0.7601	0.2248	0.0891	0.5295	**4.0050**
9	6.724	0.7696	0.2193	0.0869	0.5282	4.0036
10	7.593	0.7765	0.2153	0.0854	0.5274	4.0052
11	8.483	0.7815	0.2124	0.0842	0.5267	4.0089
12	9.392	0.7852	0.2102	0.0833	0.5262	4.0141

Table 3. ETC vs. N, $q_1 = 0.5$

N	EC	PBq_1	PBq_2	$Estartq_2$	EDS	ETC
2	1.537	0.4461	0.3970	0.1640	0.5794	4.3163
3	2.096	0.5913	0.3088	0.1276	0.5618	4.1663
4	2.704	0.6695	0.2613	0.1080	0.5523	4.0885
5	3.352	0.7176	0.2321	0.0959	0.5464	4.0434
6	4.033	0.7496	0.2127	0.0879	0.5425	4.0159
7	4.746	0.7721	0.1991	0.0822	0.5398	3.9989
8	5.487	0.7884	0.1891	0.0781	0.5378	3.9888
9	6.254	0.8007	0.1817	0.0751	0.5363	3.9834
10	7.046	0.8100	0.1761	0.0728	0.5352	**3.9814**
11	7.860	0.8171	0.1717	0.0709	0.5343	3.9818
12	8.695	0.8228	0.1683	0.0695	0.5337	3.9841

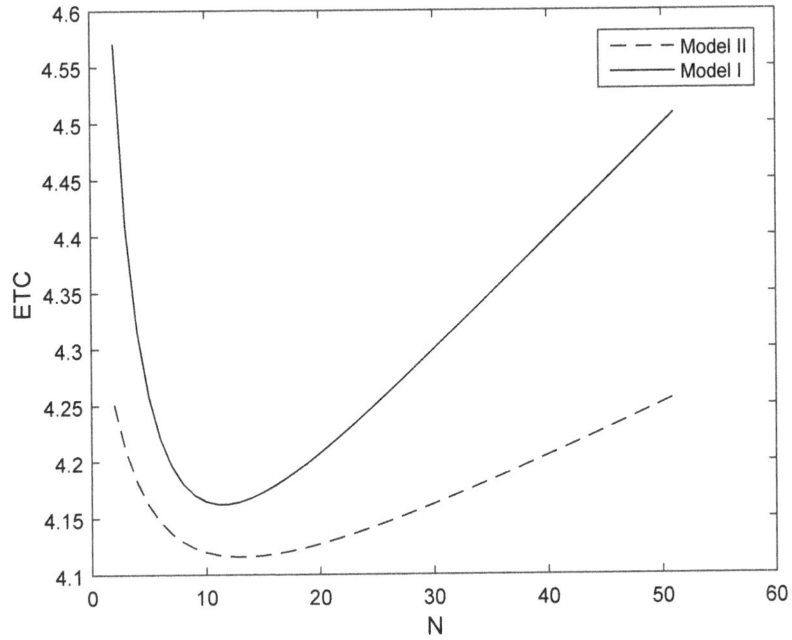

Fig. 1. ETC vs. N ($p = 0.55$; $q1 = 0.5$; $q2 = 0.7$; $\theta = 0.6$; $C_1 = 0.01$; $C_2 = 1$; $C_3 = 2$; $C_4 = 1$; $C_5 = 5$)

References

1. Alfa, A.S.: Applied Discrete-Time Queues. Springer, Heidelberg (2016). https://doi.org/10.1007/978-1-4939-3420-1

2. Anilkumar, M.P., Jose, K.P.: A discrete time *Geo*/*Geo*/1 inventory system with modified *n*-policy. Malaya J. Matematik (MJM) **8**(3), 868–876 (2020)
3. Anilkumar, M.P., Jose, K.P.: Discrete time priority queue with self generated interruption. In: AIP Conference Proceedings, vol. 2261, p. 030031. AIP Publishing LLC. (2020)
4. Anilkumar, M.P., Jose, K.P.: A Geo/Geo/1 inventory priority queue with self induced interruption. Int. J. Appl. Comput. Math. **6**(4), 1–14 (2020)
5. Chandrasekaran, V., Indhira, K., Saravanarajan, M., Rajadurai, P.: A survey on working vacation queueing models. Int. J. Pure Appl. Math. **106**(6), 33–41 (2016)
6. Drekic, S., Grassmann, W.K.: An eigenvalue approach to analyzing a finite source priority queueing model. Ann. Oper. Res. **112**(1–4), 139–152 (2002)
7. Gohberg, I., Lancaster, P., Rodman, L.: Matrix Polynomials. Classics in Applied Mathematics, Society for Industrial and Applied Mathematics (SIAM, 3600 Market Street, Floor 6, Philadelphia, PA 19104) (1982). https://books.google.co.in/books?id=KwEItnMvwbgC
8. Goswami, V.: A discrete-time queue with balking, reneging, and working vacations. Int. J. Stochastic Anal. **2014**, 1–8 (2014)
9. Grassmann, W., Tavakoli, J.: The distribution of the line length in a discrete time GI/G/1 queue. Performance Eval. **131**, 43–53 (2019)
10. Grassmann, W.K., Drekic, S.: An analytical solution for a tandem queue with blocking. Queueing Syst. **36**(1–3), 221–235 (2000)
11. Grassmann, W.K., Tavakoli, J.: A tandem queue with a movable server: an eigenvalue approach. SIAM J. Matrix Anal. Appl. **24**(2), 465–474 (2002)
12. Haverkort, B.R., Ost, A.: Steady-state analysis of infinite stochastic petri nets: comparing the spectral expansion and the matrix-geometric method. In: Proceedings of the Seventh International Workshop on Petri Nets and Performance Models, pp. 36–45. IEEE (1997)
13. Heyman, D.P.: Optimal operating policies for M/G/1 queuing systems. Oper. Res. **16**(2), 362–382 (1968)
14. Hunter, J.: Mathematical Techniques of Applied Probability, vol. ii (1983)
15. Krishnamoorthy, A., Deepak, T.G.: Modified N-policy for M/G/1 queues. Comput. Oper. Res. **29**(12), 1611–1620 (2002)
16. Lan, S., Tang, Y.: An N-policy discrete-time Geo/G/1 queue with modified multiple server vacations and Bernoulli feedback. RAIRO-Oper. Res. **53**(2), 367–387 (2019)
17. Li, J.H., Tian, N.S.: Analysis of the discrete time Geo/Geo/1 queue with single working vacation. Qual. Technol. Quant. Manage. **5**(1), 77–89 (2008)
18. Ma, Z., Wang, P., Yue, W.: Performance analysis and optimization of a pseudo-fault Geo/Geo/1 repairable queueing system with N-policy, setup time and multiple working vacations. J. Ind. Manage. Optim. **13**, 1467–1481 (2017)
19. Meisling, T.: Discrete-time queuing theory. Oper. Res. **6**(1), 96–105 (1958)
20. Mitrani, I., Chakka, R.: Spectral expansion solution for a class of Markov models: application and comparison with the matrix-geometric method. Performance Eval. **23**(3), 241–260 (1995)
21. Morse, P.M.: Queues, Inventories, and Maintenance (1958)
22. Neuts, M.F.: Matrix-geometric solutions in stochastic models: an algorithmic approach. Courier Corporation (1994)
23. Moreno, P.: A discrete-time single-server queue with a modified N-policy. Int. J. Syst. Sci. **38**(6), 483–492 (2007)
24. Sun, G.Y., Zhu, Y.J.: The Geo/Geo/1 queue model with N-policy, negative customer, feedback and multiple vacation. J. Shandong Univ. (Nat. Sci.) **2** (2010)

25. Tian, N., Ma, Z., Liu, M.: The discrete time Geom/Geom/1 queue with multiple working vacations. Appl. Math. Model. **32**(12), 2941–2953 (2008)
26. Wang, T.-Y., Ke, J.-C.: The randomized threshold for the discrete-time Geo/G/1 queue. Appl. Math. Model. **33**(7), 3178–3185 (2009)
27. Yadin, M., Naor, P.: Queueing systems with a removable service station. J. Oper. Res. Soc. **14**(4), 393–405 (1963)
28. Zhang, Z.J., Xu, X.l.: Analysis for the M/M/1 queue with multiple working vacations and N-policy. Int. J. Inf. Manage. Sci. **19**(3), 495–506 (2008)

Analysis of a Single Server Queue with Interdependence of Arrival and Service Processes – A Semi-Markov Approach

K. R. Ranjith[1(✉)], Achyutha Krishnamoorthy[2], B. Gopakumar[1], and Sajeev S. Nair[1]

[1] Government Engineering College, Thrissur, India
sajeev@gectcr.ac.in
[2] Centre for Research in Mathematics, CMS College, Kottayam, India

Abstract. In many queuing systems, the inter-arrival time and service distributions are dependent. In this paper we analyze such a system where the dependence is through a semi-Markov process. For this we assume that the arrival and service processes evolve in a finite number of phases/stages according to a Markov chain. So, the product space of the two finite sets of states (phases) is considered. The nature of transitions in the states of the combined process are such that transition rates at which the states of the combined process changes depend on the phase in which each 'marginal' process is currently in and (the phases of) the state to be visited next. We derive the stability condition and the effect of the interdependence on the stability of the system is brought out. A numerical investigation of the steady state characteristics of the system is also carried out.

Keywords: Interdependent processes · Semi-Markov process · Matrix analytic method

1 Introduction

In queueing theory literature, we notice that the problems investigated at the beginning (from the time Erlang analyzed problems in telephone systems by modeling that as a queueing problem) were considering arrival and service process to be mutually independent of each other and further inter-arrival and service times were also independent. With the introduction of the Markovian arrival/service process (MAP/MSP), the evolution within arrival/service could be modeled as dependent-two consecutive inter-arrival time/service duration are dependent. These were then extended to batch Markovian arrival and batch Markovian service processes. For a comprehensive review of such work done up to the beginning of this century, see Chakravarthy [1]. Achyutha Krishnamoorthy and Anu Nuthan Joshua extended this to consider a BMAP/BMSP/1 queue with Markov chain dependence of two successive arrival batch size and similarly that between successive service batches [2].

A. Dudin et al. (Eds.): ITMM 2020, CCIS 1391, pp. 417–429, 2021.
https://doi.org/10.1007/978-3-030-72247-0_31

The above-indicated dependence is within each component process of a queuing problem. Naturally one will be curious to know the behavior of service systems when the two-component processes are correlated; this could be between the time between the nth and the $(n+1)^{th}$ customers and the service time of the $(n+1)^{th}$ customer. There are several ways of introducing interdependence of the two processes. First, we give below a brief survey.

Conolly [3], Conolly and Hadidi [4] and Cidon et al. [5] assume that s_n, the service time of the nth arrival depends on the inter-arrival time between *n-1* and n^{th} customer. Cidon et al. [6] assume that an inter-arrival time depends on the previous service time. Conolly and Choo [7], Hadidi [8], Hadidi [9] study queueing models where the service time s_n is correlated to the inter-arrival time a_n through some density such as the bivariate exponential. Iyer and Manjunath [10] study a generalization of these models. Fendick et al. [11] study the effect of various dependencies between arrival and service processes in packet communication network queues. Combe and Boxma [12] describe how Batch Markovian Arrival Process (BMAP) can be used to model correlated inter-arrival and service times. Boxma and Perry [13] study fluid production and inventory models with dependence between service and subsequent inter-arrival time. Adan and Kulkarni [14] consider a generalization of the MAP/G/1 queue by assuming a correlation between inter-arrival and service times. They assume that the inter-arrival and service times are regulated by an irreducible discrete-time Markov chain. They derive the Laplace-Stiltjes transforms of the steady-state waiting time and the queue length distribution. Valsiou et al. [15] consider a multi-station alternating queue where preparation and service times are correlated. They consider two cases: in the first case, the correlation is determined by a discrete-time Markov chain as in Adan and Kulkarni [14] and in the second case the service time depend on the previous preparation time through their joint Laplace transform. A generalization of the G/G/1 queue with dependence between inter-arrival and service times is studied in Badila et al. [16]. In none of the above models the matrix analytic methods developed by Neuts [17–19] are used. Nor the semi-Markov approach was employed for investigation of the problem.

Sengupta [20] analyses a semi-Markovian queue with correlated inter-arrival and service times using the techniques developed in Sengupta [21]. Sengupta shows that the distributions of waiting time, time in system and virtual waiting time are matrix exponential, having phase-type representations. However, a matrix-geometric solution for the number of customers is obtained only when the inter-arrival and service times are independent. Lambert et al. [22] employ matrix-analytic methods to a queueing model where the service time of a customer depends on the inter-arrival time between himself and the previous customer. They perform the analysis, without a state-space explosion by keeping track of the age of customer in service. However, this assumption brought infinitely many parameters in to their model. Lambert et al. do not discuss the estimation part of the parameters. van Houdt [23] generalizes the above paper by presenting a queueing model where the inter-arrival times and service time are correlated, which can be analyzed as an MMAP[K]/PH[K]/1 queue for which matrix-geometric solution algorithms are available. Buchholz and Kriege [24] study a special case of van Houdt, where the dependence between inter-arrival and service times is brought in by relating phases of the arrival process with the service time distribution. More importantly, they present methods for estimating the parameters of the model.

Next, we give a slightly detailed report on Bhaskar Sengupta (Stochastic Models, 1990) [20] and B. van Houdt (Performance Evaluation, 2012) [23]. Sengupta and van Houdt have something in common in that the latter starts from the former to fill some gaps in it. Both of them delve on specific queues-the inter-arrival and service time distributions are specified. So a brief presentation of the former is sufficient. Sengupta considers the inter-arrival time sequence $\{a_n\}$ and the sequence $\{s_n\}$ duration of successive service times. Together with these, the sequence of phases where service started for successive customers and the sequence of phases from which service of customers got completed, are also taken into account. For the kth customer, while taken for service, the information, $a_1, \cdots, a_{k-1}; s_1, \cdots, s_{k-1}$ together with the service commencement phases of the first $k-1$ customers and the phases from which service completion took place for the first $k-1$ customers, are taken into consideration. Thus, too much history is required to study evolution. On closely examining the papers in the above review, we notice that the queueing models described therein could be expressed in forms like "G/G/1 queue with interdependence of arrival and service processes". In this paper, we approach the interdependence through a semi-Markov approach. For this, we need to model arrival and service processes to evolve in stages/phases. An event occurrence means either an arrival or a service completion (temporarily absorbing state). This means that we have to have finite state space first-order Markov chains to describe the evolution of the arrival and service processes. Each one has an initial probability vector and a one-step transition probability matrix. Now we take the product space of these two Markov chains (as are essential for a Markov chain). So, elements in this product space are two-dimensional objects with the first one representing the stage of current arrival and the second represents the stage in which the customer is served at present, provided there is a customer undergoing service. We use the symbol $\{(X_n \times Y_n), \text{n} \geqq 1\}$ to describe the Markov chain on the product space. The two Markov chains are independent if $P\{(X_{n+1}, Y_{n+1})|(X_n, Y_n)\} = P(X_{n+1}|X_n).P(Y_{n+1}|Y_n)$; if this equality does not hold, then the component Markov chains are interdependent. We assume that the two Markov chains to be interdependent. Transitions in this Markov chain take place according to a semi-Markov process: the sojourn time in each state (a pair) depends on the state in which it is in and the state to be visited next. On occurrence of an event, for example an arrival, we sample out the initial state from the initial probability vector of the Markov chain describing the arrival, to start the next arrival; similar explanation for the service completion and start of the next service, provided there is a customer waiting; else the server waits till the arrival of the next customer and sample out the stage to start service, from the initial probability vector of the Markov chain describing the service process.

Though we discuss a problem arising in the service system, the procedure we employ here can be applied to any branch of study where at least two processes are involved. Further, the interdependence can be group wise; those within a group are interdependent, but not between two distinct groups. Thus, the method could be applied to a wide range of study, from Economics to Medicine, Management, Commerce and even literature.

2 Interdependent Processes

We consider a queue in which arrival and service processes are related through a semi-Markov process. To this end, we assume that arrival and service of customers are in phases, as in a phase-type or MAP/MSP. There assume that we have two distinct finite state space Markov chains describing the transitions in the arrival and service processes. Each has one absorbing state-the one for the arrival process represents the occurrence of an arrival and that for the service process indicates a service completion.Assume that for the arrival process the state space of the Markov chain is $\{1, 2, \cdots, m, m+1\}$ and that for the service the state space is $\{1, 2, \cdots, n, n+1\}$ such that $m+1$ and $n+1$ are the respective absorbing states. Now consider the product space of these two sets: $\{(i,j) \mid 1 \leq i \leq m+1; 1 \leq j \leq n+1\}$ and the Markov chain on this product space as follows: Suppose that it is in state (i, j). After staying in this state for an exponentially distributed amount of time, it moves to state (i', j) or (i, j') or stays in that state itself. The sojourn time in (i, j) depends on both (i, j) and the state to be visited next. A transition with change in first coordinate represents arrival phase change with an arrival (jumping to $m+1$) or without an arrival (resulting in one of $1, 2, \cdots, m$, other than the one in which already in); similarly a transition with change in the service coordinate represents a service completion (jumping to $n+1$) or without a service completion (resulting in one of the states $1, 2, \cdots, n$, other than the one in which already in). Note that there cannot be a transition in a short interval of time $(t, t+h))$ with positive probability, when both the coordinates change. Further note that the two Markov chains are independent if and only if

$$P\{(X_{i+1}, Y_{i+1}) \mid (X_i, Y_i)\} = P\{X_{i+1} \mid X_i\}.P\{Y_{i+1} \mid Y_i\}.$$

In our case, this relation does not hold.

The idea of interdependence among random processes, evolving continuously/discrete in time, in the manner described above, was introduced by Krishnamoorthy through a series of webinars: SMARTY (Karelian Republic, August 2020); DCCN (Trapeznikov Institute of Control Sciences, Moscow, September 2020); Professor C.R. Rao Birth Centenary Talk (SreeVenkateswara University, Tirupati, September 2020), and a few others.

3 Dependence by a Row Vector

In the present study, we take two Markov Chains $\{X_i\}$ and $\{Y_i\}$ each with finite state spaces $\{1, 2, \cdots, m, m+1\}$ and $\{1, 2, \cdots, n, n+1\}$ and initial probability vectors α and β where states $m+1$ and $n+1$ are absorbing states. We assume that the state transition probabilities of the chain $\{Y_i\}$ depend on the state in which the chain $\{X_i\}$ is. We also assume that the dependence of $\{Y_i\}$ on $\{X_i\}$ is such that the rates at which the chain $\{Y_i\}$ changes its states given that the chain $\{X_i\}$ is in state k is $p_k S$, where S is an $(n+1) \times (n+1)$ matrix, $k = 1, 2, \cdots, m$.

Now consider the product $X = \{X_i \times Y_i\}$ of these chains with state space $\{(i,j)|1 \leq i \leq m+1, 1 \leq j \leq n+1, i \neq m+1, j = n+1, j \neq n+1, i = m+1\}$ with the Markovian property $P\{(X_{k+1}, Y_{k+1}) = (i', j')|(X_k, Y_k) = (i,j), (X_{k-1}, Y_{k-1}) = (i_1, j_1), \cdots, (X_0, Y_0) = (i_k, j_k)\} = P\{(X_{k+1}, Y_{k+1}) = (i', j')| (X_k, Y_k) = (i,j)\} = r_{(i,j),(i',j')}$.

Thus, $X \times Y$ is a Markov chain with temporary/instantaneous absorbing states $\{(i, n+1)|i = 1, \cdots, m\}$ and $\{(m+1, j)|j = 1, \cdots, n\}$. This chain induces a birth death process. This birth death process can be regarded as a Ph/Ph/1 queue in which the state transition times of the arrival process and that of the service process are dependent. With these assumptions, we have the following transition probabilities for the product chain:

1. $P\{(X \times Y)(t + \Delta t) = (i,j)/(X \times Y)(t) = (i,k)\} = p_i S_{kj} \Delta t, 1 \leq i \leq m, 1 \leq j, k \leq n, j \neq k$, without Type 1 or Type 2 event occurrences;
2. $P\{(X \times Y)(t + \Delta t) = (l,j)/(X \times Y)(t) = (i,j)\} = T_{il} \Delta t, 1 \leq l, i \leq m, 1 \leq j \leq n, l \neq i$, without Type 1 or Type 2 event occurrences;
3. $P\{(X \times Y)(t + \Delta t) = (i,j)/(X \times Y)(t) = (i,k)\} = p_i S_{k,n+1} \Delta t \beta_j, 1 \leq i \leq m, 1 \leq j, k \leq n$, with a Type 2 event occurrences;
4. $P\{(X \times Y)(t + \Delta t) = (l,j)/(X \times Y)(t) = (i,j)\} = T_{k,m+1} \Delta t \alpha_l, 1 \leq i, l \leq m, 1 \leq j \leq n$, with a Type 1 event occurrences;

where $T = [T_{ij}]$ is the state transition rates of the chain $\{X_i\}$. The interdependent process will have representation

$$(\alpha \otimes \beta, T \otimes I_n + J_m \otimes S),$$

where $J_m = diag(p_1, p_2, \cdot, p_m)$.

A study of the above described system is motivated by the health care models, telecommunication models etc., where the essential service has to be given to more customers within the stipulated time if the demand is high. For example, if the arrival of patients to a health care system is very high so that for giving service to all without much delay, the treatment protocol may be changed so that all the patients may get treatments required. The present scenario arising out of the COVID-19 outbreak proposes such a model. In the initial stages of the pandemic the number of patients was so small that the health care system could handle the situation easily. They are discharged from the hospitals only after three consecutive negative test results! But in the later stages when the disease is widely spread, the hospitals became crowded and the treatment protocol had to be changed so that many lives could be saved. To manage the congestion, the patients were discharged from the hospitals at a higher rate.

4 Mathematical Model

Consider a service facility where the arrival and service processes are interdependent. The assumption of dependence is motivated by the following present day scenario. In the early stages of the covid-19 outbreak, the health care system in our country took at most care in the treatment of the patients, they were kept in isolation and were sent home only after several tests. But as the number of cases increased the system had to accommodate more and more patients and so the way of treatment was changed. Here the treatment given (service) depends on the number of patients (arrival). To model such a system we use above discussed QBD as follows:

At time t, let $N(t)$ be the number of customers in the system including the one being served, $S_1(t)$ be phase of arrival and $S_2(t)$ be the phase of service. Then

$$\Omega = \{X(t) : t \geq 0\} = \{(N(t), S_1(t), S_2(t)), t \geq 0\}$$

will be a Markov chain with state space
$E = \{(0, k)|1 \leq k \leq m\} \cup \{(i,\ k,\ l)|i \geq 1, 1 \leq k \leq m, 1 \leq l \leq n\}$.

The state space of the Markov chain can be partitioned into levels $\tilde{i}$ defined as

$$\tilde{0} = \{(0,1), (0,2), ..., (0,m)\},$$

$$\tilde{i} = \{(i,1,1), (i,1,2), ...(i,1,n), .., (i,m,1), ..., (i,m,\ n)\}.$$

In the following sequel, e denotes a column matrix of 1's of appropriate order. The infinitesimal matrix of the chain is

$$Q = \begin{bmatrix} B_0 & B_1 & 0 & 0 & 0 & 0 \\ B_2 & A_1 & A_0 & 0 & 0 & 0 \\ 0 & A_2 & A_1 & A_0 & 0 & 0 \\ 0 & 0 & A_2 & A_1 & A_0 & 0 \\ . & . & . & . & . & . \\ . & . & . & . & . & . \end{bmatrix},$$

where

$$B_0 = T, B_1 = T^0 \otimes \beta \otimes \alpha; B_2 = J_m \otimes S^0,$$

and

$$A_1 = T \otimes I_n + J_m \otimes S; A_0 = T^0 \otimes \beta \otimes I_n; A_2 = J_m \otimes S^0 \otimes \alpha.$$

5 Stability Analysis

Let $A = A_0 + A_1 + A_2$
$= T^0 \otimes \beta \otimes I_n + T \otimes I_n + J_m \otimes S + J_m \otimes S^0 \otimes \alpha$
$= (T^0 \otimes \beta + T) \otimes I_n + J_m \otimes (S + S^0 \otimes \alpha)$.

If π and θ are the stationary probability vectors of $T^0 \otimes \beta + T$ and $S + S^0 \otimes \alpha$, respectively, then

$(\pi \otimes \theta)A = (\pi \otimes \theta)(T^0 \otimes \beta + T) \otimes I_n + (\pi \otimes \theta)J_m \otimes (S + S^0 \otimes \alpha)$
$= \pi(T^0 \otimes \beta + T) \otimes \theta + \pi J_m \otimes \theta(S + S^0\alpha) = 0.$

Hence, $(\pi \otimes \theta)$ is the stationary probability vector of A.

Now $(\pi \otimes \theta)A_0e = \pi T^0$ and $(\pi \otimes \theta)A_2e = (\pi P')(\theta S^0)$, where $P = [p_1, p_2, \cdots, p_n]$.

Thus, the stability condition reduces to $\pi T^0 < (\pi \mathrm{P}')\ (\theta \mathrm{S}^0)$. Hence, we have the following theorem.

Theorem 1. The Markov chain under consideration is stable if and only if $\frac{\lambda'}{\mu'} < \pi P'$ where $\lambda' = \pi T^0$ and $\mu' = \theta S^0$.

6 Steady State Analysis

The stationary probability vector z of Q is of the form $(z_0, z_1, z_1R, z_1R^2, \dots)$, where R is the minimal solution of the matrix quadratic equation

$$R^2 A_2 + RA_1 + A_0 = 0.$$

The stationary probability vector is obtained by solving the equations:

$$z_0B_0 + z_1B_2 = 0,$$
$$z_0B_1 + z_1A_1 + z_1RA_2 = 0,$$
$$z_0 + z_1(I - R)^{-1}e = 1.$$

Hence, z_1 can be determined up to a multiplicative constant using the equation

$$z_1[B_2e \otimes \beta \otimes \alpha + A_1 + RA_2] = 0.$$

7 Performance Measures

The performance of the system at stationary can be analysed using the stationary probability vector (z_0, z_1, z_1R, z_1R^2, ...). z_ne gives the probability that there are n customers in the system. Other measures like a system idle probability, expected queue length etc. follows from this. Since, our focus is to adjust the service process to control the system size even when the arrival rate is high, any information about the queue length is important.

8 Numerical Illustration

Numerical experiments are carried out to examine how the vector P influences the queue length. We chose the following values for the parameters.

$$T = \begin{bmatrix} -15 & 14 \\ 3 & -10 \end{bmatrix}, \; \alpha = \left[0.4, 0.6\right],$$

$$S = \begin{bmatrix} -7 & 1 & 2 \\ 2 & -11 & 2 \\ 2 & 2 & -10 \end{bmatrix}, \; \beta = \left[0.2, 0.3, 0.5\right].$$

In this example, the arrival of customers has two phases and the rate at which the phase of the service changes depends on which phase the arrival is in. The correlation between these two is determined by the vector P. The components of P represent the degree of dependence of the transition rates of the service with the phases of arrival. Notice that in the pattern of arrivals considered, occurrence of the event corresponding to the arrival of a customer happens in a higher rate from its second state than from the first. We found the effective service rate, ESR $= \mu'. \pi P'$, expected size $E(C)$ of the system, the server idle probability $P(I)$ and the probability $P(N > K)$ that the size of the system is enormously large (here we considered the case N is greater than 10) for various choices of P.

Table 1 shows the variation in these parameters as p_1 increases while p_2 is kept at 1. As it can be seen from the values of ESR, the service rate increases with p_1. Consequently $E(C)$ and $P(N > K)$ decrease and $P(I)$ increases. This is a result of increase in the transitions rates of the service while the arrival is in the first phase. Figure 1 depicts this increment. The fact that this dependence was linear resulted in a straight line graph. As the effective service rate increases, the system idle probability $P(I)$ also increases for obvious reason. From Fig. 2, it can be seen that $P(I)$ increases faster for small values of p_1 and slows down as $P(I)$ becomes high. The expected number of customers $E(C)$ decreases as p_1 takes higher values. Figure 3 illustrates this exponential decay. The probability of a highly populated system which is of much interest when we study healthcare models shows a similar behavior as $E(C)$. This can be seen from Fig. 4.

Table 2 helps to compare these system characteristics when $p_1 = 1$ and p_2 changes. In our example, the dependence of the transition rates of service with the second phase of arrival is determined by p_2. As the arrival of customers to the system is high in this phase, p_2 has more effect on the system characteristics than p_1. Hence, the figures corresponding to p_2 (Figs. 5, 6, 7 and 8) have the same nature of the respective figures with respect to changes in p_1 but with a greater slope.

This study illustrates how the interdependence of arrival and service processes affects the system characteristics. The system behaves according to the nature of this dependence. Thus, by making P as a control variable we can modify the system characteristics to an optimum level.

Table 1. Variation in performance measures with p_1

p_1	ESR	P(I)	E(C)	P (N > 10)
1	5.5639	0.0485	19.6305	0.5788
1.1	5.7221	0.075	12.4117	0.4262
1.2	5.8803	0.1001	9.0907	0.3173
1.3	6.0385	0.1238	7.1855	0.2388
1.4	6.1967	0.1462	5.9514	0.1817
1.5	6.3549	0.1674	5.0877	0.1397
1.6	6.513	0.1874	4.45	0.1085
1.7	6.6712	0.2063	3.9603	0.0851
1.8	6.8294	0.2242	3.5726	0.0674
1.9	6.9876	0.2412	3.2583	0.0539
2	7.1458	0.2572	2.9984	0.0434
2.5	7.9368	0.3257	2.1721	0.0166
3	8.7277	0.379	1.7322	0.0075

Table 2. Variation in performance measures with p_2

p_2	ESR	P(I)	E(C)	P (N > 10)
1	5.5639	0.0485	19.6305	0.5788
1.1	5.9621	0.1117	7.9045	0.2697
1.2	6.3603	0.1665	4.9437	0.1318
1.3	6.7585	0.2145	3.5982	0.0673
1.4	7.1567	0.2567	2.8313	0.0357
1.5	7.5549	0.2941	2.3366	0.0196
1.6	7.9531	0.3276	1.9917	0.0111
1.7	8.3513	0.3575	1.7376	0.0065
1.8	8.7495	0.3846	1.5429	0.0039
1.9	9.1477	0.4091	1.3891	0.0024
2	9.5459	0.4314	1.2645	0.0015
2.5	11.5369	0.5182	0.8836	0.0002
3	13.5279	0.578	0.6896	0

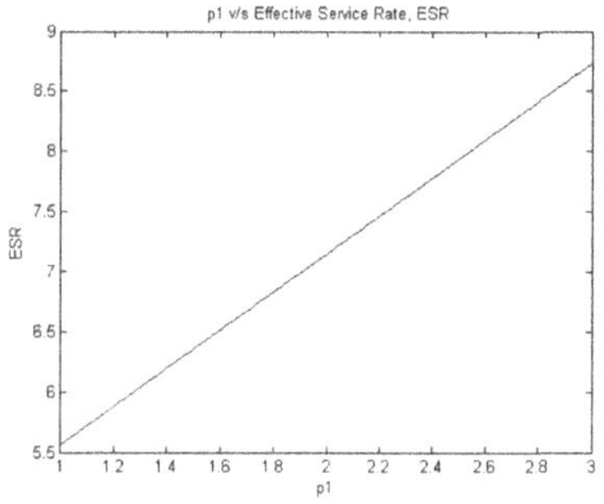

Fig. 1. p_1 vs ESR

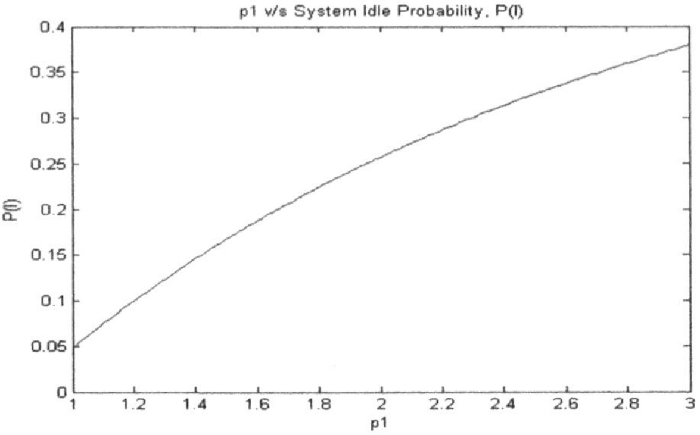

Fig. 2. p_1 vs $P(I)$

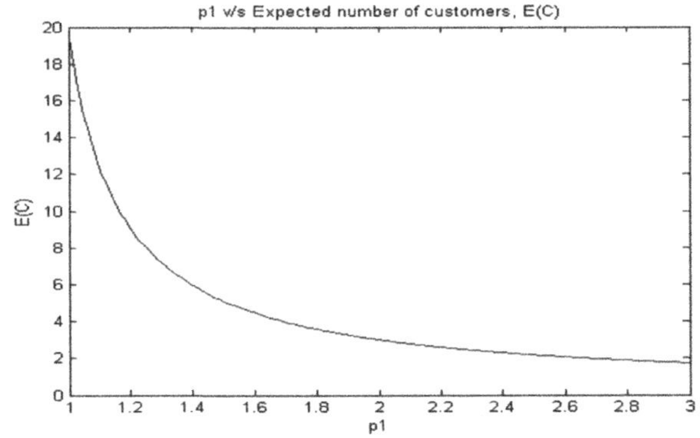

Fig. 3. p_1 vs $E(C)$

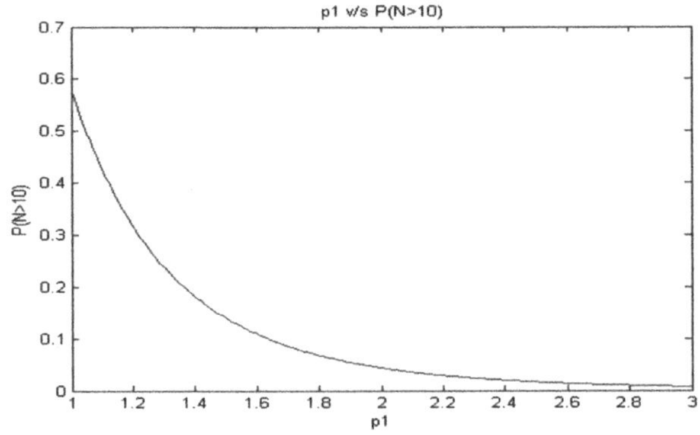

Fig. 4. p_1 vs $P(N > 10)$

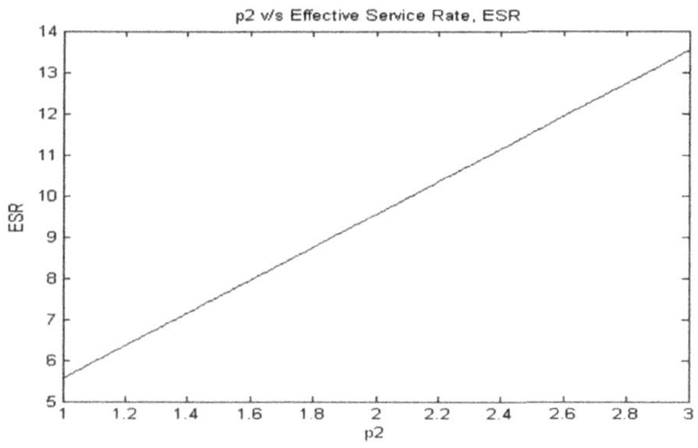

Fig. 5. p_2 vs ESR

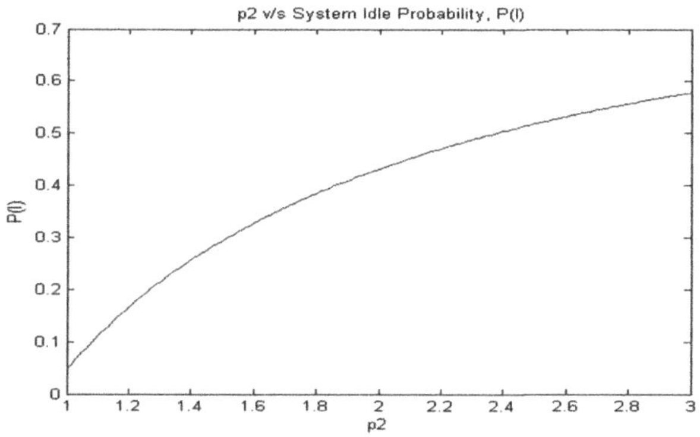

Fig. 6. p_2 vs $P(I)$

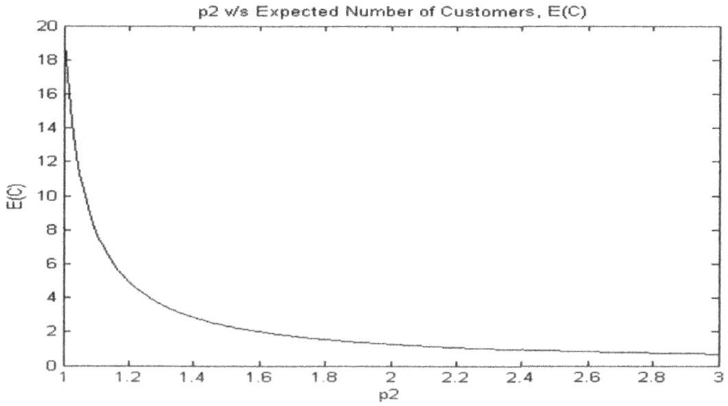

Fig. 7. p_2 vs $E(C)$

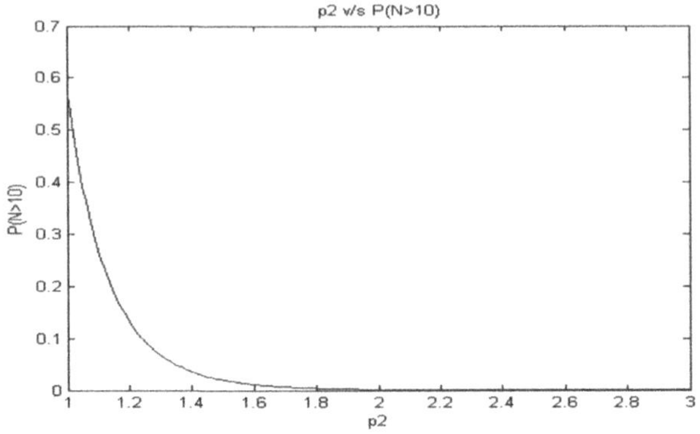

Fig. 8. p_2 vs $P(N > 10)$

9 Conclusion

In this study, we analysed a system in which the service process is dependent on the arrival process. A control over this dependency may contribute towards the system stability. In the model illustrated in the previous section, the vector P determines the degree of dependency between the two process. Thus, the numerical analysis reveals that choosing optimal values for P is possible so that the system runs efficiently. In our future work, we intent to find the service time distribution as well as the waiting time distribution of the discussed model.

References

1. Chakravarthy, S. R.: The batch Markovian arrival process: a review and future work. In: Advances in Probability and Stochastic Processes, pp. 21–49 (2001)
2. Krishnamoorthy A., Joshua A.N.: A BMAP/BMSP/1 queue with Markov dependent arrival and Markov dependent service batches. J. Ind. Manage. Optim. **3**(5) (2020). https://doi.org/10.3934/jimo.2020101
3. Conolly, B.: The waiting time for a certain correlated queue. Oper. Res. **15**, 1006–1015 (1968)
4. Conolly, B., Hadidi, N.: A correlated queue. Appl. Prob. **6**, 122–136 (1969)
5. Cidon, I., Guerin, R., Khamisy, A., Sidi, M.: On queues with inter-arrival times proportional to service times. Technical repurt. EE PUB, 811, Technion (1991)
6. Cidon, I., Guerin, R., Khamisy, A., Sidi, M.: Analysis of a correlated queue in communication systems. Technical report. EE PUB, 812, Technion (1991)
7. Conolly, B., Choo, Q.H.: The waiting process for a generalized correlated queue with exponential demand and service. SIAM J. Appl. Math. **37**, 263–275 (1979)
8. Hadidi, N.: Queues with partial correlation. SIAM J. Appl. Math. **40**, 467–475 (1981)
9. Hadidi, N.: Further results on queues with partial correlation. Oper. Res. **33**, 203–209 (1985)
10. Srikanth, K.I., Manjunath, D.: Queues with dependency between interarrival and service times using mixtures of bivariates. Stoch. Models **22**(1), 3–20 (2006)

11. Fendick, K.W., Saksena, V.R., Whitt, W.: Dependence in packet queues. IEEE Trans. Commun. **37**(11), 1173–1183 (1989)
12. Combe, M.B., Boxma, O.J.: BMAP modeling of a correlated queue. In: Walrand, J., Bagchi, K., Zobrist, G.W. (eds.) Network Performance Modeling and Simulation, pp. 177–196. Gordon and Breach Science Publishers, Philadelphia (1999)
13. Boxma, O.J., Perry, D.: A queueing model with dependence between service and interarrival time. Eur. J. Oper. Res. **128**(13), 611–624 (2001)
14. Adan, I.J.B.F., Kulkarni, V.G.: Single-server queue with Markov-dependent interarrival and service times. Queueing Syst. **45**(2), 113–134 (2003)
15. Valsiou, M., Adan, I.J.B.F., Boxma, O.J.: A two-station queue with dependent preparation and service times. Eur. J. Oper. Res. **195**(1), 104–116 (2009)
16. Badila, E.S., Boxma, O.J., Resing, J.A.C.: Queues and risk processes with dependencies. Stoch. Models. **30**(3), 390–419 (2014)
17. Neuts, M.F.: Markov chains with applications to queueing theory, which have a matrix-geometric invariant probability vector. Adv. Appl. Prob. **10**, 125–212 (1978)
18. Neuts, M.F.: Matrix Geometric Solutions for Stochastic Models. John Hopkins University Press, Baltimore (1981)
19. Neuts, M.F., Pagano, M.E.: Generating random variates from a distribution of phase type. In: Oren, T.I., Delfosse, C.M., Shub, C.M. (eds.) Winter Simulation Conference, pp. 381–387. IEEE (1981)
20. Sengupta, B.: The semi-Markovian queue: theory and applications. Stoch. Models **6**(3), 383–413 (1990)
21. Sengupta, B.: Markov processes whose steady state distribution is matrix exponential with an application to the GI/PH/1 queue. Adv. Appl. Prob. **21**, 159–180 (1989)
22. Lambert, J., van Houdt, B., Blondia, C.: Queue with correlated service and interarrival times and its application to optical buffers. Stoch. Models **22**(2), 233–251 (2006)
23. van Houdt, B.: A matrix geometric representation for the queue length distribution of multitype semi-Markovian queues. Perform. Eval. **69**(7–8), 299–314 (2012)
24. Buchholz, P., Kriege, J.: Fitting correlated arrival and service times and related queueing performance. Queueing Syst. **85**, 337–359 (2017)
25. Asmussen, S., Nerman, O., Olsson, M.: Fitting phase-type distributions via the EM algorithm. Scand. J. Stat. **23**(4), 419–441 (1996)
26. Breuer, L.: An EM algorithm for batch Markovian arrival processes and its comparison to a simpler estimation procedure. Ann. Oper. Res. **112**, 123–138 (2002). https://doi.org/10.1023/A:1020981005544
27. Horváth, G., Okamura, H.: A fast EM algorithm for fitting marked Markovian arrival processes with a new special structure. In: Balsamo, M.S., Knottenbelt, W.J., Marin, A. (eds.) EPEW 2013. LNCS, vol. 8168, pp. 119–133. Springer, Heidelberg (2013). https://doi.org/10.1007/978-3-642-40725-3_10
28. Ozawa, T.: Analysis of queues with Markovian service process. Stoch. Models. **20**(4), 391–413 (2004)
29. Whitt, W., You, W.: Using Robust queueing to expose the impact of dependence in single-server queues. Oper. Res. **66**(1), 184–199 (2017)

Author Index

Printed in the United States
by Baker & Taylor Publisher Services